Communications in Computer and Information Science   1258

Commenced Publication in 2007
Founding and Former Series Editors:
Simone Diniz Junqueira Barbosa, Phoebe Chen, Alfredo Cuzzocrea,
Xiaoyong Du, Orhun Kara, Ting Liu, Krishna M. Sivalingam,
Dominik Ślęzak, Takashi Washio, Xiaokang Yang, and Junsong Yuan

More information about this series at http://www.springer.com/series/7899

Pinle Qin · Hongzhi Wang ·
Guanglu Sun · Zeguang Lu (Eds.)

Data Science

6th International Conference
of Pioneering Computer Scientists,
Engineers and Educators, ICPCSEE 2020
Taiyuan, China, September 18–21, 2020
Proceedings, Part II

 Springer

Editors
Pinle Qin
North University of China
Taiyuan, China

Hongzhi Wang
Harbin Institute of Technology
Harbin, China

Guanglu Sun
Harbin University of Science
and Technology
Harbin, China

Zeguang Lu
National Academy of Guo Ding Institute
of Data Science
Beijing, China

ISSN 1865-0929   ISSN 1865-0937   (electronic)
Communications in Computer and Information Science
ISBN 978-981-15-7983-7   ISBN 978-981-15-7984-4   (eBook)
https://doi.org/10.1007/978-981-15-7984-4

This Springer imprint is published by the registered company Springer Nature Singapore Pte Ltd.
The registered company address is: 152 Beach Road, #21-01/04 Gateway East, Singapore 189721, Singapore

Preface

As the program chairs of the 6th International Conference of Pioneer Computer Scientists, Engineers and Educators 2020 (ICPCSEE 2020, originally ICYCSEE), it was our great pleasure to welcome you to the conference, which will be held in Taiyuan, China, September 18–21, 2020, hosted by North University of China and National Academy of Guo Ding Institute of Data Science, China. The goal of this conference is to provide a forum for computer scientists, engineers, and educators.

This conference attracted 392 paper submissions. After the hard work of the Program Committee, 98 papers were accepted to appear in the conference proceedings, with an acceptance rate of 25%. The major topic of this conference is data science. The accepted papers cover a wide range of areas related to basic theory and techniques for data science including mathematical issues in data science, computational theory for data science, big data management and applications, data quality and data preparation, evaluation and measurement in data science, data visualization, big data mining and knowledge management, infrastructure for data science, machine learning for data science, data security and privacy, applications of data science, case study of data science, multimedia data management and analysis, data-driven scientific research, data-driven bioinformatics, data-driven healthcare, data-driven management, data-driven e-government, data-driven smart city/planet, data marketing and economics, social media and recommendation systems, data-driven security, data-driven business model innovation, and social and/or organizational impacts of data science.

We would like to thank all the Program Committee members, 216 coming from 102 institutes, for their hard work in completing the review tasks. Their collective efforts made it possible to attain quality reviews for all the submissions within a few weeks. Their diverse expertise in each individual research area helped us to create an exciting program for the conference. Their comments and advice helped the authors to improve the quality of their papers and gain deeper insights.

We thank Dr. Lanlan Chang and Jane Li from Springer, whose professional assistance was invaluable in the production of the proceedings. A big thanks also to the authors and participants for their tremendous support in making the conference a success.

Besides the technical program, this year ICPCSEE offered different experiences to the participants. We hope you enjoyed the conference.

June 2020

Pinle Qin
Weipeng Jing

Organization

The 6th International Conference of Pioneering Computer Scientists, Engineers and Educators (ICPCSEE, originally ICYCSEE) 2020 (http://2020.icpcsee.org) was held in Taiyuan, China, during September 18–21, 2020, hosted by North University of China and National Academy of Guo Ding Institute of Data Science, China.

General Chair

Jianchao Zeng   North University of China, China

Program Chairs

Pinle Qin   North University of China, China
Weipeng Jing   Northeast Forestry University, China

Program Co-chairs

Yan Qiang   Taiyuan University of Technology, China
Yuhua Qian   Shanxi University, China
Peng Zhao   Taiyuan Normal University, China
Lihu Pan   Taiyuan University of Science and Technology, China
Alex kou   University of Victoria, Canada
Hongzhi Wang   Harbin Institute of Technology, China

Organization Chairs

Juanjuan Zhao   Taiyuan University of Technology, China
Fuyuan Cao   Shanxi University, China
Donglai Fu   North University of China, China
Xiaofang Mu   Taiyuan Normal University, China
Chang Song   Institute of Coal Chemistry, CAS, China

Organization Co-chairs

Rui Chai   North University of China, China
Yanbo Wang   North University of China, China
Haibo Yu   North University of China, China
Yi Yu   North University of China, China
Lifang Wang   North University of China, China

Hu Zhang	Shanxi University, China
Wei Wei	Shanxi University, China
Rui Zhang	Taiyuan University of Science and Technology, China

Publication Chair

Guanglu Sun	Harbin University of Science and Technology, China

Publication Co-chairs

Xianhua Song	Harbin University of Science and Technology, China
Xie Wei	Harbin University of Science and Technology, China

Forum Chairs

Haiwei Pan	Harbin Engineering University, China
Qiguang Miao	Xidian University, China
Fudong Liu	Information Engineering University, China
Feng Wang	RoarPanda Network Technology Co., Ltd., China

Oral Session and Post Chair

Xia Liu	Sanya Aviation and Tourism college, China

Mpetition Committee Chairs

Peng Zhao	Taiyuan Normal University, China
Xiangfei Cai	Huiying Medical Technology Co., Ltd., China

Registration and Financial Chairs

Chunyan Hu	National Academy of Guo Ding Institute of Data Science, China
Yuanping Wang	Shanxi Jinyahui Culture Spreads Co., Ltd., China

ICPCSEE Steering Committee

Jiajun Bu	Zhejiang University, China
Wanxiang Che	Harbin Institute of Technology, China
Jian Chen	ParaTera, China
Wenguang Chen	Tsinghua University, China
Xuebin Chen	North China University of Science and Technology, China
Xiaoju Dong	Shanghai Jiao Tong University, China
Qilong Han	Harbin Engineering University, China
Yiliang Han	Engineering University of CAPF, China

Yinhe Han	Institute of Computing Technology, CAS, China
Hai Jin	Huazhong University of Science and Technology, China
Weipeng Jing	Northeast Forestry University, China
Wei Li	Central Queensland University, China
Min Li	Central South University, China
Junyu Lin	Institute of Information Engineering, CAS, China
Yunhao Liu	Michigan State University, USA
Zeguang Lu	National Academy of Guo Ding Institute of Data Science, China
Rui Mao	Shenzhen University, China
Qiguang Miao	Xidian University, China
Haiwei Pan	Harbin Engineering University, China
Pinle Qin	North University of China, China
Zheng Shan	Information Engineering University, China
Guanglu Sun	Harbin University of Science and Technology, China
Jie Tang	Tsinghua University, China
Tian Feng	Institute of Software Chinese Academy of Sciences, China
Tao Wang	Peking University, China
Hongzhi Wang	Harbin Institute of Technology, China
Xiaohui Wei	Jilin University, China
lifang Wen	Beijing Huazhang Graphics & Information Co., Ltd., China
Liang Xiao	Nanjing University of Science and Technology, China
Yu Yao	Northeastern University, China
Xiaoru Yuan	Peking University, China
Yingtao Zhang	Harbin Institute of Technology, China
Yunquan Zhang	Institute of Computing Technology, CAS, China
Baokang Zhao	National University of Defense Technology, China
Min Zhu	Sichuan University, China
Liehuang Zhu	Beijing Institute of Technology, China

Program Committee Members

Witold Abramowicz	Poznań University of Economics and Business, Poland
Chunyu Ai	University of South Carolina Upstate, USA
Jiyao An	Hunan University, China
Ran Bi	Dalian University of Technology, China
Zhipeng Cai	Georgia State University, USA
Yi Cai	South China University of Technology, China
Zhipeng Cai	Georgia State University, USA
Zhao Cao	Beijing Institute of Technology, China
Richard Chbeir	LIUPPA Laboratory, France
Wanxiang Che	Harbin Institute of Technology, China
Wei Chen	Beijing Jiaotong University, China

Hao Chen	Hunan University, China
Xuebin Chen	North China University of Science and Technology, China
Chunyi Chen	Changchun University of Science and Technology, China
Yueguo Chen	Renmin University of China, China
Zhuang Chen	Guilin University of Electronic Technology, China
Siyao Cheng	Harbin Institute of Technology, China
Byron Choi	Hong Kong Baptist University, China
Vincenzo Deufemia	University of Salerno, Italy
Xiaofeng Ding	Huazhong University of Science and Technology, China
Jianrui Ding	Harbin Institute of Technology, China
Hongbin Dong	Harbin Engineering University, China
Minggang Dong	Guilin University of Technology, China
Longxu Dou	Harbin Institute of Technology, China
Pufeng Du	Tianjin University, China
Lei Duan	Sichuan University, China
Xiping Duan	Harbin Normal University, China
Zherui Fan	Xidian University, China
Xiaolin Fang	Southeast University, China
Ming Fang	Changchun University of Science and Technology, China
Jianlin Feng	Sun Yat-sen University, China
Yongkang Fu	Xidian University, China
Jing Gao	Dalian University of Technology, China
Shuolin Gao	Harbin Institute of Technology, China
Daohui Ge	Xidian University, China
Yu Gu	Northeastern University, China
Yingkai Guo	National University of Singapore, Singapore
Dianxuan Gong	North China University of Science and Technology, China
Qi Han	Harbin Institute of Technology, China
Meng Han	Georgia State University, USA
Qinglai He	Arizona State University, USA
Tieke He	Nanjing University, China
Zhixue He	North China Institute of Aerospace Engineering, China
Tao He	Harbin Institute of Technology, China
Leong Hou	University of Macau, China
Yutai Hou	Harbin Institute of Technology, China
Wei Hu	Nanjing University, China
Xu Hu	Xidian University, China
Lan Huang	Jilin University, China
Hao Huang	Wuhan University, China
Kuan Huang	Utah State University, USA
Hekai Huang	Harbin Institute of Technology, China

Cun Ji	Shandong Normal University, China
Feng Jiang	Harbin Institute of Technology, China
Bin Jiang	Hunan University, China
Xiaoyan Jiang	Shanghai University of Engineering Science, China
Wanchun Jiang	Central South University, China
Cheqing Jin	East China Normal University, China
Xin Jin	Beijing Electronic Science and Technology Institute, China
Chao Jing	Guilin University of Technology, China
Hanjiang Lai	Sun Yat-sen University, China
Shiyong Lan	Sichuan University, China
Wei Lan	Guangxi University, China
Hui Li	Xidian University, China
Zhixu Li	Soochow University, China
Mingzhao Li	RMIT University, Australia
Peng Li	Shaanxi Normal University, China
Jianjun Li	Huazhong University of Science and Technology, China
Xiaofeng Li	Sichuan University, China
Zheng Li	Sichuan University, China
Mohan Li	Jinan University, China
Min Li	South University, China
Zhixun Li	Nanchang University, China
Hua Li	Changchun University of Science and Technology, China
Rong-Hua Li	Shenzhen University, China
Cuiping Li	Renmin University of China, China
Qiong Li	Harbin Institute of Technology, China
Qingliang Li	Changchun University of Science and Technology, China
Wei Li	Georgia State University, USA
Yunan Li	Xidian University, China
Hongdong Li	Central South University, China
Xiangtao Li	Northeast Normal University, China
Xuwei Li	Computer Science College Sichuan University, China
Yanli Liu	Sichuan University, China
Hailong Liu	Northwestern Polytechnical University, China
Guanfeng Liu	Macquarie University, Australia
Yan Liu	Harbin Institute of Technology, China
Xia Liu	Sanya Aviation Tourism College, China
Yarong Liu	Guilin University of Technology, China
Shuaiqi Liu	Tianjin Normal University, China
Jin Liu	Central South University, China
Yijia Liu	Harbin Institute of Technology, China
Zeming Liu	Harbin Institute of Technology, China

Zeguang Lu	National Academy of Guo Ding Institute of Data Science, China
Binbin Lu	Sichuan University, China
Junling Lu	Shaanxi Normal University, China
Mingming Lu	Central South University, China
Jizhou Luo	Harbin Institute of Technology, China
Junwei Luo	Henan Polytechnic University, China
Zhiqiang Ma	Inner Mongolia University of Technology, China
Chenggang Mi	Northwestern Polytechnical University, China
Tiezheng Nie	Northeastern University, China
Haiwei Pan	Harbin Engineering University, China
Jialiang Peng	Norwegian University of Science and Technology, Norway
Fei Peng	Hunan University, China
Yuwei Peng	Wuhan University, China
Jianzhong Qi	The University of Melbourne, Australia
Xiangda Qi	Xidian University, China
Shaojie Qiao	Southwest Jiaotong University, China
Libo Qin	Research Center for Social Computing and Information Retrieval, China
Zhe Quan	Hunan University, China
Chang Ruan	Central South University of Sciences, China
Yingxia Shao	Peking University, China
Yingshan Shen	South China Normal University, China
Meng Shen	Xidian University, China
Feng Shi	Central South University, China
Yuanyuan Shi	Xi'an University of Electronic Science and Technology, China
Xiaoming Shi	Harbin Institute of Technology, China
Wei Song	North China University of Technology, China
Shoubao Su	Jinling Institute of Technology, China
Yanan Sun	Oklahoma State University, USA
Minghui Sun	Jilin University, China
Guanghua Tan	Hunan University, China
Dechuan Teng	Harbin Institute of Technology, China
Yongxin Tong	Beihang University, China
Xifeng Tong	Northeast Petroleum University, China
Vicenc Torra	Högskolan I Skövde, Sweden
Hongzhi Wang	Harbin Institute of Technology, China
Yingjie Wang	Yantai University, China
Dong Wang	Hunan University, China
Yongheng Wang	Hunan University, China
Chunnan Wang	Harbin Institute of Technology, China
Jinbao Wang	Harbin Institute of Technology, China
Xin Wang	Tianjin University, China
Peng Wang	Fudan University, China

Chaokun Wang	Tsinghua University, China
Xiaoling Wang	East China Normal University, China
Jiapeng Wang	Harbin Huade University, China
Qingshan Wang	Hefei University of Technology, China
Wenfeng Wang	CAS, China
Shaolei Wang	Harbin Institute of Technology, China
Yaqing Wang	Xidian University, China
Yuxuan Wang	Harbin Institute of Technology, China
Wei Wei	Xi'an Jiaotong University, China
Haoyang Wen	Harbin Institute of Technology, China
Huayu Wu	Institute for Infocomm Research, Singapore
Yan Wu	Changchun University of Science and Technology, China
Huaming Wu	Tianjin University, China
Bin Wu	Institute of Information Engineering, CAS, China
Yue Wu	Xidian University, China
Min Xian	Utah State University, USA
Sheng Xiao	Hunan University, China
Wentian Xin	Xidian University, China
Ying Xu	Hunan University, China
Jing Xu	Changchun University of Science and Technology, China
Jianqiu Xu	Nanjing University of Aeronautics and Astronautics, China
Qingzheng Xu	National University of Defense Technology, China
Yang Xu	Harbin Institute of Technology, China
Yaohong Xue	Changchun University of Science and Technology, China
Mingyuan Yan	University of North Georgia, USA
Yu Yan	Harbin Institute of Technology, China
Cheng Yan	Central South University, China
Yajun Yang	Tianjin University, China
Gaobo Yang	Hunan University, China
Lei Yang	Heilongjiang University, China
Ning Yang	Sichuan University, China
Xiaochun Yang	Northeastern University, China
Shiqin Yang	Xidian University, China
Bin Yao	Shanghai Jiao Tong University, China
Yuxin Ye	Jilin University, China
Xiufen Ye	Harbin Engineering University, China
Minghao Yin	Northeast Normal University, China
Dan Yin	Harbin Engineering University, China
Zhou Yong	China University of Mining and Technology, China
Jinguo You	Kunming University of Science and Technology, China
Xiaoyi Yu	Peking University, China
Ye Yuan	Northeastern University, China

Kun Yue	Yunnan University, China
Yue Yue	SUTD, Singapore
Xiaowang Zhang	Tianjin University, China
Lichen Zhang	Normal University, China
Yingtao Zhang	Harbin Institute of Technology, China
Yu Zhang	Harbin Institute of Technology, China
Wenjie Zhang	University of New South Wales, Australia
Dongxiang Zhang	University of Electronic Science and Technology of China, China
Xiao Zhang	Renmin University of China, China
Kejia Zhang	Harbin Engineering University, China
Yonggang Zhang	Jilin University, China
Huijie Zhang	Northeast Normal University, China
Boyu Zhang	Utah State University, USA
Jin Zhang	Beijing Normal University, China
Dejun Zhang	China University of Geosciences, China
Zhifei Zhang	Tongji University, China
Shigeng Zhang	Central South University, China
Mengyi Zhang	Harbin Institute of Technology, China
Yongqing Zhang	Chengdu University of Information Technology, China
Xiangxi Zhang	Harbin Institute of Technology, China
Meiyang Zhang	Southwest University, China
Zhen Zhang	Xidian University, China
Jian Zhao	Changchun University, China
Qijun Zhao	Sichuan University, China
Bihai Zhao	Changsha University, China
Xiaohui Zhao	University of Canberra, Australia
Peipei Zhao	Xidian University, China
Bo Zheng	Harbin Institute of Technology, China
Jiancheng Zhong	Hunan Normal University, China
Jiancheng Zhong	Central South University, China
Fucai Zhou	Northeastern University, China
Changjian Zhou	Northeast Agricultural University, China
Min Zhu	Sichuan University, China
Yuanyuan Zhu	Wuhan University, China
Yungang Zhu	Jilin University, China
Bing Zhu	Central South University, China
Wangmeng Zuo	Harbin Institute of Technology, China

Contents – Part II

Algorithm

Application

Education

Contents – Part I

Machine Learning

Network

Graphic Images

Natural Language Processing

Content-Based Hybrid Deep Neural Network Citation Recommendation Method

Leipeng Wang[1,2,3], Yuan Rao[1,2,3(✉)], Qinyu Bian[1,2,3],
and Shuo Wang[1,2,3]

[1] Shenzhen Research Institute of Xi'an Jiaotong University,
Shenzhen 518057, China
raoyuan@mail.xjtu.edu.cn
[2] Lab of Social Intelligence and Complex Data Processing, Software School,
Xi'an Jiaotong University, Xi'an 710049, China
[3] Shanxi Joint Key Laboratory for Artifact Intelligence (Sub-Lab of Xi'an
Jiaotong University), Xi'an 710049, China

Abstract. The rapid growth of scientific papers makes it difficult to query
related papers efficiently, accurately and with high coverage. Traditional citation
recommendation algorithms rely heavily on the metadata of query documents,
which leads to the low quality of recommendation results. In this paper,
DeepCite, a content-based hybrid neural network citation recommendation
method is proposed. First, the BERT model was used to extract the high-level
semantic representation vectors in the text, then the multi-scale CNN model and
BiLSTM model were used to obtain the local information and the sequence
information of the context in the sentence, and the text vectors were matched in
depth to generate candidate sets. Further, the depth neural network was used to
rerank the candidate sets by combining the score of candidate sets and multi-
source features. In the reranking stage, a variety of Metapath features were
extracted from the citation network, and added to the deep neural network to
learn, and the ranking of recommendation results were optimized. Compared
with PWFC, ClusCite, BM25, RW, NNRank models, the results of the Deepcite
algorithm presented in the ANN datasets show that the precision (P@20), recall
rate (R@20), MRR and MAP indexesrise by 2.3%, 3.9%, 2.4% and 2.1%
respectively. Experimental results on DBLP datasets show that the improvement
is 2.4%, 4.3%, 1.8% and 1.2% respectively. Therefore, the algorithm proposed
in this paper effectively improves the quality of citation recommendation.

Keywords: Citation recommendation · Recurrent neural network ·
Convolutional neural network · BERT · Deep semantic matching

1 Introduction

With the rapid increase in the number of scientific literature, researchers have found
that it is a very time-consuming task to track the latest developments in their field of
research in time and find suitable references in the process of research and writing
articles. The accumulation of literature reading is also affected by the inefficient and
backward technology of existing search papers. Traditionally, the search for related

© Springer Nature Singapore Pte Ltd. 2020
P. Qin et al. (Eds.): ICPCSEE 2020, CCIS 1258, pp. 3–20, 2020.
https://doi.org/10.1007/978-981-15-7984-4_1

research papers based on keywords and search methods based on user interests and behavioral preferences has significant drawbacks in query efficiency and recommendation coverage. In the face of a large amount of academic literature, it is an urgent problem for workers to provide an efficient and high-quality personalized citation mining and recommendation that meets their current needs.

In order to solve the above problems, two kinds of citation recommendation strategies have been proposed: global recommendation and local recommendation. Global recommendation analyzes the characteristics of titles, abstracts, and authors in the target literature to mine relevant citations, so as to obtain the relevant literature. Local recommendation is a more fine-grained recommendation. It is more complicated to make citation recommendations based on a paragraph and context of the article. This article focuses on the research of global recommendation methods.

Recently, more and more people are focusing on deep learning-based methods to study citation recommendations. Bhagavatula et al. [22] put forward the NNrank model, which uses a supervised neural network to extract the semantic information of sentences in the paper, and recommend references with metadata features. But the semantic information extracted by such a simple neural network is not very rich, and the use of metadata makes the citation biased to self-citation.

In view of the above shortcomings of the existing methods, we use the Bert model to extract the high-level semantic representation vector in the text, and use the multi-scale CNN model and the Bilstm model to obtain the local information and the sequence in-formation of the context in the sentence. Then, we match the generated text vector with deep semantics to improve the problem of insufficient text semantics in the candidate generation stage of NNrank model. Aiming at the problem that the NNrank model uses metadata to bias itself in the reordering phase, we propose a variety of different meta-path features from the citation network, add their similarities to the deep neural network to learn, optimize the ranking of the recommendation results, and finally return An important related citation recommendation list.

The main contributions of this article are as follows:

1) A new model of hybrid neural network model based on multiple strategies is proposed to extract deep semantic information and citation recommendations;
2) Use a deep neural network model to fuse multiple citation features to optimize the learning set for candidate sets to generate a high-quality academic paper citation recommendation result;
3) Perform in-depth experimental research on AAN and DBLP datasets, verifying the effectiveness of the proposed algorithm model.

The remainder of this paper is organized as follows. The second section introduces related work. The third section proposes feature extraction and task definition. The fourth section introduces specific model methods. The fifth section introduces experiments and results. The sixth section presents the conclusions of this paper.

2   Related Work

2.1   Global Citation Recommendations

Global recommendation analyzes relevant information such as titles, abstracts, and authors in the target literature to mine relevant citations, thereby obtaining comprehensive relevant literature. Global citation methods are mainly divided into three categories: collaborative filtering (CF), content-based filtering (CBF), and graph model-based citation recommendation methods.

Collaborative filtering (CF) based methods recommend citations based on ratings provided by other scientific researchers with similar research. Livne et al. [1] use collaborative filtering (CF) technology to research recommendation papers. Their work is based on citation network, which is a social network based on Citation relationship. They propose four types of collaborative filtering methods to recommend research papers, including Co-citation matching, User-Item CF, Item-Item CF and Bayesian Classifier. Yang et al. [2] proposed a collaborative filtering method based on sorting, assuming that users have similar reading interests in sorting common academic papers. Sugiyama et al. [3] extracted user research preferences from the list of published papers and cited papers, and constructed user research preferences. Users' preferences can be enhanced not only by papers published by researchers in the past, but also by papers cited by users. The content-based filtering (CBF) method uses vocabulary or topic features to determine when the paper is relevant to the needs of the researcher. Li et al. [4] proposed a conference paper recommendation method based on CBF. THE method extracts various pairwise features and applied pairwise learning to a rank model to predict papers that meet the preferences of the users.

The graph-based method considers it as a link prediction problem and solves it by using random walks. Gori et al. [6] constructed a homogenous citation graph and applied the PageRank algorithm to recommend scientific papers. Meng et al. [7] considered topics as particular nodes and build a four-layer heterogeneous publication graph, and then, they applied a random walk algorithm to recommend papers. Jardine et al. [8] extended the bias and transition probabilities of PageRank by considering topic distributions that were extracted from papers to predict scientific papers. Compared with ranking papers by whole link information on graphs, the node similarities on the sub-structures of a document network are much easier to compute, and they can reveal more explicit citation patterns. Sun et al. [9] introduced the concept of meta-path, which is a sequence of nodes in a network. They showed that meta-path-based score can obtain achievable performance for similarity search. Ren et al. [10] extracted various meta-path based features from citation graphs and proposed a hybrid model, called ClusCite, which combines nonnegative matrix factorization (NMF) with authority propagation. Guo et al. [11] extracted fine-grained co-authorship from citation graphs and recommended papers by graph-based paper ranking in a multi-layered graph. They further expanded the ranking approach with mutually reinforced learning for personalized citation recommendation. Mu et al. [12] expanded the ranking approach with mutually reinforced learning for personalized citation recommendation.

2.2   Distributed Representation of Texts

Distributed text representation refers to the method of using deep neural network algorithms to train vector representations of natural language objects (words, phrases, sentences, paragraphs, documents, etc.). Such vectors are also called text embedding vectors. Distributed representation vector is a low-dimensional dense vector learned from a large unsupervised corpus. Huang et al. [13] show that the distributed representation vector carries the semantic information of the text, which can be used as an effective expression of the text, applied to various natural language processing tasks, and achieved very excellent performance.

Bengio et al. [14] first applied distributed representation to statistical language models. This model is also known as neural network language model. In text representation learning, Word2Vec [15], Golve [16], ELMo [17] and other models use small-scale data sets to train the semantic representation of text. Relative representation methods such as bag of words have achieved certain improvements in semantic representation. Devlin et al. [18] proposed a pre training model, Bert, which can effectively obtain the rich semantic information in the text and solve the problem of multi translation of one word in the text.

In the deep semantic matching task under the domain-specific data set, establishing the relationship model between texts is the most direct way to learn distributed representation. Huang et al. [19] proposed DSSM method, which uses DNN model to represent text as low latitude semantic vector in the representation layer to predict the similarity between sentences in the matching layer, and the model has achieved good results. However, DSSM uses the bag of words model (BOW), so it loses the word order information and context information. Hu et al. [20] proposed CLSM model using CNN method, extracting context information under sliding window through convolution layer, and extracting global context information through pooling layer, so that context information can be effectively retained. However, it is difficult to retain the context information effectively. To solve the problem that CLSM model cannot capture remote context features, Palangi et al. [21] uses BiLSTM method to perform semantic matching task. The above method has achieved good results in the paired task, but in the citation recommendation task, an important problem of pairwise matching is to enumerate and calculate the relationship between each pair. Bhagavatula et al. [22] proposed the NNrank model, which uses a supervised neural network to extract the semantic information of the sentences in the paper, and recommends the references based on the characteristics of the metadata. But the semantic information extracted by such a simple neural network is not very rich.

In view of the above problems, this paper proposes a method using hybrid deep neural networks for citation recommendation tasks. Not only can the problem of insufficient rich semantic information extraction in the existing model be solved, but also various meta-paths are extracted from the citation network, which solves the problems of single citation recommendation features and optimizes the ranking of recommendation results.

3 Feature Extraction and Task Definition

3.1 Feature Extraction

This paper first uses supervised neural networks to extract valid citation features from scientific literature data. Ren et al. [10] proposed many citation features, and checked their validity in the experiments in this paper, and selected useful citation features to add to the recommended scheme in this paper. The references of academic papers contain rich meta-path information. This paper extracts meta-path features from citation heterogeneous networks and adds them to recommendation schemes to further enrich the feature information of citations in this article. In this paper, we extract 22 kinds of multi-source features from citation data, which are mainly divided into three categories: text features, metadata features, meta-path features.

Text Features: Generate embedded vectors from titles, abstracts, keywords, and meeting names in documents using the BERT model.

Metadata Features: Paper cited count, which measures the impact of the paper, users always tend to cite more articles. The number of times the author has been quoted, which measures the influence of the author. In this paper, the author's average number of citations is calculated. Author similarity Jaccard (authors), the feature of which is to calculate the Jaccard similarity of author index in citation pairs, is mainly used to measure the inclusion of similar authors in recommended papers, because these collaborators usually work in the same research field.

Meta-path Similarity: We extract various meta-path based features from the dataset. We select 15 different meta-paths, including *PAAP, PAVP, PVAP, $(PXP)^y$, $PXP \rightarrow P$, $PXP \leftarrow P$* where $X = \{A, V, T\}$, and $y = \{1, 2\}$. We choose both PathSim [23] and a random-walk based measure [24] to calculate the meta-path-based features, since PathSim can only be applied for symmetric meta-paths.

3.2 Task Definition

The citation recommendation model DeepCite proposed in this paper is defined as a two-stage task, as shown in Fig. 1.

Task 1: Candidates Generation
This paper uses a supervised hybrid neural network to train text features (title, abstract) in academic papers to generate distributed representation vectors for deep semantic matching, and then sorts the candidate sets it generates to filter and quickly generate citations for query documents d_q Set O(1000).

Task 2: Reranking
According to the candidate score $s_1(d_q, d_i)$ between the query and candidate documents in the citation candidate set O, combined with the 22 citation features (text, metadata, meta path similarity) extracted from the citation dataset, re-estimate the citation document (d_q, d_i) to sort the citation probabilities, and finally return the candidate citation with the highest citation probability for the query document d_q.

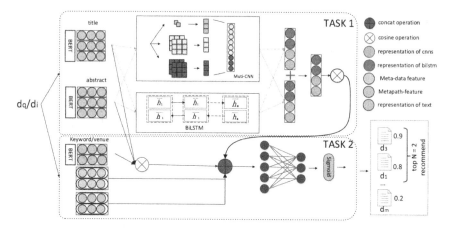

Fig. 1. The structure design of deepCite model. In task 1, we extract the text features from the title and summary to generate candidate sets through semantic matching. In task 2, we use multi-source features to reorder the candidate set and return the candidate papers with high scores for users

4   Content-Based Hybrid Deep Neural Network Citation Recommendation

In this section, we will introduce in detail how to use deep hybrid neural network to return a list of reference list top-N for users based on a given query document d_q. As shown in Fig. 1, this paper first uses the BERT multi-layer feature representation method to context embed words, then uses CNNs with different convolution kernels to extract local features of different scales, and then uses BiLSTM to strengthen the sequence relationship between words. Finally, the generated document representation vector is subjected to deep semantic matching to generate a candidate set. Further, a four-layer feedforward neural network is used to combine the candidate set score and multi-source features to learn, and finally the sigmoid function is used to output the recommended citation probability.

4.1   Candidates Generation

BERT is one of the pre-trained models with the highest performance in the learning field of NLP. The text uses the data in the scientific literature to fine-tune the original BERT model, so that the model parameters can be better adapted to the scientific literature citation recommendation field.

First, a pair of queries and candidate documents (d_q, d_i) title d[*title*] and abstract d[*abstract*] are input into the fine-tuned BERT model to generate a high-level semantic representation of the text vector $v(w_1, w_2, \ldots, w_n)$.

Where n represents the number of words, the self-attention mechanism used by the BERT model solves the problem of long-distance dependence, but indicates that the semantic information is a shallow representation, and then uses the size CNN and BiLSTM to extract richer semantic information.

Recently, CNN-based research has achieved excellent performance in the fields of text classification, named entity recognition, etc., because it can extract word-rich n-gram information to extract local features. In this paper, a grouped convolutional network is designed to extract local features at different scales. The first group is a convolutional neural network with a convolution kernel size of 1, the second group is a convolutional neural network with a convolution kernel size of 3, and the third group is also a convolutional neural network with a convolution kernel size of 3. The difference between the two groups is that they have different numbers of channels.

The sentence s_k is used as a matrix $s_k \in \aleph^{t \times m}$, where t represents the dimension of the word vector and m is the number of words in the sentence s_k. The word vector of the i-th word of the sentence s_k is expressed as $v(w_i) \in \aleph^t \cdot v(w_i : w_{i+j})$ is used to represent the word vector $[v(w_i), \ldots v(w_j)]$, the convolution kernel of the convolution operation is $W^h \in \aleph^{l \times ht}$, where h is the size of the convolution window, and the feature vector $c_i \in \aleph^l$ is generated. The definition formula is as follows:

$$e_i = f(w^h \times v(w_i : w_j) + b) \tag{1}$$

Among them, $b \in \aleph^{ht}$ has a bias term, and f is a non-linear function. In this paper, the tanh function is used. The convolution window contains all the word structures $w_i : w_h, w_2 : w_{h+1}, \ldots \ldots, w_{m-h} : w_m$ and the resulting vector is:

$$e_i = [e_1, e_2, \ldots, e_{m-h+1}] \tag{2}$$

Each convolutional neural network is activated using the ReLU function. Finally, the local features of different scales extracted by the packet convolution network are stitched together to obtain a high-level semantic representation:

$$e_{d(CNN)} = e_{d(1-CNN)} : e_{d(2-CNN)} : e_{d(3-CNN)} \tag{3}$$

CNN performs well in many tasks, but the biggest problem is the fixed field of view of filter size. On the one hand, it is impossible to model longer sequence information. On the other hand, the adjustment of filter sizes hyperparameters will be cumbersome. In natural language processing, the BiLSTM model can better express the following information. Adding the BiLSTM module to the model can strengthen the sequence relationship between words and make the words have semantic information in upper and lower languages. BiLSTM controls contextual information through forget gate, input gate, and output gate.

$$
\begin{aligned}
f_t &= \sigma(w_f[h_{t-1}, x_f] + b_f) \\
i_t &= \sigma(w_i[h_{t-1}, x_f] + b_i) \\
\tilde{C}_t &= \tanh(w_c[h_{t-1}, x_f] + b_c) \\
C_t &= f_t * C_{t-1} + i * \tilde{C}
\end{aligned}
\tag{4}
$$

Where σ is the sigmod function, w_f is the forgetting gate weight, b_f is the forgetting gate bias, i_t is the updated weight, and the tanh layer generates a new input vector. Then the output weights and output values are O_t and h_t, respectively, and the calculation formula is:

$$O_t = \sigma(w_0[h_{t-1}, x_t] + b_o)$$
$$h_t = O_t * \tanh(C_t)$$

(5)

The forward and backward outputs are stitched together as the output vector of the BiLSTM model:

$$e_{d(BiLSTM)} = ht_i^{before} : ht_i^{back}$$

(6)

Finally, the CNN convolution and BiLSTM output vectors are stitched together as the final output vector of the presentation layer:

$$e_{d[field]} = e_{d[CNN]} : e_{d[BiLSTM]}$$

(7)

In this article, title and abstract are used to represent the document vector of the paper, where μ^{title} title is a scalar model parameter:

$$e_d = \alpha^{title} \frac{e_{d[title]}}{||e_{d[title]}||_2} + \alpha^{abstract} \frac{e_{d[abstract]}}{||e_{d[abstract]}||_2}$$

(8)

This article uses cosine similarity to calculate the similarity between query and candidate document doc, that is, the candidate score.

$$s_1(d_q, d_i) = \cos ine(e_{d_q}, e_{d_i})$$

(9)

In this paper, the triples $\tau\langle d_q, d^+, d^-\rangle$ are used to establish model learning parameters, where d_q is the query document, d^+ is the referenced document, and d^- is the unreferenced document. In training, the model is trained using the Triplet Loss function to predict high cosine similarity for the (d_q, d^+) combination and low cosine similarity for the (d_q, d^-) combination.

$$loss = \max(\gamma\alpha(d^-) + s_1(d_q, d^+) + B(d^+) - s_1(d_q, d^-) - B(d^-), 0)$$

(10)

In training, using the method proposed by Bhagavatula C et al. [22], Using random negative sampling, $\alpha(d^-) = 0.3$, randomly sampling papers that are not cited by d_q. The lifting function is defined as:

$$B(d) = \frac{\sigma(\frac{d[in-citation]}{100})}{50}$$

(11)

Where σ is the sigmoid function $d[in - citation]$ is the number of times the papers have been cited in the citation dataset.

4.2 Reranking

In task 2, the text features, metadata features, and meta-path similarity features of the candidate document doc are extracted from the citation network according to the candidate set generated in task 1. Then a four-layer feedforward neural network is used to reorder the candidate set. Finally return a list of TopN citation recommendations for users.

As shown in Fig. 1, input the title, abstract, keyword, and course into the BERT to get the embedding vector, and use the cosine similarity to calculate the text feature similarity:

$$g_{field} = cosine(e_{dq[field]}, e_{di[field]}) \tag{12}$$

Then the text similarity, meta-path feature similarity features, etc. are spliced into a feedforward neural network, which is defined as follows:

$$\begin{aligned} s_2(dq, di) &= FeedForward(h) \\ h &= [g_{title}; g_{abstract}; g_{venue}; g_{keyword}; \\ &\quad Jacard(authors); doc_{citations}; aut_{citations}; \\ &\quad Metapath(PxP); s_1(d_q, d_i)] \end{aligned} \tag{13}$$

Where FeedForward is a four-layer feedforward neural network. The first three layers are the linear unit ReLU function and the last layer is the Sigmoid function. $Metapath(PxP)$ is the concatenation of the similarities of 15 different metapaths. $doc_{citations}$ refers to the number of times a candidate document should be cited. $aut_{citations}$ is the average number of times a paper publish by the candidate's author is cited. The index of $Jacard(authors)$ similarity between co-authors of the paper, $s_1(d_q, d_i)$ is the candidate set score in task 1.

5 Experiments and Results

5.1 Evaluation Metrics

To evaluate the quality of the recommendations, we use the citation information of the training papers to train our model, and the reference lists of the testing papers are used as the ground truth. Following common practice, we employ the following evaluation metrics.

- Recall
  The recall is defined as the ratio of the truly cited literature in the recommendation list to the reference of the test literature. The recall rate is an important indicator

related to the evaluation of recommendations. The formula for calculating the recall rate is as follows:

$$Recall = \frac{\sum_{d \in Q(D)} |R(d) \cap T(d)|}{\sum_{d \in Q(D)} |T(d)|}$$

- Precision
  Accuracy is the ratio of the number of documents retrieved to the total number of documents retrieved. It measures the accuracy of the retrieval system. Accuracy is an important indicator of the performance of the recommendation system. The calculation formula for accuracy is as follows:

$$precision = \frac{\sum_{d \in Q(D)} |R(d) \cap T(d)|}{\sum_{d \in Q(D)} |R(d)|}$$

- Mean Reciprocal Rank (MRR)
  MRR refers to the ranking of the standard answer in the results given by the evaluated system, taking the reciprocal as its accuracy, and averaging all the questions. The MRR calculation formula is as follows:

$$MRR = \frac{1}{|Q(D)|} \sum_{d \in Q(D)} \frac{1}{rank_d}$$

Where $rank_d$ is the position of the first correct result in test set D.
- Mean Average Precision (MAP)
  MAP is the average of a set of average accuracy rates (AP). The calculation method of AP is as follows:

$$AP_j = \sum \frac{p(i) \times pos(i)}{number\ of\ postive\ instance}$$

Where i is the ranking position in the result queue, and $P(i)$ is the accuracy of the first i results.

$$P(i) = \frac{Related\ documents}{Total\ number\ of\ documents}$$

Both MRR and MAP account for the rank of the recommended citation, and consequentially, it heavily penalizes the retrieval results when the relevant citations are returned at low rank.

5.2   Datasets

To evaluate our proposed model, we choose two bibliographic datasets, AAN and DBLP, which have different sizes of research publications in different research fields.

ANN dataset[1]: Radev et al. established the ACL Anthology Network (AAN) dataset, which contains full text information of conference and journal papers in the computational linguistics and natural language processing field. We use a subset of a 2012 release that contains 13,885 papers published from 1965 to 2012. For evaluation purposes, we divide the entire dataset into two disjoint sets, where papers published before 2012 are regarded as the training set (12,762 papers) and the remaining papers are placed in the testing set (1,123 papers).

DBLP dataset[2]: DBLP is a well-known online digital library that contains a collection of bibliographic entries for articles and books in the field of computer science and related disciplines. We use a citation dataset that was extracted and released by Tang et al. [27]. This article does not use a complete data set, but chooses a subset because some examples lack complete references. We divide papers published before 2009 into training sets (29193 papers) and papers published from 2009–2011 into Test set (2869 papers) (Table 1).

Table 1. Statistics of ANN and DBLP

		Papers	Authors	Venues	Citations
ANN	Train	12762	9766	467	68475
	Test	1123	1557	32	10437
DBLP	Train	29193	32541	751	397316
	Test	22869	3184	104	31501

5.3   Comparison with Other Approaches

In this experiment, a pre-trained BERT (12 layers, 768 hidden, 12 heads, 110M parameters) is used, and its fine-tuning is applied to the ANN and DBLP datasets. In the experiment, the length of the title of each paper is set to 50, the length of the abstract is set to 512, the maximum length of the keyword is 20, and the maximum number of citations is 100. The hidden layer of BiLSTM is 768, and the size of CNN is set to $1 \times 1, 3 \times 3, 3 \times 3$. During the training process, Adam's method was used to optimize the parameters of the model. The learning rate and dropout were set to 1e−5 and 0.55 in task 1, and set to 1e−4 and 0.5 in task 2. All experiments in this article are Linux CentOS The 7.6.1810 system is completed on the open source framework Pytorch1.0, which has an NVIDIA TITAN X graphics card (12G).

In order to verify the improvement of the performance of the evaluation index proposed by the algorithm proposed in this paper, the ANN and DBLP data sets are collected on Recall (R@20, R@40), Precision (P@10, P@20), MAP and MAP indicators. Algorithms for comparison:

- ClusCite [10]: ClusCite assumes that citation features should be organized into different groups, and each group contains its own behavior pattern to represent the

---

[1] http://clair.eecs.umich.edu/ann/.

[2] http://dblp.uni-trier.de/db/.

research interest. This method combines NMF and network regularization, to learn group and authority information for citation recommendation. To ensure fairness in the comparison, we use all extracted citation features in this paper for ClusCite.

- TopicSim: We use the original PLSA to derive the topic information; then, we recommend papers that have high topic relevance with the query.
- BM25 [25]: BM25 is a well-known ranking method for measuring the relevance of matching documents to a query based on the text. We calculate the text similarity between the papers by using both TF and IDF for BM25.
- PWFC [26]: PWFC uses a fine-grained cooperative relationship between authors to build a three-layer graph after ranking based on a random walk method.
- NNselect [22]: NNselect algorithm proposes a supervised neural network that trains the title and abstract of the paper into a low-latitude dense representation vector of text, and further uses ANN neighbors to make recommendations.
- NNRank [22]: It returns the candidate set results based on the NNselect method, and then uses a three-layer feedforward neural network to sort. In this article, we choose to use Metadata data for experiments.

The experimental results are shown in Table 2. It is possible to observe the tendency of these methods to evaluate the accuracy of the indicators. The recommendation results of the PWFC and TopSim algorithms are relatively poor, because there are certain limitations in recommending citations only through citation relationships, content relevance, and relationships between co-authors. In the experiments, the NNselect algorithm is superior to models such as TopicSim in all indicators, which indicates that text-based distributed representation is richer than text-based semantic features extracted from topic-based model features. Based on the NNselect model, NNrank uses Metadata features and deep neural networks to reorder the results of the candidate set. Experimental data shows that all index data of NNrank is better than the NNselect model, proving the important influence and effect of Metadata features on the recommendation results.

In this paper, a pre-trained model BERT is used, and its fine-tuning is applied to the ANN and DBLP scientific literature data sets. The BERT model has achieved outstanding results in multiple NLP tasks, and the text vectors it outputs are obtained using multi-scale CNN and BiLSTM models to obtain local And sequence information, and then perform deep semantic matching. Compared with the NNselect model, the DeepCite model proposed in this paper is a hybrid neural network, and its proposed semantic information is more abundant and effective. This paper uses 22 features to reorder the candidate results in task 2, and solves the problems of citation recommendation features. Experimental results prove that the DeepCite method is better than the current best model. Compared with the NNRank model and the DeepCite model in the ANN dataset, In the indicators P @ 10, P @ 20, R @ 20, R @ 40, MAP, and MRR, they increased by 2%, 2.3%, 3.9%, 3.4%, 2.4%, and 2.1%. Compared with the NNRank model in the DBLP dataset, it has increased by 2.7%, 2.4%, 4.3%, 3.2%, 1.8%, and 1.4%, respectively.

Table 2. Performance comparison between different methods on ANN and DBLP datasets

Dataset	Method	P@10	P@20	R@20	R@40	MAP	MRR
ANN	BM25	0.168	0.105	0.161	0.287	0.117	0.335
	TopSim	0.118	0.066	0.188	0.188	0.089	0.154
	PWFC	0.205	0.155	0.228	0.196	0.170	0.458
	ClusCIte	0.251	0.193	0.272	0.347	0.188	0.562
	NNselect	0.297	0.241	0.376	0.481	0.211	0.582
	NNRank	0.354	0.259	0.403	0.522	0.233	0.690
	DeepCite	**0.374**	**0.282**	**0.442**	**0.556**	**0.257**	**0.711**
DBLP	BM25	0.174	0.125	0.182	0.308	0.127	0.349
	TopSim	0.091	0.079	0.128	0.218	0.157	0.095
	PWFC	0.214	0.169	0.233	0.332	0.177	0.461
	ClusCIte	0.261	0.195	0.302	0.383	0.198	0.548
	NNselect	0.287	0.230	0.363	0.472	0.205	0.576
	NNRank	0.345	0.247	0.390	0.517	0.227	0.689
	DeepCite	**0.372**	**0.271**	**0.433**	**0.549**	**0.245**	**0.703**

5.4 Ablation Experiment Results and Analysis

In this paper, ablation experiments are used to prove the feasibility and effectiveness of each module in the DeepCite method of global citation recommendation model. In the experiment, the minimum number of citations for the papers on the ANN and DBLP datasets was set to five, and the experiments were compared with other parameters fixed.

As shown in Table 3, BERT means that in the ablation experiment, only the BERT model is used to perform semantic matching in the candidate set generation task, and then a citation is recommended for the user according to the result of the matching. The BERT_MCNN model refers to the task of generating candidate sets, inputting text vectors generated by BERT into multi-size CNNs, and then performing semantic matching, and then recommending citations for users based on the results of the matching. The BERT_BiLSTM model refers to the task of generating candidate sets, inputting text vectors generated by BERT into a bidirectional long-term and short-term memory network BiLSTM, and then performing semantic matching, and then recommending citations for users based on the results of the matching. The BERT_MCBL model refers to the task of generating candidate sets. The text vectors generated by BERT are simultaneously input into multi-size CNN and BiLSTM, and the sentence vectors of their data are further stitched for semantic matching, and then citations are recommended for users based on the matching results. In this paper, ablation experiments are used to prove the feasibility and effectiveness of each module in the candidate set generation DeepCite method, and compared with all parameters fixed.

As shown in Figs. 2, with the increase of modules, the candidate set generation module has been improved on these indicators P@10, P@20, R@20, R@40, which verifies that the model is a candidate for each module. The solution generation process has been improved, which verifies that this model has played an effective role in the

Table 3. DeepCite model ablation experiment on Ann and DBLP datasets

Dataset	Method	P@10	P@20	R@20	R@40	MAP	MRR
ANN	BERT	0.294	0.244	0.378	0.490	0.490	0.584
	BERT_MCNN	0.308	0.248	0.385	0.496	0.496	0.595
	EERT_BiLSTM	0.302	0.246	0.383	0.494	0.494	0.592
	BERT_MCBL	0.322	0.258	0.391	0.519	0.519	0.601
	DeepCite	**0.374**	**0.282**	**0.442**	**0.556**	**0.556**	**0.711**
DBLP	BERT	0.291	0.239	0.371	0.482	0.206	0.582
	BERT_MCNN	0.306	0.251	0.390	0.503	0.223	0.595
	EERT_BiLSTM	0.298	0.246	0.381	0.494	0.216	0.591
	BERT_MCBL	0.315	0.252	0.399	0.513	0.232	0.598
	DeepCite	**0.372**	**0.271**	**0.433**	**0.549**	**0.245**	**0.703**

process of candidate solution generation during the various modules. Based on the BERT model, this paper uses three sets of CNNs of different sizes to obtain the local semantic information of the text, which improves the effect of deep semantic matching. Then the BiLSTM model is used to obtain the sequence information in the text, which further improves the effect of semantic matching and generates high Quality citation candidate set.

The experimental results are shown in Table 3. Compared with the BERT model, the BERT-BL3C model proposed in task 1 in the candidate set generation stage has an accuracy rate (P@10, P@20) and a recall rate (R@20, R@40), the average reciprocal ranking (MRR), and the average accuracy rate (MAP) increased by 2.4%, 1.3%, 2.8%, 3.1%, 2.6%, and 1.6%, respectively. Increased 2.8%, 1.4%, 1.1%, 2.9%, 2.2%, 1.8% on the ANN dataset.

In Task 2, the candidate set was reordered by combining 22 citation features. Compared to BERT-BL3C, the DeepCite model has an accuracy rate (P @ 10, P @ 20) and a recall rate (R @ 20, R @) on the ANN dataset. 40), the average reciprocal ranking (MRR), the average accuracy rate (MAP) increased by 5.2%, 2.4%, 5.1%, 3.7%, 2.6%, 11%, respectively, on the DBLP dataset increased by 5.7%, 1.9%, 3.4%, 3.6%, 1.3%, 10.5%.

Meta-path Validity

In order to solve the problem thar the single feature in the traditional citation recommendation model leads to the poor effect of recommendation ranking, our adds 15 kinds of meta-path similar features to the deep neural network model to improve the ranking effect of the model. In order to prove its effectiveness, two different DeepCite models were trained on different numbers of recommended citations, one using meta-path similarity features and the other not using meta-path similarity features. The experimental results on the DBLP dataset prove that this feature effectively improves the mean to number ranking (MRR) during the reordering stage, as shown in Fig. 3.

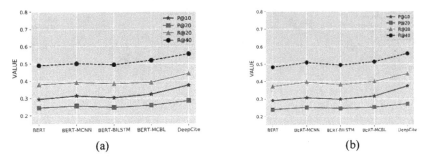

Fig. 2. Ablation experiment on DBLP and ANN datasets.

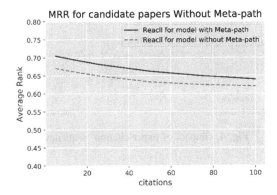

Fig. 3. MRR of predictions with varying number of candidates

Effect of Paper Citation Frequency

The minimum number of paper citations in the paper dataset used in model comparison experiments and model ablation experiments is 5. As shown in Fig. 4, we study the impact of the paper citation frequency on the performance of citation recommendation. In the ANN and DBLP datasets, the citation frequency is set between 1 and 7, respectively, at the accuracy rate (Precsion@10, Precsion@20). Experiments with recall rates (Recall @ 20, Recall @ 40). From the analysis of experimental results, as the citation frequency increases, the performance of our proposed model improves. When the citation frequency is 5, the global citation recommendation Model has the best recommendation performance. In other experiments in this paper, the reference frequency is set to 5. Generally speaking, uncited papers are not used for learning, so they can be treated as sparse data during testing. Furthermore, the experiment data can be improved according to the frequency of citations to establish a fully functional high-performance global citation recommendation model and provide users with recommendation services.

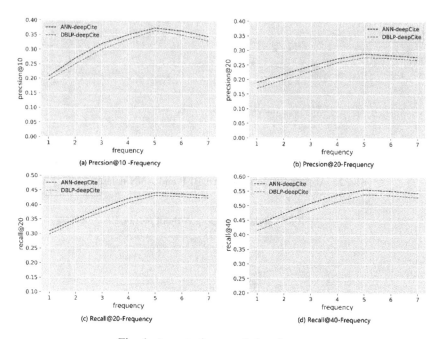

Fig. 4. Impact of paper citation frequency

6 Conclusions

This paper studies citation recommendations from two aspects: citation content and deep learning. A hybrid deep neural network model is used to extract high-level semantic representation vectors of text, and deep semantic matching is performed on query documents and candidate documents to generate candidate sets.

At the same time, the document representation vector generated by task 1 can be embedded in the vector space to improve the text matching speed. This paper extracts various effective citation features in the citation network, and uses the deep neural network to reorder the results of the candidate set in combination with the candidate set score, thereby improving the performance of citation recommendation. The experimental results show that the proposed DeepCite algorithm effectively changes the current citations. The algorithm is inefficient and the recommendation quality is low. The limit is that meta-path information of the citation network is not studied in this article. The next work will focus on the construction of complex citation networks. The deep learning technology will be used to embed the nodes and path information in the citation network.

Acknowledgment. The research work is supported by "Shenzhen Science and Technology Project" (JCYJ20180306170836595); "National key research and development program in China" (2019YFB2102300); "the World-Class Universities (Disciplines) and the Characteristic Development Guidance Funds for the Central Universities of China" (PY3A022); "Ministry of Education Fund Projects" (No. 18JZD022 and 2017B00030); "Basic Scientific Research

Operating Expenses of Central Universities" (No. ZDYF2017006); "Xi'an Navinfo Corp.& Engineering Center of Xi'an Intelligence Spatial-temporal Data Analysis Project" (C2020103); "Beilin District of Xi'an Science & Technology Project" (GX1803).

References

1. Livne, A., Gokuladas, V., Teevan, J., et al.: CiteSight: supporting contextual citation recommendation using differential search. In: Proceedings of the 37th International ACM SIGIR Conference on Research & Development in Information Retrieval, pp. 807–816 (2014)
2. Yang, C., et al.: CARES: a ranking-oriented CADAL recommender system. In: Proceedings of the 9th ACM/IEEE-CS Joint Conference on Digital Libraries, pp. 203–212. ACM (2009)
3. Sugiyama, K., Kan, M.: Scholarly paper recommendation via user's recent research interests. In: Proceedings of the 10th Annual Joint Conference on Digital Libraries, pp. 29–38. ACM (2010)
4. Li, S., et al.: Conference paper recommendation for academic conferences. IEEE Access **6**, 17153–17164 (2018)
5. Torres, R., McNee, S.M., Abel, M., et al.: Enhancing digital libraries with TechLens. In: Proceedings of the 4th ACM/IEEE-CS Joint Conference on Digital Libraries, pp. 228–236. ACM (2004)
6. Gori, M., Pucci, A.: Research paper recommender systems: a random-walk based approach, In: IEEE/WIC/ACM International Conference on Web Intelligence, pp. 778–781. IEEE (2006)
7. Meng, F., et al.: A unified graph model for personalized query oriented reference paper recommendation. In: Proceedings of the 22nd ACM International Conference on Information & Knowledge Management, pp. 1509–1512. ACM (2013)
8. Jardine, J., Teufel, S.: Topical PageRank: a model of scientific expertise for bibliographic search. In: Proceedings of the 14th Conference of the European Chapter of the Association for Computational Linguistics, pp. 501–510 (2014)
9. Sun, Y., et al.: Pathsim: meta path-based top-k similarity search in heterogeneous information networks. Proc. VLDB Endow. **4**(11), 992–1003 (2011)
10. Ren, X., et al.: ClusCite: effective citation recommendation by information network-based clustering, In: Proceedings of the 20th ACM SIGKDD International Conference on Knowledge Discovery and Data Mining, pp. 821–830 (2014)
11. Guo, L., et al.: Exploiting fine-grained co-authorship for personalized citation recommendation. IEEE Access **5**, 12714–12725 (2017)
12. Mu, D., et al.: Query-focused personalized citation recommendation with mutually reinforced ranking. IEEE Access **6**, 3107–3119 (2018)
13. Huang, E.H, Socher, R., Manning, C.D., et al: Improving word representations via global context and multiple word prototypes. In: Proceedings of the 50th Annual Meeting of the Association for Computational Linguistics: Long Papers-Volume 1, pp. 873–882. Association for Computational Linguistics (2012)
14. Bengio, Y., Ducharme, R., Vincent, P., et al.: A neural probabilistic language model. J. Mach. Learn. Res. **3**(Feb), 1137–1155 (2003)
15. Mikolov, T., Sutskever, I., Chen, K., et al.: Distributed representations of words and phrases and their compositionality. In: Advances in Neural Information Processing Systems, pp. 3111–3119 (2013)

16. Ebesu, T., Fang, Y.: Neural citation network for context-aware citation recommendation. In: Proceedings of the 40th International ACM SIGIR Conference on Research and Development in Information Retrieval, pp. 1093–1096. ACM (2017)
17. Peters, M.E., Neumann, M., Iyyer, M., et al: Deep contextualized word representations. arXiv preprint arXiv:1802.05365 (2018)
18. Devlin, J., Chang, M.W., Lee, K., et al.: BERT: pre-training of deep bidirectional transformers for language understanding. arXiv preprint arXiv:1810.04805 (2018)
19. Huang, P.S., He, X., Gao, J., et al.: Learning deep structured semantic models for web search using click through data. In: Proceedings of the 22nd ACM International Conference on Information & Knowledge Management, pp. 2333–2338. ACM (2013)
20. Hu, B., Lu, Z., Li, H., Chen, Q.: Convolutional neural network architectures for matching natural language sentences. In: Proceedings of the NIPS, pp. 2042–2050 (2014)
21. Palangi, H., et al.: Semantic modelling with long-short-term memory for information retrieval. arXiv preprint arXiv:1412.6629 (2014)
22. Bhagavatula, C., Feldman, S., Power, R., et al.: Content-based citation recommendation. arXiv preprint arXiv:1802.08301 (2018)
23. Sun, Y., Han, J., Yan, X., et al.: Pathsim: meta path-based top-k similarity search in heterogeneous information networks. Proc. VLDB Endow. **4**(11), 992–1003 (2011)
24. Lichtenwalter, R.N., Lussier, J.T., Chawla, N.V.: New perspectives and methods in link prediction. In: Proceedings of the 16th ACM SIGKDD International Conference on Knowledge Discovery and Data Mining, pp. 243–252. ACM (2010)
25. Robertson, S.E., Walker, S.: Some simple effective approximations to the 2-poisson model for probabilistic weighted retrieval. In: Croft, B.W., van Rijsbergen, C.J. (eds.) SIGIR 1994, pp. 232–241. Springer, London (1994). https://doi.org/10.1007/978-1-4471-2099-5_24
26. Guo, L., Cai, X., Hao, F., et al.: Exploiting fine-grained co-authorship for personalized citation recommendation. IEEE Access **5**, 12714–12725 (2017)
27. Tang, J., Zhang, J., Yao, L., et al.: Arnetminer: extraction and mining of academic social networks. In: Proceedings of the 14th ACM SIGKDD International Conference on Knowledge Discovery and Data Mining, pp. 990–998. ACM (2008)

Multi-factor Fusion POI Recommendation Model

Xinxing Ma, Jinghua Zhu[✉], Shuo Zhang, and Yingli Zhong

School of Computer Science and Technology,
Heilongjiang University, Harbin 150080, China
zhujinghua@hlju.edu.cn

Abstract. In the context of the rapid development of location-based social networks (LBSN), point of interest (POI) recommendation becomes an important service in LBSN. The POI recommendation service aims to recommend some new places that may be of interest to users, help users to better understand their cities, and improve users' experience of the platform. Although the geographic influence, similarity of POIs, and user check-ins information have been used in the existing work recommended by POI, little existing work considered combing the aforementioned information. In this paper, we propose to make recommendations by combing user ratings with the above information. We model four types of information under a unified POI recommendation framework and this model is called extended user preference model based on matrix factorization, referred to as UPEMF. Experiments were conducted on two real world datasets, and the results show that the proposed method improves the accuracy of POI recommendations compared to other recent methods.

Keywords: Multi-factor fusion model · Matrix factorization · Euclidean distance · Personalized recommendation

1 Introduction

1.1 A Subsection Sample

With the rapid development of the Internet, people have begun to use Internet resources and information for shopping, travel and investment activities. Due to the large amount of data, users cannot quickly screen the items they are interested in. Therefore, researchers have begun to focus on developing recommendation systems to help users make decisions [1]. Currently, location-based social networks are popular, such as Foursquare, Yelp, Gowalla, etc. They capture user ratings information and places that users often check in (e.g. restaurants, gyms, KTVs, bars, etc.) through their visit history. Recommender system can capture users' preferences from the related information and provide users with novel and interesting POIs [3, 6, 11].

POI recommendation is a new application based on location-based social network [4], which can help users explore attractive locations as well as help social network service providers design location aware advertisements for POI [8]. POI recommendation is more complicated than the ordinary recommendation system, and the

© Springer Nature Singapore Pte Ltd. 2020
P. Qin et al. (Eds.): ICPCSEE 2020, CCIS 1258, pp. 21–35, 2020.
https://doi.org/10.1007/978-981-15-7984-4_2

recommendation process will be affected by many factors. For example, the user-POIs check-in matrix is highly sparse, which is the main reason why it is difficult to capture the user's preference, the user's interest changes with time and location, and the text description of POI is incomplete and fuzzy. It has become a hot topic in the field of POI research to study these factors to improve the quality of POI recommendation. The current difficulties of POI recommendation are still data sparseness and cold start of users, which leads to impossible to accurately capture the user's preference only from a small percentage of the points of interest that we get from users [2]. In order to solve the problem of data sparseness, this paper combines the user-POIs rating matrix and the user-POIs check-in matrix, and forecasts the user's unrated POIs score using matrix factorization, which can reduce the data sparseness and facilitate the capture of user preferences. Then, the prediction matrix is combined with the user-POIs check-in matrix, which can enhance the degree of the user's favorite POIs. In this work, we also study the similarity of POIs and the geographic influence of POIs. Finally, we model four types of information in a unified POI recommendation framework and compared them experimentally. Experimental results on two real world datasets show that higher recommendation performance can be achieved by considering both user check-in behavior and user rating behavior.

The contributions of this paper are as follows:

1. User check-in behavior and user rating behavior are considered at the same time, which effectively reduces data sparseness and helps us obtain user preferences;
2. We integrate four types of information: user check-in information, user rating information, geographic influence of POIs and similarity of POIs into a unified model framework for POI recommendation on LBSNs;
3. Several experiments are conducted on two real world datasets, and the results show that our method improves the accuracy of POI recommendations compared to other recent methods.

2   Related Work

2.1   Matrix Factorization-Based Recommendations

Matrix factorization algorithm is a classic recommendation algorithm, which has good scalability and efficiency and is widely used in recommendation systems. The widely used traditional matrix factorization algorithms include Bayesian Personalized Ranking (BPR) [5], Singular Value Decomposition factorization (SVD) [10] and Probabilistic Matrix Factorization (PMF) [12]. The Bayesian personalized ranking recommendation model is based on Bayesian theory to maximize the posterior probability under prior knowledge. It realizes a recommendation system that trains multiple matrices from a user-item matrix, and a matrix represents a user's item preference to obtain the partial order relationship of multiple items to rank. Matrix singular value factorization can represent a more complex matrix by multiplying smaller and simpler sub-matrixes. These small matrices describe the important characteristics of the matrix. The idea of probability matrix factorization is that a user's preference is determined by a small number of unobserved factors. In a linear factor model, a user's preferences are modeled by linearly combining item factor vectors using specific user coefficients. However, it is often difficult to solve the problem of sparse data and cold start by simply considering the user's rating information, which will reduce the quality of recommendations. In order to solve this problem, the researchers introduced some auxiliary information or combined other algorithms in the algorithm. For example, Mohsen Jamali et al. [13] explored a matrix-based social network recommendation algorithm and added a trust propagation mechanism to the model. Gao et al. [7] integrated POI attributes, user interests and emotional indicators into a model and considered their association with check-in operations. Ren et al. [1] proposed a TGSC-PMF model, which uses text information, social information, classification information and popularity information, and effectively integrates these factors. Zhang et al. [19] model the impact of multiple tags by extracting a user tag matrix from the initial user-POIs matrix and introduce social normalization to simulate social impact. Finally, integrate multiple tags into the matrix factorization framework, and integrate social normalization and geographical impact into it.

2.2 Geographic Influence-Based Recommendation

Geographical factors are important factors in the recommendation of POI. Generally, we think that two POIs are geographically related or close to each other, then the probability of them being accessed by users at the same time is greater. In LBSN, different POIs are represented by their unique geographic locations. Users need to interact with POIs while purchasing products or entertainment, so we say that geographical influence is one of the factors that affects users' check in behavior very important factor. Some researchers have applied geographic related factors to recommend POIs. For example, Wang et al. [16] modeled the two POIs by three factors: the geographical influence of POI, the geographical sensitivity of POI, and the physical distance between POIs. Liu et al. [17] proposed a Geo-ALM model based on geographic information, which can be regarded as a fusion of geographic features and generative adversarial networks. It can learn discriminator and generator by region features and POI features of geography. Lian et al. [15] constructed a scalable and flexible framework, dubbed GeoMF++, for joint geographical modeling and implicit feedback-based matrix factorization. Some researchers have combined geographic factors with other factors and applied them to recommendation algorithms. For example, Wu et al. [18] proposed a new model to capture individual's personalized geographical constraints, combined with geographical constraints and time similarity, improved collaborative filtering model. Zhao et al. [9] created a sequence embedding rank (SEER) model for POI recommendation, and added two important factors in SEER model, i.e. time effect and geographical effect, to enhance POI recommendation system.

2.3 Similarity Measurement Method

Similarity metric learning plays an important role in practical applications. Its main idea is to learn an expected distance metric to make the distribution of similar samples more compact and the distribution of different types of samples more loose. When calculating the similarity between users/items in the recommendation algorithm, Euclidean distance and Cosine similarity are usually used for calculation. In fact, there are many other measures besides Euclidean distance and cosine similarity, such as Manhattan distance, Chebyshev distance and Mahalanobis distance. Cosine similarity and Euclidean distance are commonly used in some current recommendation algorithms to calculate the similarity between users and items, For example, He et al. [20] proposed a translation-based recommendation mode (TransRec), to predict and build the third-order interaction between users, items he previously visited and the next item in large scale. Wang et al. [21] proposed a video recommendation algorithm based on clustering and hierarchical model. The algorithm obtains similar users through clustering, and then collects the historical data of similar users' online video to form a video recommendation set, improving the performance and user experience of the recommendation system. Zhang et al. [22] used Euclidean distance instead of matrix factorization to measure the similarity between users. Two metric decomposition variables were designed in this paper, one for rating evaluation and another for personalized item ranking.

3 UPEMF Algorithm

First of all, the Gowalla and Foursquare datasets are processed using time division and multi-granularity space division to filter POIs with low visits and users who have fewer check-ins; Next, we transform the sparse rating matrix into a prediction matrix through probability matrix factorization; Again, we use the Hadamard product of the prediction matrix and the check-in matrix to form a weight matrix, and use the weight matrix to find the similarity of the POIs; At length, combining the similarity of POIs and the physical distance of POIs, the $top - k$ minimum values are obtained and recommended to users.

3.1 Symbol Description

For discussion purposes, Table 1 lists some of the important symbols used in this article.

Table 1. Related symbols and their meanings.

Symbol	Meaning
R	User-POIs rate matrix
U, V	User sets, POIs sets
I_{ij}	Indicator function
$\overline{R}$	Predicted user-POIs rate matrix
C	User-POIs check-in matrix
W	Weight matrix

(continued)

Table 1. (*continued*)

Symbol	Meaning
x, y	User-POIs check-in to a vector of two different POIs in a matrix
$\sum$	Covariance matrix
$\sigma_R^2, \sigma_U^2, \sigma_V^2$	Hyperparameter
d_{ij}	Physical distance between POI i and POI j
w	Weight value

U represents the user set, set contains m elements $U = \{u_1, u_2, \cdots, u_m\}$; V represents the POIs set, set contains n elements $V = \{v_1, v_2, \cdots, v_n\}$; $R = [R_{i,j}]_{m \times n}$ is the user-POIs check-in matrix, $R_{i,j}$ indicates the user u_i rate for POI v_j; $C = [C_{i,j}]_{m \times n}$ is the user-POIs check-in matrix, $C_{i,j}$ represents the number of times user u_i has visited the POI v_j; W is a weight matrix generated from the prediction matrix and the check-in matrix. The user's check-in matrix C and rating matrix R are very important for the POI recommendation system, which often implies the user's potential preference for POIs.

3.2 Data Preprocessing

In data preprocessing, we used two division methods: time division and multi-granularity space division. First, we use time division to divide the timestamps into four quarters. Considering that people prefer to go to the beach for vacation in summer and skiing in winter, so we divide the datasets into four types according to the timestamps of the datasets, which correspond to the check-in behavior of users in four quarters of a year. Next, We use granularity division to divide the spatial grid. There are two ways of granularity division, one is coarse-grained division and the other is fine-grained division. In brief, coarse-grained division refers to the rough division of space in the form of grid, which is conducive to capturing the overall information of the divided area; Fine-grained division pays more attention to details as local division, so through fine-grained division, we can better extract dense and discrete POIs. Finally, after time division and granularity division, we can filter out the POIs with less number of check-in times and number of check-in people in different regions in different seasons, that is, POIs with low visitor density, which can speed up the calculation speed and reduce the calculation cost.

Figure 1 shows the spatial granularity division. It can be seen from the figure that coarse-grained division is a simple division of the overall region and fine-grained division is a local division of the region.

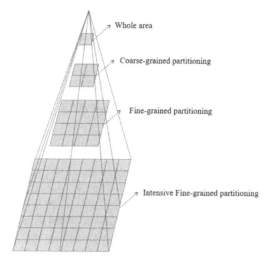

Fig. 1. Space granularity division

Figure 2 is a heat map formed by the Gowalla data set. The red areas represent areas that are popular with users, and the blue areas represent areas that users may not be interested in.

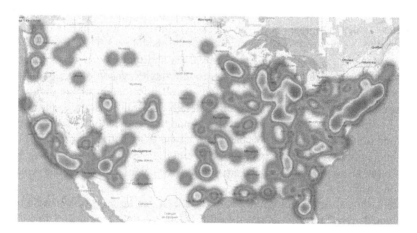

Fig. 2. Heat map formed by the Gowalla data set.

3.3 Probability Matrix Factorization Combine with Check-in Information

Probabilistic Matrix Factorization (PMF) is a classic traditional recommendation algorithm model. The idea of this model is that a user's preference is determined by a small number of unobserved factors. Since the probability matrix factorization can effectively handle large-scale data as well as sparse data and the probability matrix factorization algorithm does not spend a lot of time on the query, we use the probability matrix factorization algorithm for rate prediction. Suppose $U \in R^{m \times k}$ and $V \in R^{n \times k}$ are potential matrices of users and POIs respectively. We can calculate the predicted rate value through two potential matrices. The conditional probability distribution function of the rate matrix R is as follows:

$$p(R|U, V, \sigma_R^2) = \prod_{i=1}^{m} \prod_{j=1}^{n} [N(R_{ij}|g(U_i^T V_j), \sigma_R^2)]^{I_{i,j}^R} \tag{1}$$

$N(x|\mu, \sigma^2)$ is a Gaussian probability density function with μ as the mean and σ^2 as the variance, $g(x)$ is a Logistic function that maps the value of $U_i^T V_j$ to the $[0 - 1]$ interval. $I_{i,j}^R$ is an indicator function. If user u_i rates POI v_j, then $I_{i,j}^R = 1$, otherwise it is 0.

To prevent overfitting, it is assumed that the potential feature vectors of users and POIs follow a Gaussian distribution with a mean of 0, as follows:

$$p(U|\sigma_U^2) = \prod_{i=1}^{m} N(U_i|0, \sigma_U^2 I) \tag{2}$$

$$p(V|\sigma_v^2) = \prod_{j=1}^{n} N(V_j|0, \sigma_v^2 I) \tag{3}$$

By maximizing the loss function:

$$L(R, U, V) = \frac{1}{2} \sum_{i=1}^{m} \sum_{j=1}^{n} I_{i,j}^R (R_{ij} - g(U_i^T V_j))^2 + \frac{\lambda_U}{2} \sum_{i=1}^{m} U_i^T U_i + \frac{\lambda_V}{2} \sum_{j=1}^{n} V_j^T V_j \tag{4}$$

Finally, the prediction rate matrix $\bar{R}$ is obtained by the stochastic gradient descent method (see Fig. 3).

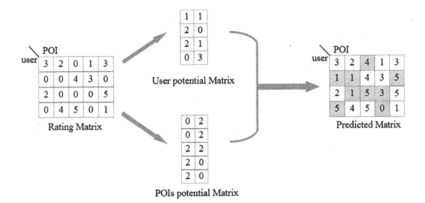

Fig. 3. Matrix factorization.

The prediction matrix $\overline{R}$ is obtained through the probability matrix factorization calculation. This method effectively solves the sparseness of the original rating matrix and fills the missing information in the original rating matrix. Different from other work, after getting the prediction matrix, we did not directly recommend it, but jointly signed in the matrix to predict the user's preference. We believe that the user's rating behavior can be combined with the user's check-in behavior. Some of the POIs that the user has checked in to have not been given a corresponding rating by the user, so we cannot know whether the user likes unrated POIs. Therefore, we combine the user-POIs rating matrix and the user-POIs check-in matrix to predict the score of the user's unrated items through matrix factorization. Then, the prediction matrix $\overline{R}$ and the check-in matrix C are used to perform Hadamard product operations, in this way, we can expand the user's preference.

For example, a certain user u_i in the data set visits the POI v_m and v_n three times and two times respectively, the user's rating of v_m is 3 and the rating of v_n is not needed. Therefore, the user's rating of v_n is 5 through prediction. We do the dot product of the user's rating and the number of check-ins. The weights are 9 and 10 respectively. The weight of the POI v_n is slightly larger, so we can judge that the user likes the POI v_n a little more (see Fig. 4).

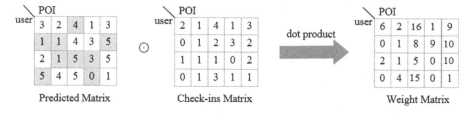

Fig. 4. Combination of prediction matrix and check-in matrix.

In the next section, we will use the weight matrix to find the similarity between POIs.

3.4   Similarities of POIs

In recommendation systems, the accuracy of recommendations is often improved by finding the similarity between users/POIs. There are many methods to calculate similarity, such as Euclidean distance, Cosine similarity, and Pearson correlation coefficient. Among them, Euclidean distance and cosine similarity are more widely used. Euclidean Distance is used to measure the absolute distance between various points in multi-dimensional space. The larger the distance, the greater the difference between individuals; Cosine Similarity uses the cosine of the angle between two vectors in space as a measure of the difference between two individuals. The smaller the angle, the more similar the two vectors are. In this article, we use distance measures to measure the differences between individuals. Euclidean distance is calculated based on the absolute values of the characteristics of each dimension. At the same time, Euclidean distance measures need to ensure that the indicators of each dimension are at the same scale level.

$$d(x,y) = \sqrt{\sum_{i=1}^{n} (x_i - y_j)^2} \tag{5}$$

The vectors $x = (v_{i1}, v_{i2}, \cdots, v_{in})$ and $y = (v_{j1}, v_{j2}, \cdots, v_{jn})$ represent different POIs vectors in the weight matrix W. We calculate the difference between POIs through Euclidean distance calculation, so that users can make personalized recommendations. However, it is not enough to consider only the differences between POIs. We also need to consider the geographical influence between POIs, so we will describe how to add geographic influence in the recommendation algorithm.

3.5   Geographic Influence of POIs

In the process of making the POI recommendation algorithm, we considered the geographical influence between the POIs. Geographic influence is the ability to capture POIs to spread visitors to other POIs. The closer the physical distance between POIs, the greater the impact between POIs. We obtain the geographic correlation between POIs by the physical distance between POIs, where $f(d_{ij})$ represents the probability that two POIs with a given physical distance d_{ij} will be accessed by the same user. The method we used to find the similarity of POIs in Sect. 3.4 is Euclidean distance, which is a representation of distance and is also geographically related. Therefore, we combine the physical distance of POIs with the similarity of POIs to get the $top - k$ minimum values and recommend POIs to users.

$$r_{ij} = wd(x,y) + (1 - w)f(d_{ij}) \tag{6}$$

$f(x) = e^x$ increases with the increase of x, which means that the physical distance between two POIs increases, so the probability of being visited by the same user will decrease. $d(x,y)$ means that the difference between POIs increases with the increase of distance. Finally, we recommend the $top - k$ minimum values to users.

4   Experiment

4.1   Data Set Description

We use two real-world datasets of the typical LBSN website Gowalla and Foursquare. Gowalla and Foursquare are two large social network sites, they allow users to check in at different locations and share their locations through the check-in. They can analyze the relevant information of the check-in data (e.g. time, space, popularity, etc.) to report a quantitative assessment of people's movement patterns. Gowalla data in the experiment was collected from the United States, and the check-in time span was from February 2009 to October 2010. We collected a total of 673,980 check-in locations for these users. The data set includes the user's social network relationship, the user's IDs, the IDs of the checked-in POIs, the latitude and longitude of each checked-in POI, and the time when the POIs checked in. In the pre-processing process of coarse-grained division, fine-grained division and time division, we filtered out regional blocks with thresholds below 10,000. In the experiment, the Gowalla dataset contains 15043 users and visited a total of 106423 POIs.

Foursquare data in the experiment was collected from the United States, and the check-in time span was from July 2011 to December 2012. A total of 386,975 check-in locations were collected. The data set includes IDs of users and POIs, geographic location and check-in time of POIs. The same as the previous data preprocessing, we filter out the area blocks with threshold value less than 10000 in the preprocessing process of coarse-grained division, fine-grained division and time division. In the experiment, the Foursquare dataset contains 9698 users who visited a total of 54216 POIs. Among them, 80% were selected as the training set, and the remaining 20% were used as the test set. The statistics of the data set are shown in Table 2.

Table 2. Datasets summary.

Information	Gowalla	Foursquare
User's number	15043	9698
POIs's number	106423	54216
Check-ins's number	673980	386975

4.2   Performance Measurements

We select the most extensive evaluation indicators in information retrieval and statistical classification: precision *precision@N* and recall *recall@N*.

$$precision@N = \frac{1}{|U|} \sum_{i=1}^{|U|} \frac{S_i(N) \cap V_j}{N} \tag{7}$$

$$recall@N = \frac{1}{|U|} \sum_{i=1}^{|U|} \frac{S_i(N) \cap V_j}{|V_j|} \tag{8}$$

$|U|$ indicates the number of users, $S_i(N)$ indicates the recommendation of N POIs to user u_i, V_j indicates the set of POIs visited by the user in the test set, and N indicates the number of recommended POIs. This article selects $N = \{5, 10, 15, 20\}$ to calculate the precision and recall values respectively, and then serves as the evaluation index of this article.

4.3   Experimental Setup

The parameter $w(0<w<1)$ mentioned in Sect. 3.5 needs to be determined through experiments. The parameter w is used to adjust the proportion of the physical distance between POIs and the similarity between POIs in the recommended results. When the value of w is larger, the distance obtained through the calculation of the physical distance of the POIs has a greater impact on the recommendation result; Conversely, the percentage of similarity of POIs obtained through the calculation of the similarity between POIs is relatively large. We finally determined the value by testing on the real-world Gowalla and Foursquare datasets.

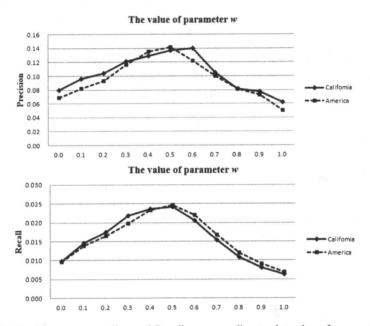

Fig. 5. Precision corresponding and Recall corresponding to the value of parameter w.

Figures 5 shows the changes in the precision and recall rates when w is $[0-1]$. As can be seen in the figure, when the value of w is about 0.5, higher precision and recall can be obtained. Therefore, in the following comparative experiment, the value of w is set to 0.5 by default.

4.4   Baseline Methods

In order to verify the effectiveness of the UPEMF model proposed in this paper, we evaluate the performance of UPEMF by comparison with the following representative POI recommendation methods.

1. PMF [11]: According to the user's historical rating data, the scores of the POIs that are not rated are predicted, and the POIs with high ratings are recommended to the user by prediction.
2. BPR [5]: The pairing method is adopted, and the item ranking is calculated by the triples formed by the user and the item. The multiple groups finally formed constitute a recommendation sequence to recommend to the user.
3. GeoMF [14]: Joint Geographical Modeling and Matrix Factorization for Point-of-Interest Recommendation, 2014): By using the spatial clustering phenomenon of user check-in behavior, the potential factors of users and POIs, the scope of users' activities and the influence area of POIs are enhanced. The spatial clustering phenomenon is captured and integrated into the matrix factorization, and the final recommendation results are obtained through the matrix factorization.
4. PACE [23]: A deep neural structure that embeds user preferences and context information. It learns the embedding of users and POIs together to predict users' preferences for POIs and various contexts related to users and POIs.
5. GeoUMF [24]: The model considers the influence of geographical factors and user factors. The objective function in the model considers the difference between the ranking generated in the recommendation model and the actual ranking in the check-in data and defines a consideration of the difference in POI access frequency in the objective function Approximation.

4.5   Performance Evaluation

Figures 6 and Figs. 7 correspond to the Gowalla dataset and Foursquare dataset respectively and show the recommended performance of $top - N(N = 5, 10, 15, 20)$ in different ways of processing the dataset.

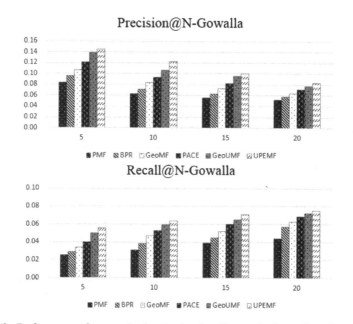

Fig. 6. Performance of our method and other baseline methods on Gowalla dataset.

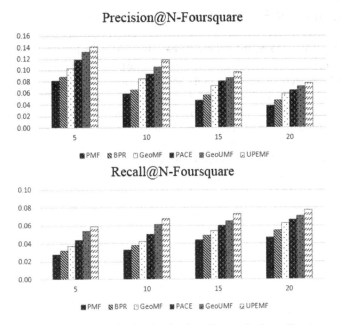

Fig. 7. Performance of our method and other baseline methods on Foursquare dataset.

According to the analysis in Fig. 6 and Fig. 7, PMF is used as a basic matrix factorization algorithm, and its precision and recall are the lowest; BPR is more precision and recall than the basic PMF algorithm, each user is regarded as a relatively independent individual to make personalized recommendation for users in BPR. The improvement of precision and recall rate proves that personalized recommendation for users can improve the quality of recommendation; GeoMF integrates the spatial clustering phenomenon into the decomposition model and increases the potential factors of users and items through the active region vector of users and the influence region vector of POIs. It can be seen from the figure that the recommendation effect of GeoMF algorithm is higher than that of the previous two algorithms, which shows that the geographical factors cannot be ignored in POI recommendation, which is conducive to improving the recommendation effect; We use the current popular deep recommendation network PACE in the baseline model to show the excellent performance of this model, and show the strong learning ability of deep neural network by comparing with BPR and PMF; GeoUMF considers the influence of geographical factors and user factors at the same time. Its performance is higher than other baseline models, which also proves the superiority of adding geographical factors and user factors; The precision and recall of the UPEMF algorithm proposed in this paper are higher than those of the previous algorithms. UPEMF combines the features of the previous three algorithms to improve the precision and recall. We can see that the pre-processing based on region and time division, and the introduction of geographical influence factors and personalized recommendations to users can significantly improve the quality of recommendations.

5 Conclusion and Future Work

In this paper, we use the relationship between user check-in behavior and user rating behavior to improve POI recommendation. We learn the connections between users and POIs and connections between POIs and POIs on LBSNs based on user check-in behavior, user rating behavior, POIs geographic influence, and POIs similarity. We model them under a unified POI recommendation framework. Our experimental results demonstrate the importance of considering both user check-in behavior and user rating behavior in explaining user behavior and improving the performance of POI recommendations on the LBSN. In the future work, we will consider how to utilize the user's comment information on POIs to learn more user preferences and habits to improve the recommendation quality.

References

1. Ren, X.: Context-aware probabilistic matrix factorization modeling for point-of-interest recommendation. Neurocomputing **1**(7), 38–55 (2017)
2. Ren, X.: Point-of-interest recommendation based on the user check-in behavior. Chin. J. Comput. **40**, 28–51 (2017)
3. Zhang, J.D., Chow, C.Y.: Point-of-interest recommendations in location-based social networks. Sigspatial Spec. **7**(3), 26–33 (2016)
4. He, J.: Inferring a personalized next point-of-interest recommendation model with latent behavior patterns. In: Proceedings of the Thirtieth AAAI Conference on Artificial Intelligence, AAAI, pp. 137–143. ACM (2016)
5. Rendle, S., Freudenthaler, C.: BPR: Bayesian personalized ranking from implicit feedback. In: Proceedings of the Twenty-Fifth Conference on Uncertainty in Artificial Intelligence, UAI, pp. 452–461. ACM (2009)
6. Gao, H.: Exploring temporal effects for location recommendation on location-based social networks. In: Proceedings of the Thirtieth AAAI Conference on Artificial Intelligence, AAAI, pp. 93–100. ACM (2016)
7. Gao, H.: Content-aware point of interest recommendation on location-based social networks. In: Proceedings of the Twenty-Ninth AAAI Conference on Artificial Intelligence, AAAI, pp. 1721–1727. ACM (2015)
8. Zhao, S.: A survey of point-of-interest recommendation in location-based social networks. In: Proceedings of the Twenty-Ninth AAAI Conference on Artificial Intelligence, AAAI, pp. 53–60. ResearchGate (2016)
9. Zhao, S., Zhao, T.: GT-SEER: geo-temporal sequential embedding rank for point-of-interest recommendation. In: 2017 International World Wide Web Conference Committee, WWW, pp. 153–162. ACM (2016)
10. Kalman, D.: A singularly valuable decomposition: the SVD of a matrix. Coll. Math. J. **1**(27), 2–23 (1996)
11. Feng, S., Li, X.: Personalized ranking metric embedding for next new POI recommendation. In: International Conference on Artificial Intelligence 2015, IJCAI, pp. 2069–2075. ACM (2015)
12. Mnih, A., Ruslan, S.: Probabilistic matrix factorization. In: Neural Information Processing Systems 2007, NIPS, pp. 1257–1264. ResearchGate (2008)

13. Jamali, M., Martin, E.: A matrix factorization technique with trust propagation for recommendation in social networks. In: Proceedings of the 2010 ACM Conference on Recommender Systems, ACM, pp. 135–142. ACM (2010)
14. Lian, D.: GeoMF: joint geographical modeling and matrix factorization for point-of-interest recommendation. In: Proceedings of the 20th ACM SIGKDD International Conference on Knowledge Discovery and Data Mining, KDD, pp. 831–840. ACM (2014)
15. Lian, D.: GeoMF++: scalable location recommendation via joint geographical modeling and matrix factorization. ACM Trans. Inf. Syst. **36**, 1–29 (2018)
16. Wang, H.: Exploiting POI-specific geographical influence for point-of-interest recommendation. In: Twenty-Seventh International Joint Conference on Artificial Intelligence, IJCAI, pp. 3877–3883. ResearchGate (2018)
17. Liu, W.: Geo-ALM: POI recommendation by fusing geographical information and adversarial learning mechanism. In: Twenty-Eighth International Joint Conference on Artificial Intelligence, IJCAI, pp. 1807–1813 (2019)
18. Wu, H., Shao, J., Yin, H., Shen, H.T., Zhou, X.: Geographical constraint and temporal similarity modeling for point-of-interest recommendation. In: Wang, J., et al. (eds.) WISE 2015. LNCS, vol. 9419, pp. 426–441. Springer, Cham (2015). https://doi.org/10.1007/978-3-319-26187-4_40
19. Zhang, Z.: Fused matrix factorization with multi-tag, social and geographical influences for POI recommendation. In: World Wide Web 2018, WWW, pp. 1135–1150. ResearchGate (2018)
20. He, R., Kang, W.: Translation-based recommendation. In: Proceedings of the Eleventh ACM Conference on Recommender Systems 2017, RECSYS, pp. 161–169. ACM (2017)
21. Wang, Y.: Collaborative filtering recommendation algorithm based on improved clustering and matrix factorization. J. Comput. Appl., 1001–1006 (2018)
22. Zhang, S.: Metric factorization: recommendation beyond matrix factorization. Inf. Retr. **6**(4), (2018)
23. Yang, C.: Bridging collaborative filtering and semi-supervised learning: a neural approach for POI recommendation. In: Knowledge Discovery and Data Mining 2017, KDD, pp. 1245–1254. ResearchGate (2017)
24. Xu, Y., Li, Y.: A multi-factor influencing POI recommendation model based on matrix factorization. In: 2018 Tenth International Conference on Advanced Computational Intelligence, ICACI, pp. 514–519 (2018)

Effective Vietnamese Sentiment Analysis Model Using Sentiment Word Embedding and Transfer Learning

Yong Huang[1,2], Siwei Liu[1(✉)], Liangdong Qu[1], and Yongsheng Li[1]

[1] College of Software and Information Security,
Guangxi University for Nationalities, Nanning, China
waaalswaaa@163.com
[2] College of Mathematics and Computer Science,
Guangxi Normal University for Nationalities, Chongzuo, China

Abstract. Sentiment analysis is one of the most popular fields in NLP, and with the development of computer software and hardware, its application is increasingly extensive. Supervised corpus has a positive effect on model training, but these corpus are prohibitively expensive to manually produce. This paper proposes a deep learning sentiment analysis model based on transfer learning. It represents the sentiment and semantics of words and improves the effect of Vietnamese sentiment analysis model by using English corpus. It generated semantic vectors through Word2Vec, an open-source tool, and built sentiment vectors through LSTM with attention mechanism to get sentiment word vector. With the method of sharing parameters, the model was pre-training with English corpus. Finally, the sentiment of the text was classified by stacked Bi-LSTM with attention mechanism, with input of sentiment word vector. Experiments show that the model can effectively improve the performance of Vietnamese sentiment analysis under small language materials.

Keywords: Sentiment analysis · Long short-term memory · Attention mechanism · Sentiment word vector · Transfer learning

1 Introduction

With the rapid development of the Internet, users generate a large number of reviews in the online community every day. These data contains a lot of useful information. For example, merchants can derive market demand for products based on customer reviews, and the government can analyze social public opinion based on comments from netizen. However, it is very expensive to analyze these data manually, so it is necessary to find an effective and automatic sentiment analysis method. Since 2000, sentiment analysis has become one of the most active research fields in natural language processing (NLP).

In 2006, the first year of deep learning. Hiton proposes a solution to the vanishing gradient problem in deep network training. A few years late, deep

---

The original version of this chapter was revised: The first sentence in section 4.2 has been revised by adding the reference [18], and the source has been added in the References section. The correction to this chapter is available at
https://doi.org/10.1007/978-981-15-7984-4_48

© Springer Nature Singapore Pte Ltd. 2020, corrected publication 2022
P. Qin et al. (Eds.): ICPCSEE 2020, CCIS 1258, pp. 36–46, 2020.
https://doi.org/10.1007/978-981-15-7984-4_3

learning technology began to be widely used in natural processing and achieved outstanding results. Deep learning model are complex in structure and highly flexible, so they have great research value. The research in this paper aims to find a deep learning model that can effectively improve the perform of sentiment classification. Our model uses deep learning models such as LSTM, transfer learning, and attention mechanisms to improve the effect.

The development of deep learning and computing technology are inseparable. The increase in the amount of computing allows computers to learn 'knowledge' autonomously from massive amounts of data. Although there is sufficient unsupervised data on the Internet, supervised data is still scarce because of the need for manual labeling, especially in minority language. Transfer learning and cross-domain learning are significant for this situation.

Using deep learning for sentiment analysis, word embedding can be said to be indispensable. Most of the existing word vectors represent the semantics of words based on context. However, when performing sentiment analysis tasks, it is obviously unreasonable not to consider the sentiment information of words in word embedding. For example, words with opposite sentiment polarity may appear in similar contexts, such as "I'm sad" and "I'm happy." Whose vectors learned from contexts will be similar. At present, there are mainly two researches on sentiment word vector, one is based on sentiment dictionaries [1,2], and the other is based on machine learning models [3,4]. The former depends too much on the dictionary, and maintaining a good sentiment dictionary is very difficult and expensive. The latter solves the problem of over-reliance on dictionaries, but how to use it to obtain word vectors is still a problem worth studying.

Based on some problems in the existing research mentioned above, this paper proposes an effective sentiment analysis model in Vietnamese. This model combines semantic and sentiment features of words in the text, while improving the problem of insufficient Vietnamese corpus through transfer learning. It consists of three parts, one is to generate the sentiment word vectors of the words, the other is using English corpus to train the shared parameters of the neural network, and finally use them to construct the neural network to classify the sentiment polarity of the Vietnamese text. Semantic vectors are generated from the context of the text, which is the same as the word vectors currently used in most sentiment analysis systems. The context representation model used in this paper is skip-gram [5]. Both deep learning and attention mechanisms are now widely used in sentiment analysis techniques [6–8]. In order to learn sentiment word vector, the study generate sentiment vectors according to the sentiment tendencies of texts by LSTM [9] and attention mechanism. After integrating sentiment vector, the information in word vector is more abundant. At the same time, through the pre-training of the English corpus, the network parameters are no longer randomly initialized. This makes our model perform better.

The main contribution of this work include:

1. This paper proposes a sentiment word vector, which represents both the context information and the sentiment information of words.
2. We show how to use English corpus to improve the effect of sentiment classification of Vietnamese through transfer learning.

The content of this paper is organized as follows. Section 2 gives a brief introduction to some research on the sentiment word embedding and sentiment analysis models in Vietnamese related to this paper. Section 3 briefly introduce mathematical principle of the sentiment word vector model proposed in this paper. The generation process of sentimental word vector is explained in detail. Section 4 use a variety of models (including the model proposed in this paper and those proposed in some other studies), and Sect. 5 to analyze the test results. The last Section summarizes the advantages and disadvantages of the proposed model, as well as the direction of improvement.

2   Relative Work

Here, we describe briefly some of the research status related to our theses.

In the research of sentiment word embedding, some scholars try to train word vectors not only retain semantics information, but also to retain sentiment information [10,11]. In this study [10], they use neural networks to train sentiment feature and semantics feature into a vector at the same.

In sentiment analysis of Vietnamese [12,13], these studies basically did not involve deep learning models. In view of problems in the above research, this paper propose an sentiment analysis model that integrates multiple features. It dose not need to rely on dictionaries, but also can effectively extract sentiment information from words. In addition, this model also uses deep learning models and transfer learning to improve model performance.

3   Sentiment Word Vector and Features Transfer

The content of this chapter is mainly the mathematical principle and model architecture of our proposed model. This paper propose to incorporate the sentiment information of sentences into word embedding. The study use the existing machine LSTM, attention mechanism and develop neural network to learn sentiment vector of the words. Sentiment Word Vector is composed of sentiment vector and semantic vector. Based on shared model parameters, the novel model use the English corpus to train the weights and biases of network as initial values for Vietnamese classification model. In order to get the final model, we need the following steps: generate the sentiment word vector; use English corpus to train the pre-train value of the network; train the Vietnamese classification model.

3.1   Semantics Vector

Our semantic vector, like most models, maps words to vectors in lower dimensions through training. By modeling the relationship between words and their context, we can get their semantic information.

The semantic vectors in this paper are all generated by Google's open source tool Word2Vec, and the text representation model is skip-gram. Some parameters of the model are set to: window = 5, min_count = 5.

3.2 Sentiment Vector

Obviously, the meaning of words will vary according to the context, so the generation of sentiment vectors should be related to the context. In addition, the word order should also be noted. Different word orders can also change the meaning of words. LSTM can capture the information of words in text according to context and word order, and overcome the problem of long dependence, which is why the proposed model use it to generate sentiment vectors.

In order to capture important information, such as words expressing emotions, attitudes, it was introduced that an attention mechanism [14] in the model to solve this problem.

The architecture of the neural network model used to generate sentimental vectors in this paper is shown in Fig. 1. As shown in the figure, the model consists of a total of four layers, the embedding layer, the LSTM layer, the attention layer and the softmax layer.

The specific calculation process of the sentiment vector is as follows.

Step 1: Use the semantic vector of the word as the initial value of its sentiment vector, and randomly initialize other parameters.

Step 2: Through step 1, the text will be transformed into a matrix of word sentiment vectors. Then, the vector of each word is used as the input of each time sequence of the LSTM. The detailed calculation process is as follows:

$$f_t = \sigma(W_f \cdot [h_{t-1}, X_t] + b_f) \tag{1}$$

$$i_t = \sigma(W_i \cdot [h_{t-1}, X_t] + b_i) \tag{2}$$

$$\widetilde{C}_t = tanh(W_C \cdot [h_{t-1}, X_t] + b_C) \tag{3}$$

$$C_t = f_t * C_{t-1} + i_t * \widetilde{C}_t \tag{4}$$

$$o_t = \sigma(W_o \cdot [h_{t-1}, X_t] + b_o) \tag{5}$$

$$h_t = o_t * tanh(C_t) \tag{6}$$

Among them, h_{t-1} is the output of the last LSTM cell, and the X_t is the sentiment vector for current input word. W_f, W_b are the weight matrix and bias matrix corresponding to the forget gate, respectively. W_i, W_C, b_i, b_c are the input gate's weight matrix and the bias matrix, respectively. W_o, b_o are the weight matrix and bias matrix of output gate.

Similarly, the output of each timing of the first layer LSTM is used as the input of the second layer LSTM, and the output $H = [h_1, h_2, \ldots, h_n]$ is obtained.

Step 3: After the sentiment features are extracted by LSTM, the obtained output is input into the attention layer, after calculating the weight of each input, they are weighted and summed. The specific calculation process is as follows, where ω is the weight matrix of softmax layer. The representation r of sentences is obtained by weighted sum of vectors in H.

$$M = tanh(H) \tag{7}$$

$$\alpha = softmax(\omega H) \tag{8}$$

$$r = H\alpha^T \tag{9}$$

The final output of the model is:

$$h^* = tanh(r) \tag{10}$$

Step 4: Enter h^* into the fully connected layer to classify sentiment polarity. W, b is the weight matrix and bias matrix of softmax layer, respectively. We set the output of the fully connected network to 1 dimension and use 0.5 as the threshold to judge the sentiment polarity s.

$$s = Sen(W \cdot h^* + b) \tag{11}$$

$$Sen(x) = \begin{cases} 0 & sigmoid(x) < 0.5 \\ 1 & else \end{cases} \tag{12}$$

The loss function of the model is as follows. s is the predicted value and z is the label value (In the label, 1 represents positive, 0 represents negative). This function selected from tensorflow, is a cross-entropy function that incorporates sigmoid and, to ensure stability and avoid overflow, the implementation uses this equivalent formulation.

$$loss = max(s, 0) - s * z + log(a + e^{-|s|}) \tag{13}$$

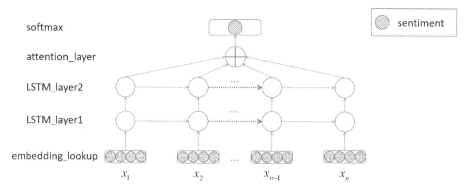

Fig. 1. The architecture for LSTM, attention mechanism and softmax layer for sentiment representation learning.

3.3   Sentiment Word Embedding Generation

After generating the semantic and sentiment vectors, we connect them to obtain the sentiment word vectors. In the composition of the sentiment word vector, the length of the semantic vector is equal to the length of the sentiment vector. As shown in Fig. 2.

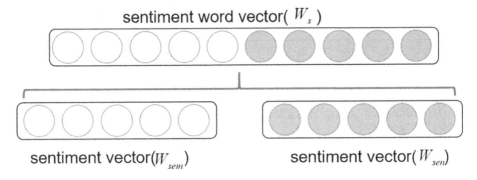

Fig. 2. The generation of sentiment word embedding.

3.4 Sentiment Classification Model Based on Transfer Learning

The model consists of two parts in total, one is a pre-training part trained using English corpus, and the other is a Vietnamese classifier based on the former. Because the architecture of the two neural networks and the size of the parameters are exactly the same, the transfer of learning can be performed by a method of sharing parameters. The model architecture is shown in Fig. 3.

As shown in Fig. 3, the input is first embedded as an sentiment word vector. Features are then extracted through the two BiLSTM layers, then weighted and summed by the attention layer, and finally classified by the fully connected layer. The model calculation details are basically the same as the sentiment word vector generation model. The difference is that results of forward and backward LSTM are stitched together as output.

Before training the network with Vietnamese text, use English corpus for training. After training in English, we use the weights and biases of the full-connected layer as initial values for the same roles in the Vietnamese sentiment classification model.

4 Experiments

In order to verify the performance of the model, our experiments used multiple sentiment analysis models to compare with our model. At the same time, the experiments also made some visualization to illustrate the motivation of using transfer learning methods.

4.1 Design

In the experiments, some classic models of machine learning and deep learning were compared with the proposed model. And ensure that everything except the model is consistent, such as the operating environment, etc. The word vectors used in all models are 106 dimensions, and the sentiment word vectors

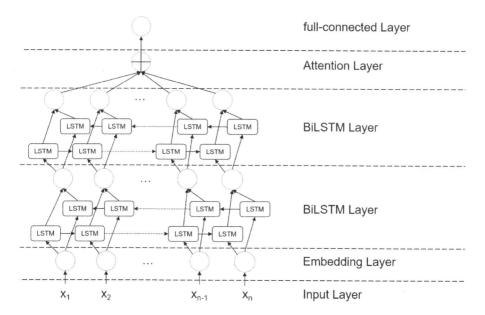

Fig. 3. The architecture for stacked BiLSTM for sentiment analysis.

are composed of 53-dimensional semantic vectors and 53-dimensional sentiment vectors. The comparison model uses SVM, TextCNN [15], stacked BiLSTM, BiLSTM+Attention [14]. In addition, we reduced the dimensionality of the data features extracted from stacked BiLSTM. The dimensionality reduction algorithm is t-SNE [16], and the results are visualized.

4.2   Dataset

Vietnamese Students' Feedback Corpus (UIT-VSFC) [18] is used for training the Vietnamese model. Also, the English corpus (we used a total of 20,000 pieces of data) uses movie reviews on IMDB [17]. The ratio of our training and test sets is 8:2. During the training process of sentiment analysis model, only the above data are used. The specific data is shown in Table 1.

Table 1. The number of data for training and testing.

Corpus	Train data	Test data
Positive data	7416	1864
Negative data	5453	1354
Total	12869	3218

4.3   Model Parameters

All the sentiment word vectors in this paper are 106-dimensional. In terms of the sentiment learning model, the length of the sentiment vector is 53 dimensions, which is half the length of the length of the total word vector. Furthermore, the learning rate of the model is 0.001, and each batch is 64 data. The number of neurons in LSTM is 128. Classification model TextCNN has convolution kernels of four sizes: 2, 3, 4, and 5, and the number of convolution kernels of each size is 128. The max length of the model input sequence is 100. In the generated sentiment word vector model, the output dimension of LSTM is 128. In the sentiment analysis model, the number of neurons in the LSTM cells in the first layer is 256, and the second layer is 128. The number of neurons in the SSWE [4] is 256 and 128, the learning rate and others are the same as above.

4.4   Experimental Results

All the results below are averaged after many tests. Results of different classification models, such as TextCNN and Bi-LSTM, are used to test the word vectors. Unless otherwise stated, word vector used in all models in the experiment are generated by word2Vec.

The accuracy (acc) and the formula for calculating the F1_score (F1) are as follows:

$$acc = \frac{TP + TN}{total} \tag{14}$$

$$F1 = \frac{2 \times P \times R}{P + R} \tag{15}$$

TP and TN are true positives and true negative, respectively. *total* is the number of the samples. The expression of P is $\frac{TP}{TP+FP}$, and R is $\frac{TP}{TP+FN}$. FP is false positives and FN is false negative.

In this experiment, we evaluated several sentiment models to show the effect of different word vectors on different models. Other classifiers not specifically Mentioned are BiLSTM with attention mechanism.

In addition, we also use BiLSTM to perform feature extraction on two language corpora, as shown in Fig. 4.

Table 2. Result of different sentiment analysis models.

Model	Result	
	Acc (%)	F1 (%)
SVM	62.06	26.82
TextCNN	84.68	81.81
BiLSTM	83.41	80.34
BiLSTM+attention	86.56	84.37
SSWE	86.29	84.33
Our model	**86.64**	**84.38**

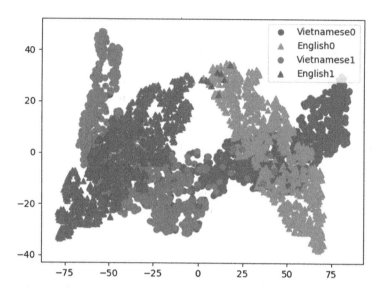

Fig. 4. Visualization of data features after dimensionality reduction.

5   Analysis

The distribution of high-level features of the two corpora after abstraction can be seen in Fig. 4. The picture contains data in two different languages, where 0 represent negative data and 1 positive. In English, it may use 'good', 'great' to express positive emotions, and in Vietnamese, it may use 'tốt', 'tuyệt'. However, after feature extraction, the data is projected onto the same feature space. At this time, the different writing of the same word in different languages is not important. In this way, in this space, the data in the source domain (English) is the same as the data in the target domain (Vietnamese), so that in the new space, the existing labeled data samples in the source domain can be better used for classification training. Because of the different fields, there still some differences in the distribution of two languages, but these can be fine-tuning by using Vietnamese corpus later. Thought pre-training and fine-tuning, cross-language training neural networks can be implemented, thereby improving the problem of sparse corpora in minority language.

It can be seen from Table 2 that the classic machine learning model SVM does not perform well in sentiment task. On the contrary, all deep learning models have achieved an accuracy rate of more than 83%. In the experiments, our model achieved better performance, which is 0.08% and 0.01% higher than BiLSTM's accuracy and F1 score, respectively. Based on the results, we speculate that the reasons for the better model proposed in this paper are as follow. The main reason is that our model effectively incorporates sentiment information into the word vector, making it more informative. Secondly, the proposed model train the sentiment vector and semantics vector separately, because it can extract and preserve the semantics and sentiment features of the text more effectively. Also, stacked BiLSTM can learn more. Last but not least, through transfer learning, we use English corpus pre-training the model to advance the effect of the it.

In general, by incorporating sentiment features into word vectors and using transfer learning method, our model effectively improves the classification perform.

6   Conclusion

It studies the incorporation of sentiment information into word vectors, and shows that transfer learning can improve performance. This paper also introduced how to improve the categorization of Vietnamese through the transfer of learning English corpus. Lastly, through experiments, we verified that our model can effectively improve the performance of Vietnamese sentiment analysis. Future work will filter the features presented by the English corpus and use a pre-training model.

Acknowledgment. This work is supported by Chinese National Science Foundation (#61763007), the higher education research project of National Ethnic Affairs Commission "Research and Practice on the Training Mode of Applied Innovative Software Talents Base on Collaborative Education and innovation" (17056), and the Innovation Team project of Xiangsihu Youth Scholars of Guangxi University For Nationalities.

References

1. Meng, S., Zhao, Y., Guan, D., Zhai, X.: A sentiment analysis method combining sentimental and semantic information. J. Comput. Appl. **39**(7), 1931–1935 (2019)
2. Xu, G., Yu, Z., Yao, H., Li, F., Meng, Y., Wu, X.: Chinese text sentiment analysis based on extended sentiment dictionary. IEEE Access **7**, 43749–43762 (2019)
3. Mullen, T., Collier, N.: Sentiment analysis using support vector machines with diverse information sources. In: Proceedings of the 2004 Conference on Empirical Methods in Natural Language Processing, pp. 412–418 (2004)
4. Tang, D., Wei, F., Yang, N., Zhou, M., Liu, T., Qin, B.: Learning sentiment-specific word embedding for Twitter sentiment classification. In: Proceedings of the 52nd Annual Meeting of the Association for Computational Linguistics (Volume 1: Long Papers), pp. 1555–1565 (2014)
5. Mikolov, T., Chen, K., Corrado, G., Dean, J.: Efficient estimation of word representations in vector space. arXiv preprint arXiv:1301.3781 (2013)
6. Xu, G., Meng, Y., Qiu, X., Yu, Z., Wu, X.: Sentiment analysis of comment texts based on BiLSTM. IEEE Access **7**, 51522–51532 (2019)
7. Yang, Z., Yang, D., Dyer, C., He, X., Smola, A., Hovy, E.: Hierarchical attention networks for document classification. In: Proceedings of the 2016 Conference of the North American Chapter of the Association for Computational Linguistics: Human Language Technologies, pp. 1480–1489 (2016)

8. Wang, Y., Huang, M., Zhu, X., Zhao, L.: Attention-based LSTM for aspect-level sentiment classification. In: Proceedings of the 2016 Conference on Empirical Methods in Natural Language Processing, pp. 606–615 (2016)

9. Hochreiter, S., Schmidhuber, J.: Long short-term memory. Neural Comput. **9**(8), 1735–1780 (1997)

10. Lan, M., Zhang, Z., Lu, Y., Wu, J.: Three convolutional neural network-based models for learning sentiment word vectors towards sentiment analysis. In: 2016 International Joint Conference on Neural Networks (IJCNN), pp. 3172–3179. IEEE (2016)

11. Collobert, R., Weston, J., Bottou, L., Karlen, M., Kavukcuoglu, K., Kuksa, P.: Natural language processing (almost) from scratch. J. Mach. Learn. Res. **12**, 2493–2537 (2011)

12. Phu, V.N., Chau, V.T.N., Tran, V.T.N., Duy, D.N., Duy, K.L.D.: A valence-totaling model for vietnamese sentiment classification. Evolving Syst. **10**(3), 453–499 (2019)

13. Nguyen-Nhat, D.-K., Duong, H.-T.: One-document training for vietnamese sentiment analysis. In: Tagarelli, A., Tong, H. (eds.) CSoNet 2019. LNCS, vol. 11917, pp. 189–200. Springer, Cham (2019). https://doi.org/10.1007/978-3-030-34980-6_21

14. Zhou, P., et al.: Attention-based bidirectional long short-term memory networks for relation classification. In: Proceedings of the 54th Annual Meeting of the Association for Computational Linguistics (Volume 2: Short Papers), pp. 207–212 (2016)

15. Kim, Y.: Convolutional neural networks for sentence classification. arXiv preprint arXiv:1408.5882 (2014)

16. Maaten, L., Hinton, G.: Visualizing data using t-SNE. J. Mach. Learn. Res. **9**, 2579–2605 (2008)

17. Maas, A.L., Daly, R.E., Pham, P.T., Huang, D., Ng, A.Y., Potts, C.: Learning word vectors for sentiment analysis. In: Proceedings of the 49th Annual Meeting of the Association for Computational Linguistics: Human Language Technologies, vol. 1, pp. 142–150. Association for Computational Linguistics (2011)

18. Van Nguyen, K., Nguyen, V.D., Nguyen, P.X.V., et al.: UIT-VSFC: Vietnamese students' feedback corpus for sentiment analysis. In: Proceeedings of the 10th International Conference on Knowledge and Systems Engineering (KSE), pp. 19–24. IEEE (2018)

Construction of Word Segmentation Model Based on HMM + BI-LSTM

Hang Zhang and Bin Wen[✉]

Yunnan Normal University, Kunming, China
ynwenbin@163.com

Abstract. Chinese word segmentation plays an important role in search engine, artificial intelligence, machine translation and so on. There are currently three main word segmentation algorithms: dictionary-based word segmentation algorithms, statistics-based word segmentation algorithms, and understanding-based word segmentation algorithms. However, few people combine these three methods or two of them. Therefore, a Chinese word segmentation model is proposed based on a combination of statistical word segmentation algorithm and understanding-based word segmentation algorithm. It combines Hidden Markov Model (HMM) word segmentation and Bi-LSTM word segmentation to improve accuracy. The main method is to make lexical statistics on the results of the two participles, and to choose the best results based on the statistical results, and then to combine them into the final word segmentation results. This combined word segmentation model is applied to perform experiments on the MSRA corpus provided by Bakeoff. Experiments show that the accuracy of word segmentation results is 12.52% higher than that of traditional HMM model and 0.19% higher than that of BI-LSTM model.

Keywords: Chinese word segmentation · HMM · BI-LSTM · Sequence tagging

1 Introduction

With the rapid development of search engines, machine translation, artificial intelligence and other fields, Chinese word segmentation has received more and more attention. In recent years, Chinese word segmentation technology has continued to develop. Chinese word segmentation methods are roughly divided into three categories: dictionary-based word segmentation algorithm, statistics-based word segmentation algorithm and deep learning-based word segmentation algorithm.

Dictionary-based word segmentation algorithm is mainly to match the words you need to split with a prepared dictionary. For example, a dictionary-based Chinese word breaker designed by Yan Liu can quickly get word breaking results [1]. However, the problem is that words that are not in the dictionary cannot be recognized, ambiguous

© Springer Nature Singapore Pte Ltd. 2020
P. Qin et al. (Eds.): ICPCSEE 2020, CCIS 1258, pp. 47–61, 2020.
https://doi.org/10.1007/978-981-15-7984-4_4

sentences cannot be understood, and the quality of word breaking results depends on the quality of dictionary construction. The algorithm based on statistics makes use of a large number of manually marked corpus covering multi domain knowledge for Chinese word frequency statistics, and then sequentially labels participle sentences [2]. Most scholars use HMM and CRF (conditional random field model) to build word segmentation model. For example, Xianjun Zhu et al. use HMM to optimize Chinese word segmentation, improve the accuracy of Chinese word segmentation [3]. This method can effectively identify the unknown words, but it needs a lot of manual marking features. The effect of word segmentation depends on the selection and setting of features. Therefore, in order to make the segmentation results more accurate, some scholars proposed to combine dictionary-based word segmentation algorithm and statistics-based word segmentation algorithm. For example, Xiaogang Zheng et al. combined dictionary-based word segmentation algorithm, word labeling-based Chinese word segmentation method and Hidden Markov word segmentation method, it can better solve Chinese ambiguity and find new login words [4]. The word segmentation algorithm based on deep learning mainly applies neural network to Chinese word segmentation, and uses neural network model to solve the problem of sequence annotation, which has high accuracy. For example, Wei Wang first uses the Bidirectional Long-Short Memory Cyclic Neural Network (Bi-LSTM) model to automatically discover the text features, then completes the Chinese word segmentation through the six word tagging set [5]. The experiment shows that it has good word segmentation effect. This method can understand sentences according to the context, but the problem is that it takes a lot of time to train the corpus, the size of training corpus also affects the result of segmentation.

To solve the problem of three kinds of word segmentation methods, this paper puts forward a Chinese word segmentation model based on the combination of HMM word segmentation model in statistics and Understanding-based neural network model. Combining the two more accurate word segmentation models, we build a combined word segmentation model based on HMM + BI-LSTM, which combines the advantages of recognizing unregistered words with the advantages of understanding sentences. The result of word segmentation is judged and the best result is chosen to combine to improve accuracy. The final experimental results show that the proposed method can recognize the words that are not in the dictionary, and can significantly improve the accuracy of the score.

2 Construction of HMM Word Segmentation Model

2.1 HMM Model

HMM is a statistical model that describes a Markov process with an unknown parameter, determines the hidden parameter, then uses the parameter for further analysis [6]. HMM is generally expressed as:

$$\lambda = (A, B, \pi) \tag{1}$$

$$A = \left[a_{ij}\right] \tag{2}$$

$$B = \left[b_j(k)\right] \tag{3}$$

Where A is the transition probability matrix, aij represents the probability of transition from state i to state j, B is the emission probability matrix, bj(k) is the probability of generating observation vk under the condition of state j, π is the initial state probability.

The three parameters of HMM for λ, state sequence I and observation sequence O can be divided into three basic problems: probability calculation problem, learning problem and decoding problem. Probability calculation problem refers to the probability of occurrence of observation sequence O under model λ. Learning problem refers to estimating the parameters of model λ by observation sequence O, so that the observation sequence P (O|λ) is the largest under the model. Decoding problem refers to solving the state sequence I with the largest conditional probability p (I|o) by model λ and observation sequence O.

2.2 HMM-Based Word Segmentation Model

The main idea of word segmentation model based on HMM is to mark the sequence of word segmentation sentences. Each Chinese character has a definite position in word-building [7]. This article sets the position state value Q of the Chinese character to {B, E, M, S}. "B" is the beginning character of the word, "m" is the middle character of the word, "e" is the end character of the word, "s" is the character of a single word. For example, the state sequence of the statement "我爱中国" is "SSBE" ("I love China". This sentence consists of four characters: "我", "爱", "中" and "国".). When the position of each Chinese character in the sentence to be segmented is determined, the result of segmentation is also determined.

HMM-based word segmentation model is to segment words in this way: Three parameters in λ are obtained by calculating the set of state values and the training corpus. Then the sentence to be segmented is taken as the observation sequence O. The maximum value of P (I|o) is calculated by λ and the observation sequence O, and the state sequence I is obtained, thus the result of segmentation is obtained, which is also a decoding problem. The segmentation flow chart is shown in Fig. 1.

Viterbi algorithm [8] is used in solving the state sequence. It mainly calculates the logarithm probability value of each initial state, then recursively calculates the logarithm probability value of each time to get an optimal path.

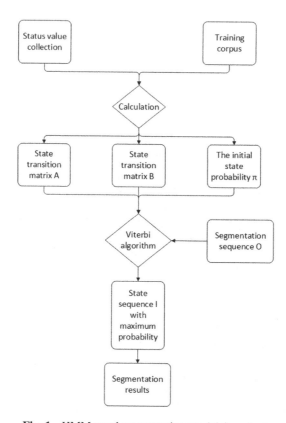

Fig. 1. HMM word segmentation model flowchart

The HMM segmentation model does not need to rely on dictionaries. It can effectively solve the problem of unrecorded word recognition by calculating and statistics. However, the HMM model is limited by strict independence assumption, so it can't predict and judge based on context information. The corpus needs a lot of manual segmentation, and the calculation and statistics take a long time.

3 Establishment of BI-LSTM Word Segmentation Model

The BI-LSTM word segmentation model is a word segmentation method based on deep learning. It mainly uses the BI-LSTM to save context information, then performs sequence labeling on the word segmentation sentence.

3.1 BI-LSTM Neural Network

BI-LSTM is a combination of forward LSTM and backward LSTM, and LSTM is an extension of recurrent neural network (RNN). RNN is a kind of recurrent neural network which takes sequence data as input, recurses in the evolution direction of sequence, and all nodes (cycle units) are linked by chain [9]. RNN has a cyclic network structure and the ability to maintain information [10]. It is mainly used in the fields of machine translation, speech recognition, and image description generation. Its main structure is shown in Fig. 2:

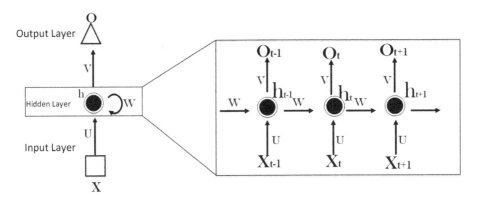

Fig. 2. RNN structure diagram

There are three layers on the left, namely the Input Layer, the Hidden Layer, and the Output Layer. On the right is an expanded view of the hidden layer. The hidden layer mainly processes information and implements the function of time memory. X is the input sample, h is the memory of the hidden layer of the sample, O is the output value, W is the weight of the input, U is the weight of the input sample at the moment, V is the output. $t-1$, t, $t+1$ are time series, f and g are activation functions, g is usually softmax or other, f is tanh or other activation functions. The calculation formula is:

$$h_t = f(U * x_t + W * h_{t-1}) \tag{4}$$

$$o_t = g(V * h_t) \tag{5}$$

From Fig. 2, it can be seen that each information passed to the back hidden layer contains the information of the front memory. However, only this line is used to record all input values, which will cause the memory value to change continuously after the time distance is continuously expanded, causing the problem of the gradient to disappear. To overcome this problem, Hochreiter and Schmidhuber proposed the Long Short-Term Memory Network (LSTM) model, which solves the problem by setting a gating unit to control the passage of information [11]. They used several gates to control the proportion of inputs to the memory cells, and the proportions forgotten from previous states [12].

The structure of LSTM is similar to RNN, except that its hidden layer is more complicated, with three more gates (forget gate, input gate, output gate) and cell state. The function of setting the gate is to control the in and out of information. The implementation method is through a sigmoid neural network layer and pairwise multiplication operation. The role of the cell state is to pass control information to the next moment. The LSTM hidden map is shown in Fig. 3:

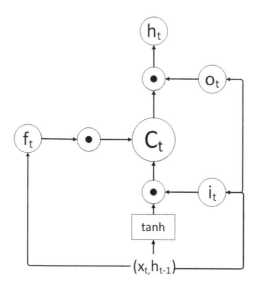

Fig. 3. LSTM structure diagram

f_t, i_t, and o_t are the forget gate, input gate and output gate at time t. X_t is the input sample, h_t is the memory of the hidden layer at time t, C_t represents the cell state at time t, and $\odot$ is the pairwise multiplication operation. The calculation process is as follows:

Step 1: The forgotten gate enter the cell through the sigmoid control information, and it will generate a f_t value from 0 to 1 according to the output h_{t-1} of the previous moment and the current input X_t (0 means completely disallowed, 1 means completely allowed by). The calculation formula is:

$$f_t = \sigma\left(W_{fx} * x_t + W_{fh} * h_t + b_f\right) \tag{6}$$

Step 2: The input gate determines the update of the information through sigmoid, and uses tanh to generate a new candidate value $\tilde{c}_t$. The calculation formula is:

$$i_t = \sigma(W_{ix}x_t + W_{ih}h_{t-1} + b_i) \tag{7}$$

$$\tilde{c}_t = g(W_{cx}x_t + W_{ch}h_{t-1} + b_c) \tag{8}$$

Step 3: Update the information, combine the results of step 1 and step 2, discard the unnecessary information and add new information. The calculation formula is:

$$c_t = f_t * c_{t-1} + i_t * \tilde{c}_t \tag{9}$$

Step 4: Determine the output of the model. First, an initial output is obtained by sigmoid control. Then use tanh to scale the $\tilde{c}_t$ value to between -1 and 1. Then, the output of the model is multiplied by the output of sigmoid one by one. The calculation formula is:

$$o_t = \sigma(W_{ox}x_t + W_{oh}h_{t-1} + b_o) \tag{10}$$

$$h_t = o_t * g(c_t) \tag{11}$$

Judging from the above structure, the LSTM model memorizes and uses the above information well, but it is not enough to link with the above information in Chinese word segmentation. Sometimes in the sentence "我爱中国", If the "中" is not associated with the previous "国", "中国" will be separated into "中\国" separately. So in this case, the same sequence can be reversely converted by using the second LSTM [13].

The BI-LSTM model is a combination of the memory results of the forward LSTM and the backward LSTM, so that it can be connected with both the above information and itself. Its main structure is shown in Fig. 4:

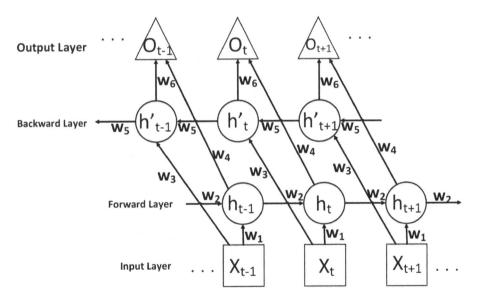

Fig. 4. BI-LSTM structure diagram

X represents the input sample, h and h′ represent the memory saved by forward LSTM and backward LSTM, O represents the output value, and W represents various weights. In the Forward layer, it is calculated forward from time 1 to time t, and the output of the hidden layer at each time is obtained and saved. In the Backward layer, the calculation is performed in the reverse direction from time t to time 1 to obtain and save the output of the hidden layer at each time. Finally, the final output is obtained by combining the results of the forward and backward layers at each time [14]. The calculation formula is:

$$h_t = f(W_1 x_t + W_2 h_{t-1}) \tag{12}$$

$$h'_t = f(W_3 x_t + W_5 h'_{t+1}) \tag{13}$$

$$o_t = g(W_4 h_t + W_6 h'_t) \tag{14}$$

3.2   BI-LSTM Word Segmentation Model

The application of BI-LSTM to Chinese word segmentation is mainly to treat the sentence with segmentation as a series of sequences, then label the sequence. This paper uses the four-digit label set {B, M, E, S}. The meaning of "BMES" is the same as that in the above HMM model word segmentation. For example, the sentence "我爱中国" is marked as "我/S 爱/S 中/B 国/E". The BI-LSTM word segmentation model is shown in Fig. 5:

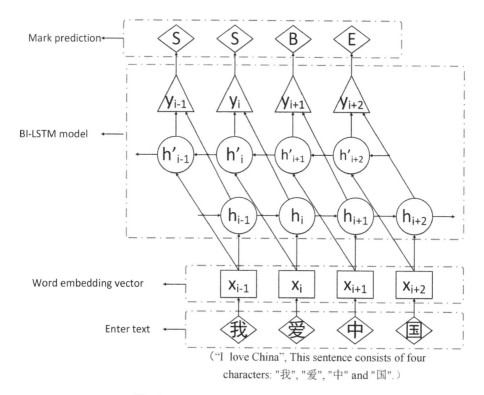

("I love China", This sentence consists of four
characters: "我", "爱", "中" and "国".)

Fig. 5. BI-LSTM word segmentation model

In this model, the natural language is first converted into symbols that can be understood by a computer, and the input text characters are vectorized. For each input $x = (x1, x2, \ldots, xn)$, a prediction sequence $(y1, y2, \ldots, yn)$ will be generated after BI-LSTM, and the score of this prediction is defined as:

$$s(x, y) = \sum_{t=1}^{n} \left(A_{y_{t-1}y_t} + \bar{y}_t \right) \tag{15}$$

A is the label weight transfer matrix, which indicates the weight of the label i to label j. The higher the value of A_{ij}, the greater the possibility of label i transferring to j [15]. $\bar{y}$ is the result matrix obtained through the neural network model.

When training, first for each training sample X, we calculate the scores S (X, y) of all possible labeled sequences y, and normalize all the scores:

$$p(y|X) = \frac{e^{s(X,y)}}{\sum_{\tilde{y} \in Y_\chi} e^{s(X,\tilde{y})}} \tag{16}$$

Then we maximize the log probability of the correct label sequence, which is to maximize the logarithm of the above formula:

$$\log(p(y|X)) = s(X, y) - \log\left(\sum_{\tilde{y} \in Y_\chi} e^{s(X,\tilde{y})} \right) \tag{17}$$

Where YX represents all possible tag sequences of X. At this point, the parameter training in BI-LSTM is also completed after the log probability is maximized. When the final prediction is made, the Viterbi algorithm is used to obtain the possible y sequence with the largest s score:

$$y^* = argmax\, s(X, \tilde{y}), \tilde{y} \in Y_X \tag{18}$$

The difference between the BI-LSTM and the HMM is that the BI-LSTM has the ability to connect to contextual information to understand semantics for word segmentation, but the complex neural network requires a lot of time to train.

4   Combination of HMM Word Segmentation Model and BI-LSTM Word Segmentation Model

The word segmentation results of the HMM word segmentation model mainly depend on the statistical results of word collocations in the corpus. When there are changes in the word segmentation field or the corpus, the word segmentation results will change accordingly, resulting in some sentences not being segmented correctly. The accuracy of the BI-LSTM word segmentation model cannot reach 100%. There are also some sentences that cannot be correctly segmented, and there may be over fitting of neural network. Therefore, we make judgments on the results of the two word segmentation and then select a combination to achieve the complementary effect of the two word segmentation result,then improve the accuracy of the word segmentation.

4.1 Dropout Technology

In order to prevent the over-fitting of the neural network, this article chooses to add the Dropout technology between the word embedding vector and the BI-LSTM. Dropout is one of the currently popular methods to avoid neural network over fitting [16]. Dropout technology is mainly to randomly select the input of the neural network, not to train each input to the neural network. As shown in Fig. 6, the left side is not added Dropout technology, the right side is added Dropout technology. This avoids the possibility that certain features only effect in a fixed input combination, and actively makes the neural network learn some common features.

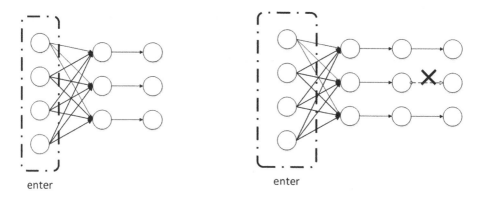

Fig. 6. Dropout technology

4.2 Construction of Word Segmentation Model Based on HMM + BI-LSTM

The result of word segmentation can be expressed as the division of words. The difference of segmentation results of different models can be expressed by the number of words. For example:

Sentence 1: "我\想\过\过\过儿\过过\的\生活" ("I want to live the life that Guoer has lived", Gooer is a personal name. However, because Guoer's Chinese name and verb "过过" are the same characters, it often leads to incorrect segmentation.).

Sentence 2: "我\想过\过\过儿\过过\的\生活".

The number of two-character vocabulary in sentence one is 4, and the number of two-character vocabulary in sentence two is 3. At this time, you can choose a better result based on the difference in the number of vocabularies. In this example, the one with more two-character vocabulary is selected, the final result is to improve the accuracy.

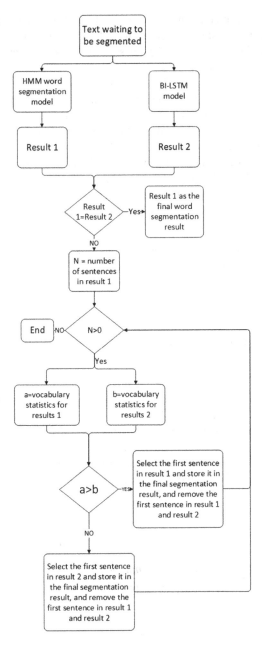

Fig. 7. Word segmentation model based on HMM + BI-LSTM

When combining the HMM and the BI-LSTM, the number of vocabularies is mainly compared from the first sentence of the two results. Then the selection is made. Finally the selected sentences are combined. The main steps are as follows:

Step 1: Use HMM segmentation model and Bi-LSTM model to segment the text to be segmented. Before using Bi-LSTM model to segment words, the text to be segmented should be vectorized and processed by dropout technology. Two results were obtained: result 1 and result 2;

Step 2: Compare the two results. If they are the same, any result will be returned. If they are not the same, step 3 will be performed;

Step 3: Count the words of result 1 and result 2 in the first sentence respectively (you can choose to count the words of three or four characters according to the situation), and get two statistics: result a and result b;

Step 4: Compare result a and result b. If result a is larger than result b, the first sentence in result 1 is selected to be stored in the final result; Otherwise, the first sentence in result 2 is selected to be stored in the final result. Delete the first sentence of result 1 and result 2, and repeat step 3 and step 4 until there are no sentences in result 1 and result 2.

The segmentation flowchart of the combined segmentation model based on HMM + BI-LSTM is shown in Fig. 7:

5 Experiments

5.1 Experiment Environment, Experiment Data and Evaluation Indicators

The main parameters of this experimental environment are: Intel(R) Core(TM) i5-8250U CPU @1.60 GHz 1.8 GHz, 8 GB memory, and Windows10 operating system. The experimental data used the MSRA Corpus provided by Bakeoff, the second international Chinese word segmentation evaluation organized by SIGHAN, with about 14 MB of data.

This experiment uses accuracy (P), recall (R), and comprehensive value index (F) to test the performance of Chinese word segmentation. The calculation formula are as follows:

$$P = \frac{Segment\ the\ correct\ number\ of\ words}{Total\ number\ of\ words\ segmented} \times 100\% \tag{19}$$

$$R = \frac{Segment\ the\ correct\ number\ of\ words}{Total\ number\ of\ words\ in\ text} \times 100\% \tag{20}$$

$$F = \frac{P \times R \times 2}{P + R} \times 100\% \tag{21}$$

5.2 Experiment Design

This experiment is implemented in python language, and the experiment is divided into three groups: the first group is using the HMM for word segmentation; the second group is using the BI-LSTM for word segmentation; the third group is to use our proposed HMM + BI-LSTM method for word segmentation. In the second group of experiments, Word2Vec provided by Gensim in the python toolkit was used to generate the word embedding vector, which was then used as input to the BI-LSTM model [17]. According to Zhang Zirui's research, when the word embedding vector dimension is 200 and dropout size is 20%, the LSTM segmentation model achieves the best performance. So the word embedding vector dimension was set to 200. The word embedding vector is then processed by the Dropout technique to avoid overfitting when the BI-LSTM model is trained. The Dropout rate is set to 0.2. In the third group of experiments, the HMM + BI-LSTM combined word segmentation method used to count the number of words with three or more characters. Because when counting the vocabulary numbers of the two results, it is found that when counting the vocabulary numbers of three words or more, the result of word segmentation is the best.

5.3 Experiment Results and Analysis

The experiment results of each segmentation model are shown in Fig. 8:

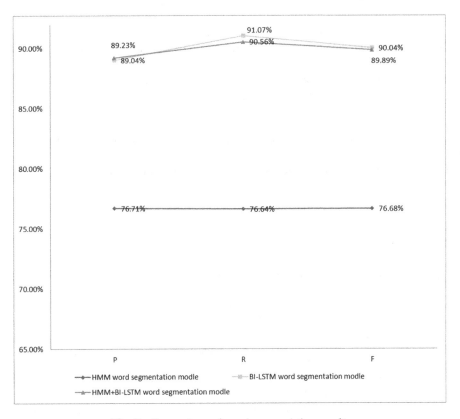

Fig. 8. Comparison of word segmentation results

The experimental results show that compared with traditional HMM model word segmentation, HMM + BI-LSTM combination model word segmentation improves the accuracy, recall and F value by 12.52%, 13.92% and 13.22%. Compared with the traditional BI-LSTM model, the HMM + BI-LSTM combination model improves the accuracy by 0.19%, but the recall rate and the F value are insufficient and need to be improved. The accuracy of the word segmentation of the HMM + BI-LSTM combination model is the highest among the three groups of experiments, which has achieved the purpose of improving the accuracy and has certain advantages.

6 Conclusion

In this paper, through the introduction and comparative analysis of traditional HMM segmentation model and Bi-LSTM segmentation model, we construct a combined segmentation model based on HMM + Bi-LSTM. This method selects and combines the HMM word segmentation model and BI-LSTM word segmentation model to form a new word segmentation result. Finally, this paper designs experiments to prove that the segmentation model built in this paper improves the accuracy of segmentation results. On the one hand, because it takes a long time to segment large text in the experiment, the next step is to explore how to improve the segmentation speed; on the other hand, both the HMM segmentation algorithm and Bi-LSTM segmentation algorithm have such disadvantage that they need to spend a lot of time in training corpus, and the combination of the two methods in this paper can not solve this problem. Thus, we will continue to solve the problem of reducing training time.

Acknowledgment. The research is supported by a National Nature Science Fund Project (61661051), Key Laboratory of Education Information of Nationalities Ministry of Education, Yunnan Key Laboratory of Smart Education, Program for innovative research team (in Science and Technology) in University of Yunnan Province, and Kunming Key Laboratory of Education Information.

References

1. Liu, Y.: Design and implementation of Chinese word segmenter based on MMSEG algorithm. Hunan University (2016)
2. Wang, W., Xu, H., Yang, W., et al.: A review of Chinese word segmentation algorithms. Group Technol. Prod. Modern. **35**(03), 1–8 (2018)
3. Zhu, X., Hong, Y., Huang, Y., et al.: Application of HMM-based algorithm optimization in Chinese word segmentation. J. Jinling Inst. Sci. Technol. **35**(03), 1–7 (2019)
4. Zheng, X., Han, L., Bai, S., Zeng, X.: A combined Chinese word segmentation method. Comput. Appl. Softw. **29**(07), 26–39 (2012)
5. Wang, W.: Bi-LSTM-6Tags-based intelligent Chinese word segmentation method. J. Comput. Appl. **38**(S2), 107–110 (2018)
6. Jiang, W., Chen, Z., Shao, D., et al.: Dynamic programming word segmentation algorithm based on domain dictionary. J. Nanjing Univ. Sci. Technol. **43**(01), 63–71 (2019)

7. Wu, S., Pan, H.: Chinese word segmentation based on hidden Markov model. Mod. Comput. (Prof. Ed.) (33), 25–28 (2018)
8. Li, R., Zheng, J.: Application research of an improved Viterbi algorithm. Comput. Eng. Des. **28**(3), 530–531 (2007)
9. Goodfellow, I., Bengio, Y., Courville, A.: Deep Learning, vol. 1, pp. 367–415. MIT Press, Cambridge (2016)
10. Ren, Z., Xu, H., Feng, S., et al.: Chinese word segmentation based on sequence labeling based on LSTM network. Appl. Res. Comput. **34**(05), 1321–1341 (2017)
11. Hochreiter, S., Schmidhuber, J.: Long short-term memory. Neural Comput. **9**(8), 1735–1780 (1997)
12. Hochreiter, S., Schmidhuber, J.: LSTM can solve hard long time lag problems. In: International Conference on Neural Information, vol. 9, pp. 473–479 (1996)
13. Huang, D., Guo, Y.: BI-LSTM-CRF Chinese word segmentation model incorporating attention mechanism. Software **39**(10), 260–266 (2018)
14. Geng, C.M., Huang, H.: Application of two-way recurrent neural network in speech recognition. Comput. Modern. (10), 1–6 (2019)
15. Jin, W., Li, W., Ji, C., et al.: Chinese word segmentation based on bidirectional LSTM neural network model. J. Chin. Inf. Process. **32**(02), 29–37 (2018)
16. Zhang, Z., Liu, Y.: Chinese word segmentation based on BI-LSTM-CRF model. J. Changchun Univ. Sci. Technol. (Nat. Sci. Ed.) **40**(04), 87–92 (2017)
17. Yao, M., Li, J., Lu, H., et al.: Chinese word segmentation based on sequence tagging of BI_LSTM_CRF neural network. Mod. Electron. Technol. **42**(01), 95–99 (2019)

POI Recommendations Using Self-attention Based on Side Information

Chenbo Yue, Jinghua Zhu$^{(\boxtimes)}$, Shuo Zhang, and Xinxing Ma

School of Computer Science and Technology, Heilongjiang University,
Harbin 150080, China
zhujinghua@hlju.edu.cn

Abstract. Point of interest (POI) recommendation is one of the most important tasks in location-based social networks (LBSN). The existing recommendation methods face two challenges: (1) the cold start problem caused by data sparsity; (2) underutilization of the abundant side information besides user-POI interaction in large-scale data. Recent research shows that a user's social relationship can be used to solve the cold start problem to some extent. The deep neural network learns users' long term and short term preferences to improve the recommendation quality. Therefore, this paper proposes a POI recommendation model called SSANet, applying side information (S) and self-attention (SA) to provide the high-satisfaction POI recommendations for users. Specifically, first, the user-POI interaction matrix were constructed by users history data to represents the user hidden representation; second, the side information includes rating scores, access frequency, social relationship, and geographic information were used to extract users preference; third, we use self-attention mechanism to learn user long term and short term preference. The experimental results on the real LBSN datasets show that the recommendation performance of the SSANet model is better than the existing POI recommendation model.

Keywords: Location-based social network · Smart computing · Point of interest · Location-based services and applications

1 Introduction

Following the express spread of smart devices and the quick decline in mobile tariffs, people are involved in a wide variety of location-related activities anytime, anywhere. With the gradual rise of LBSN platforms such as Facebook, Twitter, Foursquare, and Yelp, users can use smart terminals to check-in and rate restaurants, cinemas, parks where they have visited, and then other users can choose which places they are interested in based on these ratings and comment information. POI recommendation and further route recommendations are two of the hottest topics in recent years. In this article, we focus on POI recommendation issues.

© Springer Nature Singapore Pte Ltd. 2020
P. Qin et al. (Eds.): ICPCSEE 2020, CCIS 1258, pp. 62–76, 2020.
https://doi.org/10.1007/978-981-15-7984-4_5

The traditional POI recommendation algorithms have some deficiencies, such as the model based on matrix decomposition (MF) [1] does not consider the time interval and geographical distance impact of users check-ins. In the POI recommendation, the user's check-in records have information such as the user's check-in time and check-in location. The existing ST-RNN neural network model [2] based on RNN expansion cannot accurately simulate the temporal and spatial relationship between neighbors sign-in in user's check-in records and does not consider users long-term interest. To this end, to solve the above problems, better simulate time and space information, and make full use of side information, we propose a self-attention mechanism-based SSANet model. Firstly, the model fully considers the influence of geographical distance in the early stage. We use the Gaussian Kernel function to obtain the pairwise distance of the corresponding POI in the user's check-in records. Secondly, to further capture the user's preference for POI, we model the user's access frequency and POI rating data and assign different weights to each POI. Thirdly, we choose the node2vec model to extract the network structure features of the users social relationship and obtain the potential representation matrix. Finally, we use the self-attention mechanism to learn user long term and short term preference. Combine with the above methods, SSANet can better model Spatio-temporal information, side information and ultimately give the user a more accurate position prediction.

The main contributions of this paper include:

1. Introduce the node2vec model to extract the network structure features from a user's social relationship, and then obtain the potential representation matrix;
2. Based on the self-attention mechanism, an interest review sub-model is constructed to capture deeper internal dependencies within the sequence and learn user long term and short term preference. The model considers both user preferences and side information;
3. Experiment with three real-world datasets to verify the validity of our proposed model.

2   Related Works

In this section, we discuss related work in two ways, namely, traditional POI recommends methods and neural network POI recommend methods.

2.1   Traditional POI Recommends Methods

The POI recommendation is intended to recommend the location of interest to users [4]. Most proposed methods are based on collaborative filtering (CF) and are recommended using historical records. Some researchers use memory-based CF [5] to learn the user's preferences. Other work uses the matrix decomposition model to recommend. In [6–8,11], the researchers found that check-ins can be considered as implicit feedback and applied weighted regularization matrix decomposition [9] to simulate implicit feedback data. Other researchers believe that the recommendation problem is the task of pairwise ranking. In [3,20], researchers used Bayesian personalized ranking loss [14] to learn the pairwise preferences of POI. [19] proposes a new gravity model called LORE, to exploit the Spatio-temporal sequential influence on POI recommendations. [18] An approach called GeoSoCa, proposes a kernel estimation method with adaptive bandwidth, which simulates the geographic correlation between POIs, and models the check-in frequency of user's friends and user's preferences for POI categories as power-law distributions respectively, to represent the correlation between social relationships and POI categories.

2.2   Neural Network POI Recommend Methods

The neural network is not only used for feature learning to simulate various features of users or projects but also as a core recommendation model to simulate nonlinear, complex interactions between users and projects. [17,21] further improved it by autoregressive methods. A model called ST-RNN was proposed, which take into account the spatial and temporal context to simulate user behavior for the next location recommendation. [10] proposes a geographical information based adversarial learning model (Geo-ALM), combining geographic features and generative adversarial networks to learn the discriminator and generator interactively, and finally improve recommended performance. [13] proposes a translation-based recommender framework (STA) that considers spatiotemporal contexts to model the third-order relationship among users, POIs, and spatiotemporal contexts. [15] proposed a deep neural network structure called PACE for POI recommendation, which use the smoothness of semi-supervised learning to mitigate the sparseness of collaborative filtering. [16] A neural network model called JNTM was used to jointly model the social network structure and the user's trajectory behavior. [22] proposes an Adversarial POI Recommendation (APOIR) model, using a recommender and a discriminator to learn the distribution of user latent preference. [23] proposes a topic-enhanced memory network (TEMN), to utilize the strengths of both the global structure of latent patterns and local neighborhood-based features in a nonlinear way, to enhance the performs of POI recommendations. [24] proposed a Time-LSTM model and two variants that were equipped with time gate in LSTM to simulate the time interval in next POI recommendation.

3   POI Recommendation Model - SSANet

This section first introduces the proposed POI recommendation model SSANet, and then elaborate this model on two main modules: (1) information embedding module; (2) information interaction module.

3.1   Problem Definition

Define $U = \{u_1, u_2, u_3, \cdots, u_M\}$ as a set of M users, $P = \{p_1, p_2, p_3, \cdots, p_N\}$ as a set of N POIs. For user u, has a series of historical POI check-in records, from t_1 to t_{i-1} denote as $P_i^u = \left\{ p_{t_1}^u, p_{t_2}^u, p_{t_3}^u, \cdots, p_{t_{i-1}}^u \right\}$, which $P_{t_k}^u$ represent the POI p that the user u access at the moment t_k. The goal of we purposed model SSANet is to recommend a list that include top k POIs that have never accessed for user to access at the moment t_i. Specifically, user's higher predicted score for the unvisited POI p_j indicate that user u want to have a higher probability to access the POI p_j at the time t_i. Based on the predicted score, we can recommend top k POIs to user.

3.2 Architecture

The proposed architecture is shown in Fig. 1. It consists of two modules: the information embedding module and the information interaction module. For the information embedding module, we introduce it from two parts. First, we use the input represent as multi-hot encoding to model the user-POI check-in sequence to generate a potential representation matrix. Second, in the side information extraction part, we extract the POI geographic location information through the Gaussian Kernel function and normalize the user's rating score and POI access frequency through softmax. Based on the network representation technology (node2vec) pre-training strategy, the network structure feature extraction from the user's social relationship information. In the information interaction module, based on the self-attention mechanism, deeper internal dependencies within the sequence are captured and we can further learn user long term and short term preference, then we use three bottleneck layers to aggregate the information we learned and get the final prediction.

3.3 Information Embedding Module

User Represents Embedding. Given the user's check-in records, our goal is to learn the potential representation of the POI sequence to promote POI recommend accuracy. In SSANet, we design a transfer matrix to map the user's Spatio-temporal intention and potential features of POI in the feature space. The input is a multi-hot user preference vector show as: $P_i^u = \left\{ p_{t_1}^u, p_{t_2}^u, p_{t_3}^u, \cdots, p_{t_{i-1}}^u \right\}$, Where $p_{t_k}^u$ is 1 indicate that the user u have accessed POI p at time t_k. The projection process can be shown as follows:

$$h^u = f_1 \left(W_u P_i^u + b_u \right) \tag{1}$$

h^u represent the potential representation vector of user u, $W_u \in \mathbb{R}^{N \times N}$, b_u denote the weight matrices and bias vectors, respectively.

POI Spatio-Temporal Information Embedding. In the LBSN, there is a physical distance between user and POI, POI and POI, which is different from other recommendation tasks. In the user's check-in records, users' appearance usually limited to a few specific areas, which is a well-known geographical clustering phenomenon in user's check-in activity [6]. From this phenomenon, it can

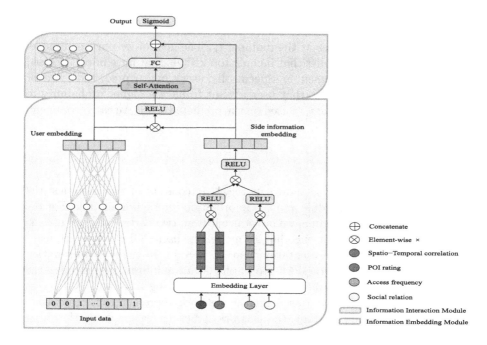

Fig. 1. The structure of SSANet.

be inferred that user prefers to access unvisited POI which near the POI they have visited before, and the degree of user preference depend on the attributes and distances of two points. If the attributes are the same, the closer distance between two points, the greater the probability that the unvisited POI is accessed by user. To combine the POI geographic distance properties, we use the Gaussian Kernel function to extract neighbor-aware influence for the visited POI, as shown below:

$$k\left(p_{t_i}, p_{t_j}\right) = \exp\left(-\frac{\left\|p_{t_i} - p_{t_j}\right\|^2}{2\sigma^2}\right) \tag{2}$$

where p_{t_i} and p_{t_j} are the geographic coordinates of two POIs that user have visited. The value range of the Gaussian Kernel is $k\left(p_{t_i}, p_{t_j}\right) \in [0, 1]$. Finally, the Gaussian Kernal value vector $k \in \mathbb{R}^N$ can be obtained by calculating the pairwise Gaussian Kernal values of each POI pair.

POI Rating Score and Access Frequency Embedding. We can obtain the user's rating score and access frequency through preprocess the LBSN dataset. Softmax normalization is used to scale its value to between (0,1), and the sum of scoring probability and access frequency probability of all check-in POIs of each user is 1, respectively. As a result, we can more easily characterize the user's preference for the POI they have visited. As shown below:

$$r\left(p_j|p_i\right) = \frac{\exp\left(p_i\right)}{\sum_{j=1}^{N}\exp\left(p_j\right)} \tag{3}$$

$$q\left(p_j|p_i\right) = \frac{\exp\left(p_i\right)}{\sum_{j=1}^{N}\exp\left(p_j\right)} \tag{4}$$

The probability value vector $r \in \mathbb{R}^N$, $q \in \mathbb{R}^N$ can be obtained by calculating the rating score probability value and the access frequency probability value of each user, respectively.

User's Social Relationship Embedding. When learning the user's social relationship representation, we try the deepwalk and node2vec models for training respectively. The experimental results show that the node2vec model performs better. This is because node2vec uses a next node sampling method. This method can follow a certain probability selection to the next node, even if the probability is equal, is much better than the random method of deepwalk. So in this paper, we choose the node2vec model to extract the network structure features of the user relationship. For each user, we use node2vec to generate m random walk sequences of length n (where m and n are the parameters we specify) and then apply the Skip-Gram language model with hierarchical softmax on these random walk sequences. And finally, find users who are related to the current user. In this way, a user's embedding is used to maximize the probability of seeing its neighbors in the sequence. Thus, we can get a potential representation matrix of the user's social relationship.

Node2vec describe feature learning problem in the network as a maximum likelihood optimization problem. Given a graph $G = (U, E)$, containing all users, no direction, no weight.

$$f : U \rightarrow \mathbb{R}^N \tag{5}$$

Let f be the mapping function of the user node to the feature representation, $f \in |U| * N$, where N is the dimension represented by the feature.

$$N_s(u) \subset U \tag{6}$$

The above formula is defined as the network neighborhood of the user node generated by the neighborhood sampling strategy S. We seek to optimize the following objective function, maximize the logarithmic probability of observing the network neighborhood $N_S(u)$, which is conditional on its characteristic representation $f(u)$.

Define the possibilities and give two conditions that constitute the objective function to be optimized:

$$\max_{f} \sum_{u \in U} \log \Pr\left(N_s(u)|f(u)\right) \tag{7}$$

Conditional independence:

$$\Pr\left(N_s(u)|f(u)\right) = \prod_{n_i \in N_S(u)} \Pr\left(n_i|f(u)\right) \tag{8}$$

The symmetry between nodes:

$$\Pr\left(n_i|f(u)\right) = \frac{\exp\left(f\left(n_i\right) \cdot f(u)\right)}{\sum_{v \in U} \exp(f(v) \cdot f(u))} \tag{9}$$

The final objective function:

$$\max_f \sum_{u \in U} \left[-\log Z_u + \sum_{n_i \in N_S(u)} f\left(n_i\right) \cdot f(u) \right] \tag{10}$$

From this, we can get the potential representation vector $s \in \mathbb{R}^N$ of the user's social relationship.

Side Information Representation. In this section, to fully extract POI geographical distance information, POI access frequency, POI rating score information, and user's social relationship; to deal with the complex interaction between user-POI, and further POI recommendation, we integrate the above side information as follows:

$$h = f_2(ReLu(k \odot r) \odot ReLu(q \odot s)) \tag{11}$$

h is a latent representation vector of side information, and $\odot$ is an element-wise point multiplication.

3.4   Information Interaction Module

In the user's check-in records, there should be some POIs that are more representative than other POIs that directly reflect the user's preferences. These more representative POIs should make more contributions to the user's hidden representation to express user preferences. Thus, we introduce a self-attention mechanism to learn user long term and short term preference. For the reason of using self-attention, we consider three aspects compared with RNN, which is a popular neural network that used for POI recommendation in recent years. First, when the input sequence l is smaller than the dimension d, the time complexity of each layer self-attention has an advantage over RNN; when l is larger than d, each POI only calculates the attention with a limited r POIs. Second, the multi-attention mechanism does not depend on the calculation of the previous moment, so it better than RNN because of its parallelism. Third, when self-attention learns the potential relationship between each POI and other POI separately, regardless of how far they are, the maximum path length is only 1, which can capture long-distance dependencies.

The purpose of this mechanism is to learn the embedding weighted sum, then form the user's further hidden representation by integrating users embedding representation and the side information of the POI, to capture the dependencies between POIs within the sequence and learn user long term and short term preference.

We aggregate the POI hidden representation with the side information as follows:

$$h_{tc} = f_2 \left(W_{tc} \left(h^u \right)^T h + b_{tc} \right) \tag{12}$$

The importance based on self-attention that we learned can be expressed as follows:

$$h_{att} = att \left(W_{att} \left(h^u \right)^T h_{tc} + b_{att} \right) \tag{13}$$

Finally, the final predicted value is obtained through three bottleneck layers as follows:

$$h_1 = f_1 \left(W_1 h_{att} + b_1 \right) \tag{14}$$

$$h_2 = f_1 \left(W_2 h_1 + b_2 \right) \tag{15}$$

$$h_3 = f_1 \left(W_3 h_2 + b_3 \right) \tag{16}$$

$$y_{pred} = f_3 \left(W_4 h_3 + h + b_4 \right) \tag{17}$$

h_1, h_2, h_3 are the output vectors of the three bottleneck layers, h is the latent representation vector of side information, and y_{pred} is the final prediction value we want.

3.5 Neural Network Training

For POI recommendation tasks, we consider combine regularization terms, the objective function of the proposed model is shown as follows:

$$L = smooth_{L_1}(Y - X) + \lambda \left(\|W_{tc}\|_2^2 + \|W_{att}\|_2^2 \right) \tag{18}$$

in which

$$smooth_{L_1} \left(y_{pred} - x \right) = \begin{cases} 0.5 \left(y_{pred} - x \right)^2 & (a) \\ |y_{pred} - x| - 0.5 & otherwise \end{cases} \tag{19}$$

where λ is the regularization parameter, W_{tc} and W_{att} are the parameters learned by the aggregate layer and attention layer, respectively. (a) is a judgment condition: $|y_{pred} - x| < 1$.

By minimizing the objective function, the partial derivatives for all parameters can be calculated by gradient descent with backpropagation. We use Adam optimizer to automatically adjust the learning rate during the training process.

4 Experiment

In this section, we evaluate our proposed model on three real-world datasets.

4.1 Datasets Description

We evaluate our proposed model on three large real-world LBSN datasets (Foursquare[1], Gowalla[2] and Yelp[3]). The check-in records in the datasets include user ID, POI ID, POI latitude and longitude, timestamps, POI rating score, and POI access frequency. The user's rating score for the check-in POIs is between 1–5, and the timestamps are represented in UNIX format. After pre-processing, the relevant details of the datasets are shown in Table 1.

Table 1. Datasets summary.

Datasets	Foursquare	Gowalla	Yelp
Number of users	24941	18737	30887
Number of POIs	28593	32510	18995
Total check-ins	1196248	1278274	860888
Density	0.1677%	0.2098%	0.1467%

4.2 Evaluation Metrics

In this article, we use Precision@k, Recall@k, F1-score@k, and MAP@k to evaluate our model with other models. For each user, Precision@k indicates the percentage of locations in the top k recommended POIs user visited, and Recall@k indicates the percentage of access locations that may appear in the top k recommended POIs. F1-score is the weighted harmonic average of Precision and Recall, it combines the results of Precision and Recall. When F1-score is high, the method is more effective. MAP@k is the average accuracy of the recommended top k POIs, where the average accuracy is the average of the accuracy values of all post-rank POIs.

4.3 Experimental Setup

In the experiment, f_1 is the *tanh* function, f_2 is the *relu* function, and f_3 is the *sigmoid* function. The learning rate and regularization parameters are set to 0.001, 0.001 respectively. The batch size is 128 in the Gowalla dataset, 256 in Foursquare and Yelp, the number of attention dimension α is 256 in Foursquare and Gowalla, 128 in Yelp. The fully connected layer is set to [N, 256, 64, 256, N]. The dropout rate is set to 0.5. Our experimental equipment is configured as follows: CPU as i7 8700k, the memory as 48G, GPU as Nvidia GeForce GTX 1080Ti. We run our model through the pytorch 1.2.0 framework.

---

[1] https://sites.google.com/site/yangdingqi/home/foursquare-dataset.
[2] http://snap.stanford.edu/data/loc-gowalla.html.
[3] https://www.yelp.com/dataset/challenge.

4.4   Baseline Methods

To demonstrate the effectiveness of our approach, we compare it to the following POI recommendation models.

1. **WRMF** [9]: weighted regularized matrix factorization, which minimizes the squared error loss by assigning both observed and unobserved check-ins with different confidential values based on matrix factorization.
2. **BPRMF** [14]: Bayesian personalized ranking, which optimizes the ordering of the preferences for the observed locations and the unobserved locations.
3. **RankGeoFM** [7]: ranking-based geographical factorization, which is a ranking-based matrix factorization model that learns user's preference rankings for POIs, as well as include the geographical influence of neighboring POIs.
4. **PACE** [15]: preference and context embedding, a deep neural architecture that jointly learns the embeddings of users and POIs to predict both user preference over POIs and various contexts associated with users and POIs.
5. **SAE-NAD** [12]: a model with a self-attention encoder and neighbor-aware decoder for implicit feedback.
6. **APOIR** [22]: a recommender and a discriminator to learn the distribution of user latent preference.

4.5   Performance Comparison

The performance comparison between our model and the baseline model in different datasets are shown in Fig. 2, 3, 4. The top-K in the abscissa represents the top k POI we want to recommend. The index of ordinate is the evaluation index we used in the experiment.

Observing our model, first of all, our proposed model achieves the best performance in most of the evaluation metrics on three datasets. Take the Foursquare dataset for example (Fig. 2): (1) SSANet achieves 4.6% on Precision@5, 11.2% on Recall@5, 9.0% on F1-score@5 and 9.3% on MAP@5 over SAE-NAD which performs the second best; (2) SSANet achieves 15.8% on Precision@5, 18.1% on Recall@5, 18.4% on F1-score@5, 24.1% on MAP@5 over RankGeoFM which performs best in the traditional methods. From this we can see that our model performance has improved significantly, this is because we make full use of side information to learn hidden feedback in our model, and after that, we use the latest neural network method self-attention to further capture the dependencies inside the POI sequence and learn user long term and short term preference. In the traditional recommend method, the performance of RankGeoFM is optimal, which is the same as PACE performance in most experimental results, even surpass the latest deep learning method based on neural network, APOIR. This is because RankGeoFM uses a ranking-based matrix decomposition model, which is a good way to learn the user's preference for geographic location. The performance of the SAE-NAD model is second only to the model we proposed. This is because the SAE-NAD model only considers the geographical location information of POI when learning user implicit feedback, but not make full use of side

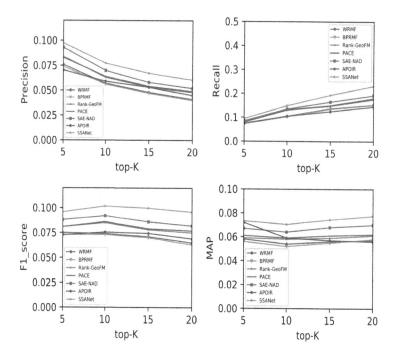

Fig. 2. The comparison of performance on Foursquare.

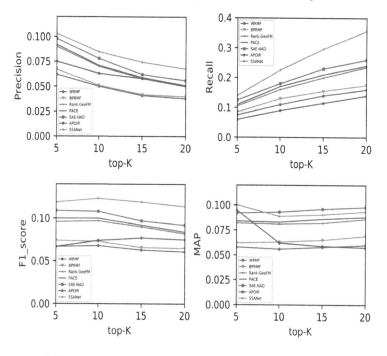

Fig. 3. The comparison of performance on Gowalla.

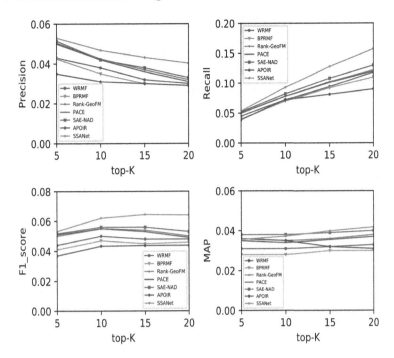

Fig. 4. The comparison of performance on Yelp.

information of POI, and also not consider the user's social relationship when recommend. The latest model APOIR's comparative experimental results are not satisfactory, which shows that simply using an adversarial network to train for POI recommendation cannot well learn the potential intention representation in the user check-in sequence.

4.6 Cold Start Experiment

In POI recommendation, the cold start problem indicates that how to design a recommendation system and make users satisfied with its recommendation results without a large amount of check-in data.

Table 2. Cold start performance on F1-score.

	F1@5	F1@10	F1@15	F1@20
No social relation	0.0519	0.0609	0.0627	0.0619
With social relation	0.0532	0.0622	0.0646	0.0642

In this paper, we use the user's social relationship to solve the cold start problem to some extent. Because Yelp is the only dataset that contains the

Table 3. Cold start performance on MAP.

	MAP@5	MAP@10	MAP@15	MAP@20
No social relation	0.0344	0.0360	0.0384	0.0401
With social relation	0.0356	0.0372	0.0399	0.0418

user's social relationship data, so we experiment with cold start performance on the Yelp dataset. We use the node2vec model to extract network structure features from the user's social relationship. The experiment shows that, compare with that without considering the social relation of the user, then add the social relation, the performance of F1-score improves 3.7% when recommending the top 20 POIs, and the performance of MAP improves 0.2% when recommending the top 20 POIs, shown in Table 2, 3.

4.7 Sensitivity Analysis

In this section, we analyze the impact of change the number of attention dimensions α with F1-score on performance. We choose to experiment on Foursquare and Yelp datasets.

Table 4. Sensitivity analysis on foursquare.

	$\alpha = 16$	$\alpha = 32$	$\alpha = 64$	$\alpha = 128$	$\alpha = 256$
F1@5	0.0586	0.0879	0.0886	0.0870	0.0959
F1@10	0.0634	0.0903	0.0927	0.0933	0.1018
F1@15	0.0613	0.0862	0.0882	0.0899	0.0998
F1@20	0.0576	0.0814	0.0834	0.0849	0.0959

Table 5. Sensitivity analysis on yelp.

	$\alpha = 16$	$\alpha = 32$	$\alpha = 64$	$\alpha = 128$	$\alpha = 256$
F1@5	0.0511	0.0524	0.0505	0.0538	0.0532
F1@10	0.0590	0.0603	0.0597	0.0620	0.0622
F1@15	0.0610	0.0622	0.0613	0.0639	0.0646
F1@20	0.0609	0.0617	0.0609	0.0639	0.0642

When the attention dimension α is relatively small, it is difficult to fully express the relationship between users, the user-POI interaction in the user's check-in records and the side information of POI. When the attention dimension α increases, the performance of the model is significantly improved. As shown in Table 4, 5, the experiment prove that on the Foursquare dataset, when the number of attention dimension α is 256, the experimental results are optimal. On the Yelp dataset, when the number of attention dimensions α is 128 or 256, the experiment can reach the most excellent effect.

5   Conclusion and Future Work

We propose a POI recommendation model based on LBSN, called SSANet. Considering user preference and POI side information, users are recommended with highly satisfied top k POIs. Firstly, the user-POI embedding representation is constructed according to the user's check-in records and side information, the nonlinear user-POI relationship is learned. Secondly, a self-attention mechanism is introduced to capture the dependency between the internal user-POI of the model, and users' long term and short term preferences are further learned. The experimental results of three real-world datasets show that our model's recommendation performance is better than the existing POI recommendation model. In the future work, we will consider the picture information of POI, and extract similar features in the picture through a deep neural network to further improve the recommendation performance.

References

1. Koren, Y., Bell, R., Volinsky, C.: Matrix factorization techniques for recommender systems. Computer **42**(8), 30–37 (2009)
2. Liu, Q., Wu, S., Wang, L., et al.: Predicting the next location: a recurrent model with spatial and temporal contexts. In: Thirtieth AAAI Conference on Artificial Intelligence (2016)
3. Cheng, C., Yang, H., Lyu, M.R., et al.: Where you like to go next: Successive point-of-interest recommendation. In: Twenty-Third International Joint Conference on Artificial Intelligence (2013)
4. Bao, J., Zheng, Y., Wilkie, D., et al.: Recommendations in location-based social networks: a survey. GeoInformatica **19**(3), 525–565 (2015)
5. Bobadilla, J., Ortega, F., Hernando, A., et al.: Recommender systems survey. Knowl.-Based Syst. **46**(Complete), 109–132 (2013)
6. Li, H., Ge, Y., Hong, R., et al.: Point-of-interest recommendations: Learning potential check-ins from friends. In: Proceedings of the 22nd ACM SIGKDD International Conference on Knowledge Discovery and Data Mining, pp. 975–984 (2016)
7. Li, X., Cong, G., Li, X.L., et al.: Rank-geofm: a ranking based geographical factorization method for point of interest recommendation. In: Proceedings of the 38th International ACM SIGIR Conference on Research and Development in Information Retrieval, pp. 433–442 (2015)
8. Lian, D., Zhao, C., Xie, X., et al.: GeoMF: joint geographical modeling and matrix factorization for point-of-interest recommendation. In: Proceedings of the 20th ACM SIGKDD International Conference on Knowledge Discovery and Data Mining, pp. 831–840 (2014)

9. Hu, Y., Koren, Y., Volinsky, C.: Collaborative filtering for implicit feedback datasets. In: 2008 Eighth IEEE International Conference on Data Mining, pp. 263–272. IEEE (2008)
10. Liu, W., Wang, Z.J., Yao, B., et al.: Geo-ALM: POI recommendation by fusing geographical information and adversarial learning mechanism. In: Proceedings of the 28th International Joint Conference on Artificial Intelligence, pp. 1807–1813. AAAI Press (2019)
11. Liu, Y., Wei, W., Sun, A., et al.: Exploiting geographical neighborhood characteristics for location recommendation. In: Proceedings of the 23rd ACM International Conference on Conference on Information and Knowledge Management, pp. 739–748 (2014)
12. Ma, C., Zhang, Y., Wang, Q., et al.: Point-of-interest recommendation: exploiting self-attentive autoencoders with neighbor-aware influence. In: Proceedings of the 27th ACM International Conference on Information and Knowledge Management, pp. 697–706 (2018)
13. Qian, T., Liu, B., Nguyen, Q.V.H., et al.: Spatiotemporal representation learning for translation-based POI recommendation. ACM Trans. Inf. Syst. (TOIS) **37**(2), 1–24 (2019)
14. Rendle, S., Freudenthaler, C., Gantner, Z., et al.: BPR: Bayesian personalized ranking from implicit feedback. arXiv preprint arXiv:1205.2618 (2012)
15. Yang, C., Bai, L., Zhang, C., et al.: Bridging collaborative filtering and semi-supervised learning: a neural approach for poi recommendation. In: Proceedings of the 23rd ACM SIGKDD International Conference on Knowledge Discovery and Data Mining, pp. 1245–1254 (2017)
16. Yang, C., Sun, M., Zhao, W.X., et al.: A neural network approach to jointly modeling social networks and mobile trajectories. ACM Trans. Inf. Syst. (TOIS) **35**(4), 1–28 (2017)
17. Zhang, F., Yuan, N.J., Lian, D., et al.: Collaborative knowledge base embedding for recommender systems. In: Proceedings of the 22nd ACM SIGKDD International Conference on Knowledge Discovery and Data Mining, pp. 353–362 (2016)
18. Zhang, J.D., Chow, C.Y.: GeoSoCa: exploiting geographical, social and categorical correlations for point-of-interest recommendations. In: Proceedings of the 38th International ACM SIGIR Conference on Research and Development in Information Retrieval, pp. 443–452 (2015)
19. Zhang, J.D., Chow, C.Y.: Spatiotemporal sequential influence modeling for location recommendations: a gravity-based approach. ACM Trans. Intell. Syst. Technol. (TIST) **7**(1), 1–25 (2015)
20. Zhao, S., Zhao, T., Yang, H., et al.: STELLAR: spatial-temporal latent ranking for successive point-of-interest recommendation. In: Thirtieth AAAI Conference on Artificial Intelligence (2016)
21. Zheng, Y., Tang, B., Ding, W., et al.: A neural autoregressive approach to collaborative filtering. arXiv preprint arXiv:1605.09477 (2016)
22. Zhou, F., Yin, R., Zhang, K., et al.: Adversarial point-of-interest recommendation. In: The World Wide Web Conference, pp. 3462–34618 (2019)
23. Zhou, X., Mascolo, C., Zhao, Z.: Topic-enhanced memory networks for personalised point-of-interest recommendation. In: Proceedings of the 25th ACM SIGKDD International Conference on Knowledge Discovery & Data Mining, pp. 3018–3028. ACM (2019)
24. Zhu, Y., Li, H., Liao, Y., et al.: What to do next: modeling user behaviors by time-LSTM. In: IJCAI, vol. 17, pp. 3602–3608 (2017)

Feature Extraction by Using Attention Mechanism in Text Classification

Yaling Wang and Yue Wang[(⊠)]

School of Information, Central University of Finance and Economics,
Beijing, China
yuelwang@163.com

Abstract. In recent years, machine learning technology has made great success in fields of computer vision, natural language processing, speech recognition and so on. And machine learning model has been widely used in face recognition, automatic driving, malware detection, intelligent medical analysis and other practical tasks. In this paper, attention mechanism is proposed to combine with LSTM model to extract features in text classification. The results show that, on the one hand, LSTM + Attention can improve classification performance; on the other hand, by sorting by word weights generated by the attention layer, we find some meaningful word features, however, its recognition performance is not good. Some possible reasons were analyzed and it was found that attention mechanism sometimes misjudges wrong word features, resulting from these wrong words often appearing at the same time with meaningful word features.

Keywords: LSTM · Attention · Feature extraction · Interpretability · Text classification

1 Introduction

In recent years, machine learning technology has made great success in computer vision, natural language processing, speech recognition and other fields. Machine learning model has been widely used in face recognition, automatic driving, malware detection, intelligent medical analysis and other important practical tasks. In some scenarios, the performance of machine learning model is even better than that of human learning model. Although machine learning is superior to human in many meaningful tasks, its performance and application are also questioned due to the lack of interpretability [1]. For ordinary users, the machine learning model, especially the deep neural network working model, is like a black box, giving it an input and feeding back a decision result. No one can know exactly the decision basis behind it and whether the decision it makes is reliable or not. However, in practical tasks, especially in security sensitive tasks, the lack of interpretability in DNN may bring serious threats to many practical applications. For example, the lack of an interpretable automatic medical

This work is supported by National Defense Science and Technology Innovation Special Zone Project (No. 18-163-11-ZT-002-045-04).

© Springer Nature Singapore Pte Ltd. 2020
P. Qin et al. (Eds.): ICPCSEE 2020, CCIS 1258, pp. 77–89, 2020.
https://doi.org/10.1007/978-981-15-7984-4_6

diagnosis model may lead to wrong treatment plans for patients. In addition, recent studies have shown that DNN itself faces a variety of security threats, that is, maliciously constructed antagonistic samples are prone to make DNN model classification errors [2], and their vulnerability to anti samples is also lack of interpretability. Therefore, the lack of interpretability has become a practical task of machine learning, which is also one of the main obstacles to further development and application of machine learning in business. In order to improve the interpretability and transparency of machine learning model, establish the trust relationship between users and decision-making model, and eliminate the potential threat of models in actual deployment and application, in recent years, academia and industry have conducted extensive and in-depth research, and put forward a series of interpretability methods of machine learning model.

This paper explores using the attention mechanism to explain the results of the machine learning model, and find out which parts of the text are more critical to the results. In order to find out the application value of interpretable results, the output results of attention mechanism are analyzed.

The experimental results show that the attention module can not only improve the performance of classification model, but also get some feature words. Compared with human experience, the recognition accuracy of feature words from attention mechanism recognition is 34%. Further analysis of these words shows that the recognition performance of attention mechanism is limited, and attention mechanism sometimes misjudges some feature words.

The organizational structure of the paper is as follows: Sect. 2 introduces the work of interpretability and the attention mechanism. Section 3 briefly introduces the model and mechanism used in the experiment. In Sect. 4, we introduce the process of the experiment, take the car industry data as an example, apply the neural network prediction model and interpretable method, and show the results. In Sect. 5, based on the analysis of the interpretable results, we study whether the results of attention mechanism can really find the important words for classification. Section 6 summarizes the results of this paper.

2   Related Work

In the data mining and machine learning scenarios, there are many methods for text classification [15] and feature selection [14]. It's also very important to explain the model and the results. Interpretability is defined as the ability to interpret or present understandable terms to humans [3]. In essence, interpretability is the interface between human and decision model. It is not only the precise agent of the decision model, but also related to the human understanding [4]. Guidotti R et al. reviewed the interpretable aspects of machine learning models, introduced the source, purpose and significance of interpretable problems, classified interpretable methods, and finally pointed out the development direction of interpretable methods.

Malhotra et al. [13] introduced long term short-term memory (LSTM) networks have been shown to be particularly useful for learning sequences containing long-term patterns of unknown length because of their ability to maintain long-term memory. In such a network, the more hidden layers can learn the higher level of time character-istics. The prediction error is modeled as multivariate Gaussian distribution, which is used to evaluate the possibility of abnormal behavior.

In terms of specific interpretable algorithms, an effective method is to introduce attention mechanism [7]. Attention mechanism originates from the study of human cognitive neuroscience. In cognitive science, due to the bottleneck of information processing, human brain can choose a small part of useful information from a large amount of input information intentionally or unintentionally to focus on processing, while ignoring other visible information, which is the attention mechanism of human brain [8]. In the case of limited computing power, attention mechanism is an effective way to solve the problem of information overload. By determining the input part that needs attention, the limited information processing resources are allocated to more important tasks. In addition, attention mechanism has good interpretability, and attention weight matrix directly reflects the areas of interest in the decision-making process of the model.

Some scholars apply interpretable methods to the datasets. Bahdanau et al. [10] introduced attention mechanism into machine translation based on Encoder-Decoder architecture, effectively improving the performance of "English French" translation. In the encoding stage, machine translation model uses Bi-RNN to encode source language into vector space. In the decoding stage, the attention mechanism assigns different weights to the hidden state of the decoder, which allows the decoder to selectively process different parts of the input sentence in each step of generating French trans-lation. Finally, by visualizing the attention weights, users can clearly understand how words in one language rely on words in another language for correct translation. Yang et al. [11] introduced the hierarchical attention mechanism into the task of text clas-sification. At the same time, attention weight quantifies the importance of each word, which can help people clearly understand the contribution of each word to the final emotion classification results.

The research of interpretability of machine learning model has made a lot of achievements. By evaluating the interpretable methods, we can measure their perfor-mance. However, in the application of practical problems, the results of interpretable methods are not completely understandable and acceptable to human beings. Therefore, this paper uses attention mechanism as an interpretable method to examine practical problems.

3   The Model

3.1   LSTM

The full name of LSTM is Long Short-Term Memory, which is a kind of RNN (Recurrent Neural Network). Due to its design features, LSTM is very suitable for modeling temporal data, such as text data. BiLSTM is the abbreviation of Bi-directional Long Short-Term Memory, which is composed of forward LSTM and backward LSTM. Both of them are often used to model context information in NLP tasks.

At the moment t, the LSTM model consists of the input word X_t, the cell state C_t, the temporary cell state $\tilde{C}_t$, the hidden layer state h_t, the forgetting gate f_t, the memory gate i_t, and the output gate o_t. The calculation process of LSTM can be summarized as follows: through forgetting and memorizing new information in cell state, useful information can be transmitted for subsequent calculation, while useless information can be discarded, and hidden layer state h_t will be output at each time step, in which forgetting, memory and output are calculated by forgetting gate f_t calculated by hidden layer state h_{t-1} at last time and current input X_t, memory gate i_t, output gate o_t to control. The overall framework is shown in Fig. 1.

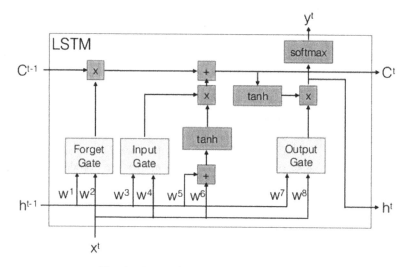

Fig. 1. Overall framework of LSTM model

The hidden state h in the recurrent neural network stores the historical information, which can be regarded as a kind of memory. In a simple cyclic network, the hidden state is rewritten every time, so it can be regarded as a kind of short-term memory. In neural network, long-term memory can be regarded as time network parameters, which implies the experience learned from training data, and the update cycle is much slower than short-term memory. In the LSTM network, the memory unit C can capture some key information at a certain time, and save the key information for a certain time interval. The life cycle of information stored in memory unit C is longer than that of short-term memory h, but it is much shorter than that of long-term memory, so it is called long-term short-term memory.

3.2   LSTM + Attention in Text Classification

Attention mechanism can recognize the important information in the data and calculate the weight score of each word in the input data sequence. The frame of LSTM + Attention model is as follows (Fig. 2).

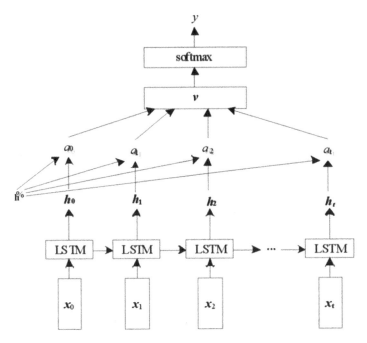

Fig. 2. LSTM + Attention model

The input sequence in the figure is the vector representation of each word after a segment of text segmentation, which represents $x_0, x_1, \ldots, x_t,$. Each input is passed into the LSTM unit, and the output $h_0, h_1, \ldots, h_t,$ of the corresponding hidden layer is obtained. Here, attention is introduced into the hidden layer to calculate the probability distribution value of attention $a_0, a_1, \ldots, a_t,$ allocated by each input.

The calculation process of the probability distribution of attention value is as follows (Fig. 3).

In RNN, at time i, if we want to generate y_i words, we can know the output value H_{i-1} of hidden layer node $i - 1$ at time $i - 1$ before target generates y_i, and our purpose is to calculate the importance value of words $x_1, x_2, \ldots, x_m$ in the input sentence when generating y_i, then we can use the hidden layer node state H_{i-1} of target output sentence $i - 1$ to compare with the RNN hidden layer node state h_j corresponding to each word in the input sentence source, that is, through the function $F(h_j, H_{i-1})$ to obtain the alignment possibility of the target word y_i and each input word, this F function may adopt different methods in different algorithms, and then the output of the function f is normalized by *Softmax* to obtain the attention distribution probability distribution value that conforms to the value range of the probability distribution a_{ij}. It represents the weight score of the word j in the data i.

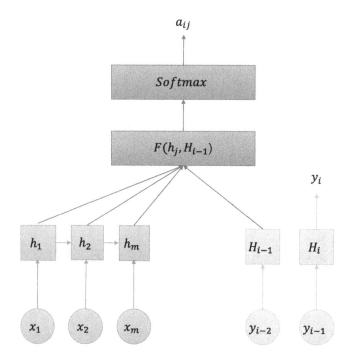

Fig. 3. The generation process of attention distribution coefficient

4   Classification and Attention

4.1   Experiment Preparation

The experimental programming language in this work is Python 3, the development tool is Jupyter notebook, and the deep learning framework used is Keras.

The experimental dataset of this paper comes from the positive and negative comment data of 100 automobile brands. The data includes users' evaluation of each brand product in the automobile industry, as well as their corresponding labels (positive and negative).

In this experiment, the LSTM + Attention model is used to predict whether the text content of the comment is positive or negative, and attention mechanism is introduced to find out which parts of the texts are more important for classification. Note that the data set is a Chinese text set, which we translated into English during the presentation (Table 1).

Chinese text data needs to be transformed into tensor before it can be input into neural network. The data processing steps of this experiment are as follows:

(1) Participle. The original whole text content is divided into Chinese words by using the Jieba word segmentation device. In this experiment, the exact pattern is used for word segmentation.

Table 1. Illustration of data sample

Text	Label
English: @China government website: FAW Volkswagen recalled part of Suteng which had safety hazards due to rear suspension. After the recall, lining plate was added, which did not fundamentally solve the safety hazards. Since the recall, the public acknowledged that there was a problem with Suteng, that is, the rear axle may be broken	Negative
English: The sales data of BYD in March 2015 is as follows. Both S7 and Qin performed well. Congratulations	Positive

(2) Index conversion. After all the text data is segmented, a dictionary is constructed. Each Chinese word corresponds to an index value. The index value list is the dictionary number sorted according to the most frequent occurrence of words. The size of the dictionary can be set to 2000, that is, only the first 2000 most frequent words in the training data are retained. After the dictionary system is built, the word list can be converted into an index list, and the length of the list is set to 20, that is, a comment text contains 20 words.

(3) Vectorization. There are two conversion methods: one is to fill the list with the same length, and then convert the list into an integer tensor with the shape of (number of data, length of comment text); the other is to encode the list with one hot code, and convert it into a vector composed of 0 and 1, and then the network can process the data. The latter method was used in this experiment.

At the same time, the Chinese labels "negative" and "positive" are converted to 0 and 1 respectively.

4.2   Experimental Design

In this paper, the LSTM + Attention model is used to classify Chinese comment texts. In order to evaluate the performance of the classification model, the classification results are measured by accuracy, AUC and F1.

In order to illustrate the influence of attention mechanism on the classification results, the classification results of the LSTM + Attention model are compared with some base line models.

A simple stack with a relu activated full connection layer (dense), known as a multi-layer perceptron (MLP). Its network structure is relatively simple, usually as the first experimental model.

For the classification model, two key architectures need to be determined: how many layers the network has and how many hidden units each layer has.

In the process of training neural network, we need to determine the loss function and optimizer. In binary classification, "binary_crossentropy" can be used as loss function. In terms of optimizer selection, "rmsprop" is often a good enough choice.

In order to verify the above settings, the original data is divided into training set, verification set and test set according to the ratio of 7:2:1. In this experiment, the batch size was set to 512, and the epoch of model training was 20.

4.3   Experimental Result

Classification Results. Table 2 shows the comparison results of the overall average accuracy value, AUC value and F1 value of different classification models in the test set.

Table 2. Comparison of average classification results of different classification models

Classification models	Accuracy	AUC	F1
MLP	90%	0.94	0.89
CNN	88%	0.92	0.88
LSTM	94%	0.96	0.91
LSTM + Attention	98%	0.98	0.93

It can be seen from the results in Table 2 that on the same dataset, with the trained word vectors as the text features, the classification effect of LSTM + Attention classification model is better than the other classification models. The reason is that attention calculates the probability distribution value of each word in the text, which can effectively extract the key words in the text, and the key words play an important role in the semantic expression of the text, which shows that attention plays a certain role in improving the performance of text classification.

Interpretability Results. Table 3 shows the recognition result of attention mechanism of a comment instance. The attention layer generates weight scores for each word in each comment sentence.

Table 3. Recognition result of attention mechanism

Input	True label	Prediction	Score
Volkswagen recall part of Suteng which has the safety problems in the rear suspension	Negative	Negative	[0.032, 0.271, 0.002, 0.002, 0.056, 0.026, 0.087, 0.176, 0.132, 0.005, 0.022, 0.089]

The input data is "Volkswagen recall part of Suteng which has the safety problems in the rear suspension", and the label predicted by the model is "negative". The attention module generates importance scores for each word after segmentation, and highlights that the words that play an important role in this comment sentence to negative prediction are "recall", "hidden danger" and "problem".

5   Discussion on Feature Extraction by LSTM + Attention

5.1   Feature Extraction by LSTM + Attention

The extraction method is as follows: the attention level generates the weight score of each word in each sentence. In the positive and negative categories, we traverse the dataset respectively, store the weight score of each word in each sentence in a matrix, and then calculate the average weight score of each word. The word list is sorted according to word score values, and the top k words are the extracted as important features.

Through this method, we get the most important 50 words in positive and negative categories. Part of the results is shown in Table 4 and 5.

Table 4.   Negative word list (part)

Rank	Word	Score
1	**Spontaneous combustion**	0.198
2	Powerful	0.194
3	On the road	0.193
4	**Garbage**	0.189
5	Insurance	0.185
6	Environmental protection	0.159
7	**Malfunction**	0.149

Table 5.   Positive word list (part)

Rank	Word	Score
1	Xinbaolai	0.259
2	Engine oil	0.188
3	Germany	0.162
4	They	0.159
5	2003	0.152
14	**Solve**	0.094
29	**Know**	0.067

From the positive and negative results, attention mechanism recognizes some words that can be understood by human beings, such as the words "fault " and "crisis" for the negative sample, and "solve" and "know" for positive the positive sample.

However, some words are incomprehensible. For example, the word "powerful" has positive meaning in the human mind, but it is assigned to words that are important for negative evaluation. Besides, the meaning of some words can't be understood intuitively by human beings like "on the road" for negative evaluation.

In order to improve the interpretability of the remaining recognized words (making it easier to be understood by humans), we designed a filtering algorithm to find the context information of these words, form bigram phrases, and check whether the interpretability is enhanced.

Algorithm 1. The filtering algorithm to find bigrams

Input: every comment sentence in dataset; top k important words and scores generated in Section 5.1

Output: Top k important words and bi-gram phrases

1:    scan the sentences with top k words

2:    group the top k words with the words just before and after them and form **bigrams** in the sentences

3:    **if** both words in a bigram **exist in** the word list of top k

4:    add the bigram phrase into word list

5:    importance score of a bigram = $word_1\ socre + word_2$ score

6: Sort by importance scores

7: return the top 50 words and bi-gram phrases

The word list is updated by the filtering algorithm. Table 6 shows some of new phrases in the Top 50 word list.

Table 6. New top50 word list (part)

Phrase	Score	Class
Really bad	**0.375**	**Negative**
Rights protection incident	**0.244**	**Negative**
Sudden recall	**0.153**	**Negative**
Quality is	0.125	Positive
Related forums	0.180	Positive
Really	0.136	Negative
Rights protection	**0.119**	**Negative**
Quality	0.065	Positive
Forums	0.081	Positive

Through the recognition of bigram phrases, some important words that cannot be understood by human beings are explained. For example, in a single word, it is not clear why the word "really" is important for a negative classification. However, through the combination of words, it is found that "really" is often accompanied by "bad", and "bad" is a negative word with obvious emotion, so the word "really" is also recognized as an important negative word by the attention module.

5.2   Examining the Problems of Feature Extraction

In order to judge the performance of attention mechanism, we use the external emotion dictionary, which includes the words of positive and negative emotions. And we manually recognize some specific positive and negative words in car industry besides emotional dictionary.

In our negative word list, we find 10 words in the emotion dictionary and 7 words of negative events in the automotive industry, and the precision (top 50) of negative word list is 34% (17/50). Similarly, in our positive word list, there are 2 words in emotion dictionary and 5 words for positive events, so the precision is 14% (7/50). The results of feature extraction for negative categories are better than positive ones. The reason may be that there are obvious indicators in the negative categories.

Anyway, it can be seen from the evaluation results that attention mechanism can identify some human understandable and interpretable results. But the precision value is not high.

Section 5.1 mentioned that there are some strange results in feature extraction. We try to find some reasons by looking at the original data and roughly divided into several categories.

- Why does the seemingly positive word "powerful" appear in the negative vocabulary? By looking for sentences with the word "powerful", we find that this word is a negative modifier in some ironic comments! For example, the original sentence means that the vehicle is of poor quality and easy to catch fire (it is "powerful" in this sense).
- Why are some car categories such as "Japanese cars" important for negative classification? This is due to the negative comments on the brand in the data set. Most people think Japanese cars are not good. There are also some car parts such as "clutch" in the feature word list, most of which indicate that there is a problem with this part.
- Why is the seemingly neutral word "smell" important for negative classification? This is because the word "peculiar smell" is often used in car news, which is a negative evaluation of cars. There are also conjunctions such as "after all". Because the word often represents a negative turn in the context, it is recognized as a characteristic word of a negative class.
- In addition, there are still some words in the result of feature extraction that we can't understand the emotional meaning intuitively such as "on the road". By looking up the sentences containing the word "on the road", we find that "on the road" often appears in the same sentence describing some accident.

To sum up, we can see that attention mechanism sometimes misjudges some wrong words because these wrong words often appears at the same time with the real import words.

6 Conclusion

Firstly, this paper introduces attention mechanism into LSTM model, and finds that attention module can improve classification performance significantly. Then it inputs the recognition results of attention mechanism, sorts out and orders the words by its important scores, and outputs the 50 words that are the most important to predict the positive and negative categories respectively. By introducing the external emotion dictionary, calculating the precision value and evaluating the performance of attention mechanism, it is found that attention mechanism can successfully extract some important words, but its performance is limited. Although some bigram phrases can be found by the filtering algorithm, but the performance is still limited. And the reason is discussed and draws the conclusion that attention mechanism sometimes misjudges some wrong words because these wrong words often appears at the same time with the real important words. The reason is that the attention mechanism cannot recognize word features satisfactorily without knowing semantic meaning of words.

Through the interpretable results, it can provide a new idea for the artificial selection of features. Printing the original data containing the important feature words and the importance scores of the words can be used to manually identify the true meaning of the feature words in the sentences. Adding these words to the existing artificial features is an iterative process of manually extracting the important feature words. Through the above methods, the truly important characteristic words can be added to the list of characteristic words.

References

1. Ibrahim, M., et al.: Global explanations of neural networks. In: AAAI/ACM Conference ACM (2019)
2. Li, J., Ji, S., Du, T., et al.: Text bugger: generating adversarial text against real-world applications. In: Proceedings of the 26th Annual Network and Distributed Systems Security Symposium. ISOC, Reston (2019)
3. Doshi-Velez, F., Kim, B.: Towards A Rigorous Science of Interpretable Machine Learning (2017)
4. Riccardo, G., et al.: A survey of methods for explaining black box models. ACM Comput. Surv. **51**(5), 1–42 (2018)
5. Sebastian, B., et al.: On pixel-wise explanations for non-linear classifier decisions by layer-wise relevance propagation. PLoS ONE **10**(7), e0130140 (2015)
6. Ribeiro, M.T., Singh, S., Guestrin, C.: "Why Should I Trust You?": Explaining the Predictions of Any Classifier (2016)
7. Vaswani, A., Shazer, N., Parmar, N., et al.: Attention is all you need. In: Proceedings of the 31st International Conference on Neural Information Processing Systems. Curran Associates Inc, Red Hook (2017)
8. Xu, K., et al.: Show, attend and tell: neural image caption generation with visual attention. In: Computer EnCE, pp. 2048–2057 (2015)
9. Arras, L., et al.: "What is relevant in a text document?": an interpretable machine learning approach. PLoS ONE **12**(8), e0181142 (2017)

10. Bahdanau, D., Cho, K., Bengio, Y.: Neural machine translation by jointly learning to align and translate. arXiv preprint arXiv:1409.0473 (2014)
11. Yang, Z., et al.: Hierarchical attention networks for document classification. In: Proceedings of the 2016 Conference of the North American Chapter of the Association for Computational Linguistics: Human Language Technologies (2016)
12. Poerner, N., Roth, B., Schütze, H.: Evaluating neural network explanation methods using hybrid documents and morphological agreement (2018)
13. Malhotra, P., et al.: Long short term memory networks for anomaly detection in time series. In: European Symposium on Artificial Neural Networks (2015)
14. Deng, X., et al.: Feature selection for text classification: a review. Multimedia Tools Appl. **78**(3), 3797–3816 (2019)
15. Kowsari, et al.: Text classification algorithms: a survey. Information **10**(4) (2019)
16. Xie, J., et al.: Chinese text classification based on attention mechanism and feature-enhanced fusion neural network. Computing **102**(3), 683–700 (2019). https://doi.org/10.1007/s00607-019-00766-9
17. Liu, G., Guo, J.: Bidirectional LSTM with attention mechanism and convolutional layer for text classification. Neurocomputing **337**, 325–338 (2019)
18. Tong, X., et al.: Semantic correlations loss: improving model interpretability for multi-class classification. In: 2019 IEEE International Conference on Big Data (Big Data). IEEE (2019)

Event Extraction via DMCNN in Open Domain Public Sentiment Information

Zhanghui Wang, Le Sun[✉], Xiaoguang Li, and Linteng Wang

School of Information, Liaoning University, Shenyang, China
952528760@qq.com
http://www.lnu.edu.cn

Abstract. Event extraction (EE) is a difficult task in natural language processing (NLP). The target of EE is to obtain and present key information described in natural language in a structured form. Internet opinion, as an essential bearer of social information, is crucial. In order to help readers quickly get the main idea of news, a method of analyzing public sentiment information on the Internet and extracting events from news information is proposed. It enables users to quickly obtain information they need. An event extraction method was proposed based on Chinese language public opinion information, aiming at automatically classifying different types of public opinion events by using sentence-level features, and neural networks were applied to extract events. A sentence feature model was introduced to classify different types of public opinion events. To ensure the effective retention of text information in the calculation process, attention mechanism was added to the semantic information, and an effective public opinion event extractor was trained through CNN and LSTM networks. Experiments show that structured information can be extracted from unstructured text, and the purpose of obtaining public opinion event entities, event-entity relationships, and entity attribute information can be achieved.

Keywords: Text processing · Attention mechanism · Syntax analysis · Neural network

1 Introduction

With the increasing number of Internet users, people can post, obtain and transmit information through the Internet. While different editors report different news, they will describe the news from hundreds of perspectives, so it is difficult for the public to find the relationship between events. Public opinion information reflects a social attitude, most of which is the public's development of social

This work was supported by National Natural Science Foundation of China under Grant (No. 61802160); Doctoral Start-Up Fund of Liao-ning Province (No. 20180540106); Liao-ning Public Opinion and Network Security Big Data System Engineering Laboratory (No. 04-2016-0089013).

P. Qin et al. (Eds.): ICPCSEE 2020, CCIS 1258, pp. 90–100, 2020.
https://doi.org/10.1007/978-981-15-7984-4_7

events or ideas generated by certain aspects of social managers. It covers the public attitudes, opinions, and emotions on various events. A large amount of public sentiment information is generated every day on the Internet, so the extraction of public opinion information is very indispensable. Event extraction is a difficult task in information extraction. The ultimate goal is to find event triggers that can represent the type of event and its parameters. Event trigger words can accurately express events. The event trigger word detection task is an important step in event extraction. This verb is often used to refer to collisions between objects. Therefore, knowing the part-of-speech of a word is greatly helpful for the event extraction process. For example, consider the following sentence:

S1: 金融危机冲击中国市场。(The financial crisis hit the Chinese market.)

S2: 军地协作解决官兵涉法问题。(Military and local cooperation to solve the law-related problems of officers and soldiers.)

In S1, '冲击 (hit)' represents a trigger of type Finance. However in S2, '解决 (sove)' expresses event type Military. If we have the word part of speech as a priori knowledge of the selection trigger, we can reasonably speculate that '冲击' is a trigger for the type of Finance, not the Military. This information is obtained at the lexical level. And there are multiple grammatical structures in the Chinese language and the same word can be used in the description of different event types. For instance, consider the following sentences:

S3: 香港股市暴涨，恒指飙升近千点。(The Hong Kong stock market soared, and the HSI soared nearly a thousand points.)

S4: 农田被淹、河水暴涨，尤溪出现入汛以来最强降雨。(The farmland was flooded and the river water skyrocketed. Youxi experienced the strongest rainfall since flooding.)

In S3, '暴涨' (skyrocket) is a trigger of type Finance. While the same word is the Society trigger in another sentence. It can be seen that the same word has different meanings in different situations and Chinese grammar is more complicated than other languages. Through the above Chinese language characteristics, this paper proposes a method combining statistical methods and neural networks to extract Chinese public opinion events.

We have made improvements in the following areas:

1) We propose a new expression of word vector, which consists of word information, part of speech and special identification entity information.
2) To determine the integrity of semantic information, we adopt the method of the encoder to decoder in the process of encoding the sentence, and use the encoder as the result of the sentence encoding.
3) We combine neural networks with attention mechanisms to extract event parameters. When different event parameters are extracted, different weights are assigned to different words according to the semantic meaning of the words.

2   Related Work

Most current event extraction technologies are based on pattern matching methods and neural network-based models. Pattern recognition mainly uses a large number of manual identifiers to count the trigger words in the text. This method needs a lot of manual annotation and time. Pattern matching based methods can achieve better results in specific areas. And there is some work based on machine learning methods that transform the recognition of event categories and event elements into classification problems, and their cores lie in the construction of classifiers and the selection of features. For the first time, Chieu et al. (2002) introduced the maximum entropy model for event recognition in event extraction, which enabled the extraction of lecture announcements and personnel management events [1]. Hector Llorens et al. (2010) used the CRF model for semantic role labeling and applied it to TimeML event extraction to improve system performance [2]. The recent development of event extraction is the aspect of building a neural network model, the model can extract event triggers and event parameters by training. Zhiheng Huang et al. (2015) employed bidirectional LSTM-CRF models for sequence Tagging [3]. Zae Myung Kim et al. (2016) proposed using a cyclic neural network to extract events [4]. In the meantime, Thien Huu Nguyen et al. (2016) proposed a method for extracting joint events through recurrent neural networks [5]. Due to the less trainable data, Kang Liu et al. (2017) proposed a method to automatically generate text data based on features [6]. This method is implemented by filtering and expanding the trigger word. Lishuang Li et al. (2018) formulated to use Bi-GRU to extract biological events [7].

3   Methodology

In this paper, the public opinion event extraction task is realized by two stages of Long Short-Term Memory Networks (LSTM) coding classification and convolution extraction. First, we classify the text in the first stage, we encode news texts via LSTM and classify the encoded results by event type. In order to make the encoding result contain complete semantic information, we adopt the encoder-decoder structure to control the encoded results. The second stage is to identify the event information, which extracts the event trigger words and event arguments words from the classification result through the Attention mechanism and the DMCNN. The integrity and accuracy of the trigger word extraction will directly affect the quality of the event extraction.

3.1   Text Vectorization

Text information is not conducive to the understanding of the computer. In order to facilitate computer understanding and calculation, the text information is converted into a vector data form. The vector expresses semantic information, which is beneficial to the training of neural networks. First of all, we take the word segmentation operation on Chinese text. Because Chinese grammar is complicated, different segmentation positions in sentences may produce different semantic words. In this article, we use Jieba to tokenize text. After the word segmentation, the sentence is represented by the sequence $s = [w_1, w_2, ..., w_n]$, where w_i is the i-th word in the sentence and n is the length of the sentence. In this paper, text embedding inputs include word vectors, part-of-speech vectors, and entity vectors to ensure accurate judgment of event types.

Word Embedding. The word embedding is to express word information with vectors. Currently, in NLP tasks, word vector has two representations. One hot representation defines the vocabulary size as the vector dimension. The definition and labeling of vectors is based on the position of the vocabulary in the vocabulary. For example: In Baghdad, the American tank fired on the Palestine Hotel. The vector of the word 'fired' is $[0, 0, 0, 0, 1, 0, 0, 0, 0]$, the method is simple to understand. But if the text dimension very large, it will result in a very large feature space, and this representation results in the similarity of any two words are zero. Distributed representation is a continuous vector representing words as low-dimensional. This expression is to reduce the dimensionality of each word to a shorter word vector through neural network training. All the trained words form the vector space, and the relationship between words is also studied in the word space. So this method saves computational cost and there can contain more semantic information. In this paper, we use the word2vec model proposed by Mikolov et al. to train the word vector, and the skip-gram model is used to obtain the vector in the corpus. Finally, we turn it into a matrix $E = [e_1, e_2, ..., e_n]$, where e_i is the word-initial embedding of i-th word.

Part of Speech Vector. Part of speech (POS) is an essential part of word information. Part of speech is the grammatical category of each word in the corpus, which can express the features of the semantic information of the text. Part of speech information can provide useful information for the next trigger word detection task and event arguments acquisition task. For example, verbs or nouns can be used as event triggers, adjectives can't. Therefore, the information is beneficial to improve the accuracy of event extraction. The paper uses the Jieba tokenizer to preprocess the news text and perform POS tagging on the processed word sequence. There are 39 types of POS features in Jieba, including adjectives, nouns, verbs, etc. We construct a 39-dimensional feature dictionary, and turn it into a matrix $PV = [pv_1, pv_2, ..., pv_n]$, where each word corresponds to a feature vector.

Entity Vector. Words may contain deeper meanings, such as names of places, humans, etc. In the process of analysis, the semantic information of certain words is divided into three categories: entity class, time class, and number class. There are 7 sub-categories, names of people, places, Organization name, time, date, currency, and percentage. We organize and encode the entity information from 001 to 111 representing 7 subcategories, 000 indicates that the word does not belong to these four categories. The entity vector matrix is $EV = [ev_1, ev_2, ..., ev_n]$. Finally, the above three forms of vectors are stitched as semantic word vectors of public opinion text information, and then input to the neural network: $V = [E, PV, EV]$

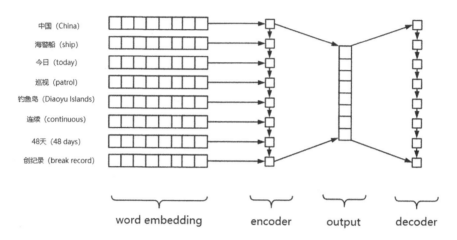

中国 (China)
海警船 (ship)
今日 (today)
巡视 (patrol)
钓鱼岛 (Diaoyu Islands)
连续 (continuous)
48天 (48 days)
创纪录 (break record)

word embedding   encoder   output   decoder

Fig. 1. Text Vectorization. In order to classify the text, we encode the text information, in which the encoding and decoding process use LSTM.

Text Encoding. Text information is sequence-based information. Simple feedforward neural networks can't capture the meaning of sequences well, which will result in the loss of semantic information. RNN (Recurrent Neural Network) can solve this problem. The input to RNN contains not only the input of this step but also the output information of the previous step. However, RNN has poor memory ability when processing long sequence information. LSTM (Long Short-Term Memory) is a long-term and short-term memory network. It is a recursive neural network with memory function, suitable for sequence processing with long distance but containing important information. The word vector is stitched by word embedding, part of speech vector and entity vector. It is sent to LSTM to extract the key information, and the encoded result obtained is regarded as a sentence vector. The model receives a sequence $V = [v_1, v_2, ..., v_n]$. The input is calculated as follows:

$$v_{1:n} = v_1 \oplus v_2 \oplus ... \oplus v_n \qquad (1)$$

In order to keep the meaning of the sentence in the coded information, as shown in Fig. 1, we use the encoder-to-decoder to encode the text. And the encoder output is calculated as follows:

$$E, h_t = LSTM(v_{1:n}, h_{t-1}) \qquad (2)$$

And the decoder output is calculated as follows:

$$D, h_t = LSTM(E, h_{t-1}) \tag{3}$$

We train the network by calculating the cross-entropy of the decoder results and initial inputs.

3.2 Text Categorization

From the above analysis, all vectors are projected in a vector space, and the information is numerous and complicated. It is not conducive to the extraction and conversion of the information. So the paper adopts a multi-dimensional spatial classification algorithm, which classifies all texts into 34 event types. The multi-dimensional space classification algorithm uses the KNN (K-Nearest Neighbor). The input information of the KNN algorithm is a text vector, the text vector is calculated in a feature space, and the output information is the category of the text information. KNN algorithm classifies event text according to the distance between instances and instances. By this method, the news text is divided into 34 categories, such as military, financial, sports and so on.

3.3 Event Extraction

In this section, we build a network model combining attention mechanism and DMCNN network, which can extract event triggers and parameters from text. When extracting event information, the neural model pays different attention to words with different parts of speech based on grammatical attributes. The whole process of the model is to introduce the attention mechanism into information extraction, the word vector is the overall input of the network model, and the recognition process takes DMCNN to classify the information. Our model is shown in Fig. 2.

Attention Vector. The attention vector a_i is calculated as follows:

$$a_i = \alpha_i \times v_i \tag{4}$$

$$\alpha = softmax(\beta) \tag{5}$$

$$\beta_i = selu(pv_i \cdot w^c) \tag{6}$$

where $\alpha = [\alpha_1, \alpha_2, ..., \alpha_n]$ and $\beta = [\beta_1, \beta_2, ..., \beta_n]$ are vectors, w^c is a weight vector.

Trigger Detection. The sample data has been divided into 34 event types. Select a type of event sample set and extract the prediction trigger. The next step is to convolve the word vector, segment the sentence with the position of the predicted trigger word, and maximize each part of the pool. The obtained

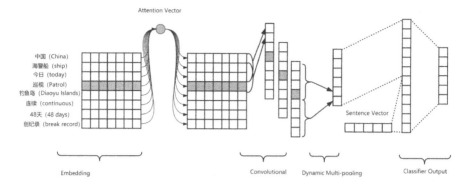

Fig. 2. Event Extraction. The trained word vector is labeled according to the information after the word segmentation, and then the required information is extracted from the labeled word vector.

result is spliced with the encoder result obtained by LSTM, and input to the fully connected layer to classification. The purpose of the convolutional layer is to capture the combined semantics of the sentence. In calculation, let $a_{i:i+m}$ represent the concatenation of $a_i, a_{i+1}, ..., a_{i+m}$. The function c_i is generated by the following operators:

$$c_i = f(w \cdot a_{i:i+m-1} + b) \tag{7}$$

where w is a filter, b is a bias term and f is a non-linear. To capture more semantic features, we use multiple convolution operations. Use the following definition for the filter $W = w_1, w_2, ..., w_n$, the calculation formula is:

$$c_{ji} = f(w_j \cdot a_{i:i+m-1} + b_j) \tag{8}$$

where j represents the filter. The predictive trigger word divides the text into two segments. In order to capture important semantic information, we use the dynamic multi-pooling method to obtain the result. The pooling calculation is as follows:

$$p_{ji} = max(c_{ji}) \tag{9}$$

Map the above coding result and pooling result to the feature vector $T = [P, E]$. The feature vectors are input to a fully connected network for classification.

$$O = w_s \cdot T + b_s \tag{10}$$

Event Arguments Detection. The event argument extraction is similar to the event trigger extraction process. The event arguments are mainly nouns, adjectives, and a few are verbs, so we need to pay more attention to these types of words when training. Moreover, in most cases, the parameters need to be extracted multiple times to avoid losing information.

3.4 Training

We determine all parameters of the network model as $theta$ = $[v, w^c, W, b, w_s, b_s]$. Suppose we are given an input example s and the output information of the model is O. We calculate the softmax of the output as:

$$p(i|x, \theta) = \frac{e^{o_i}}{\sum_{q=1}^{n} e^{o_q}} \tag{11}$$

Suppose (x_i, y_i) is an example of the model, where y is the output information of the model and x is the input information of the model, then the objective function is as:

$$J(\theta) = \sum_{1}^{N} \log p(y^{(i)}|x^{(i)}, \theta) \tag{12}$$

During model training, we use stochastic gradient descent and the Adadelta update rule to optimize the objective function.

4 Experiments

4.1 Experimental Datasets and Evaluation Indicators

We utilized the network public opinion data corpus as our dataset. The data set is obtained by us sorting out the network data. The data set size is 15.3M, which contains 34 event types. There are 1800 documents in the data set. We randomly select 50 for testing, 50 documents verification results and the remaining 1,700 as training sets. We use Precision (P), Recall (R) and F measure (F) as the yardstick.

$$P = \frac{Number\ of\ correctly\ extracted\ events}{Number\ of\ all\ extracted\ events} \tag{13}$$

$$R = \frac{Number\ of\ correctly\ extracted\ events}{Total\ number\ of\ true\ events} \tag{14}$$

$$F = \frac{2 \times P \times R}{P + R} \tag{15}$$

4.2 Training Curves

We use precision and dataset size to measure the accuracy of training.

The left figure shows that with the increase of the dataset, the accuracy of the experimental results improves faster; but when the dataset is 1,700 documents, the experimental results enter a convergence period. As we can see, when the dataset is already large enough, the experimental results do not change as the dataset increases. The right figure shows the training results between P and R. After combining the attention mechanism in information extraction, the accuracy and completeness of event extraction has been greatly improved (Fig. 3).

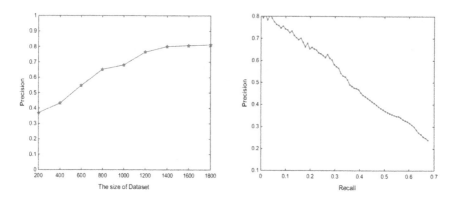

Fig. 3. The training curves

4.3   Comparative Experiment

Table 1 shows the overall performance of the above method in CEC data set. For the extraction of Chinese events, we use the following method as a comparison experiment:

1) Xia's DNN model is by [8], which introduces deep neural networks and word embeddings into the recognition process.
2) Hong's cross-entity is by [9], which improves the accuracy of information extraction by using cross entity theory.
3) Li's structure is by [10], which predicts event information through a structure based on global features.
4) Chen's DMCNN is by [11], which extracts events based on dynamic multi-pooling convolutional neural networks.

Table 1. The Experimental result on blind test data.

Methods	Trigger			Argument			Argument role		
	P	R	F	P	R	F	P	R	F
DNN	61.1	67.8	64.3		N/A			N/A	
Hong's cross-entity		N/A		53.4	52.9	53.1	51.6	45.5	48.3
Joint event extraction	76.9	65.0	70.4	69.8	47.9	56.8	64.7	44.4	52.7
DMCNN	80.4	67.7	73.5	68.8	51.9	59.1	62.2	46.9	53.5
Our model	81.2	67.3	73.6	69.6	51.6	59.2	63.4	45.7	53.1

It can be seen from the above table that the method performs well and verifies the feasibility of our method. Since our training set divides text information into 34 categories, including military, natural disasters, and health, we take single-category texts of terrorist attacks, earthquakes, and food poisoning in CEC to conduct comparative experiment analysis on the above three category models. The methods performed in the texts of terrorist attacks are general, but in the texts of earthquakes, the methods perform well, indicating that our natural disaster training set contains more information about earthquakes, which also verifies the feasibility of our method.

5   Conclusion

This paper proposes an event extraction method based on the attention mechanism, which can extract valuable information to the user from the open domain news text. The model uses a combination of neural networks and statistics. The model does not have a complicated preprocessing process and adds the semantic information of the word itself to the analysis. Experiments show that the method is effective. Future work will demonstrate and improve performance. We will use the extracted information to build a public opinion knowledge map.

References

1. Chieu, H.L., Ng, H.T.: A maximum entropy approach to information extraction from semi-structured and free text. In: Proceedings of the 18th National Conference on Artificial Intelligence, pp. 786–791 (2002)
2. Llorens, H., Saquete, E., Navarro-Colorado, B.: TimeML events recognition and classification: learning CRF models with semantic roles. In: COLING, pp. 725–733 (2010)
3. Huang, Z., Xu, W., Yu, K.: Bidirectional LSTM-CRF models for sequence tagging. CoRR abs/1508.01991 (2015)
4. Kim, Z.M., Jeong, Y.-S.:: TIMEX3 and event extraction using recurrent neural networks. In: BigComp, pp. 450–453 (2016)
5. Nguyen, T.H., Cho, K., Grishman, R.: Joint event extraction via recurrent neural networks. In: HLT-NAACL, pp. 300–309 (2016)
6. Chen, Y., Liu, S., Zhang, X., Liu, K., Zhao, J.: Automatically labeled data generation for large scale event extraction. In: ACL, vol. 1, pp. 409–419 (2017)
7. Li, L., Wan, J., Zheng, J., Wang, J.: Biomedical event extraction based on GRU integrating attention mechanism. BMC Bioinform. **19-S**(9), 177–184 (2018)
8. Xia, Y., Liu, Y.: Chinese event extraction using deep neural network with word embedding. CoRR abs/1610.00842 (2016)
9. Hong, Y., Zhang, J., Ma, B., Yao, J.-M., Zhou, G., Zhu, Q.: Using cross-entity inference to improve event extraction. In: ACL, pp. 1127–1136 (2011)
10. Li, Q., Ji, H., Huang, L.: Joint event extraction via structured prediction with global features. In: ACL, vol. 1, pp. 73–82 (2013)
11. Chen, Y., Liheng, X., Liu, K., Zeng, D., Zhao, J.: Event extraction via dynamic multi-pooling convolutional neural networks. In: ACL, vol. 1, pp. 167–176 (2015)
12. Chen, Y., Liu, S., He, S., Liu, K., Zhao, J.: Event extraction via bidirectional long short-term memory tensor neural networks. In: Sun, M., Huang, X., Lin, H., Liu, Z., Liu, Y. (eds.) CCL/NLP-NABD -2016. LNCS (LNAI), vol. 10035, pp. 190–203. Springer, Cham (2016). https://doi.org/10.1007/978-3-319-47674-2_17

13. Zeng, X., Zeng, D., He, S., Liu, K., Zhao, J.: Extracting relational facts by an end-to-end neural model with copy mechanism. In: ACL, vol. 1, pp. 506–514 (2018)
14. Zeng, X., He, S., Liu, K., Zhao, J.: Large scaled relation extraction with reinforcement learning. In: AAAI, pp. 5658–5665 (2018)
15. Liu, X., Huang, H., Zhang, Y.: Open domain event extraction using neural latent variable models. In: ACL, vol. 1, pp. 2860–2871 (2019)
16. Li, W., Cheng, D., He, L., Wang, Y., Jin, X.: Joint event extraction based on hierarchical event schemas from FrameNet. IEEE Access **7**, 25001–25015 (2019)
17. Zheng, S., Cao, W., Xu, W., Bian, J.: Doc2EDAG: an end-to-end document-level framework for Chinese financial event extraction. CoRR abs/1904.07535 (2019)
18. Kuila, A., Chandra Bussa, S., Sarkar, S.: A neural network based event extraction system for Indian languages. In: FIRE (Working Notes), pp. 291–301 (2005)
19. Li, L., Huang, M., Liu, Y., Qian, S., He, X.: Contextual label sensitive gated network for biomedical event trigger extraction. J. Biomed. Inform. **95**, 103221 (2019)
20. Zhang, Y., Liu, Z., Zhou, W.: Event recognition based on deep learning in Chinese texts. PLoS ONE **11**, e0160147 (2016)
21. Romadhony, A., Widyantoro, D.H., Purwarianti, A.: Utilizing structured knowledge bases in open IE based event template extraction. Appl. Intell. **49**(1), 206–219 (2018). https://doi.org/10.1007/s10489-018-1269-0
22. Yang, H., Chen, Y., Liu, K., Xiao, Y., Zhao, J.: DCFEE: a document-level Chinese financial event extraction system based on automatically labeled training data. In: ACL (4), pp. 50–55 (2018)
23. Liu, J., Chen, Y., Liu, K., Zhao, J.: Event detection via gated multilingual attention mechanism. In: AAAI, pp. 4865–4872 (2018)
24. Ji, G., Liu, K., He, S., Zhao, J.: Distant supervision for relation extraction with sentence-level attention and entity descriptions. In: Proceedings of the Thirty-First AAAI Conference on Artificial Intelligence AAAI, pp. 3060–3066 (2017)

Deep Neural Semantic Network for Keywords Extraction on Short Text

Chundong She[1], Huanying You[1], Changhai Lin[2(✉)], Shaohua Liu[1],
Boxiang Liang[1], Juan Jia[1], Xinglei Zhang[1], and Yanming Qi[3]

[1] Laboratory for Cyber Physical Systems, School of Electronic Engineering,
Beijing University of Posts and Telecommunications, Beijing, China
{scd,yyou88627,liushaohua,liangboxiang,jiajuan,
sunzhongkai}@bupt.edu.cn
[2] Suge Communication Technology Co., Ltd., Yibin, Sichuan, China
411469879@qq.com
[3] Sichuang Jiuzhou Video Technology Co. Ltd., Mianyang, Sichuan, China
28047273@qq.com

Abstract. Keyword extraction is a branch of natural language processing, which plays an important role in many tasks, such as long text classification, automatic summary, machine translation, dialogue system, etc. All of them need to use high-quality keywords as a starting point. In this paper, we propose a deep learning network called deep neural semantic network (DNSN) to solve the problem of short text keyword extraction. It can map short text and words to the same semantic space, get the semantic vector of them at the same time, and then compute the similarity between short text and words to extract top-ranked words as keywords. The Bidirectional Encoder Representations from Transformers was first used to obtain the initial semantic feature vectors of short text and words, and then feed the initial semantic feature vectors to the residual network so as to obtain the final semantic vectors of short text and words at the same vector space. Finally, the keywords were extracted by calculating the similarity between short text and words. Compared with existed baseline models including Frequency, Term Frequency Inverse Document Frequency (TF-IDF) and Text-Rank, the model proposed is superior to the baseline models in Precision, Recall, and F-score on the same batch of test dataset. In addition, the precision, recall, and F-score are 6.79%, 5.67%, and 11.08% higher than the baseline model in the best case, respectively.

Keywords: Semantic similarity · Semantic network · Short text · Keywords extraction

C. She, H. You, C. Lin and S. Liu—These authors contributed to the work equally and should be regarded as co-first authors.

© Springer Nature Singapore Pte Ltd. 2020
P. Qin et al. (Eds.): ICPCSEE 2020, CCIS 1258, pp. 101–112, 2020.
https://doi.org/10.1007/978-981-15-7984-4_8

1 Introduction

Keyword extraction technology has exhibited their potential on improving many natural language processing (NLP) and information retrieval (IR) tasks, such as document indexing [1], text summarization [2, 3], text categorization [4, 5], opinion mining [6], knowledge base [7] and Question Answering [8]. Although many automatic frameworks for extracting keywords have been proposed [9–13], there exist more factors need to be considered for extracting keywords for short texts, for examples, the length of the text is relatively short, the most prevalent word frequency is 1, the number of structured features required for conversion to machine learning is much limited, and the size of the text-rank word network is also small. Thus, the existing methods suitable for long texts are not directly applicable to extract keywords from the short text. In this paper, rather than generate quantified features to extract keywords, we try to extract keywords according to the similarity between the text and words. Based on this idea, we propose a deep neural semantic network to map the short text and the candidate keyword to the same semantic space for extracting keywords. First, we learn the corresponding relationship between the text and the keywords to obtain the same semantic space of the text and the key words, then map the text and the words to the trained semantic space at the same time and get their respective semantic vectors, and then the similarity between the original text and the keyword extracted is calculated to evaluate the effect. Finally, the keywords are extracted according to the similarity ranking.

The paper is organized as follows. In Sect. 2, we review the related work of keyword extraction and semantic representation of words. In Sect. 3, we describe our model structure and corresponding formula derivation. In Sect. 4, we introduce our data, model training, and inference of the model. In Sect. 5, we show our model results and analyze the reason why our model outperforms baseline models. In Sect. 6, we draw our conclusion and outlook for future work.

2 Related Work

In general, existing approaches for the keyword extraction problem can be divided into three classes. The first is to extract keywords based on statistics represented by TF-IDF (Term Frequency-Inverse Document Frequency) [14]. TF-IDF is an unsupervised learning method. It calculates the weight of all candidate words and then selects k large weighted words as keywords. The second is to extract the keywords from the construction word network represented by text-rank [15]. It derived from Google's Page-rank and more concerned the importance of each word in the word-network. The third is to extract keywords based on the machine learning methods [16–21]. An algorithm was proposed using Character n-Grams instead of Word n-Grams, tokenization tools and external stop words list to extract Trend Keywords on Thai language in social media platform [22]. Dynamic N-gram (DNG) has been proposed to extract the Keywords of Chinese News [23]. The semantic similarity between words and sentences has also been used to extract important keywords from documents [24]. Corresponding machine learning methods could transform unsupervised learning methods into supervised learning solving a binary classification problem. Instead of calculating the weight of each candidate word, multiple features of candidate words are generated by certain rules, and then the candidate is directly judged by the model.

Recently, deep learning approaches have been the focus of much research on a wide variety of NLP and IR problems. The main feature of these approaches is that during the learning process, they discover not only the mapping from representation to output but also the representation itself, which removes the empirical process of manual feature engineering characteristic of traditional supervised methods, and greatly improves adaptability in new tasks and domains. The click log of search users has been used to calculate the similarity model between query and documents [25]. Firstly, the query and document are mapped to a vector space at the same time through a deep neural semantic network. Then the similarity between query and document is calculated. Finally, the best matching document is given according to the rank of similarity. A text representation (word-embedding) is the basis of using a deep neural network in NLP. Neural network [26] has been used to do language model work, trying to get low-dimensional, dense vector representations of words. A simplified model Word2Vec has been proposed [27], which includes two models, CBOW (Continuous bag-of-words) and Skip-gram. The former is used to predict the probability of the appearance of target words through window context, and the latter uses the target words to predict the probability of each contextual word in the window. Le and Mikolov extended the idea of Word2Vec and put forward the core idea of Doc2Vec, which regards document vector as "context" [28]. The word vector and document vector can be obtained by the Doc2Vec algorithm. The deep neural network [29] based on the transformer continues the pre-training and micro-adjustment idea in the field of CV (Computer Vision) and uses a large number of corpus and huge training network to train the corpus, which can provide semantic feature vectors for multiple NLP tasks. In this paper, we use the BERT model [29] to obtain the text representation vector as the initial vector of the task.

3   The Model

3.1   Deep Neural Semantic Network Framework

The work of our model is divided into two parts. In the first stage, the short text and the candidate word can be mapped into a vector space, the obtained vector can be regarded as the semantic vector. In this stage, the model learn the corresponding relationship between the text and the keywords, and also obtain the semantic vector space of the text and the keywords at the same time. Receiving the semantic vectors of the text and words when we put them to the deep neural semantic network, respectively. In the second phase of the model, the similarity between the short text vector and each candidate term vector representation is calculated, and the k candidates with the highest similarity are considered as the keywords of the text. The vector used to calculate semantic similarity is the last semantic vector obtained in the first phase.

The structure of the deep neural semantic network is shown in Fig. 1. The model mainly includes four layers. (1) In the input layer, the input is the short text and the semantic word; (2) In the semantic feature layer, we use the BERT as the semantic feature extraction model so as to get the semantic feature vector with 768 dimensions, which advantage is that the BERT has already a certain semantic representation ability so that it can help us learn the semantic vector of the text better; (3) The residual network layer is used to fine-tune the semantic feature vectors to make the semantic feature vectors more suitable for our keyword extraction task to obtain the final semantic vector; (4) The semantic representation layer serves as our semantic vector.

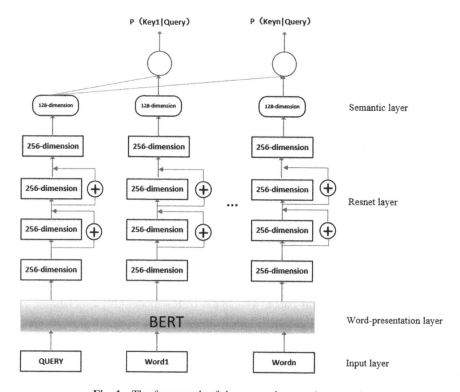

Fig. 1. The framework of deep neural semantic network.

3.2   Deep Neural Semantic Network Formulation

As shown in Fig. 1, the main function of the deep neural semantic network is to map the text and keyword to a semantic vector space. First, we obtain the initialization semantic vector of 768-dimensional text and candidate words by the BERT, then we learn and reduce the dimension of the initialization semantic vector through the residual network to obtain the vector that can represent short text and candidate words.

In this paper, we take the semantic vector of initialization as x, and the vector of output as y, l_i, $(i = 1, , \ldots, N - 1)$, l_i is the hidden layer, W_i is the weight matrix of layer i, b_i is the deviation of layer i, thus,

$$l_1 = W_1 x + b_1 \tag{1}$$

$$l_i = f(W_i l_{i-1} + b_i) + l_{i-1}, \ (i = 1,, \ldots N - 1) \tag{2}$$

$$y = f(W_{N-1} l_{N-1} + b_i) \tag{3}$$

$$f(x) = \frac{1 - e^{-2x}}{1 + e^{-2x}} \tag{4}$$

We use $\{(t_1, k_1), (t_2, k_2), \ldots\}$ to represent the set of short text and corresponding keyword pairs. k_i is a keyword under a text t_i. Assume that each text-keyword pair is independent of each other, the joint probability of this set is

$$\prod_i P(k_i | t_i) \tag{5}$$

Where $P(k_i | t_i)$ denotes the conditional probability. The expression of the joint probability is converted into the maximum likelihood function. We use the softmax function to establish a probability model, that is,

$$p(k|t; \vec{w}) = \frac{\exp(R(t, k; \vec{w}))}{\exp(R(t, k; \vec{w})) + \sum_{k' \in K^-} \exp(R(t, k'; \vec{w}))} \tag{6}$$

Where K^- denotes the set of irrelevant keywords under t, $R(t, k; \vec{w})$ is an equation about the parameter of the similarity measure between short text and words. In this paper, the cosine distance is used as the similarity measure.

$$R(t, k; \vec{w}) = \frac{\vec{y}(t)^T \vec{y}(k)}{||\vec{y}(t)|| \bullet ||\vec{y}(k)||} \tag{7}$$

Where $\vec{y}(t) = \alpha(q, \vec{w}^{(t)})$ and $\vec{y}(k) = \beta(k; \vec{w}^{(k)})$ are the semantic vector of the text and the candidate, respectively. $\vec{w}^{(t)}$, $\vec{w}^{(k)}$ are the parameter representations corresponding to the text and the candidate word. Finally, we can get the loss function of the model.

$$\begin{aligned} L(\vec{w}) &= -\sum_i \ln p(k_i | t_i) \\ &= -\sum_i \ln \frac{\exp(R(t, k; \vec{w}))}{\exp(R(t, k; \vec{w})) + \sum_{k' \in K^-} \exp(R(t, k'; \vec{w}))} \end{aligned} \tag{8}$$

Where K_i^- denotes a non-keyword sampling set of short text. In this paper, the number of negative samples is 4. Therefore, our model needs to solve parameters $\vec{w}$, α and β to minimize the loss function.

4   Experiments

In this section, we will introduce the training data, test data, distribution of the number of keywords, model training, and model inference.

4.1   Datasets

We have tagged a total of 20,000 sentences and 66,950 keywords. 16,000 sentences are taken as a training set, while 4000 sentences are used as a test set. Our training sample form is {short text, keyword, negative sample set}. If there are N keywords in the short text, N training samples are generated, where the negative sample set contains 4 words (non-keywords). Specifically, the negative sample selected from tagged keywords set

Table 1. The sample of the training dataset

Short text	Keywords	Negative sample set
巧手折纸星星收纳盒，漂亮简单又实用 (Folding paper into star-shaped storage boxes with dexterity that are beautiful, simple and practical.)	收纳盒 (storage box)	{治愈系，枸杞，纪实摄影，养生} (Genre of Healing, goji berries, documentary photography, nourish)
	折纸(Folding paper	{挪威，砒霜，养生，治愈系} (Norwegian, arsenic, nourish, Genre of Healing, wolfberry)
	星星 (star)	{枸杞，治愈系，折纸，收纳盒} (goji berries, Genre of Healing, folding paper, storage box)
枸杞虽然可以滋补养生，瞎吃却比砒霜还要毒 (Although goji berries are nourish good for health, eating them blindly is more poisonous than arsenic.)	砒霜 (arsenic)	{枸杞，挪威，星星，折纸} (goji berries, Norwegian, star, folding paper)
	枸杞 (goji berries)	{收纳盒，纪实摄影，治愈系，折纸} (storage box, documentary photography, Genre of Healing, folding paper)
	养生 (nourish)	{收纳盒，星星，治愈系，挪威} (storage box, star, Genre of Healing , Norwegian)
	治愈系 (Genre of Healing)	{砒霜，折纸，枸杞，星星} (arsenic, Folding paper, goji berries, star)
治愈系：置身于时间与时空的流逝中，挪威纪实摄影 (Genre of Healing: immersed in the passage of time and space - Norwegian documentary photography.)	挪威 (Norwegian)	{折纸，养生，枸杞，收纳盒} (Folding paper, nourish, goji berries, storage boxes)
	纪实摄影 (documentary photography)	{星星，养生，砒霜，收纳盒} (star, nourish, arsenic, storage boxes)

(except keywords set) with a certain probability that decided by the frequency of tagged keywords, the greater the frequency, the greater the probability. A set of samples is shown in Table 1.

There are 3 short texts, 9 tagged keywords, and the corresponding negative sample set generated from the tagged keywords excepted its keywords set with a probability decided by frequency.

Our test sample form is {short text, candidates, keywords}. Unlike the training sample, there is no negative sample set, the candidates of a sample are sub-words of short text, keywords are regarded as the label of short text. A set of samples is shown in Table 2.

There are 5 short texts, 19 tagged keywords, and the corresponding candidates for each short text. The number of tagged keywords set change from 3 to 6, the candidates are sub-words of short text, and the number of candidates relates to the short text, the word segmentation algorithm. Most the number of the keywords for the short text is 3

Table 2. The sample of the test dataset

Short text	Candidates	Keywords
熊出没：趁熊二睡觉强哥想偷偷把小熊徽章偷回来(*Boonie Bears*: While Bramble was asleep, Logger Vick wanted to steal the bear emblem back, secretly.)	{熊出没, 熊二, 睡觉, 强哥, 想, 偷偷, 小熊, 徽章, 偷回来} (*Boonie Bear*, Bramble, sleep, Logger Vick, want, secret, bear, emblem, steal back)	{熊出没, 熊, 徽章} (Boonie Bear, bear, bear emblem)
实拍印度农民采摘未成熟的青香蕉制作香蕉片全过程 (Filming the whole process of Indian farmers unposedly picking immature green bananas to make dried banana chips.)	{实拍, 印度, 农民, 采摘, 未成熟, 青, 香蕉, 制作, 片, 全过程} (Film, Indian, farmers, pick, immature, green, banana, make, chip, the whole process)	{香蕉, 采摘, 印度, 过程, 成熟, 农民} (banana, pick, Indian, process, mature, farmer)
小恐龙独自流浪跟着大型恐龙能找食物真担心它被踩到 (The little dinosaur wanders alone with the big dinosaur so it can find food. Really worried about it being stepped on.)	{恐龙, 流浪, 跟着, 大型, 找, 食物, 真, 担心, 踩} (dinosaur, wander, with, big, find, food, really, worry, step)	{恐龙, 流浪, 食物} (dinosaur, wander, food)
王者荣耀:哪件装备出的人最多官方公布装备购买率排行 (*Honor of Kings*: What's the equipment that most people sell? The official announced the ranking of equipment purchase rate.)	{王者, 荣耀, 哪件, 装备, 官方, 公布, 购买率, 排行} (king, honor, what, equipment, official, announce, purchase rate, ranking)	{王者荣耀, 装备, 官方} (honor of king, equipment, official)
智能手机也能3D全息投影了,超酷的 (The smartphone can also project a 3D holography. Super cool.)	{智能手机, 3D, 全息, 投影, 超酷} (smartphone, 3D, holography, project, cool)	{全息投影, 3D, 智能手机} (project with holography, 3D, smartphone)

Table 3. The distribution for the number of keywords in the test dataset

The number of keywords	3	4	5	6
The number of samples	3240	480	230	50
Sample ratio	64.80%	9.60%	4.60%	1%

but some are not equal to 3, and the distribution of the number of keywords for test data-set is shown in Table 3.

4.2 Model Training

The model structure used in this paper is shown in Fig. 1. During the training process, to make the model converge faster, we add a Batch Normalization layer to each layer of the residual network and use the Tanh function (formula 4) as our activation function. The softmax_cross_entropy function is used as the objective function, and the Adam function is used as the optimizer. Firstly, we put the samples into the encode interface of BERT to get the 786-dimensional initial text feature vector of text and words, and then feed the initial text feature vector to the deep neural semantic network and get 128-dimensional semantic vectors suitable for keyword extraction. Then we calculate the semantic similarity between short text and positive and negative samples. In this paper, we use cosine distance as the measure of semantic similarity. Finally, the loss function of the model is obtained, and the model is trained with the goal of minimizing the loss function.

4.3 Inference

When performing keyword prediction, we need to calculate the similarity between the short text and each word, so we should get all the sub-words of the short text in the first step of the prediction phase, and then feed the text and sub-words to BERT to get the initial text vector. After that, through the deep neural semantic network to get the final semantic vector and calculate the semantic similarity, and finally select the top-k words we need as our final keywords according to the ranking of the semantic similarity.

5 Results and Analysis

To verify the validity of the model, we selected the classic keyword extraction model TF-IDF and Text-Rank model as the compared model. Besides, since the initial feature vector obtained by BERT can also express semantics to a certain degree. We added a method of directly using BERT to extract keywords to compare with our model. In this paper, we use precision, recall, and F-score as the evaluation indicators. The deep neural semantic network to the test dataset and list comparisons of experimental results are shown in Table 4 and Fig. 2. In Table 4, we can see that as the number of extracted keywords increases, the accuracy of the model decreases. By contrary, the recall and the F-score increase. The results of the deep neural semantic network proposed in this paper on the precision, recall, and F-score are better than the baseline model. In the case of extracting 3 keywords, the precision of the model is higher than TextRank by 6.79%. In the case of extracting 4 keywords, the F-score of the model is 5.67% higher than Text-Rank. In the case of extracting 5 keywords, the model's recall is higher than Text-Rank 11.08%. Also, we can see that the effect of extracting keywords through BERT is similar to the baseline model, indicating that BERT's semantic feature vector can be used to extract keywords directly, but the deep neural semantic network model proposed is still better.

Table 4. Performance comparison between four algorithms on the test dataset.

Method	Top3 candidates			Top4 candidates			Top5 candidates		
	Precision	Recall	F-score	Precision	Recall	F-score	Precision	Recall	F-score
TF-IDF	43.50%	37.37%	40.21%	41.88%	47.94%	44.70%	39.28%	55.57%	46.02%
TextRank	40.24%	31.89%	35.59%	40.97%	41.29%	41.13%	40.07%	47.55%	43.86%
Bert	43.6%	37.45%	40.29%	42.46%	48.62%	45.44%	39.20%	55.88%	46.08%
DNSN	**47.03%**	**40.39%**	**43.46%**	**43.84%**	**50.20%**	**46.80%**	**41.13%**	**58.63%**	**48.34%**

As we can see, the proposed model is superior to the basic model for three reasons:

1) TextRank and TF-IDF belong to the method of unsupervised learning. They calculate the importance of each candidate word in the entire text through statistical co-occurrence relationships, while our model calculates the semantic similarity between the text and words through a supervised learning method to extracted keywords.

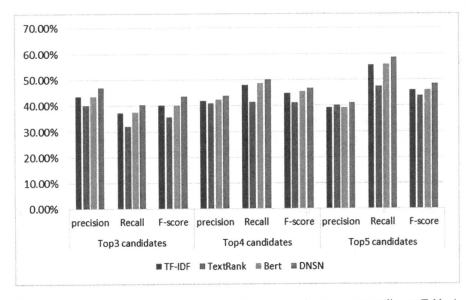

Fig. 2. Performance comparison of four algorithms on test dataset corresponding to Table 4.

2) The similarity between text and words is more suitable for keyword extraction of short text, the deep neural semantic network-based method can effectively learn the representation of words and information of text in high-dimensional semantic space.

3) The introduction of a BERT can reduce the impact of insufficient corpus and tedious text preprocessing methods. Compared with BERT method, the proposed method has better effect. We argue that the proposed model adjusts and learns the initial semantic vector of BERT, making the final semantic vector more suitable for keyword extraction task.

6   Conclusion

In this paper, we proposed to use a deep learning network i.e., DNSN, to simultaneously map short texts and candidate words to the same semantic space, and extract keywords by calculating the semantic similarity between short texts and words. We first use BERT to obtain the initial semantic feature vector, then use the deep neural semantic network to optimize the semantic feature vector to obtain a semantic vector that is more suitable for keyword extraction, and finally extract the keywords by the ranked similarity between the short text and the words. We designed three sets of comparative experiments to verify the effectiveness of the model. The results show that our model is better than the baseline models in precision, recall, and F-score. In future work, we will apply the model to extract keywords for the paragraph.

Acknowledgment. This research was partly supported by the Major Program of National Natural Science Foundation of China (Grant Nos. 91938301), the National Defense Equipment Advance Research Shared Technology Program of China (41402050301-170441402065), and the Sichuan Science and Technology Major Project on New Generation Artificial Intelligence (2018GZDZX0034).

References

1. Frank, E., Paynter, G.W., Witten, I.H., Gutwin, C., NevillManning, C.G.: Domain-specific keyphrase extraction. In: Proceedings of 16th International Joint Conference on Artificial Intelligence, pp. 668–673 (1999)
2. Pal, A.R., Maiti, P.K., Saha, D.: An approach to automatic text summarization using simplified lesk algorithm and wordnet. Int. J. Control Theory Comput. Model. **3** (2013)
3. Zhang, Y., Zincir-Heywood, N., Milios, E.: World wide web site summarization. Web Intell. Agent Syst. **2**, 39–53 (2004)
4. Hulth, A., Megyesi, B.B.: A study on automatically extracted keywords in text categorization. In: Proceedings of the 21st International Conference on Computational Linguistics and the 44th Annual Meeting of the Association for Computational Linguistics, pp. 537–544 (2006)
5. Özgür, A., Özgür, L., Güngör, T.: Text categorization with class-based and corpus-based keyword selection. In: Yolum, p, Güngör, T., Gürgen, F., Özturan, C. (eds.) ISCIS 2005. LNCS, vol. 3733, pp. 606–615. Springer, Heidelberg (2005). https://doi.org/10.1007/11569596_63

6. Berend, G.: Opinion expression mining by exploiting keyphrase extraction. In: Proceedings of the 5th International Joint Conference on Natural Language Processing, pp. 1162–1170 (2011)

7. Yang, Z., Nyberg, E.: Leveraging procedural knowledge base for task-oriented search. In: Proceedings of the 38th International ACM SIGIR Conference on Research & Development in Information Retrieval. ACM (2011)

8. Liu, R., Nyberg, E.: A phased ranking model for question answering. In: Proceedings of the 22nd ACM International Conference on Information & Knowledge Management, CIKM 2013, New York, NY, USA, pp. 79–88 (2013)

9. Riloff, E., Lehnert, W.: Information extraction as a basis for high-precision text classification. ACM Trans. Inf. Syst. (TOIS) **12**(3), 296–333 (1994)

10. Witten, I.H., Paynter, G.W., Frank, E., Gutwin, C., NevillManning, C.G.: KEA: practical automatic keyphrase extraction. In: Proceedings of the 4th ACM Conference on Digital Libraries, DL 1999, New York, NY, USA, pp. 254–255 (1999)

11. Turney, P.D.: Learning algorithms for keyphrase extraction. Inf. Retrieval **2**(4), 303–336 (2000)

12. Medelyan, O., Perrone, V., Witten, I.H.: Subject metadata support powered by Maui. In: Hunter, J., Lagoze, C., Lee Giles, C., Li, Y.-F. (eds.) JCDL, pp. 407–408. ACM (2010)

13. Litvak, M., Last, M.: Graph-based keyword extraction for single-document summarization. In: Proceedings of the Workshop on Multisource Multilingual Information Extraction and Summarization, MMIES 2008, pp. 17–24. Association for Computational Linguistics, Stroudsburg (2008)

14. Salton, G., Wong, A., Yang, C.S.: A vector space model for automatic indexing. Commun. ACM **18**(11), 613–620 (1975)

15. Mihalcea, R., Tarau, P.: TextRank: bringing order into texts. In: Proceedings of the 2004 Conference on Empirical Methods in Natural Language Processing, Barcelona, Spain, 25–26 July 2004, pp. 404–411 (2004)

16. Zhang, C.: Automatic keyword extraction from documents using conditional random fields. J. Comput. Inf. Syst. **4**, 1169–1180 (2008)

17. Chen, Y., Yin, J., Zhu, W., Qiu, S.: Novel word features for keyword extraction. In: Dong, X.L., Yu, X., Li, J., Sun, Y. (eds.) WAIM 2015. LNCS, vol. 9098, pp. 148–160. Springer, Cham (2015). https://doi.org/10.1007/978-3-319-21042-1_12

18. Ardiansyah, S., Majid, M.A., Zain, J.M.: Knowledge of extraction from trained neural network by using decision tree. In: Proceedings of the International Conference on Science in Information Technology, Balikpapan, Indonesia, 26–27 October, pp. 220–225 (2016)

19. Witten, I.H., Paynter, G.W., Frank, E., Gutwin, C., Nevill-Manning, C.G.: KEA: practical automatic keyphrase extraction. In: Proceedings of the ACM Conference on Digital Libraries, Berkeley, CA, USA, pp. 254–255 (1999)

20. Zhou, C., Li, S.: Research of information extraction algorithm based on hidden Markov model. In: Proceedings of the International Conference on Information Science and Engineering, Hangzhou, China, pp. 1–4 (2010)

21. Kanis, J.: Digging language model – maximum entropy phrase extraction. In: Sojka, P., Horák, A., Kopeček, I., Pala, K. (eds.) TSD 2016. LNCS (LNAI), vol. 9924, pp. 46–53. Springer, Cham (2016). https://doi.org/10.1007/978-3-319-45510-5_6

22. Ousirimaneechai, N., Sinthupinyo, S.: Extraction of trend keywords and stop words from Thai Facebook pages using character n-grams. Int. J. Mach. Learn. Comput. **6**, 589–594 (2018)

23. Lin, R.-X., Yang, H.-L.: Apply the dynamic n-gram to extract the keywords of Chinese news. In: Ali, M., Pan, J.-S., Chen, S.-M., Horng, M.-F. (eds.) IEA/AIE 2014. LNCS (LNAI), vol. 8482, pp. 398–406. Springer, Cham (2014). https://doi.org/10.1007/978-3-319-07467-2_42

24. Hasan, H.M.M., Sanyal, F., Chaki, D.: A novel approach to extract important keywords from documents applying latent semantic analysis. In: 2018 10th International Conference on Knowledge and Smart Technology (KST), Chiang Mai, pp. 117–122 (2018). https://doi.org/10.1109/kst.2018.8426144

25. Huang, P.S., et al.: Learning deep structured semantic models for web search using clickthrough data. In: Proceedings of the 22nd ACM International Conference on Information & Knowledge Management. ACM (2013)

26. Bengio, Y., Schwenk, H., Senécal, J.S., Morin, F., Gauvain, J.L.: Neural probabilistic language models. In: Holmes, D.E., Jain, L.C. (eds.) Innovations in Machine Learning. STUDFUZZ, vol. 194, pp. 137–186. Springer, Heidelberg (2006). https://doi.org/10.1007/3-540-33486-6_6

27. Mikolov, T., Chen, K., Corrado, G., et al.: Efficient estimation of word representations in vector space, arXiv preprint arXiv:1301.3781 (2013)

28. Le, Q.V., Mikolov, T.: Distributed representations of sentences and documents. In: Proceedings of the 31st International Conference on Machine Learning, pp. 1188–1196 (2014)

29. Devlin, J., et al.: BERT: pre-training of deep bidirectional transformers for language understanding (2018)

Application of News Features in News Recommendation Methods: A Survey

Jing Qin[1(✉)] and Peng Lu[2]

[1] Northeastern University, Shenyang 110169, China
annyproj@126.com
[2] Criminal Investigation Police University of China, Shenyang 110035, China
.

Abstract. In recent years, many traditional news websites developed corresponding recommendation systems to cater to readers' interests and news recommendation systems are widely applied in traditional PCs and mobile devices. News recommendation system has become a critical research hotspot in the field of recommendation system. As News contains more text information, it is more helpful to improve the recommendation effect to obtain the content related to news features (location, time, events) from the news. This survey summarizes news features-based recommendation methods including location-based news recommendation methods, time-based news recommendation methods, events-based news recommendation methods. It helps researchers to know the application of news features in news recommendation methods. Also, this suvery summarizes the challenges faced by the news recommendation system and the future research direction.

Keywords: News recommendation · News features · Recommendation system

1 Introduction

Since the concept of recommendation system was put forward in 1992 [14], recommendation system applies to different fields such as news (e.g., GroupLens, Toutiao news, Sina News, Google News), e-commerce (e.g., Amazon, eBay, Tmall, Jingdong Mall), movies (e.g., MovieLens), music (e.g., Last.com). As news contains a large number of texts and has strong real-time performance, the research on news recommendation methods has become a hot research field in recommendation methods. The number of papers retrieved in DBLP (https://dblp.uni-trier.de/) shows that the number of papers published in the direction of news recommendation has increased from less than 10 papers per year before 2010 to 59 papers in 2019, and the research results also publish in relevant international conferences, such as WWW, SIGIR, IJCAI.

The item that news recommendation system recommended is news, which consists of 6 elements, such as a person, time, spot, event, cause, and process of occurrence, with a robust real-time nature. In the existing news recommendation

© Springer Nature Singapore Pte Ltd. 2020
P. Qin et al. (Eds.): ICPCSEE 2020, CCIS 1258, pp. 113–125, 2020.
https://doi.org/10.1007/978-981-15-7984-4_9

methods, the six elements of news applied to the research of three news features (location, event, time). To facilitate researchers to understand the application of combining news features in news recommendation methods, this paper focusing on the application of news features in news recommendation methods. This paper summarizes the news features-based news recommendation methods in recent years, including location-based news recommendation methods [3, 4, 7, 18, 23, 31, 34, 39], time-based news recommendation methods [9–11, 20, 22, 28, 30, 32, 35, 36, 38, 41, 43], and event-based news recommendation methods [17, 24, 26, 33, 40, 42].

2   News Features-Based News Recommendation Methods

In recent years, researchers have proposed many news recommendation methods based on news features. These research results accurately analyzed the news content mainly from the following points, such as the place where the news happened, the time when the news took place, and events.

The location-based news recommendation methods, we focus on using the topic model to mine news related to location information. In the time-based news recommendation methods, we focus on the real-time news recommendation and session-based news recommendation from news timeliness. In the events-based news recommendation method, news events obtain from news content and search the relevance of news events, notably the recommendation method based on news text analysis.

2.1   Location-Based News Recommendation Methods

The research hotspot of location-based news recommendation methods is the news recommendation based on the news recommendation using the location-based topic model and the user's current location or a specific range of content in the current location.

Topic models widely used in news recommendation systems in recent years mainly include PLSA (Probabilistic Latent Semantic Analysis) [16], LDA (Latent Dirichlet Allocation) [5], ESA (Explicit Semantic Analysis) [12, 13]. Topic models usually use vectors to represent documents. PLSA is based on likelihood principle. Compare to LSA (Latent Semantic Analysis) [8], PLSA allows deal with polysemous words. LSA only detects synonyms. LDA is a three-level hierarchical Bayesian model. In LDA, each topic has its word distribution (i.e., Polynomial distribution), and the parameters of each word distribution according to Dirichlet distribution [5]. ESA uses concepts in Wikipedia to represent natural language text as vector representations with semantic concepts [12, 13]. Besides, some models improved on the topic model described above [7, 31, 34].

The topic model used in the location-based news recommendation system includes the STPM (Spatial Topical Preference Model) model proposed by [31], which is an improved model based on the LDA model. The ELSA (Explicit Localized Semantic Analysis) model proposed by [34] is an improved model based on the ESA model. The CLSA (Clustering-based Localized Semantic Analysis) proposed by [7], the ALSA (Auto encoder-based Localized Semantic Analysis), and DLSA (Deep Localized Semantic Analysis), all of them improved based on ELSA model.

The ELSA model [34] is the general model of the location-based news recommendation model, and the recommendation process is three-step. First, a document with geographic information related to the location of the user obtained from a third-party platform such as Twitter. Then, the location-related information in the two types of documents found to generate a topic distribution model. Finally, the topic distribution model graded with the documents with geographic information and news information, respectively, by using a scoring function to generate top-k recommendation items. Among them, CLSA model and ALSA model pays more attention to minimizing the distance in clustering that has nothing to do with the local target news which distinguishes users. In the DLSA model, the similarity between the news that happened locally and the news took place in the user's location maximized and the unrelated news similarity minimized. Therefore, the DLSA model is more suitable for location-based news recommendation than CLSA and ALSA models.

The DLSA model is improved on the ELSA model and is optimal among the three models proposed by [7]. Three deep neural network models are added to ELSA to generate feature representation of news and location, and tan function used as the activation function. Finally, the similarity between vectors is calculated by cosine similarity to generate top-k news articles for the user. Experiments on Twitter dataset [1] show that DLSA is better than ELSA, CLSA, and ALSA in MRR (mean reciprocal rank) and MAP (mean average precision) metrics. The main advantage and disadvantage of location-based models (STPM, ELSA, DLSA) in news recommendations shown in Table 1.

The GeoFeed (i.e., a location-aware news feed system) is proposed by [3] to recommend news for users with the help of the location information of users' friends and their current location information in social networks. The GeoRank (i.e., location-aware news feed ranking system) is proposed by [4]. It uses the user's location information to the news recommendation and pays more attention to the similarity of news in space, time, and personal preferences. [18] proposed a news recommender system, which considered users' current GPS (Global Positioning System) location with a 50 km radius.

In addition to the location-based topic model used in news recommendation listed in the above, the user's location information included as part of the user's profiles in [23,39] and news are recommended to the user using the user's preferences for the corresponding location on the social network (e.g., Twitter, Facebook).

2.2   Time-Based News Recommendation Methods

Time is also critical in news recommendations. The research of related time-based news recommendation methods includes real-time news recommendation and session-based news recommendation.

Table 1. The main advantage and disvantage of location-based models (STPM, ELSA, DLSA) in news recommendations

Model	Base model	Main advantage	Main disadvantage
STPM	LDA	1. Having considered the user's interest preference in different locations;	1. Not considering the geographical context of a location;
		2. Avoid the sparsity of the locational infomation	2. Cold-start problem
ELSA	ESA	1. Considered the geographic context of a location;	1. Not considering users' personal preferences;
		2. The same geographial category have different topic distributions	2. Cold-start problem
DLSA	ELSA	1. The topic space is lower dimensional than ELSA;	1. Not considering users' presonal preferences;
		2. Avoid the topic space sparsity and redundancy problems	2. Not considering the news contextual informations

Real-Time News Recommendation Methods. For news, the implementation of real-time recommendation is to maximize the satisfaction of users'needs for the timeliness of news content. The problem to be solved in a real-time recommendation method is to provide personalized news recommendations for users according to their feedback in a short period. Therefore, the real-time recommendation method pays more attention to the short-time information feedback to the user, the model of the user portrait mainly according to the user's current browsing web pages, according to the user's click behavior to obtain the user, with the help of third-party knowledge base, or the use of time window to obtain the user's interest.

[28] shows that the optimal combination recommendation method is dynamically selected to ensure that the user can obtain the most satisfactory news content in a short time according to the real-time request of the user. [11] proposed that trends and temporal relevance of news to recommend news to users according to the semantics. The models used in the news content analysis part include keywords and named entities, which include the news related to the event, time, location, and other information. The part of trend analysis relies in part on statistics from Google Trends. [9] proposed to make use of the implicit feedback of the users to the news click to obtain the news of interest to the users, to use the Map-Reduce mechanism, and through the fast parallel processing way to obtain the most popular news quickly. [30] proposed a real-time news recommendation based on the daily news stream. The daily news model

transformed into a specific event, an event chain constructed, and the information events determined from the event chain. [2] proposed to make use of the implicit feedback of the users to the news click to obtain the news of interest to the users, and to use the Map-Reduce mechanism and through the high-speed parallel processing way to obtain the most popular news quickly. [22] proposed a function called Match the News, which is an extended function in the Firefox browser. It is to recommend articles for users according to the relevant articles on the web page they are currently visiting. The extended function uses the weight function based on BM25H [21] for keyword extraction.

Except for focusing on semantic and contextual information, the real-time news recommendation method also focuses on the user's feedback, updating the model incrementally in a time window to reduce computational time. CTR (click-through rate) [11] metric mainly used to evaluate the real-time methods, but very little public data is available for real-time recommendation news methods. Therefore, most of the real-time recommended news methods based on contest datasets (e.g., NEWSREEL [6]).

Session-Based News Recommendation Methods. Because most of the users browsing news are unregistered users, there is some difficulty in obtaining the past behavior of users. Therefore, to better recommend the news of interest to users, researchers regard the relevant records of browsing news in session as the user's preference and recommend the news of interest to users accordingly. User browsing records in session reflect users' short-term interest preferences to some extent, so it can also alleviate the cold-start problem of new users in news recommendations. In addition to the traditional referral list, the researchers used user behavior records in session to predict the user's next click behavior. The session-based news recommendation framework mainly includes CHAMELEON [36], URecSYS [43], StreamingRec [20]. Also, most session-based news recommendation methods adopt deep learning methods.

[37] proposed an instance of the CHAMELEON framework [36], in which CNN (Convolution Neural Network) model and LSTM (Long Short-Term Memory) model added. The CNN model is added to ACR (Article Content Representation) module of CHAMELEON framework to generate news content vector efficiently, and LSTM model is added to NAR (Next-Article Recommendation) module to generate session vector representation, which is more conducive to improve the accuracy of news recommendation. The structure of ACR module and NAR module in CHAMELEON framework and the instance of CHAMELEON Framework shown in Fig. 1, Fig. 2 respectively. In [38], a CHAMELEON framework based proposed a hybrid method to deal with item cold-start problems.

URecSYS framework [43] also applied the user's session, and through users click-stream dataset to find recommendation rules. The framework focus on the discover recommendation rules and rules-based recommend to users. The recommend rule includes article-level recommendation rule discovery and topic-level recommendation rule discovery. StreamingRec framework [20] implements

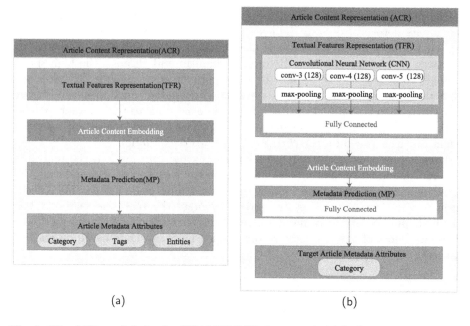

Fig. 1. The ACR module in the CHAMELEON framework (a) [36], The ACR module in the instance of CHAMELEON framework (b) [37]

some session-based news recommendation method mainly include GRU4Rec [15], SkNN [19], V-SkNN [29]. Experiments on Outbrain and Plista datasets [25] show that V-SkNN is better than GRU4Rec and SkNN in MRR and F1 metrics. Also, SkNN is better than GRU4Rec in MRR and F1 metrics.

[35] proposed that based on the hybrid method of content-based and collaborative filtering, short-term interests in the user session. The LDA [5] model is used in news text analysis to classify news by topic. [41] proposed a Neural Network-based model DeepJONN (Deep Joint Neural Networks). The RNN (Recurrent Neural Network) method used to obtain the user's click patterns and association features from the clickstreams in the session. The CNN model used to analyze the news content, and the text content expressed as a vector form in character-level coding. [10] proposed that the recommendation model combines the short-term and long-term reading behaviors of users and proposed medium-term reading behavior. When analyzing news content, the characteristics of news are considered, including popularity, timeliness, and user's reading behavior. User sessions are used to analyze and partition situations in which users read too long or too short a time when reading consecutive events. [32] proposed two recommendation models of deep neural networks. One based on RNN model, which used to obtain user click records and user browsing history information in the session, and the other based on CNN model, which used to obtain the user's long-term interest.

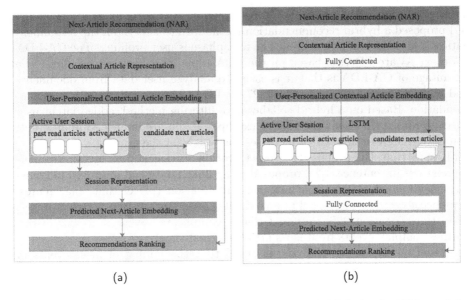

Fig. 2. The NAR module in the CHAMELEON framework (a) [36], The NAR module in the instance of CHAMELEON framework (b) [37]

It can be seen from the analysis results that the session-based news recommendation methods are a supplement to the user's historical selection items. Through the news sequence selected in the user session, it not only can obtain the relevance of news context selection but also used in the analysis of the user's long-term and short-term interests. Deep Learning methods widely used in session-based news recommendation methods. The RNN model is suitable for sequence analysis in the session, the LSTM model is suitable for long-term and short-term interest analysis, and the CNN model is suitable for text analysis and long-term interest mining.

2.3 Events-Based News Recommendation Method

The event-based news recommendation method includes a method of recommending news events to users as items [42], a recommendation method of extracting people from news events [33], and a method of recommending news for users according to their interest in news events [17].

[42] proposes a hybrid recommendation method to recommend news events to users as granularity. In this recommendation method, it established an event model to enable users to recommend reports related to news events based on personalized requirements. [33] proposed a framework RePAIR (Recommend Political Actors In Real-time From News Websites), it made recommendations for politicians in current affairs news. RePAIR framework used ACE (Automatic Content Extraction) method and a frequent based actor ranking algorithm. The recommendation process of the RePAIR framework includes scraped

data, extraction data, political Actor extraction, and recommendation model. [17] proposed a hybrid recommendation method. In this method, the recommendation is divided into two phases. The first phase is user profiling. The CA-LDA (Context-Aware LDA) is based on the LDA topic model in user profiling. The advantage of CA-LDA is that it is easier to analyze user data from documents and more meaningful than doc2vec [27] or LDA. The second phase is a recommendation. Based on a hybrid collaborative filtering method, added latent user offset to record the user's rating. For example, if you are interested in an event, you have a high rating, which increases the weight of the event.

Besides that, [40] proposed a 'News Curation Service' system by using semantic relations in sentences. [24] proposed using linked data and improved to transform news event information into meta-information (place, time, tag, genre). Sentence structures are stored in linked data, including subject, activity, object, date, time, location. [26] proposed that a complicated event processor distinguishes complex events from news articles in real-time.

Based on the above analysis, it noted that the mining of news events in news recommendation helps to improve the accuracy of recommendation results, and the semantic analysis of news content still focused on the recommendation method. However, in the real-time recommendation model, the time-consuming problem is still a problem to be solved.

3   Prospects for News Recommendations

News recommendation system is a research hotspot in the recommendation system, more and more news websites have begun to use recommendation technology, such as Xinhua News Agency, Sina,Yahoo. Because news is different from other recommendation objects, news itself contains a large amount of information and frequently updates, and it is more timely, affected by user location and time factors. Although there have been a lot of research results in the field of news recommendation, there are still many important and difficult problems in the news recommendation algorithm, which are worthy of in-depth study by researchers.

3.1   The Analysis of News Content Takes a Long Time

Effective analysis of news content is more conducive to news clustering and recommending interesting news for users. However, due to the large amount of text involved in news content, although the existing machine learning and deep learning methods can greatly improve the speed of text analysis, it is still an important challenge to real-time news algorithm. In addition, semantic analysis, synonym analysis and other operations must be carried out on the text content, but these operations also involve some corpus and cross-language thesaurus.

3.2 The Quality of News Cannot Be Guaranteed

Under the trend of amount of online news is ever-increasing, the detection of false news has become a hot issue in recent years, and the related academic research has increased since 2017. In addition, due to the emergence of 'headline party' in recent years, the quality of news content has also been affected to a certain extent. Therefore, in the process of news recommendation, it is the focus of study that not only to ensure the accuracy and efficiency of the recommendation results, but also to ensure the quality of news.

3.3 The News Classification System Is Too Difficult to Realize Cross-Language Classification

The classification granularity of the existing news classification system is coarse-grained, such as domestic, international, military, cultural, or sports, entertainment, current affairs and so on. The classification of news is very important to the cold-start of news recommendation, so the fine-grained classification of news classification system is another hot issue in news recommendation. In addition, because news content involves different languages, it needs cross-language semantic recognition when classifying news, so it still faces the problem of time-consuming too much in semantic recognition.

3.4 Privacy Protection in News Recommendation

In the news recommendation algorithm, the use of a large number of users' reading history information and users' location information will bring out the problem of user privacy disclosure. In recent years, user privacy protection is also a key issue in the research of recommendation algorithm. It is also a difficult problem to provide users with personalized news recommendation while protecting their privacy. Although some researchers use differential privacy, mixed privacy protection and other privacy protection methods, they will affect the accuracy of the recommendation algorithm to some extent. Therefore, personalized, accurate and efficient news recommendation algorithm for users is still an urgent problem to be solved in the recommendation algorithm research.

3.5 Lack of Standards to Measure Diversity and Serendipity in News Recommendation

Most of the news recommendation methods focus on the interests of users, so that users only immerse themselves in the news of interest, and do not pay attention to other news that may also be interested in. Therefore, to improve the diversity and serendipity of news recommendation can better expand the horizons of reading for users, and at the same time also can explore more interests of users, in order to improve the user's reading experience. Because news content differs from other recommendation object, it is not only the diversity of news headlines, but also includes the events involved in news and the relevance between news events.

Therefore, the existing standards for evaluating the diversity and surprise degree of news items are not fully applicable to news recommendation methods, and the diversity and serendipity degree of recommendation methods will become an important issue in the future research in the field of news.

4 Conclusions

The emergence of the news recommendation system reduces the time for users to find news, and can timely and accurately recommend the news of interest to users. The research of the news recommendation system is a hot spot in the research field of recommendation systems. This paper summarizes the application of news features in news recommendation methods. Then, this paper discussed the key and challenging issues in news recommendation methods and hoped to be of all help to researchers in related fields.

References

1. Abel, F., Gao, Q., Houben, G.-J., Tao, K.: Analyzing user modeling on Twitter for personalized news recommendations. In: Konstan, J.A., Conejo, R., Marzo, J.L., Oliver, N. (eds.) UMAP 2011. LNCS, vol. 6787, pp. 1–12. Springer, Heidelberg (2011). https://doi.org/10.1007/978-3-642-22362-4_1
2. Bailey, D., Pajak, T., Clarke, D., Rodriguez, C.: Algorithms and architecture for real-time recommendations at news UK. In: Bramer, M., Petridis, M. (eds.) SGAI 2017. LNCS (LNAI), vol. 10630, pp. 264–277. Springer, Cham (2017). https://doi.org/10.1007/978-3-319-71078-5_23
3. Bao, J., Mokbel, M.F., Chow, C.: Geofeed: a location aware news feed system. In: 2012 IEEE 28th International Conference on Data Engineering, pp. 54–65 (2012)
4. Bao, J., Mokbel, M.F.: Georank: an efficient location-aware news feed ranking system. In: 21st SIGSPATIAL International Conference on Advances in Geographic Information Systems, SIGSPATIAL 2013, Orlando, FL, USA, pp. 184–193 (2013)
5. Blei, D.M., Ng, A.Y., Jordan, M.I.: Latent Dirichlet allocation. J. Mach. Learn. Res. **3**, 993–1022 (2003)
6. Brodt, T., Hopfgartner, F.: Shedding light on a living lab: the CLEF NEWSREEL open recommendation platform. In: Fifth Information Interaction in Context Symposium, IIiX 2014, Regensburg, Germany, 26–29 August 2014, pp. 223–226 (2014)
7. Chen, C., Lukasiewicz, T., Meng, X., Xu, Z.: Location-aware news recommendation using deep localized semantic analysis. In: Candan, S., Chen, L., Pedersen, T.B., Chang, L., Hua, W. (eds.) DASFAA 2017. LNCS, vol. 10177, pp. 507–524. Springer, Cham (2017). https://doi.org/10.1007/978-3-319-55753-3_32
8. Deerwester, S.C., Dumais, S.T., Landauer, T.K., Furnas, G.W., Harshman, R.A.: Indexing by latent semantic analysis. JASIS **41**(6), 391–407 (1990)
9. Domann, J., Lommatzsch, A.: A highly available real-time news recommender based on apache spark. In: Jones, G.J.F., et al. (eds.) CLEF 2017. LNCS, vol. 10456, pp. 161–172. Springer, Cham (2017). https://doi.org/10.1007/978-3-319-65813-1_17
10. Epure, E.V., Kille, B., Ingvaldsen, J.E., Deneckère, R., Salinesi, C., Albayrak, S.: Recommending personalized news in short user sessions. In: Proceedings of the Eleventh ACM Conference on Recommender Systems, RecSys 2017, Como, Italy, pp. 121–129 (2017)

11. Ficel, H., Haddad, M.R., Zghal, H.B.: Large-scale real-time news recommendation based on semantic data analysis and users' implicit and explicit behaviors. In: Advances in Databases and Information Systems - 22nd European Conference, ADBIS 2018, Budapest, Hungary, pp. 247–260 (2018)

12. Gabrilovich, E., Markovitch, S.: Overcoming the brittleness bottleneck using Wikipedia: enhancing text categorization with encyclopedic knowledge. In: Proceedings, The Twenty-First National Conference on Artificial Intelligence and the Eighteenth Innovative Applications of Artificial Intelligence Conference, Boston, Massachusetts, USA, 16–20 July 2006, pp. 1301–1306 (2006)

13. Gabrilovich, E., Markovitch, S.: Computing semantic relatedness using Wikipedia-based explicit semantic analysis. In: IJCAI 2007, Proceedings of the 20th International Joint Conference on Artificial Intelligence, Hyderabad, India, pp. 1606–1611 (2007)

14. Goldberg, D., Nichols, D.A., Oki, B.M., Terry, D.B.: Using collaborative filtering to weave an information tapestry. Commun. ACM 35(12), 61–70 (1992)

15. Hidasi, B., Karatzoglou, A.: Recurrent neural networks with top-k gains for session-based recommendations. In: Proceedings of the 27th ACM International Conference on Information and Knowledge Management, CIKM 2018, Torino, Italy, 22–26 October 2018, pp. 843–852 (2018)

16. Hofmann, T.: Probabilistic latent semantic indexing. In: SIGIR 1999: Proceedings of the 22nd Annual International ACM SIGIR Conference on Research and Development in Information Retrieval, Berkeley, CA, USA, pp. 50–57 (1999)

17. Hsieh, C., Yang, L., Wei, H., Naaman, M., Estrin, D.: Immersive recommendation: news and event recommendations using personal digital traces. In: Proceedings of the 25th International Conference on World Wide Web, WWW 2016, Montreal, Canada, pp. 51–62 (2016)

18. Ingvaldsen, J.E., Özgöbek, Ö., Gulla, J.A.: Context-aware user-driven news recommendation. In: Proceedings of the 3rd International Workshop on News Recommendation and Analytics (INRA 2015) co-located with 9th ACM Conference on Recommender Systems (RecSys 2015), Vienna, Austria. pp. 33–36 (2015)

19. Jannach, D., Ludewig, M.: When recurrent neural networks meet the neighborhood for session-based recommendation. In: Proceedings of the Eleventh ACM Conference on Recommender Systems, RecSys 2017, Como, Italy, pp. 306–310 (2017)

20. Jugovac, M., Jannach, D., Karimi, M.: StreamingRec: a framework for benchmarking stream-based news recommenders. In: Proceedings of the 12th ACM Conference on Recommender Systems, RecSys 2018, Vancouver, BC, Canada, pp. 269–273 (2018)

21. Karkali, M., Plachouras, V., Stefanatos, C., Vazirgiannis, M.: Keeping keywords fresh: a BM25 variation for personalized keyword extraction. In: 2nd Temporal Web Analytics Workshop, TempWeb 2012, Lyon, France, 16–17 April 2012, pp. 17–24 (2012)

22. Karkali, M., Pontikis, D., Vazirgiannis, M.: Match the news: a firefox extension for real-time news recommendation. In: The 36th International ACM SIGIR conference on research and development in Information Retrieval, SIGIR 2013, Dublin, Ireland, pp. 1117–1118 (2013)

23. Kazai, G., Yusof, I., Clarke, D.: Personalised news and blog recommendations based on user location, Facebook and Twitter user profiling. In: Proceedings of the 39th International ACM SIGIR Conference on Research and Development in Information Retrieval, SIGIR 2016, Pisa, Italy, pp. 1129–1132 (2016)

24. Khrouf, H., Troncy, R.: Hybrid event recommendation using linked data and user diversity. In: Seventh ACM Conference on Recommender Systems, RecSys 2013, Hong Kong, China, pp. 185–192 (2013)
25. Kille, B., Hopfgartner, F., Brodt, T., Heintz, T.: The Plista dataset. In: Proceedings of the 2013 International News Recommender Systems Workshop and Challenge, NRS 2013, pp. 16–23. Association for Computing Machinery, New York (2013)
26. La Fleur, A., Teymourian, K., Paschke, A.: Complex event extraction from real-time news streams. In: Proceedings of the 11th International Conference on Semantic Systems, pp. 9–16. ACM, New York (2015)
27. Le, Q.V., Mikolov, T.: Distributed representations of sentences and documents. In: Proceedings of the 31th International Conference on Machine Learning, ICML 2014, Beijing, China, 21–26 June 2014, pp. 1188–1196 (2014)
28. Lommatzsch, A.: Real-time news recommendation using context-aware ensembles. In: de Rijke, M., et al. (eds.) ECIR 2014. LNCS, vol. 8416, pp. 51–62. Springer, Cham (2014). https://doi.org/10.1007/978-3-319-06028-6_5
29. Ludewig, M., Jannach, D.: Evaluation of session-based recommendation algorithms. User Model. User-Adapt. Interact. **28**, 331–390 (2018). https://doi.org/10.1007/s11257-018-9209-6
30. Lyu, L., Fetahu, B.: Real-time event-based news suggestion for Wikipedia pages from news streams. In: Companion of the The Web Conference 2018 on The Web Conference 2018, WWW 2018, Lyon, France, pp. 1793–1799 (2018)
31. Noh, Y., Oh, Y., Park, S.: A location-based personalized news recommendation. In: International Conference on Big Data and Smart Computing, BIGCOMP 2014, Bangkok, Thailand, pp. 99–104 (2014)
32. Park, K., Lee, J., Choi, J.: Deep neural networks for news recommendations. In: Proceedings of the 2017 ACM on Conference on Information and Knowledge Management, CIKM 2017, Singapore, pp. 2255–2258 (2017)
33. Solaimani, M., Salam, S., Khan, L., Brandt, P.T., D'Orazio, V.: Repair: recommend political actors in real-time from news websites. In: 2017 IEEE International Conference on Big Data, BigData 2017, Boston, MA, USA, pp. 1333–1340 (2017)
34. Son, J.W., Kim, A., Park, S.: A location-based news article recommendation with explicit localized semantic analysis. In: The 36th International ACM SIGIR conference on research and development in Information Retrieval, SIGIR 2013, Dublin, Ireland, pp. 293–302 (2013)
35. Sottocornola, G., Symeonidis, P., Zanker, M.: Session-based news recommendations. In: Companion of the The Web Conference 2018 on The Web Conference 2018, WWW 2018, Lyon, France, pp. 1395–1399 (2018)
36. de Souza Pereira Moreira, G.: CHAMELEON: a deep learning meta-architecture for news recommender systems. In: Proceedings of the 12th ACM Conference on Recommender Systems, RecSys 2018, Vancouver, BC, Canada, pp. 578–583 (2018)
37. de Souza Pereira Moreira, G., Ferreira, F., da Cunha, A.M.: News session-based recommendations using deep neural networks. In: Proceedings of the 3rd Workshop on Deep Learning for Recommender Systems, DLRS@RecSys 2018, Vancouver, BC, Canada, pp. 15–23 (2018)
38. de Souza Pereira Moreira, G., Jannach, D., da Cunha, A.M.: On the importance of news content representation in hybrid neural session-based recommender systems. CoRR abs/1907.07629 (2019)
39. Tiwari, S., Pangtey, M.S., Kumar, S.: Location aware personalized news recommender system based on Twitter popularity. In: Gervasi, O., et al. (eds.) ICCSA 2018. LNCS, vol. 10963, pp. 650–658. Springer, Cham (2018). https://doi.org/10.1007/978-3-319-95171-3_51

40. Yoko, R., Kawamura, T., Sei, Y., Tahara, Y., Ohsuga, A.: News recommendation based on semantic relations between events. In: Workshop and Poster Proceedings of the 4th Joint International Semantic Technology Conference, JIST, Chiang Mai, Thailand, pp. 128–131 (2014)
41. Zhang, L., Liu, P., Gulla, J.A.: A deep joint network for session-based news recommendations with contextual augmentation. In: Proceedings of the 29th on Hypertext and Social Media, HTBaltimore, MD, USA, pp. 201–209 (2018)
42. Zhen-dong, N., Shuai, W., Shi-hang, W., Jie, C.: Distributed news event hybrid recommendation approach. Trans. Beijing Inst. Technol. **37**(7), 721–726 (2017)
43. Zihayat, M., Ayanso, A., Zhao, X., Davoudi, H., An, A.: A utility-based news recommendation system. Decis. Support Syst. **117**, 14–27 (2019)

Security

New Lattice-Based Digital Multi-signature Scheme

Chunyan Peng[✉] and Xiujuan Du

Qinghai Normal University, Xining 810008, China
7456911@qq.com

Abstract. The traditional digital multi-signature schemes are mostly based on large integer factorization and the discrete logarithm, which cannot be secured in quantum environment. The paper presents a lattice-based multi-signature scheme that can resist the quantum attack using the hardness of average-case short integer solution problem (SIS). Multi-signature includes the simultaneous signature and sequential signature. The paper describes respectively the key generation, multi-signature generation and multi-signature verification of the two types of schemes. Moreover, experimental results prove that the digital multi-signature scheme based on lattice is especially efficient and secure to multi-signature generation.

Keywords: Lattice · Multi-signature · Simultaneous signature · Sequential signature

1 Introduction

The development of digital signature technology has a very wide range of applications in many security fields, such as identity signature right, data integrity, non-repudiation and so on [1, 2]. The requirements of these different applications are obviously different, so there are many variations in the way of digital signature. If more than one user needs to sign the same message together in the electronic transactions and other activities under the network environment, which is digital multi-signature that we often call [3, 4]. Multi-signature is a special kind of digital signature scheme that can allow a group of signers to produce a compact signature cooperatively on a common message. The multi-signature scheme has been provided in [5], a group of signers (denoted by U) can jointly produce a compact signature which can be verified very similarly to the ordinary signature scheme on a common message m. All of potential signers are denoted by $U_1, U_2, \ldots, U_N$. It has already been proved that both the shortest vector problem (SVP) [6, 7] and the closest vector problem (CVP) [8] are NP-hard problems [9, 10] by theoretical analysis in lattice theory problem. And so far, there is no any algorithms that can be used to decipher them by quantum computer. Therefore, the study of cryptosystem based on lattice problem construction has gradually become a domestic problem in the post quantum era if the cryptosystem can resist to the quantum computer attacks. It has attracted the attention of many experts and scholars to research the typical post quantum encryption mechanism of public key cryptography based on lattice problem.

© Springer Nature Singapore Pte Ltd. 2020
P. Qin et al. (Eds.): ICPCSEE 2020, CCIS 1258, pp. 129–137, 2020.
https://doi.org/10.1007/978-981-15-7984-4_10

The remainder of this paper is organized as follows. In Sect. 2, the related work along with motivation is presented. In Sect. 3, the detailed explanation of preliminaries is discussed. Section 4 presents our proposed simultaneous multi-signature scheme and Sect. 5 presents our proposed sequential multi-signature scheme. Section 6 analyzes the correctness and security of the new lattice-based digital multi-signature scheme. Section 7 concludes the paper.

2 Related Works

The earliest definition of multi-signature was proposed by K. Itakura and K. Nakamura [11] in 1983. After N signers sign on a common message one by one, they will get a signature. The length of the signature is a certain constant, so this signature method is called digital multi-signature. Based on different cooperation modes or different opportunities of signers, the kind of digital signature can be divided into two types: simultaneous multi-signature and sequential multi-signature. When verifying the multi-signatures, you need to test and verify the public keys of N Users. And it must be guaranteed that the amount of computation cannot be increased with the number of users during the process of verifying the public keys. In recent years, cryptographers have also used discrete logarithm, large integer decomposition and other related mathematical theories to construct many famous digital multi-signature schemes. For example, the two scholars, M. Bellare and G. Neven also have proposed a 3-rounds interactive multi-signature scheme by using the ordinary public key model at CCS2006 Conference [12]. The authors J. H. Ji and R. J. Zhao have provided a new digital multi-signature scheme based on the Schnorr scheme, and have proved the security of the sequential multi-signature scheme in [13]. However, most of these signature schemes are based on large integer decomposition and discrete logarithm problems [14, 15]. With the continuous development of quantum computers, these problems that may be solved in polynomial time obviously cannot meet the needs of digital security signature technology [16, 17]. Therefore, it is necessary to design a multi-signature scheme based on the public key cryptosystem. And the proposed scheme can resist all kinds of attacks although using the quantum computer. It is also an urgent problem to be solved in the field of password security research in the post quantum era.

Ajtai has put forward a scheme of cipher construction based on NP-hard problem of lattice in 1996 [18]. This mechanism opened up a new direction of public key cryptosystem based on lattice. In 2008, the authors Gentry etc. have constructed the first verifiable security signature mechanism by using the original image sampling threshold function based on lattice problem, which is also a public key encryption scheme based on identities of signers [19]. This is a research achievement with milestone significance for the application of lattice theory in cryptography. At the same time, it has appeared an unprecedented peak in lattice public key cryptosystem field. In 2010, Cash etc. have implemented the digital signature scheme named bonsai tree model by using the technology of lattice expansion, which does not need the support of random oracle technology [20]. In this paper, we have provided a secure digital multi-signature based on lattice theory and have designed two kinds of new multi-signature schemes based on lattice problem, one is simultaneous signature, the other is sequential signature. Since lattice is proved to be secure in quantum environment, the proposed scheme has higher security and better ability to resist attacks than other schemes. Compared with the traditional polynomial number theory problems [21], the multi-signature scheme can reduce the computation of signature generation by using simple operations such as small integer modular addition and modular multiplication, as well as the original image sampling algorithm. At the same time, it has the characteristics of simple calculation and higher efficiency.

3 Preliminaries

Lattice is a kind of linear structure. It is an algebraic system based on partial order set. Most of the operations on lattice are linear operations. Because of the different objects of lattice theory, it is usually defined and analyzed in different forms. In the field of computer information security and engineering, the definition of lattice is as follows.

Definition 1. $v_1, \ldots, v_n \in \mathbb{R}^m$ is set to be a set of linearly independent vectors. The generated lattice L refers to the set of vectors $v_1, \ldots, v_n$ which composed of linear combination of vector $v_1, \ldots, v_n$, and the coefficients used are all in $\mathbb{Z}$, that is

$$L = \{n_1 v_1 + n_2 v_2 + \cdots + n_i v_i | n_1, n_2, \cdots, n_i \in \mathbb{Z}\} = \left\{ \sum_{i=1}^{d} n_i v_i | n_i \in \mathbb{Z} \right\},$$ such a set L

is called lattice. Any set of linearly independent vectors that can generate a lattice L is called the basis of the lattice L, and the number of vectors in the set is called the dimension of the lattice L, denoted as $\dim(L) = n$.

Definition 2. Given $q, m, n \in \mathbb{Z}$, and matrix $A \in \mathbb{Z}_q^{n \times m}$, $u \in \mathbb{Z}_q^n$, q-module is defined as the following formula (1).

$$\begin{aligned}
\Lambda_q(A^T) &= \{x \in Z^m : x = A^T s \bmod q, s \in Z^n\} \\
\Lambda_q^{\perp}(A) &= \{x \in Z^m : Ax = 0 \bmod q\} \\
\Lambda_q^u(A) &= \{x \in Z^m : Ax = u \bmod q\}
\end{aligned} \tag{1}$$

To find the shortest non-zero vector problem in a lattice is the shortest vector problem SVP, and to find the vector closest to the specified non lattice vector in a lattice is the closest vector problem CVP. Both SVP and CVP are the most basic NP-hard problems in lattice theory.

Definition 3. In the lattice L, finding a shortest non-zero vector $v \in L$ aims to minimize the Euclidean norm $||v||$, which is called the shortest vector problem.

Definition 4. Given a vector $w \in \mathbb{R}^m$ that is not in lattice L, and looking for a vector $v \in L$ that is closest to w, therefore there is a vector $v \in L$ that can minimize Euclidean norm $||w - v||$, which is called the closest vector problem.
    A group of high-quality bases of a lattice can be obtained by the method of lattice-based reduction. The basic hard problems in the above two lattices can be transformed into other approximate hard problems by a certain lattice-based reduction algorithm.

The common approximate problem can be solved by using the small integer solution problem SIS, which is defined as the following Definition 5.

Definition 5. Given a integer q, a matrix $A \in \mathbb{Z}_q^{n \times m}$, a real number β, we can find a non-zero vector $e \in \mathbb{Z}^m$ to satisfy $Ae = 0 \bmod q$ and $||e|| \leq \beta$. Here, q, m, β can be represented polynomial that generated with n, and A is randomly and evenly distributed.

Since the matrix A is selected randomly and uniformly, the SIS problem of small integer problems is an NP-hard problem when selecting appropriate parameters. The theory has been proofed by Micciancio and Regev [22].

Theorem 1. In [18], suppose n is a security parameter, q is a prime number, let $m \geq 2n \log q$, when satisfying $A \in \mathbb{Z}_q^{n \times m}$, $s \geq \omega \sqrt{\log m}$ and $e \leftarrow D_z^m$, s has a great probability to make the statistics of $u = Ae \bmod q$ on $\mathbb{Z}_q^n$ be distributed evenly.

The theorem shows that if the input is the points sampled by the discrete Gaussian distribution on the lattice, then the random and uniform matrix vector product should also be uniformly distributed in the vector space produced by the matrix.

Theorem 2. In [19, 23], suppose n as a parameter of the algorithm, for any prime number $q = poly(n)$, and a integer $m \geq 5n \log q$, there is a probability polynomial algorithm TrapGen(1^n), here, 1^n is an input and its output is the statistic of a full rank vector set T which is close to uniformly distributed matrix $A \in \mathbb{Z}_q^{n \times m}$ and $\Lambda^\perp(A)$. Meanwhile, the vector set satisfy $||T|| \leq L \approx m^{1+\varepsilon}, \varepsilon > 0$.

In [19], the author has provided the construction process of the original image sampling gate boundary function. The construction of the gate boundary function is described as follows.

(1) According to Theorem 2, the matrix $A \in \mathbb{Z}_q^{n \times m}$ and the vector set T can be obtained, A is a public key, and T is private key and it is also gate boundary in the following function $f_A(\cdot)$.

(2) We can define a one-way gate boundary function: $f_A(x) = Ax(\bmod q)$, among them, $x \in D_n = \left\{ x \in \mathbb{Z}_q^m, ||x|| \leq s\sqrt{m} \right\}$ and the vector x satisfies the distribution condition of D_z^m, s.

(3) To give another vector $u \in \mathbb{Z}_q^m$, and select vectors $t \in \mathbb{Z}_q^m$ by using the method of linear algebra, which can satisfy the equation $At = u(\bmod q)$. Then using the gate boundary T, the center t, offset s outputs a vector v of discrete Gaussian distribution on in the Lattice $\Lambda^\perp(A)$, we can compute the original image value $\sigma = t + v$ to make $A\sigma = u(\bmod q)$ is established.

In the original image sampling function mentioned above, the gate boundary base T directly determines the norm of the output vector σ. Only known the gate boundary basis T, the owner of T can get an original image of the vector $u \in \mathbb{Z}_q^n$ which satisfies $\sigma \in D_n$ at $f_A(x) = Ax(\bmod q)$. In this paper, we use this principle to construct the simultaneous multi-signature scheme and the sequential multi-signature scheme.

4 Simultaneous Multi-signature Scheme

Simultaneous multi-signature is sometimes called broadcast digital multi-signature scheme. It means that the source of the message sends the message to each signer of signature at the same time, and then the signature message is collected by a signature collector who is selected jointly by all of signers. The collector sorts the signature message and sends it to the verifier. Finally, the signature verifier tests and verifies if the signature message is valid or not. Now suppose there are N signers of $U_1, U_2, \ldots, U_N$ who will sign for the same message m simultaneously.

4.1 Key Generation

Input n is a security parameter, q, m, s are all the function of n, c is constant, set $q = poly(n)$, $m \geq cnlogq$, the parameter $s \geq m^{1+\varepsilon} \cdot \omega(logm)^{1/2}$. Suppose $U_i (1 \leq i \leq N)$ is the IDs of N signers, then generate N random uniform matrix $A_i \in \mathbb{Z}_q^{n \times m} (1 \leq i \leq N)$ and a set of gate boundary bases $S_i \in \mathbb{Z}_q^{n \times m}$ on the lattice $\Lambda_q^{\perp}(A_i)$ by using TrapGen (1n). To generate another matrix $B_i \in \mathbb{Z}_q^{n \times m} (1 \leq i \leq N)$ and a set of short bases $T_i \in \mathbb{Z}_q^{n \times m}$ of the lattice $\Lambda_q^{\perp}(B_i)$ by using short base sampling algorithm of matrix A_i, which satisfies $B_i A_i^{\perp} = 0$. $(A_i, B_i)_{i=1}^{N}$ is the public key and T_i is the private key for all signers $U_i (1 \leq i \leq N)$.

4.2 Simultaneous Multi-signature

If there are N users $U_i (1 \leq i \leq N)$ will sign for the same message m simultaneously, the steps of multi-signature are as the following.

(1) Users $U_i (1 \leq i \leq N)$ select a vector $r \leftarrow \{0, 1\}^n$ by using random oracle $H_1 : \{0, 1\}^* \rightarrow \mathbb{Z}_q^{n \times k}$, and then calculate $h_i = H_1(m; r) \in \mathbb{Z}_q^{n \times k}$.
(2) To select the vector $t_i \in \mathbb{Z}_q^m$ randomly and uniformly while satisfying $B_i t_i = h_i (mod \ q)$.
(3) For each signer $U_i (1 \leq i \leq N)$ can use the center $-t_i$, the parameter s and the original image sampling algorithm to extract vectors $v_i \leftarrow D_{\Lambda^{\perp}, s, -t_i}$ and find out $\sigma_i = t_i + v_i$. Then, check whether it meets the requirements $||\sigma_i|| \leq s\sqrt{m}$, if not, the signer can extract the vector v_i again and make $B_i \sigma_i = h_i (mod \ q)$ established.
(4) To calculate $z_i = A_i^{\perp} s_i + \sigma_i (mod \ q)$, then to construct a NIWI proof π which is provided in [24], and the output is the signature $e = (r, z_1, z_2, \ldots, z_N, \pi)$.

Now the collector can send the result e to the verifier, e is the generated message after simultaneous multi-signatures of all of signers $U_1, U_2, \ldots, U_N$.

4.3 Verification of Simultaneous Multi-signature

After the verifier receives the multiple signature results $e = (r, z_1, z_2, \ldots, z_N, \pi)$, it is necessary to verify whether π is correct firstly. If π is correct, and the number of public keys exactly equals the signed message e in z_i, and meanwhile each signer $U_i (1 \leq i \leq N)$

can satisfy the equation $A_i z_i = h_i (mod\ q)$, then the verification is successful and the output is 1, otherwise the verification is failure and then output is 0.

5 Sequential Multi-signature Scheme

In the sequential digital multi-signature scheme, the first thing is that all senders start to negotiate the sequence of signature for a message together. Secondly, the message is sent to the first signer in order. When the first signer has finished the signature, the message is sent to the second signer in the order of negotiation, and so on, until the last signer completes signing and delivers it to the verifier. In the process of sequential signature, except for the first signer, each other signer will verify the validity of the previous signature after receiving the signature message. If the signature is valid, then the signer sends the signature message to the next, and the signature will be going on. If the signature is invalid, then the signer refuses to sign the message again, and the signature will be terminated. When the verifier receives the signature message, he will verify whether the signature is valid or not. If the verification is valid, the multi-signature message is valid; otherwise, the multi-signature message is invalid. Similar to the simultaneous multi signature mechanism, we set N signers $U_1, U_2, \ldots, U_N$ to sign the same message m in a sequence. Because the key generation process is the same as the simultaneous multi-signature scheme, the detailed process of sequential multi-signature is discussed directly.

5.1 Sequential Multi-signature

Suppose there are N signers $U_i (1 \leq i \leq N)$ who sign a same message m in a predetermined order. The signature process of the first signer U_1 is as the following steps.

(1) To select a vector $r \leftarrow \{0, 1\}^n$ randomly, use the random oracle $H_1 : \{0, 1\}^* \rightarrow \mathbb{Z}_q^{n \times k}$, and then calculate: $h_1 = H_1(m; r) \in \mathbb{Z}_q^{n \times k}$.

(2) To select the vector $t_1 \in \mathbb{Z}_q^m$ randomly and uniformly, while satisfying $B_1 t_1 = h_1 (mod\ q)$.

(3) To take $-t_1$ as the center, and use the parameter s and the original image sampling algorithm to extract vectors $v_1 \leftarrow D_{A^\perp, s, -t_1}$, can be found out $\sigma_1 = t_1 + v_1$. Then to check whether the inequality is satisfied with $\|\sigma_1\| \leq s\sqrt{m}$ and the equation $B_1 \sigma_1 = h_1 (mod\ q)$ is established.

(4) To calculate $z_1 = A_1^\perp s_1 + \sigma_1 (mod\ q)$, the first signer U_1 sends r, z_1, σ_1 which are the result of signature to the second signer U_2.

Then the signature process of each rest signer $U_i (i > 1)$ is as the following steps.

(1) The signer U_i receives $B_i z_i = h_i (mod\ q)(i \geq 1)$ from the last signer U_{i-1} and verifies if it is true or not. If the equation is false, the signature is terminated; if it is true, then continue with the following steps.

(2) Randomly select a vector $r \leftarrow \{0, 1\}^n$, using random oracle $H_1 : \{0, 1\}^* \rightarrow \mathbb{Z}_q^{n \times k}$ to calculate $h_i = H_1(m; r) \in \mathbb{Z}_q^{n \times k}$.

(3) To select a vector $t_i \in \mathbb{Z}_q^m$ random uniformly, while satisfying $B_i t_i = h_i (mod\ q)$.

(4) To take $-t_i$ as the center, use the parameter s and the original image sampling algorithm to extract vectors $v_i \leftarrow D_{A^{\perp}, s, -t_i}$, can be found out $\sigma_i = t_i + v_i$. Then to check whether the inequality is satisfied with $||\sigma_i|| \leq s\sqrt{m}$ and the equation $B_i \sigma_i = h_i (mod\ q)$ is established.

(5) To calculate $z_i = A_i^{\perp} s_i + \sigma_i (mod\ q)$, and send the result of signature $r, z_1, \ldots, z_i, \sigma_1, \ldots, \sigma_i$ to the next signer. If the next signer is $U_i (i = N)$, and then construct a NIWI proof π and send the result $(r, z_1, \ldots, z_N, \sigma_1, \ldots, \sigma_N, \pi)$ to the verifier.

5.2 Verification of Sequential Multi-signature

After the verifier receives multiple signatures $(r, z_1, \ldots, z_N, \sigma_1, \ldots, \sigma_N, \pi)$, and the receiver starts to verify if the number of public keys N is the number of z_i in the signature e. Meanwhile, for each $U_i (1 \leq i \leq N)$, the equation $B_i z_i = h_i (mod\ q)$ must be established. If the equation is true, the output is 1 which indicates that the signature is valid; otherwise, the output is 0 which indicates that the signature is invalid.

6 Security Analysis

6.1 Correctness

If the signers $U_i (1 \leq i \leq N)$ and the verifier comply with the signature rules, the verifier will always believe that the multi-signature that has generated above are correct. Suppose $(r, z_1, z_2, \ldots, z_N, \sigma_1, \sigma_2, \ldots, \sigma_N, \pi)$ is the signature of a legal group member, then π is a correct proof. Secondly, for any i in $1 \leq i \leq N$, $B_i z_i = B_i (A_i^{\perp} s_i + \sigma_i) = B_i \sigma_i = h_i (mod\ q)$ is established. The verification is successful, so it can be proved that the signature scheme is effective.

6.2 Security

The multi-signature mechanism is based on lattice theory, which can successfully resist the security initiative attack of quantum computer. Theoretical analysis shows that the new scheme is also secure against passive attack, and its security mainly depends on two factors. The first one is the difficulty of NP-hard problems to obtain the private key and the public key $(A_i, B_i)_{i=1}^{N}$ of multi-signature. The second one is the difficulty of forging part of multi-signature to make the signatures satisfying the verification for the adversary. In any case, it is not feasible for any adversary to obtain part of the private key for every adversary. When computing $z_i = A_i^{\perp} s_i + \sigma_i (mod\ q)$ and finding out σ_i by z_i, it is impossible to solve the SIS problem on lattice. In addition, the multi-signature key is generated by the signature key of each signer. If other signers want to forge the final signature message, they need to know the signature key. This is contrary to the zero-knowledge proof proposed by G. keygen in [24], which means that we must solve the problem of the nearest vector, which is obviously not feasible in calculation.

7 Conclusion

With the rapid development of quantum computer in the new era [25, 26], the traditional signature scheme based on the big integer decomposition problem is proved to be not secure, but SIS has not found its decoding algorithm in the quantum computing environment. As far as we know, this paper constructs two different types of digital multi-signature schemes based on lattice theory for the first time, and these two multi-signature schemes are able to resist quantum attacks. We have compared it with the traditional multi-signature system in the article [27]. The signature in [27] uses techniques of ring signatures and bilinear pairings. These operations need to take more time and computation. But the signature we provided includes most of operations such as simple modular addition and modular multiplication, so which needs less time and is more efficiency. At the same time, the validity and security of multi-signature schemes are also discussed. In the future work, we need to optimize the signature length, storage and calculation costs.

Acknowledgements. This work is supported by Qinghai Office of Science and Technology (No. 2019-ZJ-7086, No. 2018-SF-143), the National Social Science Foundation of China (No. 18XMZ050), the Key Laboratory of IoT of Qinghai (No. 2020-ZJ-Y16).

References

1. Kaur, R., Kaur, A.: Digital signature. In: 2012 International Conference on Computing Sciences (ICCS). IEEE (2012)
2. Bellare, M., Miner, S.K.: A forward-secure digital signature scheme. In: Wiener, M. (ed.) CRYPTO 1999. LNCS, vol. 1666, pp. 431–448. Springer, Heidelberg (1999). https://doi.org/10.1007/3-540-48405-1_28
3. Mitomi, S., Miyaji, A.: A multisignature scheme with message flexibility, order flexibility and order verifiability. In: Dawson, E.P., Clark, A., Boyd, C. (eds.) ACISP 2000. LNCS, vol. 1841, pp. 298–312. Springer, Heidelberg (2000). https://doi.org/10.1007/10718964_25
4. Harna, L., Jian, R.: Efficient identity-based RSA multisignatures. Comput. Secur. **27**(1), 12–15 (2008)
5. Yi, L.-J., Bai, G.-Q., Xiao, G.-Z.: Proxy multi-signature: a new type of proxy signature schemes. Acta Electronica Sin. **29**, 569–570 (2001)
6. Micciancio, D.: The shortest vector in a lattice is hard to approximate to within some constant. In: The 39th Annual Symposium on Foundations of Computer Science, Los Alamitos, pp. 92–98. IEEE Computer Society (1998)
7. Khot, S.: Hardness of approximating the shortest vector problem in lattices. J. ACM **52**(5), 789–808 (2005)
8. Dinur, I.: Approximating SVP_∞ to within almost-polynomial factors is NP-hard. In: Bongiovanni, G., Petreschi, R., Gambosi, G. (eds.) CIAC 2000. LNCS, vol. 1767, pp. 263–276. Springer, Heidelberg (2000). https://doi.org/10.1007/3-540-46521-9_22
9. Boas, P.V.E.: Another NP-complete partition problem and the complexity of computing short vectors in lattices (1981)
10. Ajtai, M.: The shortest vector problem in L2 is NP-hard for randomized reductions. In: The 30th Annual ACM Symposium on Theory of Computing, pp. 10–19. ACM, New York (1998)

11. Itakura, K., Nakamura, K.: A public-key cryptosystem suitable for digital multisignatures. NEC Res. Dev. **71**, 1–8 (1983)
12. Bellare, M., Neven, G.: Multi-signatures in the plain public-key model and a general forking lemma. In: 13th ACM Conference on Computer and Communications Security (CCS2006), Alexandria, USA, 30 October–3 November (2006)
13. Ji, J., Zhao, R.: New digital multi signature system. J. Comput. Sci. **20**(6), 533–538 (1997)
14. Wang, J.F., Wang, S., Liu, Y.H.: Chinese characters knapsack public key encryption algorithm research. Appl. Mech. Mater. **198–199**, 1518–1522 (2012)
15. Huang, M.-D., Raskind, W.: Signature calculus and discrete logarithm problems. In: Hess, F., Pauli, S., Pohst, M. (eds.) ANTS 2006. LNCS, vol. 4076, pp. 558–572. Springer, Heidelberg (2006). https://doi.org/10.1007/11792086_39
16. Shor, P.W.: Scheme for reducing decoherence in quantum computer memory. Phys. Rev. A **52**(4), R2493 (1995)
17. Shor, P.W.: Polynomial-time algorithms for prime factorization and discrete logarithms on a quantum computer. Siam Rev. **41**(2), 303–332 (1999)
18. Ajtai, M.: Generation hard instances of lattice problems. Complex. Comput. Proofs Quad. di Mat. **13**, 1–32 (2004)
19. Gentry, C., Peikert, C., Vaikuntanathan, V.: Trapdoors for hard lattices and new cryptographic constructions. In: Proceedings of STOC, Victoria, pp. 197–206. ACM (2008)
20. Cash, D., Hofheinz, D., Kiltz, E., Peikert, C.: Bonsai trees, or how to delegate a lattice basis. In: Gilbert, H. (ed.) EUROCRYPT 2010. LNCS, vol. 6110, pp. 523–552. Springer, Heidelberg (2010). https://doi.org/10.1007/978-3-642-13190-5_27
21. Ljujic, Z.: Problems in additive number theory. Dissertations & Theses Gradworks (2011)
22. Micciancio, D., Regev, O.: Worst-case to average-case reductions based on gaussian measures. SIAM J. Comput. **37**(1), 267–302 (2007)
23. Ajtai, M.: Generating hard instances of the short basis problem. In: Wiedermann, J., van Emde Boas, P., Nielsen, M. (eds.) ICALP 1999. LNCS, vol. 1644, pp. 1–9. Springer, Heidelberg (1999). https://doi.org/10.1007/3-540-48523-6_1
24. Micciancio, D., Vadhan, S.P.: Statistical zero-knowledge proofs with efficient provers: lattice problems and more. In: Boneh, D. (ed.) CRYPTO 2003. LNCS, vol. 2729, pp. 282–298. Springer, Heidelberg (2003). https://doi.org/10.1007/978-3-540-45146-4_17
25. De Raedt, H., Jin, F., Willsch, D.: Massively parallel quantum computer simulator, eleven years later. Comput. Phys. Commun. **237**, 47–61 (2018)
26. Potapov, V., Gushanskiy, S., Guzik, V., Polenov, M.: The computational structure of the quantum computer simulator and its performance evaluation. In: Silhavy, R. (ed.) CSOC2018 2018. AISC, vol. 763, pp. 198–207. Springer, Cham (2019). https://doi.org/10.1007/978-3-319-91186-1_21
27. Tonien, D., Susilo, W., Safavi-Naini, R.: Multi-party Concurrent Signatures. In: Katsikas, S. K., López, J., Backes, M., Gritzalis, S., Preneel, B. (eds.) ISC 2006. LNCS, vol. 4176, pp. 131–145. Springer, Heidelberg (2006). https://doi.org/10.1007/11836810_10

Research on Online Leakage Assessment

Zhengguang Shi[✉], Fan Huang, Mengce Zheng, Wenlong Cao, Ruizhe Gu,
Honggang Hu, and Nenghai Yu

Key Laboratory of Electromagnetic Space Information,
University of Science and Technology of China, Hefei 230027, China
szg10086@mail.ustc.edu.cn

Abstract. Leakage assessment is the most common approach applied
for assessing side-channel information leakage and validating the
effectiveness of side-channel countermeasures. Established evaluation
approaches are usually based on Test Vector Leakage Assessment
(TVLA) that deployed in a divide and conquer flow with offline computations, which causes two apparent shortcomings in required memory and time. In this paper, a lightweight framework of online leakage
assessment is proposed. The problems were analyzed and the evaluation
approach was further validated with a Field Programmable Gate Array
(FPGA). The experimental results show that it can implement online
processing on newly collected data, and instantly stop to give the result
when detecting credible leakage. The online leakage assessment can significantly economize on memory and time. It has good performance when
there is limited memory or real-time evaluations are needed.

Keywords: Leakage assessment · Online processing · Efficient
evaluation · Side-channel countermeasures

1 Introduction

Side-Channel Analysis (SCA) attack is a high issue in the crypto community that
violates the security and reliability of hardware implementations. The fundamental of SCA is to utilize the relationship between the side-channel information,
e.g. power consumption, and intermediate values during an existing hardware
implementation running, for the phenomenon that power consumptions of some
certain operations may significantly differ when the device processes 0 and 1,
and these differences may be relative to secret keys [12].

Evoked by the demand of security, researchers proposed some countermeasures, such as hiding [15] and masking [7]. The key idea is to eliminate the difference or to break the relationships between power traces and intermediate values,

The authors would like to thank Information Science Laboratory Center of USTC for
the hardware/software services. This work was supported by National Natural Science
Foundation of China (Nos. 61972370 and 61632013), Fundamental Research Funds for
Central Universities in China (No. WK3480000007).

P. Qin et al. (Eds.): ICPCSEE 2020, CCIS 1258, pp. 138–150, 2020.
https://doi.org/10.1007/978-981-15-7984-4_11

so that the traces are not lucrative to the adversary any more. A new problem occurs subsequently that how to verify whether a countermeasure works [10] because some countermeasures still have statistical vulnerabilities [6,16]. The most common approach is Leakage Assessment, which is a statistical way aiming to detect the similarity of two data sets.

The concept of Leakage Assessment was raised by Coron et al. in 2001 [5]. They also put forward a basic evaluation idea that success confirms the existence of leakage, while failure cannot refuse it but indicates that the leakage is not significantly detected with the given confidence. Researcher then developed Leakage Assessment with different approaches. In 2011, Gilbert et al. proposed a testing methodology that based on t tests for the first time [10]. It's easy to carry out and can validate side-channel resistance at any certain order. In 2013, Becker et al. proposed Test Vector Leakage Assessment (TVLA) in practice [3] that employs t tests to perform 2 non-specific Fixed-VS.-Random tests and 896 specific Fixed-VS.-Fixed tests between two data sets collected with elaborate inputs. Since then, it was widely accepted as a reference for its standard evaluation flow and sensitivity to leakage.

In 2015, Moradi et al. concluded the leakage assessment methodology [14] and proposed a guideline that detects leakage at any order with increment computations. Based on previous work, Reparaz et al. proposed Fast Leakage Assessment [13] in 2017 that uses distribution histograms of samples instead of the whole data set to accelerate the computations. In addition, the work of Ding et al. [8] gives some guidance in practical evaluations that can speed up the evaluation process and enhance the detecting capability. Considering that t tests can assert the leakage's presence but lack support of asserting the leakage's absence and the evaluation result is affected by the sample size[9], Bache et al. proposed Confident Leakage Assessment[1] based on confidence intervals to complete t tests. Then, they extended the approach to an evaluation framework [2].

Our Contribution. In this paper, we pay attention to single-variate evaluations based on t tests. Established evaluation approaches using t tests, e.g. the approaches proposed in [14] and [13], are usually based on TVLA that deployed in a divide and conquer flow:

1. collect a certain amount (denoted as m) of power traces with an acquisition device and temporarily store them in the local memory. The number m depends on the scale of each trace and the memory size of the device;
2. send the collected data to a computer and store them on disks;
3. repeat the first two steps until the total number of traces reaches a predefined number N_t ($N_t = c \cdot m$ where c represents the number of repetitions);
4. process the data to perform offline t tests on the computer.

While it has some advantages such as flexible control on each stage and the CPU on the desktop having a high performance, the flow leads to two apparent shortcomings. One is **required memory** since it needs plenty of memory to store the collected data on both the employed acquisition device and the computer.

Another one is **required time** since the data processing cannot begin until all traces are collected and sent to the computer. Thus, if the evaluation device has limited memory or if we want real-time evaluation results in case of, e.g. on-chip leakage monitoring, these kinds of approaches are not suitable. In addition, if the device leaks, we will waste time on dealing with excess data since we cannot exactly forecast the number of required traces to identify the leakage.

From this point of view, we focus on the advantages of online processing and propose a frame of Online Leakage Assessment approach. It can collect power traces and online process newly collected data to perform t tests. When it detects credible leakage, it will instantly stop and give the result. We validated our approach with an FPGA and compared its performance with the approaches proposed in [14] and [13]. According to the experiments, it outperforms two approaches on required memory and time a lot.

Outline. In Sect. 2, we give the notations used in this paper as well as some preliminaries about our work. In Sect. 3, we give the core method and the procedure of the online approach. Furthermore, we implemented our approach as well as those in [14] and [13] for comparisons to validate its performance. Then, we describe the experimental setup in Sect. 4. In Sect. 5, we show the advantages of the proposed online approach according to numerical experimental results. Finally, we conclude our work in Sect. 6.

2   Preliminaries

In this section, we provide the notations used in this paper and introduce some related works.

2.1   Notations

We shall collect two data sets for evaluations, $D^f = \{T_i^f\}_{i=1}^{N_t}$ and $D^r = \{T_i^r\}_{i=1}^{N_t}$, where $T_i^f = \{\tau_{i,j}^f\}_{j=1}^{N_p}$ and $T_i^r = \{\tau_{i,j}^r\}_{j=1}^{N_p}$. The notations about collected data are defined as follows.

- D^f (resp. D^r) denotes the data set of fixed-input traces (resp. random-input traces).
- T_i^f (resp. T_i^r) denotes the ith collected trace of D^f (resp. D^r).
- $\tau_{i,j}^f$ (resp. $\tau_{i,j}^r$) denotes the jth point (time sample) on T_i^f (resp. T_i^r).
- N_t denotes the number of traces collected for both D^f and D^r.
- N_p denotes the number of points (time samples) on each collected trace.
- Q denotes the number of quantization bits of each point, i.e. $\tau_{i,j}^f$ or $\tau_{i,j}^r$.

D^f (resp. D^r) can be regarded as a matrix of N_t rows and N_p columns. Its ith row is denoted as $D_{i,}^f$ (resp. $D_{i,}^r$), i.e. T_i^f (resp. T_i^r), and its jth column is denoted as $D_{,j}^f$ (resp. $D_{,j}^r$). In the fast leakage assessment, we need to generate two families of histograms for D^f and D^r, which are denoted as H^f and H^r. Similarly, $H^f_{i,}$ (resp. $H^r_{i,}$) and $H^f_{,j}$ (resp. $H^r_{,j}$) denote the ith row and the jth column of H^f (resp. H^r), respectively. The parameters used for t tests are defined as follows.

- M^f_j (resp. M^r_j) denotes the mean of $D^f_{,j}$ (resp. $D^r_{,j}$) or $H^f_{,j}$ (resp. $H^r_{,j}$).
- S^f_j (resp. S^r_j) denotes the variance of $D^f_{,j}$ (resp. $D^r_{,j}$) or $H^f_{,j}$ (resp. $H^r_{,j}$).

2.2   TVLA Based on t Tests

As mentioned above, the fundamental aim of evaluation is to identify whether two sets of power consumptions that collected from the same operation while processing different values are significantly different or not. The most popular approach is using the TVLA procedure with Welch's t tests to quantify their similarities. Basically, a t test usually takes the assumption that both two sets are drawn from the same population as the null hypothesis and gives a probability to validate whether to accept the hypothesis or not.

Concretely, let $D_0 = \{\tau_{i,j}{}^f\}_{i=1}^{N_0}$ with $j = j_0$ and $D_1 = \{\tau_{i,j}{}^r\}_{i=1}^{N_1}$ with $j = j_0$, i.e. $D^f_{,j_0}$ and $D^r_{,j_0}$, denote two sets to be test. Let $M_0 = M^f_{j_0}$ (resp. $M_1 = M^r_{j_0}$) and $S_0 = S^f_{j_0}$ (resp. $S_1 = S^r_{j_0}$) stand for the mean and the variance of D_0 (resp. D_1), respectively. Using Welch's two-tailed t tests, the t-statistic and the degree of freedom v can be computed as

$$t = \frac{M_0 - M_1}{\sqrt{\frac{S_0}{N_0} + \frac{S_1}{N_1}}}, \quad v = \frac{(\frac{S_0}{N_0} + \frac{S_1}{N_1})^2}{\frac{(\frac{S_0}{N_0})^2}{N_0-1} + \frac{(\frac{S_1}{N_1})^2}{N_1-1}}. \tag{1}$$

If we get $|t| > 4.5$, we will reject the null hypothesis with a confidence of 0.9999 without considering the degree of freedom, i.e. the device under test (DUT) leaks.

As mentioned above, Test Vector Leakage Assessment (TVLA) is widely regard as a reference for its standard evaluation flow and sensitivity to leakage. It employs t tests to perform two categories of tests [3]. The non-specific Fixed-VS.-Random tests are recommended by the writers because according to the experiments, they can identify leakage using fewer traces than the specific Fixed-VS.-Fixed tests by orders of magnitude. Hence, they are more sensitive to leakage.

We modified Fixed-VS.-Random tests and applied it to our work. We collected N_t fixed-input traces for D^f and N_t random-input traces for D^r. Then we perform only one **Test** with D^f and D^r that t tests between $D^f_{,j}$ and $D^r_{,j}$ for each $1 \leq j \leq N_p$. This will decrease the reliability of the evaluation results comparing to performing two independent **Tests** in [3], but simplify the evaluation process and improve the efficiency a lot.

2.3 Fast Leakage Assessment

Considering the traditional approach for computing mean and variance will traverse the whole data set several times, Reparaz et al. proposed Fast Leakage Assessment to perform efficient computations. In this paper, we implement it for a comparison with our Online Leakage Assessment approach.

The core idea of Fast Leakage Assessment is "counter". For traces consisting of N_p Q-bit-integer points (time samples), we create a family of histograms (i.e. an array) $H^f[2^Q][N_p]$ (resp. $H^r[2^Q][N_p]$) for fixed-input set D^f (resp. random-input set D^r) that each element corresponds to a certain sample value among $[0, 2^Q - 1]$, and its value represents how many times the sample value occurs at this time sample. In detail, when receiving a newly collected trace T_i^f, for each $\tau_{i,j}^f$, we increase $H^f[\tau_{i,j}^f][j]$ by 1. When collecting stage finishes, we can compute the mean and variance at time sample j as:

$$M_j^f = \frac{1}{\sum_{i=0}^{2^Q-1} H^f[i][j]} \sum_{i=0}^{2^Q-1} i \cdot H^f[i][j], \tag{2}$$

and

$$S_j^f = \frac{1}{(\sum_{i=0}^{2^Q-1} H^f[i][j]) - 1} \sum_{i=0}^{2^Q-1} \left(i - M_j^f\right) \cdot \left(i - M_j^f\right) \cdot H^f[i][j]. \tag{3}$$

The M_j^r and S_j^r can be achieved with H^r in a similar way.

3 Methodology

In this section, we give the core method and the procedure of online approach. To verify its advantages, we implement three designs for comparisons.

3.1 Online Computation

In the online approach, when collecting a new sample value, we need to instantly process it to detect leakage. This prompts us to search for an efficient computation approach that can avoid a mass of redundant computations and ensure numerical stability [11]. In [14], Moradi et al. take an iterative approach using increment computations and give us inspirations.

The approach is detailed in [4]. Let X denote a set consisting of N elements and each element is denoted as x_i ($1 \le i \le N$). Let M and S denote the mean and the variance of X. According to the definitions, the mean M and the variance S can be computed as

$$M = \frac{1}{N} \sum_{i=1}^{N} x_i, \quad S = \frac{1}{N-1} \sum_{i=1}^{N} (x_i - M)^2.$$

Let $M_k = \frac{1}{k}\sum_{i=1}^{k} x_i$ denote the mean of first k elements of X and we get $M_{k-1} = \frac{1}{k-1}\sum_{i=1}^{k-1} x_i$ for $1 < k \leq N$. Then, we have

$$kM_k = \sum_{i=1}^{k-1} x_i + x_k = (k-1)M_{k-1} + x_k = kM_{k-1} + x_k - M_{k-1}.$$

Hence, we can achieve M_k iteratively as

$$M_k = \begin{cases} x_k, & \text{if } k = 1, \\ M_{k-1} + \frac{1}{k}(x_k - M_{k-1}), & \text{if } 1 < k \leq N. \end{cases} \tag{4}$$

As for S, we define $S' = \sum_{i=1}^{N}(x_i - M)^2$ and hence $S = \frac{1}{N-1}S'$. Then we can manipulate S' to $S' = \sum_{i=1}^{N} x_i^2 - \frac{1}{N}(\sum_{i=1}^{N} x_i)^2$. Similarly, we get $S'_k = \sum_{i=1}^{k} x_i^2 - \frac{1}{k}(\sum_{i=1}^{k} x_i)^2$ and $S'_{k-1} = \sum_{i=1}^{k-1} x_i^2 - \frac{1}{k-1}(\sum_{i=1}^{k-1} x_i)^2$ for $1 < k \leq N$. Then, we have

$$kS'_k = k\left(\sum_{i=1}^{k-1} x_i^2 + x_k^2\right) - \left(\sum_{i=1}^{k-1} x_i + x_k\right)^2$$

$$= k\sum_{i=1}^{k-1} x_i^2 + kx_k^2 - \left(\sum_{i=1}^{k-1} x_i\right)^2 - 2x_k\sum_{i=1}^{k-1} x_i - x_k^2$$

$$= k\left(\sum_{i=1}^{k-1} x_i^2 - \frac{1}{k-1}\left(\sum_{i=1}^{k-1} x_i\right)^2\right) + \left(\frac{1}{k-1}\right)\left(\sum_{i=1}^{k-1} x_i\right)^2$$

$$+ (k-1)x_k^2 - 2x_k\sum_{i=1}^{k-1} x_i$$

$$= kS'_{k-1} + (k-1)M_{k-1} + (k-1)x_k^2 + 2x_k(k-1)M_{k-1}$$

$$= kS'_{k-1} + (k-1)(x_k - M_{k-1})^2.$$

Hence, we can achieve S'_k iteratively as

$$S'_k = \begin{cases} 0, & \text{if } k = 1, \\ S'_{k-1} + \frac{k-1}{k}(x_k - M_{k-1})^2, & \text{if } 1 < k \leq N. \end{cases} \tag{5}$$

In our design, the procedures of online computing M_j^f and S_j^f with a new sampling point $\tau_{i,j}^f$ are detailed in Algorithm 1. M_j^r and S_j^r can be achieved with $\tau_{i,j}^r$ in the same way. Then, we can perform t tests instantly. The whole procedure of the online approach is shown in Algorithm 2. Note that in our experiments, we set both N_0 and N_1 as N_t to simplify the evaluation process, i.e. we collect a random-input trace each after collecting a fixed-input trace.

3.2   Distinct Designs and Comparisons

In order to quantify the performance of the online approach, we implement three different designs, i.e. Design_c, Design_f, and Design_o, for comparisons in required memory and time. Their explanations and flows are given as follows.

Algorithm 1. The procedure of online computing mean and variance

---

Input: $\tau_{i,j}{}^{f}$
Output: Current $M_j{}^{f}$ and $S_j{}^{f}$
  1: **if** $i = 1$ **then**
  2:     $M_j{}^{f} \leftarrow \tau_{i,j}{}^{f}$
  3:     $S_j{}^{f} \leftarrow 0$
  4:     $S'_j{}^{f} \leftarrow 0$
  5: **else**
  6:     $lastM \leftarrow M_j{}^{f}$
  7:     $lastS \leftarrow S_j{}^{f}$ ▷ reserve the last M and S
  8:     $M_j{}^{f} \leftarrow lastM + \frac{1}{i}(\tau_{i,j}{}^{f} - lastM)$
  9:     $S'_j{}^{f} \leftarrow lastS + \frac{i-1}{i}(\tau_{i,j}{}^{f} - lastM)^2$
10:     $S_j{}^{f} \leftarrow \frac{1}{i-1}S'_j{}^{f}$
11: **end if**

Algorithm 2. The procedure of $Design_o$

Input: $\left\{T_i{}^{o}\right\}_{i=1}^{2N_t}$, $type$.   ▷ original collected power traces
Output: t values.
  1: **Initial:** $flag \leftarrow 0, N_0 \leftarrow 0, N_1 \leftarrow 0$
  2: **while** $flag = 0$ **do**   ▷ leakage is not identified
  3:    **if** $type = 0$ **then**   ▷ collect a fixed-input trace
  4:       $N_0 \leftarrow N_0 + 1$ ▷ the number of fixed-input traces
  5:       **for** $j = 1 \rightarrow N_p$ **do**
  6:          measure the voltage as $\tau_{N_0,j}{}^{f}$
  7:          update $M_j{}^{f}$ and $S_j{}^{f}$ with Algorithm 1
  8:       **end for**
  9:    **else**   ▷ collect a random-input trace
10:       $N_1 \leftarrow N_1 + 1$ ▷ the number of random-input traces
11:       **for** $j = 1 \rightarrow N_p$ **do**
12:          measure the voltage as $\tau_{N_1,j}{}^{r}$
13:          update $M_j{}^{r}$ and $S_j{}^{r}$ with Algorithm 1
14:       **end for**
15:    **end if**
16:    **if** $N_0 > 1$ **and** $N_1 > 1$ **then**
17:       **for** $j = 1 \rightarrow N_p$ **do**
18:          perform t tests on the FPGA with Eq. 1
19:          **if** $|t| > 4.5$ **then**   ▷ leakage is identified
20:             $flag \leftarrow 1$
21:             lighten an LED on the FPGA   ▷ give results
22:             **break**
23:          **end if**
24:       **end for**
25:    **end if**
26: **end while**

Design$_c$. Common approach proposed by Schneider and Moradi in [14], which employs offline increment computations:
1. collect N_t fixed-input traces $\{T_i^f\}_{i=1}^{N_t}$, N_t random-input traces $\{T_i^r\}_{i=1}^{N_t}$ (each with N_p points) and send them to a computer,
2. store the traces as D^f and D^r on the computer,
3. compute means and variances using Eq. (4) and Eq. (5) on the computer,
4. perform t tests using Eq. (1) on the **computer**.

Design$_f$. Fast Leakage Assessment approach proposed by Reparaz et al. in [13]:
1. collect N_t fixed-input traces $\{T_i^f\}_{i=1}^{N_t}$, N_t random-input traces $\{T_i^r\}_{i=1}^{N_t}$ (each with N_p points) and send them to a computer,
2. store the traces as two families of histograms H^f and H^r,
3. compute means and variances using Eq. (2) and Eq. (3) on the computer,
4. perform t tests using Eq. (1) on the **computer**.

Design$_o$. Online Leakage Assessment approach proposed in this paper:
1. collect a new fixed-input trace $T_i^f = \{\tau_{i,j}^f\}_{i=1}^{N_p}$ and instantly update $\{M_j^f\}_{j=1}^{N_p}$ and $\{S_j^f\}_{j=1}^{N_p}$ using Eq. (4) and Eq. (5),
2. collect a new random-input trace $T_i^r = \{\tau_{i,j}^r\}_{i=1}^{N_p}$ and instantly update $\{M_j^r\}_{j=1}^{N_p}$ and $\{S_j^r\}_{j=1}^{N_p}$ using Eq. (4) and Eq. (5),
3. for each $1 \le j \le N_p$, perform t test using Eq. (1) on the **FPGA**,
4. repeat the first three steps until the leakage is identified (or i reaches a pre-defined number N_t).

4   Implementation

We implement three designs on an FPGA and a computer. In this section, we first introduce the experimental setups then give the design flow on them.

The designs are implemented on a Xilinx Virtex-7 XC7VX485T-2FFG1761 FPGA, also called VC707 board (see Fig. 1). There is a dual 12-bit, 1 mega-samples per second (MSPS) ADC module called XADC on the board [17] which can be used for sampling. In our experiment, we use the XADC to collect the data with dedicate external differential analog input V_P, V_N channel. The XADC is set working in single channel, bipolar input and continuous sampling mode and enable averaging by writing to the control registers.

In the conversion stage, the XADC always produces a 16-bit result in the 16-bit status registers, the 12 MSBs of which correspond to the 12-bit conversion data and the 4 LSBs are used to improve resolution through averaging or filtering. Notes that when bipolar input is enabled, the maximum input range of the differential analog input $(V_P - V_N)$ is ± 0.5 V. By default, the XADC uses the on-board reference voltage, thus the LSB size in volts is equal to $1\text{V}/2^{12}$ or $1\text{V}/4096 = 244\mu\text{V}$. To collect data, we programmed an AMS101 card working with the FPGA to emulate power traces that collected from an CW303 board running an unprotected AES for simulations. The original power traces are denoted as $\{T_i^o\}_{i=1}^{N_t}$ where $T_i^o = \{\tau_{i,j}^o\}_{j=1}^{N_p}$.

Design$_o$ is implemented completely on the FPGA while Design$_c$ and Design$_f$ collect traces on the FPGA and perform processing on a computer. Here, we use

Fig. 1. The VC707 board, ① represents the AMS101 card plugging on the XADC header, ② represents the UART module, ③ represents LEDs.

the UART module on the FPGA to send collected traces to the computer. In Design$_o$, when it detects leakage, an LED on the FPGA will be lighten to give the result. The computer we use is a desktop with an Intel Core i5-4460 CPU @ 3.2 GHz. On the desktop, we implement the rest of Design$_c$ and Design$_f$ using Visual Studio 2017. All codes are running in Release versions.

To implement the designs, we built an SoC on the VC707 board referring to [18] using Vivado 2013.3, the structure of which is illustrated in Fig. 2. Then we developed functions over the SoC using Xilinx SDK 2013.3 so that the FPGA can satisfy the requirements of collecting power traces, processing and sending data.

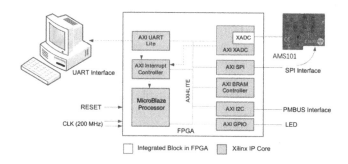

Fig. 2. The structure of the SoC built on VC707.

Since Design$_c$ and Design$_f$ perform offline computations on the desktop, we pre-process the received data with Visual Studio 2017, i.e. convert the received hexadecimal numbers to decimal integers, and generate D^f, D^r for Design$_c$ and H^f, H^r for Design$_f$. At last, we perform t tests for each time sample.

5   Performance

In this section, we study the advantages of the online approach. As mentioned in Sect. 3.2, we implemented three designs to compare their performances on required memory and time. For each issue, we will do theoretical analysis and give conclusions according to the experimental results.

5.1   Required Memory

According to the flows of three designs, $Design_c$ and $Design_f$ both need memory to store traces or histograms, while $Design_o$ does not need it. Hence, the online approach should consume less memory than the former two designs. In our design, each sampling point in the FPGA is a 16-bit unsigned integer but only 12 MSBs are valid. We assume the FPGA stores the whole 16-bit sampling values, and sends every 256 sampling values to the desktop due to its limited memory (indeed it can be much larger, but it does not matter to the conclusion), so it costs memory of at most 512 bytes to store them on the FPGA. On the desktop, each point on traces and each element of histograms are stored as 32-bit (4-byte) integers. Each mean and variance value are stored as single-precision floating points on both the desktop and the FPGA, so that each consumes 4 bytes. Thus, the required memory in three designs are described in Table 1.

Table 1. Required memory and an illustrated example of three designs

Design type	Main storage	Required memory	An example using $N_t = 1024, N_p = 1024$
$Design_c$	D^f, D^r $M_j^f, M_j^r, S_j^f, S_j^r$ 256 points	$(8N_t \cdot N_p + 16N_p)\,\mathrm{B}$ on the desktop 512 B on the FPGA	8 MB on the desktop 512 B on the FPGA
$Design_f$	H^f, H^r $M_j^f, M_j^r, S_j^f, S_j^r$ 256 points	$(8 \cdot 2^{12} \cdot N_p + 16N_p)\,\mathrm{B}$ on the desktop 512 B on the FPGA	32 MB on the desktop 512 B on the FPGA
$Design_o$	$M_j^f, M_j^r, S_j^f, S_j^r$	none on the desktop $(16N_p)\,\mathrm{B}$ on the FPGA	None on the desktop 16 KB on the FPGA

From above, we can conclude that the online approach required much less memory than compared approaches, e.g. it saves more than 99% of total required memory when $N_t = 1024$ and $N_p = 1024$. Although the required memory of $Design_o$ depends on N_p, usually it's limited in practical evaluations, because N_p is determined by the sample rate of the employed acquisition device and the running period of the encryption algorithm on the DUT. Hence, our approach fit the case when there is limited memory available for evaluations.

5.2   Required Time

Since our approach can do online processing on collected data and doesn't contain sending and pre-processing stage, it should have better performance on required time. We performed two groups of experiments and study how the required time of three designs differ with N_p and N_t, respectively. For each pair of N_p and N_t of two groups, we repeated the experiment for 10 times, recorded each runtime values of the whole running process (denoted as t^{total}) and took the average values for analysis.

We take t^{total} as the most intuitive indicator to show the performance of all three designs in Fig. 3(a) and Fig. 3(b). We can see that t^{total} of Design$_o$ grows apparently slower than other two designs over both N_t and N_p. Since the required time of collecting each point is expected to be fixed, we also recorded the required time of the collecting stage (denoted as $t^{collect}$). Then, we subtract it from t^{total} to achieve $\tilde{t}^{total} = t^{total} - t^{collect}$ and take a view of $\tilde{t}^{total}$. The results shown in Fig. 3(c) and Fig. 3(d) give a more intuitive comparison.

In addition, we employed three designs to evaluate the original power traces separately. We empirically set $N_t = 512$ (because $v \approx 2 \cdot N_t$ should be greater than 1000 [14]) for Design$_c$ and Design$_f$ and doubled it if leakage detection failed, while just ran Design$_o$ as Algorithm 2 shows. N_p is set as 1024. Finally all three designs identified leakage successfully. Although all three designs could have

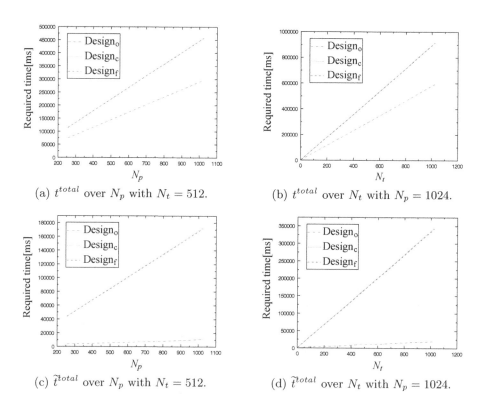

(a) t^{total} over N_p with $N_t = 512$.

(b) t^{total} over N_t with $N_p = 1024$.

(c) $\tilde{t}^{total}$ over N_p with $N_t = 512$.

(d) $\tilde{t}^{total}$ over N_t with $N_p = 1024$.

Fig. 3. The experimental results of t^{total} and $\tilde{t}^{total}$ of three designs.

succeeded at $N_t = 527$, only $Design_o$ stopped at $N_t = 527$ immediately while $Design_c$ and $Design_f$ needed to deal with 1024 power traces and as a result, they spent effort and time on excess data.

Hence, we can conclude that $Design_o$ outperforms other two designs on required time a lot. On one hand, $Design_o$ does not need the sending stage and the pre-processing stage, while in [14], Moradi et al. proposed that the low speed of communication between the PC and the DUT (or the acquisition device) is usually the bottleneck of evaluation efficiency, thus our approach can significantly economize on required time. On the other hand, when working in practical, it will stop immediately when identifying leakage, thus it can avoid spending effort on dealing with excess data.

6   Conclusion

In this paper, we focus on the advantages of online processing and analyze the shortcomings of established divide and conquer evaluation flows. Then we propose a framework of Online Leakage Assessment approach and validate it with an FPGA. It can perform online processing while collecting power traces, and it will stop instantly and give the result when it detects leakage. According to our experiments, it can lower the required memory and runtime a lot comparing to offline approaches.

Considering that the online approach will lead to impacts on the reliability of the evaluation result (see Sect. 2.2), the approach is suitable for the cases where efficient evaluations and real-time reactions are needed (for example, on-chip leakage monitoring), since it's fast and can give real-time evaluation results. In addition, the approach is lightweight and easily implemented.

Furthermore, we recommend to make the online approach cooperate with offline approaches when collecting a new trace. It first performs real-time online evaluations, then sends data collected to the computer to perform offline evaluations if the leakage is not identified. On one hand, the online approach can give real-time guidance to help offline approaches estimate the magnitude of N_t. On the other hand, if there exist fatal errors in collecting power traces, it still retains opportunities to distinguish and discard them during offline evaluations.

References

1. Bache, F., Plump, C., Güneysu, T.: Confident leakage assessment – a side-channel evaluation framework based on confidence intervals. In: 2018 Design, Automation Test in Europe Conference Exhibition (DATE), pp. 1117–1122 (2018)
2. Bache, F., Plump, C., Wloka, J., Güneysu, T., Drechsler, R.: Evaluation of (power) side-channels in cryptographic implementations. it - Inform. Technol. **61**(1), 15–28 (2019). https://www.degruyter.com/view/journals/itit/61/1/article-p15.xml
3. Becker, G.: Test vector leakage assessment (TVLA) methodology in practice. In: International Cryptographic Module Conference (2013). http://icmc-2013.org/wp/wp-content/uploads/2013/09/goodwillkenworthtestvector.pdf

4. Chan, T.F., Golub, G.H., Leveque, R.J.: Algorithms for computing the sample variance: analysis and recommendations. Am. Stat. **37**(3), 242–247 (1983)

5. Coron, J.-S., Kocher, P., Naccache, D.: Statistics and secret leakage. In: Frankel, Y. (ed.) FC 2000. LNCS, vol. 1962, pp. 157–173. Springer, Heidelberg (2001). https://doi.org/10.1007/3-540-45472-1_12

6. De Cnudde, T., Bilgin, B., Gierlichs, B., Nikov, V., Nikova, S., Rijmen, V.: Does coupling affect the security of masked implementations? In: Guilley, S. (ed.) COSADE 2017. LNCS, vol. 10348, pp. 1–18. Springer, Cham (2017). https://doi.org/10.1007/978-3-319-64647-3_1

7. De Cnudde, T., Reparaz, O., Bilgin, B., Nikova, S., Nikov, V., Rijmen, V.: Masking AES with $d+1$ shares in hardware. In: Gierlichs, B., Poschmann, A.Y. (eds.) CHES 2016. LNCS, vol. 9813, pp. 194–212. Springer, Heidelberg (2016). https://doi.org/10.1007/978-3-662-53140-2_10

8. Ding, A.A., Chen, C., Eisenbarth, T.: Simpler, faster, and more robust T-test based leakage detection. In: Standaert, F.-X., Oswald, E. (eds.) COSADE 2016. LNCS, vol. 9689, pp. 163–183. Springer, Cham (2016). https://doi.org/10.1007/978-3-319-43283-0_10

9. Ding, A.A., Zhang, L., Durvaux, F., Standaert, F.-X., Fei, Y.: Towards sound and optimal leakage detection procedure. In: Eisenbarth, T., Teglia, Y. (eds.) CARDIS 2017. LNCS, vol. 10728, pp. 105–122. Springer, Cham (2018). https://doi.org/10.1007/978-3-319-75208-2_7

10. Gilbert Goodwill, B.J., Jaffe, J., Rohatgi, P., et al.: A testing methodology for side-channel resistance validation. In: NIST non-invasive attack testing workshop (2011). http://csrc.nist.gov/news_events/non-invasive-attack-testing-workshop/papers/08_Goodwill.pdf

11. Higham, N.J.: Accuracy and Stability of Numerical Algorithms. Society for Industrial and Applied Mathematics, Philadelphia, 2nd edn. (2002). ISBN: 0898715210

12. Kocher, P., Jaffe, J., Jun, B.: Differential power analysis. In: Wiener, M. (ed.) CRYPTO 1999. LNCS, vol. 1666, pp. 388–397. Springer, Heidelberg (1999). https://doi.org/10.1007/3-540-48405-1_25

13. Reparaz, O., Gierlichs, B., Verbauwhede, I.: Fast leakage assessment. In: Fischer, W., Homma, N. (eds.) CHES 2017. LNCS, vol. 10529, pp. 387–399. Springer, Cham (2017). https://doi.org/10.1007/978-3-319-66787-4_19

14. Schneider, T., Moradi, A.: Leakage assessment methodology. In: Güneysu, T., Handschuh, H. (eds.) CHES 2015. LNCS, vol. 9293, pp. 495–513. Springer, Heidelberg (2015). https://doi.org/10.1007/978-3-662-48324-4_25

15. Tiri, K., Verbauwhede, I.: A logic level design methodology for a secure DPA resistant ASIC or FPGA implementation. In: Proceedings Design, Automation and Test in Europe Conference and Exhibition, France, vol. 1, pp. 246–251 (2004)

16. Wang, A., Zhang, Yu., Tian, W., Wang, Q., Zhang, G., Zhu, L.: Right or wrong collision rate analysis without profiling: full-automatic collision fault attack. Sci. China Inf. Sci. **61**(3), 1–11 (2017). https://doi.org/10.1007/s11432-016-0616-4

17. Xilinx: UG480-7 series FPGAs and Zynq-7000 SoC XADC dual 12-bit 1 MSPS Analog-to-Digital Converter. https://www.xilinx.com/support/documentation/user_guides/ug480_7Series_XADC.pdf

18. Xilinx: UG960-7 series FPGA AMS targeted reference design. https://www.xilinx.com/support/documentation/boards_and_kits/ams101/2013_3/ug960-7series-ams-trd-user-guide.pdf

Differential Privacy Trajectory Protection Method Based on Spatiotemporal Correlation

Kangkang Dou[1(✉)] and Jian Liu[2]

[1] Jiangsu Automation Research Institute, Lianyungang 222061, China
douk2008@163.com
[2] Vivo Mobile Communications (Shenzhen) Co, Ltd., Shenzhen 518000, China

Abstract. Location-based services provide service and convenience, while causing the leakage of track privacy. The existing trajectory privacy protection methods lack the consideration of the correlation between the noise sequence, the user's original trajectory sequence, and the published trajectory sequence. And they are susceptible to noise filtering attacks using filtering methods. In view of this problem, a differential privacy trajectory protection method based on spatiotemporal correlation is proposed in this paper. With this method, the concept of correlation function was introduced to establish the correlation constraint of release track sequence, and the least square method was used to fit the user's original track and the overall direction of noise sequence to construct noise candidate set. It ensured that the added noise sequence has spatiotemporal correlation with the user's original track sequence and release track sequence. Also, it effectively resists attackers' denoising attacks, and reduces the risk of trajectory privacy leakage. Finally, comparative experiments were carried out on the real data sets. The experimental results show that this method effectively improves the privacy protection effect and the data availability of the release track, and it also has better practicability.

Keywords: Location-Based Services · Trajectory privacy protection · Spatiotemporal correlation · Cross-correlation constraints · Noise candidate set

1 Introduction

Location-Based Services (LBS) are getting real-time acquisition of geographic location information of mobile users by mobile social networks, wireless communication technology, and intelligent sensor positioning equipment, and then provide users with the required information and services. The main services of LBS include map services, social services, and navigation services, such as Google Maps, Facebook, Weibo, Keep, etc. While users are enjoying the services provided by LBS, a large amount of trajectory data will be formed according to the changes of mobile users' real-time location, which contains different kinds of spatiotemporal information and personal privacy (such as home address, workplace, religious beliefs, living laws, social relations Wait). The problem of information protection of trajectory data has been the main issue to be solved urgently in the field of privacy protection and LBS.

© Springer Nature Singapore Pte Ltd. 2020
P. Qin et al. (Eds.): ICPCSEE 2020, CCIS 1258, pp. 151–166, 2020.
https://doi.org/10.1007/978-981-15-7984-4_12

With mobile users paying more attention to personal trajectory privacy information and the related service providers' demand for high-quality services, solving the lack of differential protection methods, poor privacy protection effects, and data availability generally existing in trajectory privacy protection methods are main jobs. In addition, keeping the dynamic balance between the trajectory privacy protection effect and data availability is also important. Experts and scholars have carried out privacy protection on sensitive location points on the trajectory, which means they only considered the situation of single-time LBS. In this way, the method achieves the effect of trajectory privacy protection. However, because of the spatiotemporal correlation in the trajectory data, even if all sensitive locations in the trajectory are anonymously processed, the attacker can still use relative knowledge and technical methods to mine data and analysis them to obtain private information.

In this paper, a differential privacy trajectory protection method based on spatiotemporal correlation is proposed, which not only uses the correlation function, considers the time cross-correlation of the published trajectory, but also uses the least square method to fit the original users' trajectory and all direction of the noise sequence to construct a noise candidate set. The method ensures the spatiotemporal correlation between noise sequence, the user original trajectory sequence, and release trajectory sequences. The method effectively prevents attackers from using filtering methods to remove noise and improves the privacy protection effect and data utilization rate.

Section 2 of this paper introduces the work related to the privacy protection of tracks. Section 3 proposes a trajectory privacy protection method based on spatiotemporal correlation and introduces it specifically. In Sect. 4, the author provides the process of experiments and the analysis of the results. Finally, the author summarizes the full works and predicts future research works.

2 Related Work

At present, most of the trajectory privacy protection methods are derived from the location privacy protection methods. As a new research hotspot, the related research results in recent years are also very rich. The existing track privacy protection methods at home and abroad can be divided into the following categories according to different encryption ideas:

(1) Generalization method

The core idea of generalization is to generalize the location and trajectory data of users and generalize the original location points or routes to regions or planes, thereby reducing the probability of attackers acquiring user information. Zhangjialei et al. Put forward an improved path k-anonymity privacy protection method, which makes use of a certain time node to constrain the time needed to search k-1 similar paths, and effectively improves the operation efficiency of the path k-anonymity privacy protection method [1]. In order to resist the track similarity attack, Jia Junjie and others proposed an (k, e) - anonymity algorithm based on the l-difference method. The algorithm uses the slope of the track to measure the sensitivity of the track, so that the slope difference value of K tracks is greater than or equal to e, ensuring that the anonymity track after generalization meets the l-difference principle [2]. In Ref. [3], Bayesian Stackelberg game is applied to track privacy protection, and the attack mode and defense formal representation of attackers are transformed into game model. A user based track privacy protection method is proposed.

(2) Inhibition method

The core idea of suppression method is to selectively publish the user's original track data and restrict the release of some sensitive information, so as to achieve the purpose of track privacy protection.

In order to balance the effect of privacy protection and data availability, researchers have proposed many feasible methods. In Ref. [4], in order to prevent the leakage of user identity information in track data, a track privacy protection method based on the inhibition method is proposed to eliminate the correlation of privacy data. In Ref. [5], a (k, c) - L trajectory privacy protection model is proposed. In the process of anonymous processing of trajectory data, sensitive points are suppressed to improve data availability.

(3) Disturbance method

The core idea of the disturbance method is to generate a certain amount of false tracks for each track in the user's original track, and then replace the user's original track with the false track to hide the real track information or the user's original trajectory data can be disturbed to generate a group of false trajectory directly, which can be mixed with the original trajectory to achieve the effect of privacy protection.

In Ref. [6], a privacy protection method of pseudo trajectory based on temporal and spatial correlation is proposed. By calculating the time accessibility of each position point in the pseudo trajectory and the overall direction of the pseudo trajectory, the temporal and spatial correlation between the pseudo trajectory and the original trajectory is ensured. In Ref. [7], a data publishing algorithm is proposed to add bounded noise to the user's original track, which can effectively improve the effect of track privacy protection. Dong Yulan and others proposed a path privacy protection method to meet user-defined requirements. Considering the leakage probability of each location point in the original track, the similarity degree of the track and the frequency of service request, they generated a set of false tracks to protect the track privacy [8].

(4) Differential privacy protection method

The main idea of differential privacy protection method is to add random noise to the user's original trajectory to achieve the effect of privacy protection. It is a strong privacy protection method. Because it has strict mathematical definition and ignores the background knowledge of the attacker, in recent years, there are many related researches and widely used.

In Ref. [9], combines the statistical principle with differential privacy, adds Laplacian noise of different degrees to the original track according to different differential privacy budget, and improves the privacy protection effect and data availability. In Ref. [10], a post mapping plane Laplacian mechanism is proposed, which transforms the problem of location publishing into the problem of finding the optimal mapping location point, effectively reducing the information loss. Sun Kui et al. Put forward an optimized differential privacy data publishing method [11], which generalizes the original trajectory data, adopts the exponential mechanism to select the optimization scheme, and carries on the specialized iteration to it, then uses the decision tree algorithm to divide the equivalence classes, and finally adds the disturbing noise to the original trajectory to form the final publishing data. In Ref. [12], a differential privacy trajectory protection method based on spatiotemporal correlation is proposed to dynamically calculate the privacy level of each location point in the trajectory data, and then calculate the probability of privacy leakage caused by the data release of each location point to other location points according to the Markov probability matrix, so as to ensure that the final released data meet the spatiotemporal correlation. In Ref. [13], in order to make the noise sequence and the user's original track sequence have correlation, the idea of autocorrelation function is introduced into the differential privacy protection method to ensure that the autocorrelation function of the noise sequence and the original track sequence are the same, and effectively resist the attacker's attacks such as filtering.

The above four track privacy protection methods are compared in terms of privacy protection effect, data availability and method operation efficiency, and the results are shown in Table 1.

Table 1. Comparison of advantages and disadvantages of four ways to protect track privacy.

Methods of track privacy protection	Advantages of the method	Disadvantages of the method
Generalization method	Simple implementation and high data availability	Low efficiency in method operation
Inhibition method	Good privacy protection	Poor data availability
Disturbance method	Simple implementation and high efficiency of method operation	Poor data availability
Differential privacy protection method	Good privacy protection	High difficulty in implementation, low efficiency in method operation, poor data availability

3   Differential Privacy Trajectory Protection Method Based on Spatiotemporal Correlation

In the trajectory privacy protection method proposed in this paper, by setting the power spectral density of the filter and the impulse response, the Laplace noise transformed from the Gaussian white noise is added to the real user trajectory. Noise candidate set constraints ensure that the noise sequence and the user's original trajectory sequence and the final release trajectory sequence meet the constraints of spatiotemporal correlation, and provide high-intensity trajectory privacy protection.

3.1 Constraints on Cross-Correlation of Published Trajectory Sequences

In order to solve the problem that the release trajectory sequence does not meet the timing correlation, this paper introduces the concept of correlation function and uses the function idea to describe the timing correlation between trajectories.

Definition 1. Related functions: For signals $x(s)$, $y(t)$. The correlation function indicates the degree of correlation between the values of s and t at any two different times. The similarity between two signals is measured by correlation coefficient. Expressed as:

$$\rho_{xy} = \frac{COV(X,Y)}{\sqrt{D(X)}\sqrt{D(Y)}} = \frac{\sigma_{xy}}{\sigma_x \sigma_y} \tag{1}$$

Definition 2. Autocorrelation function: Autocorrelation function: The autocorrelation function is the cross-correlation of the signal itself, indicating the dependence between t1 and t2 of the same sequence at different times, that is, describing the value of the random signal x(t) between time t and time $(t + \tau)$ Relevance. Expressed as:

$$R^x_{(t_1,t_2)} = E[x(t_1), x(t_2)] \tag{2}$$

The autocorrelation of continuous functions can be expressed as:

$$R^x_{(t_1,t_2)} = x(\tau) * x^*(-\tau) = \int_{-\infty}^{+\infty} x(t+\tau)x^*(t)dt = \int_{-\infty}^{+\infty} x(t)x^*(t-\tau)dt \tag{3}$$

Definition 3. Cross-correlation function: Auto-correlation is a special case of cross-correlation. The cross-correlation function represents the dependence between different sequences at different times t1 and t2, that is, describes the degree of correlation between the random signals x (t) and y (t) at time t and $(t + \tau)$. Expressed as:

$$R^{(x,y)}_{(t_1,t_2)} = E[x(t_1), y(t_2)] \tag{4}$$

The cross-correlation of continuous functions can be expressed as:

$$R^{(x,y)}_{(t_1,t_2)} = \int_{-\infty}^{+\infty} x^*(t)y(t+\tau)dt \tag{5}$$

Through the concepts of autocorrelation function and cross-correlation function mentioned above, specific definitions can be given to the temporal correlation.

Definition 4. Timing correlation: If the differential privacy protection method is adopted, the autocorrelation function RX' (τ) of the noise sequence added to the user trajectory is equal to the autocorrelation function RX (τ) of the user's original trajectory sequence, i.e. RX' (τ) = RX (τ), it means that the added noise sequence and the user's original trajectory have timing correlation, that is, when the noise sequence is added to the user's original trajectory to form a release trajectory sequence, the release trajectory sequence is also time-dependent. In theory, an attacker cannot identify the user's original trajectory sequence by performing noise removal attacks from the perspective of timing correlation.

In the signal system, the input of independent signals and the processing of the filter can form a non-independent signal sequence. Through the definition of the autocorrelation function, the consistency of the user's original trajectory sequence and the added noise sequence at different times can be defined by the autocorrelation function, but the new sequence formed by the two after being processed by the filter does not necessarily meet the timing correlation Sex. Literature [14] proved that if the cross-correlation function of the release trajectory sequence formed by the filter processing at any two moments is the superposition of the auto-correlation functions of the original trajectory sequence and the noise sequence at these two moments, then the release trajectory sequence with temporal cross-correlation.

Definition 5. The cross-correlation constraint of the release trajectory sequence: add a noise sequence to the user's original trajectory sequence, and form a release trajectory sequence after being processed by the filter. If the cross-correlation function of the release trajectory sequence at any two consecutive times is the sum of the autocorrelation functions of the user's original trajectory sequence and the added noise sequence at these two times, it means that the release trajectory sequence has time series correlation.

The cross-correlation constraints of the published trajectory sequence can be expressed as:

$$R^{(Z_i, Z_{i+1})}_{(t_i, t_{i+1})} = R^{X}_{(t_i, t_{i+1})} + R^{Y}_{(t_i, t_{i+1})} \tag{6}$$

Where Z is the post-trajectory sequence formed after processing, X is the original user trajectory sequence, and Y is the added noise sequence.

3.2   Noise Candidate Set

The cross-correlation constraint of the published trajectory sequence can ensure that the user's trajectory privacy will not be leaked by the attacker through the removal of noise in the time series, but the added noise still exists in the geographical space as a logical location point or not. The possibility of reaching a point. In order to satisfy the spatial correlation between the user trajectory sequence and the added noise sequence, this paper proposes the concept of noise candidate set, that is, if the noise added at the sensitive position of the user's original trajectory is in the noise candidate set, the noise is in the actual geographic location Is the reachable point, when all the noise meets this condition, it can be explained that the added noise sequence satisfies the spatial correlation, and then the overall direction of the noise sequence and the original trajectory is checked. If the overall direction of the two meets the requirements, the combined The release trajectory sequence cannot be removed by the attacker from the perspective of spatial correlation, resulting in privacy leakage.

The noise candidate set proposed in this paper uses the least squares method to fit the original direction of the user and the overall direction of the noise sequence, so as to ensure that the added noise sequence and the overall direction of the user's original trajectory are similar, so that the attacker cannot noise the noise sequence through the direction of movement filter. The direction of movement of the user's original trajectory can be expressed as slope:

$$l = \hat{b} = \frac{\sum_{i=1}^{n} x_i y_i - n\bar{x}\bar{y}}{\sum_{i=1}^{n} x_i^2 - n\bar{x}^2} = \frac{\sum_{i=1}^{n} (x_i - \bar{x})(y_i - \bar{y})}{\sum_{i=1}^{n} (x_i - \bar{x})^2} \tag{7}$$

Among them, $\bar{x} = \frac{1}{n}\sum_{i=1}^{n} x_i, \bar{y} = \frac{1}{n}\sum_{i=1}^{n} y_i$.

In order to ensure that the overall direction of the noise sequence is similar to the overall direction of the user's original trajectory, the start and end positions of the noise sequence need to be determined first, so the starting point and the end point of the user's original trajectory are used as sensitive points to generate a noise candidate set StartSet = {start1, start2,..., startn} and EndSet = {end1, end2,..., endn}, and select the starting and ending noise position points respectively, so that the slope l2 of the overall direction of the noise sequence is close to the slope l1 of the original direction of the original trajectory and the noise position Meet differential privacy. In addition, in order to ensure that any two adjacent position points in the noise sequence can be reached within the release time interval and the movement speed of the noise sequence and the original trajectory are similar, it is necessary to ensure that the start and end noise position points are reachable within the time interval, so the noise starting and ending points should also meet:

$$\begin{cases} dis = \left| dis\langle start^d, end^d \rangle - dis\langle start^{real}, end^{real} \rangle \right| \leq \delta_{dis\_all} \\ time = \left| time\langle start^d, end^d \rangle - time\langle start^{real}, end^{real} \rangle \right| \leq \delta_{t\_all} \end{cases} \tag{8}$$

Among them, $dis\langle start^d, end^d \rangle$ Point from the starting position $start^d$ to end point end^d moving distance, δ_{dis_all} threshold to limit the distance traveled, $time\langle start^d, end^d \rangle$ represents the time required to move from the starting position point $start^d$ to the end position point end^d, and δ_{t_all} 为 is the threshold value to limit the moving time.

After determining the starting and ending noise position points, it is necessary to add noise to each sensitive position point of the original trajectory one by one to form a noise sequence, so the noise candidate set of each position point needs to be determined. When generating the i-th ($2 \le i \le n-1$) position of the noise sequence, the Manhattan distance r + random (random is a random number) between the i−1 and i-th position of the user's original trajectory for the radius, make the circle with the i−1 position of the noise sequence as the center of the circle. On both sides of the line where l2 is located, a position point is selected as a noise candidate position point every interval $\theta°$ to form a noise candidate set until the angle between the grid edge and the line where l2 lies reaches the threshold. Then, randomly select the noise position in the noise candidate set, which can not only ensure the randomness of the added noise position, but also ensure that the noise position in the middle of the noise sequence is close to the whole, and will not cause leakage due to the position of individual points away from the whole.

The schematic diagram of noise candidate set is shown in Fig. 1. Among them, the shaded part is the noise candidate set, where $r = \sqrt{(x_i - x_{i-1})^2 + (y_i - y_{i-1})^2}$, $\theta = \arccos\left(1 - \frac{d^2}{r^2}\right)$, d is the side length of the grid.

When constructing the added noise sequence, you can take points from the shadow area where the sensitive position points in the user's original trajectory are located, and anonymize the sensitive position points to form a noise sequence, so that the noise sequence and the user's original trajectory sequence and release the trajectory sequence satisfies the spatial correlation.

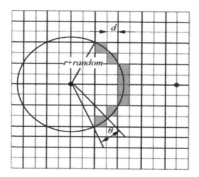

Fig. 1. Noise candidate set

After constructing the noise sequence according to the above steps, the slope 12 of the overall moving direction and the time reachability and moving distance of any two adjacent position points need to be compared and checked. For the specific check method, see formula (6)–(8), where the threshold is preset by the system. If the two comparison conditions are met, it means that the noise sequence has a spatial correlation with the user's original trajectory sequence. If any of the conditions are not met, the noise sequence needs to be regenerated.

3.3 Differential Privacy Trajectory Protection Method Based on Spatiotemporal Correlation

This paper combines the cross-correlation constraints of published trajectory sequences and the limitations of noise candidate sets, and proposes a differential privacy trajectory protection method based on spatiotemporal correlation, it is used to solve the problem of poor spatial-temporal correlation between the user's original trajectory, noise sequence and published trajectory. If any noise location point of the noise sequence is in the noise candidate set and the location point of any two consecutive moments of the release trajectory sequence satisfies the cross-correlation constraint, it can be guaranteed that the release trajectory sequence and the user's original trajectory sequence and noise sequence satisfy the time and space Correlation makes it impossible for an attacker to remove noise through the space-time relationship, thereby achieving the security of user trajectory privacy.

According to Definition 5, if at any two consecutive adjacent times, the original trajectory sequence and the noise sequence satisfying the auto-correlation are superimposed, the resulting release trajectory sequence satisfies the cross-correlation constraint. In this paper, on the basis of literature [15], four groups of white Gaussian noise X1, X2, X3, X4 (λ = 1, 2, 3, 4), through $Lap(2\ \lambda^2)$ Laplace transform, Get a set of Laplace noise sequences, $X = X_1^2 + X_2^2 - X_3^2 - X_4^2$. According to the literature [16], if the autocorrelation function of the noise sequence $R_N(\tau)$ Autocorrelation function with user's original trajectory sequence $R_O(\tau)$ satisfy relationship $R_O(\tau) = 8R_N(\tau)^2$, then the noise sequence and the original trajectory sequence satisfy the timing correlation. Therefore, on the basis of the above, this paper generates a Laplacian noise sequence that satisfies the autocorrelation constraint and is in the noise candidate set, and then adds the noise sequence to the original trajectory sequence according to the cross-correlation constraint condition to form a spatiotemporal correlation. The release trajectory sequence plays the role of privacy protection.

In the implementation of the algorithm, the user's original trajectory sequence can be divided into shorter time and frequency stable sequences at a certain time interval, and then the signal processing, filtering method and other methods are used to disturb and add noise, and finally form a space-time correlation Publish the trajectory sequence. The specific execution process is shown PPSC Algorithm.

Algorithm: Differential privacy trajectory protection method based on spatiotemporal correlation PPSC($O,\varepsilon_i,T,traj_{\text{noise}}$)

Input: User original trajectory sequence O, differential privacy budget parameter set ε, time
interval T, noise candidate set $traj_{\text{noise}}$

Output: Publish trajectory sequence R

1. $\forall \varepsilon_i \in \varepsilon$

2. $\exists X_j \in X(j = 1,2,3,4)$, according to $X_j \sim N(0, \sqrt{2\lambda^2})$, generate Gaussian white noise

$$L(D) = f(D) + Laplace(\frac{\Delta f}{\varepsilon_i}) \ Pr[x] = \frac{1}{2\lambda} exp^{\frac{|x-\mu|}{\lambda}}$$

//According to differential privacy budget parameters ε_i generate 4 groups of white
Gaussian noise X1, X2, X3, X4

3. IF $X_j \in traj_{\text{noise}}$ THEN CONTINUE; //Determine whether it belongs to the noise can-

didate set

4. $R_O(\tau) = calculateR(O, \tau)$; // Calculate the autocorrelation function of the user's origi-

nal trajectory, formula (5)

5. $R_N(\tau) = \sqrt{\frac{R_O(\tau)}{8}}$ // Calculate the autocorrelation function of the noise sequence to be
added

6. According to the autocorrelation function of the noise sequence $NR_N(\tau)$, get X_j gaussian white noise converted to new correlation $X_j'(j=1,2,3,4)$, $X_j' \sim N(0,2\lambda^2)$// The filter is $h(\tau) = \sqrt{\frac{R_O(\tau)}{16\pi N}}$

7. $X = X_1^2 + X_2^2 - X_3^2 - X_4^2$ //Synthetic Laplace noise sequence

8. $R=O+X$ //According to the cross-correlation constraints, the noise sequence and the

user's original trajectory are superimposed to generate a release trajectory sequence R

9. RETURN R

---

4 Experiment and Result Analysis

4.1 Experimental Data and Operating Environment

This article uses Geolife and Gowalla two real data sets for extraction as experimental
data. The Geolife data set contains 17621 tracks, which is composed of the real track
data of 182 users from April 2007 to August 2012 in Beijing. A total of 4 attributes for
coordinate latitude extraction; The Gowalla data set contains the check-in data of
15116 users who checked in using mobile social applications in California from

February 2009 to October 2010. Similarly, this article extracted user numbers, time stamps, longitude coordinates, A total of 4 attributes of coordinate latitude are used as the experimental data set.

Hardware: Intel core i5-8300H processor, 16 G memory, graphics card NVIDIA GTX1050ti.

Experimental Design and Indicators

In the experiment, set the differential privacy budget parameter $\varepsilon \in [0.1,1]$, the step size is 0.1; the differential privacy protection model parameter $\zeta \in [0.1,1]$, the step size is 0.1; And randomly set the privacy level $S \in [0,1]$ of the preset sensitive position set, with a step size of 0.1.

In order to compare and verify the effectiveness of the method proposed in this paper, the differential privacy trajectory protection method PPSC based on spatiotemporal correlation was compared with the method proposed in reference 14 (abbreviated DP-UR) and the method proposed in Ref. [17] (abbreviated P-refix) Comparative Experiment.

Regarding the trajectory privacy protection method, the main evaluation indicators considered in this article are three, namely: privacy protection effect, data availability and operation efficiency.

4.2   Experimental Results and Analysis

Results and analysis of differential privacy trajectory protection method based on spatiotemporal correlation.

(1) Privacy protection effect

Figure 2(a) and (b) show the impact of differential privacy budget parameters on privacy protection on two data sets. The analysis shows that the larger the differential privacy budget parameter, the lower the interference level to the user's original trajectory, so the privacy protection effect gradually decreases. Among the three methods, P-refix has the worst privacy protection effect, because the P-refix method adds random noise to the user's original trajectory according to the differential privacy budget parameter, and the amount of noise will directly affect the quality of privacy protection. Because this method does not consider the spatiotemporal correlation between the user's original trajectory and the noise sequence, it is easy for an attacker to filter the noise to recognize the user's real trajectory, so the privacy protection effect of this method is poor. The DP-UR method and the PPSC method both consider the timing correlation between the original user trajectory and the noise sequence in the trajectory privacy protection, so the privacy protection effect is better. It can be seen that with the change of the differential privacy budget parameter ε, the privacy protection effect of the PPSC method can reach up to about 95%, and the worst can reach about 70%. Under the same differential privacy budget parameter ε conditions, the PPSC method has an average privacy protection effect higher than 3% compared to the DP-UR method; the PPSC method has an average privacy protection effect higher than 10% higher than the P-refix method In contrast, in terms of privacy protection, the PPSC method proposed in this paper has significantly improved.

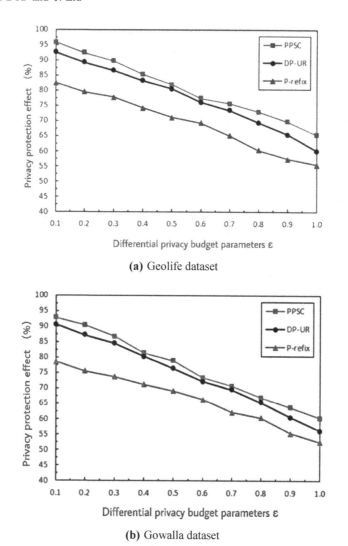

(a) Geolife dataset

(b) Gowalla dataset

Fig. 2. The impact of differential privacy budget parameters on privacy protection

(2) Data availability

Figure 3(a) and (b) respectively reflect the impact of differential privacy budget parameters on data availability on the two data sets. It can be seen that under the same differential privacy budget parameters ε, the data availability of the PPSC method is higher. This is because the P-refix method adopts the method of adding random noise to the user's original trajectory, and the amount of added noise is uncontrollable, so it will cause a large amount of information loss in the original trajectory and low data availability. Although the DP-UR method considers adding different noises according to the different privacy levels of different location points on the trajectory, when calculating the privacy level of different location points, only the distance between the location points is considered, and the user's travel method is not used. Taking into account, the definition of privacy level is not precise enough. The PPSC method proposed in this paper comprehensively considers the Manhattan distance between different locations on the original trajectory and the way users travel, and obtains the privacy level of the connected sensitive locations, and then adds reasonable and differential noise to disrupt. So the information loss of the user's original track is less, and the data availability of the release track provided to third parties is higher. Similarly, the results of the three methods on the Geolife dataset are better than the Gowalla dataset.

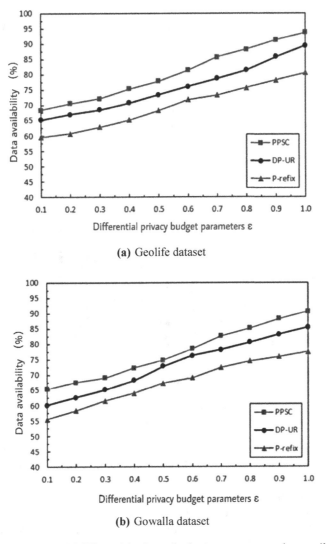

(a) Geolife dataset

(b) Gowalla dataset

Fig. 3. The impact of differential privacy budget parameters on data availability

(3) Operating efficiency

Figure 4 shows the effect of the size of the Geolife dataset on the efficiency of the method. It can be seen from the figure that the running time of the three trajectory privacy protection methods increases with the increase of the data set, showing an effect of approximately linear growth. This is because as the data set increases, the time that the machine will process the data will increase. Compared with the three methods, the running time of the PPSC method proposed in this paper is slightly higher than the other two methods, that is, the operating efficiency is slightly lower than the other two methods. This is because when the PPSC method protects the original trajectory of the user, it first calculates the privacy level of all the location points passed by the user trajectory, which not only provides privacy protection for the preset sensitive location set, but also provides privacy protection for the connected sensitive location points. And calculate the personalized differential privacy budget parameters ε according to the different privacy levels, and then add the different Laplace disturbance noise to the original trajectory. When adding noise, the timing correlation between the user's original trajectory and the noise sequence also needs to be considered, so that the noise sequence satisfies the noise candidate set limit and the published sequence satisfies the cross correlation constraint. Therefore, the PPSC method has more calculations and restrictions in the process of trajectory privacy protection, and the method running time is higher than the other two methods, that is, the method running efficiency is lower.

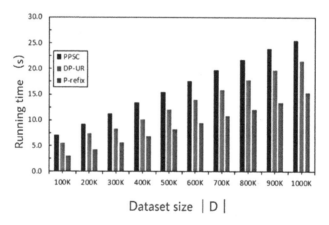

Fig. 4. Effect of data set size | D | on running time

According to the data analysis of the experimental results, it can be seen that with the change of the data set size | D |, the longest running time of the PPSC method takes about 25 s and the shortest takes about 7.5 s. Under the condition of the same data set size | D |, the PPSC method is compared with the DP-UR method, the average running time is about 2 s longer; compared with the P-refix method, the PPSC method has an average running time about 5 s longer. In practical experience, the reduction in the operating efficiency of the PPSC method will not cause a large impact on users.

5   Conclusion

In the context of location-based services, this paper aims at the problems that exist in the current track privacy protection methods that are easy to be cracked by inference attacks, and studies the existing track privacy protection model from the perspective of time-space correlation, and proposes a track privacy protection method with better privacy protection effect and higher data availability. The main ideas of this paper include putting forward the idea of noise candidate set, and the combination with the constraint mechanism of publishing track cross-correlation, proposing a differential privacy track protection method based on the time-space correlation. It ensure that the added noise sequence, the user's original track sequence, and the final distribution track sequence are consistent in time-series correlation, and spatial correlation and logic indistinguishability further improve the effect of privacy protection. The PPSC algorithm needs a certain amount of calculation when using the time constraint and spatial correlation of trajectory to build noise candidate set. In the face of the arrival of big data era and the protection of mass trajectory privacy data, how to ensure the effect of privacy protection, on the premise of enhancing the practicability of the method is an important direction of follow-up research.

References

1. Zhang, J., Zhong, C., Fang, B., Ding, J., Jia, Y.: An improvement of track privacy protection method based on k-anonymity. Intell. Comput. Appl. 9(05), 250–252+256 (2019)
2. Jia, J., Huang, H.: A trajectory (k, e)-anonymous algorithm against trajectory similarity attacks. Comput. Eng. Sci. 41(05), 828–834 (2019)
3. Shen, H., Bai, G.: Protecting trajectory privacy: a user-centric analysis. J. Netw. Comput. Appl. 82, 128–139 (2017)
4. Chen, R.: Privacy-preserving trajectory data publishing by local suppression. Inf. Sci. 232(3), 83–97 (2013)
5. Sui, P., Li, X.: A privacy-preserving approach for multimodal transaction data integrated analysis. Neurocomputing 253, 56–64 (2017)
6. Lei, K., Li, X.: Dummy trajectory privacy protection scheme for trajectory publishing based on the spatiotemporal correlation. J. Commun. 37(12), 156–164 (2016)
7. Li, M., Zhu, L.: Achieving differential privacy of trajectory data publishing in participatory sensing. Inf. Sci. 400(8), 1–13 (2017)
8. Dong, Y., Pi, D.: Novel trajectory privacy preserving mechanism based on dummies. Comput. Sci. 44(8), 124–128 (2017)
9. Zhu, W., You, Q., Yang, W.: Trajectory privacy preserving based on statistical differential privacy. J. Comput. Res. Dev. 54(12), 2825–2832 (2017)
10. Hou, Y.: Research on trajectory information protection mechanism and its application based on differentiated privacy. Chongqing University of Posts and Telecommunications (2019)
11. Sun, K., Zhang, Z., Zhao, C.: An enhanced differential privacy data release algorithm. Comput. Eng. 43(4), 160–165 (2017)

12. Wu, Y., Chen, H., Zhao, S.: Differentially private trajectory protection based on spatial and temporal correlation. Chin. J. Comput. **41**(2), 311–324 (2018)

13. Wang, H., Xu, Z., Xiong, L.: CLM: differential privacy protection method for trajectory publishing. J. Commun. **38**(6), 1–12 (2017)

14. Hu, Z.: Research on personalized trajectory privacy protection method based on location service. Harbin Engineering University (2019)

15. Cho, E., Myers, S.A., Leskovec, J.: Friendship and mobility: user movement in location-based social networks. In: 17th ACM SIGKDD International Conference on Knowledge Discovery and Data Mining, pp. 1082–1090. ACM, New York (2011)

16. Chen, R., Fung, B.C.M.: Differentially private transit data publication: a case study on the montreal transportation system. In: 18th ACM SIGKDD International Conference on Knowledge Discovery and Data Mining, pp. 213–221. ACM, New York (2012)

A New Lightweight Database Encryption and Security Scheme for Internet-of-Things

Jishun Liu[1], Yan Zhang[1], Zhengda Zhou[2(✉)], and Huabin Tang[2]

[1] The Third Research Institute of Public Security of Ministry of China, Shanghai, China
{liujs, zhangyan}@mctc.org.cn
[2] Zhejiang ZhiBei Tech Co. Ltd, Hangzhou, Zhejiang, China
{zhouzd, tanghb}@zhibeitech.com

Abstract. Internet-of-Things (IoT) extends the power of Internet and bring tremendous opportunity to academia and industry. However the security and data privacy challenges become major obstacles for its adoption and deployment. To address these issues, an encryption and security scheme is proposed for a lightweight database which is suitable for embedded systems with limited storage and computing resources. The scheme encrypts data are in both storage and used memory. So it can prevent sensitive data leakage from untrusted applications, zero-day-vulnerability and malicious attacks for the lightweight database. The prototype of the proposed scheme was presented and the feasibility and effectiveness was evaluated. The experimental results demonstrate the scheme is practical and effective.

Keywords: Internet of Things · Data encryption · Lightweight database

1 Introduction

Internet of Things (IoT) is a technology trend that makes more and more physical object and mini-devices connect to the internet and create many new opportunities which help people live and work smarter [1]. The connected physical objects and mini-devices in IoT bridge the gap between physical and digital world to improve the quality and productivity of people's life, society and industries. Most IoT applications successfully integrate data-driven insights into many application scenarios and make utilities and public services more efficient and convenient. Meanwhile, the security and data privacy issues become major obstacles for in the adoption and deployment of IoT.

Because IoT is a nascent market, many pioneers in this area put much efforts in getting the products and solutions to market quickly, rather than in taking sufficient considerations to the security and data privacy issues. With the increasing involvement of IoT in people's lives and industries, concerns of potential security breaches are raised. In 2016, the European Union approved a new data privacy law called the General Data Protection Regulation (GDPR), which is considered the world's most stringent data protection law. The Cybersecurity Law of the People's Republic of China officially came into force on June 1, 2017,which serves as a "Basic Law" for cybersecurity legislation.

The original version of this chapter was revised: The figures in Table 1. on p. 173 have been corrected. The correction to this chapter is available at https://doi.org/10.1007/978-981-15-7984-4_47

© Springer Nature Singapore Pte Ltd. 2020, corrected publication 2020
P. Qin et al. (Eds.): ICPCSEE 2020, CCIS 1258, pp. 167–175, 2020.
https://doi.org/10.1007/978-981-15-7984-4_13

Therefore security and data privacy are the critical research issues which have received a great deal of attentions, and the open issues which require more efforts [2, 3].

Since the IoT involved a tremendous number of end devices and sensors which are designed and implemented by resource-constrained embedded systems, the solutions and techniques for data protection must be considered under its limited capabilities in the terms of computing power, memory, storage and bandwidth [4]. We proposed a practical encryption and security scheme for lightweight databases on the embedded system, aimed at protecting data from hacking and misusing [5, 6].

The remainder of this paper is organized as follows. The next section discusses further motivation for our research and provides some background introductions. Section 3 introduces our proposal to address data protection on the embedded system for IoT applications. Section 4 introduces the design and implementation of our scheme and discusses the experimental results. Finally, Sect. 5 presents the conclusions and outlines the directions for future research.

2 Relate Works

The security and data privacy in the field of Internet of Things (IoT) is a critical issue for IoT applications to make a tremendous number of end devices and sensors connected and share information with each other. As IoT devices and sensors are normally implemented with embedded systems having limited computing and memory capacity, it is a challenge to prevent IoT applications from data breaches and being targeted for cyber-attacks. The cryptographic techniques are considered the solutions to secure IoT applications. In [7], a framework is introduced for the benchmarking of lightweight block ciphers on a multitude of embedded platforms. [8] presents existing state-of-the-art advances reported on the cryptographic solutions for industrial IoT. [9] present information security related challenges in IoT application such as object Identification, object authentication, data privacy, lightweight cryptosystems, security protocol, etc.

Databases have a very important role to play in handling IoT data adequately and securely, and the various threats pose challenges to the database security in terms of confidentiality, integrity and access control [10]. [11] presented a practical encrypted relational DBMS, which proposed an SQL-aware encryption strategy that maps SQL operations to encryption schemes, an adjustable query-based encryption method and an onion encryption method which can adjust encryption level of each data item based on user queries. In [12], an end-to-end encrypted database and an end-to-end encrypted query protocol are proposed to enable clients to operate on (search, sort, query, and share) encrypted data without exposing encryption keys or cleartext data to the database server. [13] proposed a practical and functionally rich database system that encrypts the data only with semantically secure encryption schemes. [14] proposed a database called EnclaveDB, which uses trusted execution environments such as SGX enclaves to guarantee confidentiality and integrity with low overhead. The proposed database utilized a small trusted computing base to implement the in-memory storage and query engine, the transaction manager and the pre-compiled stored procedures. In addition, the proposed database employed an efficient protocol for checking integrity and freshness of the database log, which supported concurrent, asynchronous appends and truncation.

Some related works focused on the query execution over encrypted data in database on the cloud servers or DAAS (Data as a service). [15] proposed a parallel query execution methodology using multithreading technique up to 6 threads to improve the performance of the data encryption. [16] proposed a secure and scalable scheme which utilized hierarchical cubes to encode multi-dimensional data records and constructed a secure tree index on top of such encoding to achieve sublinear query time. [17] proposed a privacy-preserving framework without online trusted third parties, which employed a grid-based location protection method and an efficient task assignment algorithm.

Most related works focused on databases running on server, and lightweight databases running on IoT embedded system are often only deployed OS (Operation System, such as Linux) hardening mechanisms to improve data security.

3   A New Encryption and Security Scheme

Considering the limitations of the computing power and memory of IoT devices, we proposed a new encryption and security scheme for lightweight database which can improve data security on embedded system platforms. The proposed scheme is decoupled from any specific database system, and we implemented the prototype for SQLite. SQLite is a widely-used lightweight embedded SQL database engine in IoT applications. The following parts will present the proposed scheme in detail.

3.1   Threats and Security Requirements

The threats to lightweight databases in IoT applications are different from the major streams databases, such as Oracle, SQL Server and MySQL, which are designed to run on the backend servers. As adopted for distributed deployment normally, the lightweight databases security solutions are commonly involved with numerous distributed embedded devices, instead of one or a cluster of servers in the local server rooms or data centers. Moreover, the embedded platforms normally have too limited computational and memory resources to employment the regular information security schemes, furthermore, the framework and architecture of the embedded platforms are various, which makes the lightweight database security issue more complex. Considering the differences, the proposed scheme adopted a lightweight solution to realize database security with low system overhead. The implementation of the scheme can be decoupled from the specific operation systems and databases and support cross-platform deployments.

The most common database threats include excessive privileges, database injection attacks, storage media exposure, exploitation of vulnerabilities, unmanaged sensitive data, etc. The proposed scheme addresses all these threats to lightweight database in IoT applications.

Firstly, though many proposed cryptosystems and security protocols are considered secure and robust, they may not be suitable for embedded devices. Furthermore, physical security of the embedded devices cannot be guaranteed secured in many practical scenarios, which make the sensitive data protection more difficult. Secondly, due to constrained resources on embedded devices, lightweight databases ignored the authentication procedure, or applied weak authentication schemes. Thus, the absent or weak authentication vulnerability causes malicious users or devices to access database and to steal, delete and tamper data by posing a legal user or device. Thirdly, because of the variety of embedded devices, the number of the hardware and software vulnerabilities are more than the traditional databases. The traditional security products and mechanisms such as IPS, antivirus are not applicable in many IoT application scenarios. Thus, it is a challenge to prevent the various vulnerabilities exposing to hackers or malicious users.

3.2   Techniques and Contributions

The proposed scheme considers the architecture of an IoT application that stores sensitive data in the lightweight databases, which require to ensure confidentiality and integrity of data with constrained resources. Figure 1 illustrates the typical IoT architecture and the threat model for the proposed scheme. The end devices are the embedded platforms that integrated sensors and actuators, which are able to sense the environment, collect and preprocess the data, and passed the data to the application servers on cloud or backend systems. The end devices always locates on untrusted environment and runs the risk of malware and vulnerability exploitations, thus the sensitive data stored in lightweight databases is in danger of data breaches and tampering. In addition, the application servers gather, store, process and analyze massive volumes of data by data analytics engines and machine learning mechanisms, and provide the data services to clients. Thus, the attacker are able to steal the data form the end devices assume the identity of legitimate legal application server.

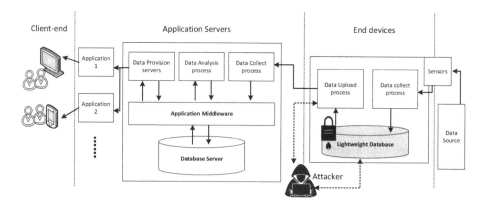

Fig. 1. Overview of IoT architecture and the threat model. The proposed scheme targets attackers to the lightweight database on the end devices. The shaded part depicts the protected databased and the lock depicts the encrypted sensitive data.

As shown in Fig. 1, the goal of the proposed scheme is to ensure the confidentiality and integrity of data in an lightweight database on the end devices of IoT applications. The two types of attacker are considered in the threat model, the servers-side attacker and the devices-side attacker. The server-side attacker assumes the identity of legitimate legal application server by reverse analysis of service API, communication protocol and authentication mechanism, in order to get access to the end devices. The device-side attacker scans and exploits the vulnerabilities of the end devices and gets the access to the sensitive data.

Considering the two types of attacker in the threat model, the proposed scheme provided decoupled data encryption and hybrid authentication for lightweight databases to meet the security requirements in the practical IoT application scenarios. We employed encryption techniques to provide the data confidentiality and integrity in both storage and in used memory. The scheme can be easy to implement for diversified hardware platforms and customized operating systems. Considering the constrained resources, the scheme support various cryptographic algorithms, especially lightweight cryptographic algorithms. Furthermore, the scheme provided a hybrid authentication mechanism, which supports machine-to-human and machine-to-machine interactions interfaces.

3.3   Decoupled Data Encryption

In order to ensure data confidentiality and integrity, we proposed a decoupled data encryption scheme for lightweight databases, which can realize the data encryption regardless of operation systems and databases. The data encryption mechanism is implemented by modifying the codes in the standard JDBC driver JAR file, which provides a complete set of interfaces that allows for IoT applications to access databases. As the application programs normally interact with databases through JDBC driver, data encryption mechanism can be realized in a transparent way, avoiding both the code changes for application programs and database engines.

As shown in Fig. 2, the modified JDBC driver is deployed between the application and database, where all the sensitive data can be encrypted and decrypted automatically and transparently to applications. In this way, the sensitive data in the database is always encrypted, in both the database's storage and memory. Meanwhile, the modified JDBC driver also provides Integrity and legality checking, which can help detect the data tampering and code injections.

Figure 2 illustrates the typical data processing flow between the application program and the lightweight database engine. In step 1 and 2, the application program sends the access request to the database via the JDBC driver, and our modified code is triggered to verify the identification, to check the integrity of the target data and to complete the initial work for data encryption. Section 3.3 in this paper will present more details of authentication mechanism. The step 3–5 is shown the typical data access flow. In the step 3, a query is sent by application program, and the modified JDBC driver parses the SQL command and complete the related preprocess. For example, if the SQL command key words such as "select" is found, the driver can gain the targeted field data of the database and initialize the decryption operations. And if the SQL command key words such as "update" or "insert" is found, the driver can gain

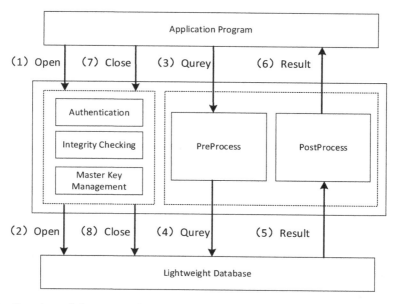

Fig. 2. Overview of the proposed scheme. Steps (1)–(8) illustrate the typical process flow including open (connect), sent queries, receiving results and closing (unconnected).

the targeted field data needed to encrypt and complete encryption operation. After preprocessing, the modified JDBC driver forwards the SQL commands to the database in the step 4 and receives the returned result from the database in the step 5. Then, the modified JDBC driver parses returned data and completes postprocess. For example, if the returns of the "select" query is found, the driver can gain the targeted field data of database and complete decryption operation according to security configuration.

Because the modified JDBC driver is compatible with the most cryptographic algorithms, the regular lightweight cryptographic algorithms [7] can integrate into the scheme easily. Additionally, because of the datatype restrictions from the databases or applications, format-preserving encryption (FPE) [18, 19] is introduced to our scheme.

3.4   Hybrid Authentication

In order to build trust in IoT devices and data, we proposed a hybrid authentication mechanism in the scheme, which supports both machine-to-human and machine-to-machine interactions.

As shown in Fig. 2, when the application program launches the access request to the database via the JDBC driver, the proposed authentication mechanism is triggered. In a machine-to-human mode, we employ a salted challenge-response authentication and a password-based encryption in the scheme. In a machine-to-machine mode, we employ a authenticated identity-based cryptography. The user's password or machineID is stored securely in the IoT devices during the user initialization or registration, which employs the salted password hashing algorithms. The challenge-response mechanism is employed to prevent the Man-in-the-Middle attacks and replay attacks.

If authentication process does not complete successfully, the application program cannot gain access to the database. If someone accesses to the database by bypassing the modified JDBC driver, the sensitive data is stored in ciphertext which avoiding data breaching. The password-based encryption is employed to create strong secret keys, which are used to encrypted the sensitive data in database.

Besides, the proposed scheme is compatible with TEE or HSM(hardware security module), which uses the hardware protected cryptographic keys and accelerated cryptographic operations to strengthen encryption practices.

4 Experimental Results

To evaluate the performance of the proposed scheme, we performed experiments on the Android (8.1.0). mobile platform with one 2.0 GHz Qualcomm Snapdragon 625 8-core processor, 4 GB RAM and 64 GB ROM. As the proposed scheme is decoupled from any particular database, we implemented our prototype for SQlite 3.28.0 and the JDBC driver version is sqlite-jdbc-3.27.2.1.jar. We ran the test program to access SQLite file via the modified JDBC drivers. The experiments used the symmetric encryption algorithms (AES 128 in ECB mode). We did not use the AES hardware accelerating and relied on software AES. The Definition of the target table for the test as follows.

```
CREATE TABLE COMPANY(
    ID INTEGER PRIMARY KEY AUTOINCREMENT,
    NAME TEXT NOT NULL,
    MOBILE1 CHAR(50) NOT NULL,
    MOBILE2 CHAR(50),
    TELEPHONE  CHAR(50),
    ADDRESS CHAR(50)
);
```

In the experiment, we evaluated the latency introduced by the proposed scheme. The test program launched "INSERT" or "SELECT" operation repeatedly to SQLite database and recorded the execution time. The "INSERT" statement adds one or more rows into a table which bring data encryption process, and the "SELECT" statement retrieves data from a single table which bring data decryption process.

Table 1. Comparison of the execution time with and without encryption (Millisecond).

Work mode	100	1000	10000	Avg
"INSERT" common mode	1348.9	10745.3	94360.4	9.591
"INSERT" security mode	1409.1	11057.2	95102.5	9.691
"SELECT" common mode	88.5	317.4	2707.2	0.280
"SELECT" security mode	149.3	543.3	4835.4	0.498

Table 1 tabulates the execution time of "INSERT" or "SELECT" operation with and without encryption when the number of iteration is set as 100,1000 and 10000 respectively. In Common Mode, the modified JDBC drivers works without encryption and In Security Mode, the modified JDBC drivers works with encryption. For "INSERT" operation, the average execution time per instruction is 9.591 ms in the Common Mode and 9.691 ms in the Security Mode, which increased only 1% with data encryption process. For "SELECT" operation, the average execution time per instruction is 0.280 ms in the Common Mode and 0.498 ms in the Security Mode, which increased 77.99% with data decryption process.

5 Conclusion and Future Work

We proposed a new encryption and security scheme for lightweight database which is suitable for embedded systems with limited storage and computing resources. The proposed scheme focused on data security issues and provided decoupled data encryption and hybrid authentication for lightweight databases to meet the security requirements in the practical IoT application scenarios. We presented the design and implementation of the prototype for the proposed scheme and evaluated the feasibility and effectiveness. The future work will be implemented to make encryption more effective and efficient by exploiting the capabilities and characteristic of hardware and employing the new lightweight cryptographic algorithms.

Acknowledgments. This work was supported by National Key Research and Develop Plan of China (2018YFF0215601-3).

Reference

1. Gubbi, J., Buyya, R., Marusic, S., et al.: Internet of Things (IoT): a vision, architectural elements, and future directions. Fut. Gener. Comput. Syst. **29**(7), 1645–1660 (2013)
2. Atlam, H.F., et al.: Integration of cloud computing with internet of things: challenges and open issues. In: IEEE International Conference on Internet of Things & IEEE Green Computing & Communications & IEEE Cyber. IEEE (2018)
3. Hou, J., Qu, L., Shi, W.: A survey on internet of things security from data perspectives. Comput. Netw. **148**, 295–306 (2019)
4. Kouicem, D.E., Bouabdallah, A., Lakhlef, H.: Internet of things security: a top-down survey. Comput. Netw. **141**, 199–221 (2018)
5. Alsmadi, I., et al.: Web and Database Security. In: Practical Information Security (2018)
6. Top Ten Database Security Threats. http://www.imperva.com/downloads/TopTenDatabaseSecurityThreats.pdf
7. Daniel, D., et al.: Triathlon of lightweight block ciphers for the Internet of things. J. Cryptogr. Eng. **9**, 283–302 (2018)
8. Raymond, K.K., Stefanos, G., Park, J.H.: Cryptographic solutions for industrial internet-of-things: research challenges and opportunities. In: IEEE Transactions on Industrial Informatics, p. 1 (2018)

9. Zhang, Z.K., et al.: IoT security: ongoing challenges and research opportunities. In: 2014 IEEE 7th International Conference on Service-Oriented Computing and Applications (SOCA). IEEE (2014)
10. Basharat, I., Azam, F.: Database security and encryption: a survey study. Int. J. Comput. Appl. **47**, 28–34 (2012)
11. Popa, R.A., et al.: CryptDB: processing queries on an encrypted database. Commun. ACM **55**(9), 103–111 (2012)
12. Egorov, M., Wilkison, M.L.: ZeroDB white paper (2016)
13. Poddar, R., Boelter, T., Popa, R.A.: Arx: an encrypted database using semantically secure encryption. Proc. VLDB Endow. **12**(11), 1664–1678 (2019)
14. Priebe, C., Vaswani, K., Costa, M.: EnclaveDB: a secure database using SGX. In: 2018 IEEE Symposium on Security and Privacy (SP). IEEE (2018)
15. Ahmad, A., et al.: Parallel query execution over encrypted data in database-as-a-service (DaaS). J. Supercomput. **75**(4), 2269–2288 (2019)
16. Wu, S., Li, Q., Li, G., Yuan, D., Yuan, X., Wang, C.: ServeDB: secure verifiable and efficient range queries on outsourced database. In: IEEE 35th International Conference on Data Engineering (ICDE) 2019, pp. 626–637 (2019)
17. Yuan, D., Li, Q., Li, G., Wang, Q., Ren, K.: PriRadar: a privacy-preserving framework for spatial crowdsourcing. IEEE Trans. Inf. Forensics Secur. **15**, 299–314 (2020)
18. Liu, Z., Jia, C., Li, J.-W.: Research on the format-preserving encryption techniques. J. Softw. **23**(1), 152–170 (2012)
19. Dworkin, M.J.: SP 800-38G Recommendation for Block Cipher Modes of Operation: Methods for Format-Preserving Encryption. National Institute of Standards & Technology (2016)

Research on MLChecker Plagiarism Detection System

Haihao Yu[1], Chengzhe Huang[1], Leilei Kong[2(✉)], Xu Sun[1], Haoliang Qi[2], and Zhongyuan Han[2]

[1] Heilongjiang Institute of Engineering, Harbin 150400, China
[2] Foshan University, Foshan 528000, China
kongleilei1979@gmail.com

Abstract. Plagiarism detection system plays an essential role in education quality improvement by helping teachers to detect plagiarism. Using a number of measures customized to determine occurrences of plagiarism is the most common approach for plagiarism detection tool. It is simple and effective, while it lacks flexibility when applied in more complicated situations. This paper proposes the MLChecker, a smart plagiarism detection system, to provide more flexible detection tactics. An automatic plagiarism dataset construction method was exploited in MLChecker to dynamically update the plagiarism detection algorithms according to different detection tasks. And the full-process quality management functions were also provided by MLChecker. The result shows that the detection accuracy is raised by 56%. Compared with traditional plagiarism detection tools, MLChecker is with higher accuracy and efficiency.

Keywords: Plagiarism · Plagiarism detection system · Plagiarism dataset · MLChecker

1 Introduction

With the rapid development of Internet technology, students can quickly obtain plagiarism sources using search engines, machine translation, etc. The use of these tools has made plagiarism easier and more serious [1]. Plagiarism detection is a useful and powerful anti-plagiarism tool to help teachers evaluating the students' learning outcomes. Plagiarism detection system plays an essential role in education quality improvement [2].

But as the use of the plagiarism detection system, students continuously implement anti-plagiarism technologies to avoid the plagiarism detection. They evade plagiarism detection using synonym substitution, syntactic modification, sentence reduction, combination, reorganization, conceptual generalization, and specialization. It poses a severe challenge to the plagiarism detection system. The algorithm of plagiarism detection needs to be improved continuously.

On the other hand, plagiarism corpus is the research foundation of plagiarism detection technology. Which reflecting the real plagiarism plays an essential role in the analysis of plagiarism rules and the construction of plagiarism detection algorithm [3]. However, the plagiarists will not easily agree to the application of the rewritten text,

© Springer Nature Singapore Pte Ltd. 2020
P. Qin et al. (Eds.): ICPCSEE 2020, CCIS 1258, pp. 176–181, 2020.
https://doi.org/10.1007/978-981-15-7984-4_14

which makes the construction of real plagiarism corpus to improve the performance of plagiarism detection more and more difficult.

To resolve the problems of various plagiarism means and the lack of plagiarism corpus, we developed MLChecker, a plagiarism detection system based on machine learning to detect the documents generating during the teaching procedure, such as the experiments reports, the course papers, etc. The MLChecker innovatively implements the following functions.

1. Adaptive plagiarism detection for different plagiarism types. The MLChecker uses an adaptive detection algorithm to dynamically identify different plagiarism types, including low obfuscation plagiarism with simple modification and high obfuscation plagiarism with interpretation modification [4].
2. Automatic acquisition of plagiarized corpus. The MLChecker uses a plagiarized corpus acquisition algorithm based on the natural annotation, and high-quality paraphrase plagiarism corpus is obtained automatically from the documents submitted by students multiple times on the same topic.
3. Automatic update of plagiarism algorithm. The MLChecker continuous training its model by using the real plagiarism cases and updates the detection algorithm automatically.

2 Frameworks and Core Algorithms

Figure 1 shows the plagiarism detection framework of MLChecker, DPI (Deep Paraphrase Identification) realizes adaptive plagiarism detection for different plagiarism types, PCC (Plagiarism Corpus Constructor) realizes the automatic acquisition of plagiarized corpus and provides the data to DPI for training the models.

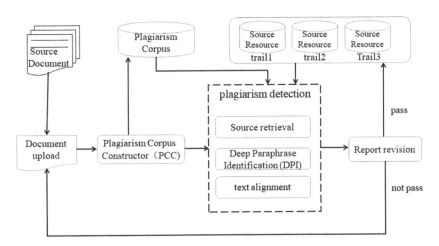

Fig. 1. Plagiarism detection framework of MLChecker

Figure 2 shows the primary function of MLChecker. For students, the system mainly implements the functions of student report uploading, inspection details viewing, downloading, etc. For teachers, the system mainly implements the functions of establishing courses and experiments, uploading basic information of students, and setting parameters. For leaders, the system mainly implements statistical analysis and query functions.

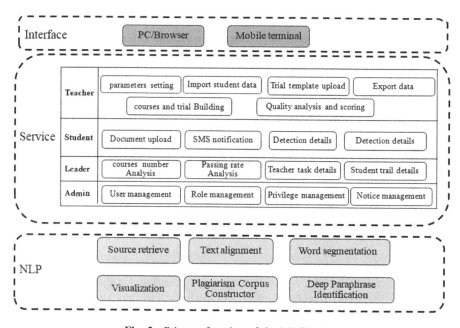

Fig. 2. Primary function of the MLChecker

The following Fig. 3 gives an examples of the student detection report, and Fig. 4 shows are results page of detection progress and qualification rate.

文本复制检测报告单(全文标明引文)

ADBD2019R_201910281638251572251905387   检测时间：   2019-10-28 16:38:25

检测文献： c:/profile/corporapath/~_____v-HBase编程开发-3303\txt\20161847-2.txt
作　　者： ▓▓▓▓
检测范围： 201617534.txt
　　　　　 201617553.txt
　　　　　 201617711.txt
　　　　　 201617771.txt
　　　　　 201618222.txt
提交顺序： 6

1.全文(总7643字)   69.5538%

被抄袭文件名	文章总长度	重复字数	重复比例
1.201617553.txt	8197	848	10.3452%
2.201618222.txt	7355	5028	68.3617%

相似文档： 201618222.txt
相似文档内容：

(HDFS默认工作目录格式为/user/hadoop)c.检查文件是否存在d.上传本地文件到HDFS系统e.追加到文件末尾的指令f.覆盖原有文件的指令从HDFS中下载指定文件，如果本地文件与要下载的文件名称相同，则自动对下载的文件重命名；将HDFS中指定文件的内容输出到终端中；显示HDFS中指定的文件的读写权限、大小、创建时间、路径等信息；给定HDFS中某一个目录，输出该目录下的所有文件的读写权限、大小、创建时间、路径等信息，如果该文件是目录，则递归输出该目录下所有文件相关信息；提供一个HDFS内的文件的路径，对该文件进行创建和删除操作。如果文件所在目录不存在，则自动创建目录；提供一个HDFS的目录的路径，对该目录进行创建和删除操作。创建目录时，如果目录文件所在目录不存在则自动创建相应目录；删除目录时，由用户指定当该目录不为空时是否还删除该目录；创建文件删除文件向HDFS中指定的文件追加内容，由用户指定内容追加到原有文件的开头或结尾；追加在后面追加在前面删除HDFS中指定的文件；删除HDFS中指定的目录，由用户指定目录中如果存在文件时是否删除目录；在HDFS中，将文件从源路径移动到目的路径。编程实现以下任务：将HDFS中指定文件的内容输出到终端中（对应shell命令第3题

Fig. 3. Plagiarism detection report

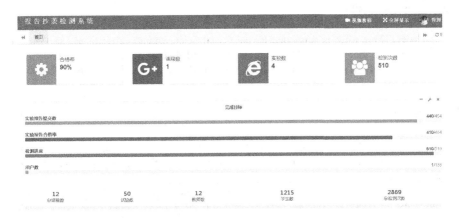

Fig. 4. Detection progress and qualification rate

The system has provided the services for 50 experimental courses and 1215 students. In Table 1, we take three experimental courses to indicate the changes of plagiarism ratio of student's experimental reports and the improvements of teacher's working efficiency after using MLchecker.

Table 1. Plagiarism ratio and efficiency

Courses	Method	Number of students	Inspection times (s)/document	Increase times	Plagiarism ratio	Increase rate	Accuracy
Mapreduce	MLChecker	110	1	8	10%	40%	95.5%
	Manual	110	9		50%		43%
Requirement analysis	MLChecker	111	3	4	6%	36%	96%
	Manual	111	12		42%		36%
J2EE SSM frame	MLChecker	197	2	5	8%	38%	95.8%
	Manual	197	10		46%		39%

Seen from Table 1, after using MLChecker, the average inspection times of teachers decreased by nearly six times, the ratio of plagiarism decreased by 38%, and the accuracy of plagiarism detection increased by 56%.

3   Conclusion

We report the MLChecker, a plagiarism detection system based on statistical machine learning. MLChecker can detect the different plagiarism types by a dynamical plagiarism detection algorithm. Besides, it realizes the full-process teaching quality management. Result of application indicates that the average detection time of a report is only 7 s, the accuracy rate of the system is nearly 96%, and the system can detect 10,000 reports per day. The system can meet the needs of teaching and research activities in universities.

Acknowledge. This work is supported by the Social Science Fund of Heilongjiang Province (No. 18TQB103), the National Natural Science Foundation of China (No. 61806075 and No. 61772177), and the Natural Science Foundation of Heilongjiang Province (No. F2018029).

References

1. Haneef, I., Nawab, A., Muhammad, R., Munir, E.U., Bajwa, I.S.: Design and development of a large cross-lingual plagiarism corpus for Urdu-English language pair. Sci. Program. **2019**, 1–11 (2019)
2. Kruchinin, S.V., Bagrova, E.V.: Anti-plagiarism system "Antiplagiat HEI". Implementation into Russian education quality improvement. In: 2019 International Conference Quality Management, Transport and Information Security, Information Technologies, pp. 622–624. IEEE (2019)

3. Thaiprayoon, S., Palingoon, P., Trakultaweekoon, K.: Design and development of a plagiarism corpus in Thai for plagiarism detection. In 2019 11th International Conference on Knowledge and Systems Engineering, pp. 1–5. IEEE (2019)
4. Kong, L.L., Han, Z.Y., Qi, H.L., Yang, M.Y.: Source retrieval model focused on aggregation for plagiarism detection. Inf. Sci. **503**, 336–350 (2019)
5. Kong, L.L.: Research on Plagiarism Detection Modeling based on Statistical Machine Learning. Doctor, Harbin Engineering University (2017)
6. Baba, K.: Filtering documents for plagiarism detection. In: Soldatova, L., Vanschoren, J., Papadopoulos, G., Ceci, M. (eds.) DS 2018. LNCS (LNAI), vol. 11198, pp. 361–372. Springer, Cham (2018). https://doi.org/10.1007/978-3-030-01771-2_23
7. Foltýnek, T., Meuschke, N., Gipp, B.: Academic plagiarism detection: a systematic literature review. ACM Comput. Surv. **52**(6), 1–42 (2019)
8. Pajic, E., Ljubovic, V.: Improving plagiarism detection using genetic algorithm. In: International Convention on Information and Communication Technology Electronics and Microelectronics, pp. 571–576 (2019)
9. Alsallal, M., Iqbal, R., Palade, V., Amin, S., Chang, V.: An integrated approach for intrinsic plagiarism detection. Fut. Gener. Comput. Syst. **96**, 700–712 (2017)
10. Kong, L., Han, Z., Haoliang, Q.I., Zhimao, L.U.: A ranking-based text matching approach for plagiarism detection. In: IEICE Transactions on Fundamentals of Electronics, Communications and Computer Sciences, pp. 799–810 (2018)

Algorithm

Research on Global Competition and Acoustic Search Algorithm Based on Random Attraction

Xiaoming Zhang[1(✉)], Jiakang He[1,2], and Qing Shen[1]

[1] School of Information Engineering, Beijing Institute of Petrochemical
Technology, Beijing 102617, China
zhangxiaoming@bipt.edu.cn
[2] School of Information Science and Technology,
Beijing University of Chemical Technology, Beijing 10029, China

Abstract. In view of complex optimization problems with high nonlinearity and multiple extremums, harmony search algorithm has problems of slow convergence speed, being easy to be stalled and difficult to combine global search with local search. Therefore, a global competitive harmonic search algorithm based on stochastic attraction is proposed. Firstly, by introducing the stochastic attraction model to adjust the harmonic vector, the harmonic search algorithm was greatly improved and prone to fall into the local optimum. Secondly, the competitive search strategy was made to generate two harmony vectors for each iteration and make competitive selection. The adaptive global adjustment and local search strategy were designed to effectively balance the global and local search capabilities of the algorithm. The typical test function was used to test the algorithm. The experimental results show that the algorithm has high precision and the ability to find the global optimum compared with the existing algorithms. The overall accuracy is increased by more than 50%.

Keywords: Harmony search algorithm · Random attraction model · Global competition · Global optimal

1 Introduction

In recent years, many researchers have proposed many intelligent optimization algorithms to solve numerical function optimization problems. Valdez devised a new method using fuzzy logic to dynamically adapt some parameters of the particle swarm optimization (PSO) algorithm, such as inertia weights and learning factors [1]. A three-objective differential evolution (DE) method to solve MMOPs (The multimodal optimization problems) was proposed by Yu and The third objective constructed by niche technology insensitive to niche parameters greatly improved population diversity [2]. Li introduced a new gene recombination operator (GRO) into the artificial bee colony (ABC) and solved the problem of slow convergence rate of ABC algorithm [3]. Chakri introduced directional echolocation into the standard bat algorithm to enhance its exploration and development ability, and it solved the problem of premature convergence that can occur under certain conditions [4].

© Springer Nature Singapore Pte Ltd. 2020
P. Qin et al. (Eds.): ICPCSEE 2020, CCIS 1258, pp. 185–196, 2020.
https://doi.org/10.1007/978-981-15-7984-4_15

Inspired by the experience of musicians in music creation, Z.W. Geem proposed the harmony search algorithm. Compared with other optimization algorithms, harmony search (HS) algorithm has more advantages. Its algorithm is simple and easy to be implemented with fewer adjustable parameters [5], and it has better search capability and robustness [6]. Due to these advantages, HS algorithm has been widely applied [7–10].

First of all, HS algorithm inevitably has some defects. The characteristic of HS algorithm is to use random search instead of gradient based search, which will lead to the impact of the initial solution of the algorithm on the overall algorithm effect, that is, the randomness is very large. Secondly, HS algorithm lacks guidance information, and its search efficiency is relatively low. Moreover, HS algorithm is prone to blind search in the late stage of evolution, and it cannot effectively adjust the structure of solution vectors. The high-quality solutions it produces may cause the algorithm to produce 'premature' convergence or fall into local optimization. In view of the above problems, researchers made a series of improvements to the HS algorithm. Quan-Ke Pan improved the harmony search mechanism and proposed an adaptive global harmony search algorithm (SGHS). Experimental results showed that by was significantly better than the basic HS algorithm [11]. Xiang proposed an improved global best harmony search algorithm, called IGHS algorithm, by using artificial colony algorithm improved random creation for reducing global best harmony search (GHS) stochastic and avoiding premature convergence of the algorithm. Through using the opposite learning, and differential evolution to improve the search ability, the experimental results show that IGHS is far superior to the basic harmony search (HS) algorithm and GHS algorithm [12]. In literature [13], a global competitive harmony search (GCHS) algorithm is proposed. By establishing a competitive search mechanism, two harmony vectors are generated in each iteration and competitive selection is made. The adaptive global adjustment and local learning strategies are designed to balance the local and global search capabilities of the algorithm. Guo proposed a hybrid meta-heuristic method through mixed harmonic search (HS) and firefly algorithm (FA), namely HS/FA, which combined the advantages of the two algorithms and achieved good experimental results [14]. An improved differential HS (IDHS) was proposed by Wang. It proposed a new creation process by integrating the ideas in the differential evolutionary algorithm and achieved good results [15].

Although the variants of the previously proposed HS algorithms are well designed, they still have some disadvantages, such as slow convergence when dealing with high-dimensional problems, and premature and sharp convergence of harmonic causes the algorithm to stall and fall into the local optimal problem. In this paper, it aims to improve the HS algorithm to solve the contradiction between fast convergence and premature fusion, balance the focus of global search and local exploration, and propose a harmony search (RAGCHS) algorithm with random attraction global competition strategy. By providing some exploration guidance information through random attraction, the diversity of harmony vectors in the memory bank is improved, and the premature fusion phenomenon is avoided while the convergence speed is improved. Through global competition, reverse learning and other strategies, the problem is solved to balance the global search and local search.

2 HS and Its Variants

In this section, the original HS is introduced to better understand its procedure and principles.

The harmony search (HS) algorithm compares Musical Instruments to design variables in optimization problems. The number of Musical Instruments can be regarded as the dimension of design variables. The harmony of Musical Instruments' tones is equivalent to a solution vector of optimization problems. The algorithm firstly generates a certain amount of harmony as the initial liberation into the harmony memory bank (HM), searches for the new harmony in HM with probability HMCR as the new solution, and searches for the new solution in the possible range of variables outside HM with probability 1-hmcr. The algorithm then generates a local disturbance to the new solution with a probability PAR. It determines that whether the objective function value of the new solution is better than the worst solution in HM. If so, replace it. Then it will iterate until a predetermined number of iterations (Tmax) is reached.

2.1 HS Flow

The basic steps of the HS algorithm are as follows.

Step 1: initialize the algorithm parameters and determine HM size (HMS), HCMR, PAR, BW and maximum number of iterations (Tmax).

Step 2: initialize the harmony memory bank, randomly generate HMS harmonies and put them into the harmony memory bank. The form of the harmony memory bank is as follows:

$$HM = \begin{bmatrix} x^1 & f(x^1) \\ x^2 & f(x^2) \\ \vdots & \vdots \\ x^{HMS} & f(x^{HMS}) \end{bmatrix} \tag{1}$$

Step 3: a new harmony is generated, generate a new harmony $x^i = (x_1, x_2, \ldots, x_N)$, and generate each component of the new harmony x_i (i = 1, 2, ..., N) by learning the harmony memory bank, tuning, and random selection of elements.

Step 4: update the harmony memory bank. Evaluate the new harmony in step 3. If it is better than the one with the worst fitness in HM, update the new harmony in HM to replace the harmony with the worst fitness.

Step 5: check if the algorithm termination condition is met, repeat step 3 and step 4 until the number of iterations reaches Tmax.

3 RAGCHS

3.1   The Overall Train of Thought

The algorithm proposed in this paper is a variant algorithm of harmony search. By improving the imperfect part of harmony algorithm and adding the excellent idea and strategy of other intelligent optimization algorithms, the algorithm of random attraction global competitive harmony search is proposed.

In the analysis of the improvisation process in harmony search, it is found that in the HS algorithm, the value of BW largely determines the optimization accuracy of the algorithm, and the selection of BW is an inevitable problem in the HS algorithm. Many HS variants use other methods to replace the tone modulation mechanism in the original method [12, 13, 16]. Therefore, we use a novel improvisation method instead of the original tone bandwidth modulation.

The stochastic attraction model, firefly algorithm (FA), was proposed by professor Yang of the university of Cambridge in 2008. It mainly takes the firefly's flash as signal system to attract other fireflies. FA uses the full-attraction model (FAM) for population evolution, that is, each firefly moves towards a brighter firefly. Assuming the population size is N, the evolution mode of the full attraction model is adopted. The brightest firefly needs to move 0 times, and the darkest firefly needs to move $n - 1$ times, that is, the maximum total number of times of the population evolution generation needs to move is $Mfull = N * (n - 1)/2$. This model enables fireflies to obtain more search opportunities, but the search efficiency is very low, and in the face of complex problems are also prone to shock phenomenon. Therefore, some researchers proposed a stochastic attraction model, in which each firefly moves only once to a randomly selected firefly whose brightness is higher than its own, and the population evolution generation only needs to move $n - 1$ times, thus simplifying the model and saving computing resources. Excellent optimization results have been achieved [17, 18]. In this paper, the random attraction model is introduced into the harmony search algorithm to provide certain guidance information for tone modulation, which improves the blind search problem of HS algorithm to some extent.

In this paper, a concept of quality harmony is proposed, the harmony in the harmony memory is calculated, and the harmony with high fitness is set as quality harmony. The good harmonies here will be used in the random attraction process.

The traditional HS algorithm only generates a new harmony in each iteration and compares it with the worst harmony in the harmony memory bank. If the harmony is better than the worst harmony, it replaces it, which leads to its relatively low search efficiency. And the search range is small, easy to fall into the local optimal. In this paper, competitive selection mechanism is added to generate two new harmonies at each iteration, one focusing on global adjustment with a large search range, and the other focusing on local search to accelerate the approach to global optimal. To some extent, the competitive selection mechanism not only speeds up the convergence speed but also avoids the rapid convergence and premature stagnation of the algorithm, which balances the global search and the local search.

3.2 Algorithm Design

The traditional harmony algorithm and most of its variants use probabilistic random-ness in the generation of harmony memory. In order to improve the quality of initial harmony, a reverse learning strategy is introduced in the initialization of harmony memory bank. Reverse learning refers to obtaining the reverse solution from the original solution and comparing and evaluating it. Studies have proved that the con-vergence rate of the reverse learning strategy used in population initialization is far faster than that of the traditional pure random strategy [19]. Itself as follows, first even randomly generated a batch of the original harmonic vectors, according to each of the first harmonic vectors generated a reverse harmonic vectors, fitness function is used to calculate the fitness, will be the first harmonic vectors and reverse and acoustic vector comparison, choose fitness higher harmonic vectors, into the ultimate harmony memory Banks.

In the process of improvisation of new harmony, random attraction is realized by random selection of high-quality harmony, and its formula is as follows

$$r = \left\| x_g - x \right\| \tag{2}$$

$$x_{new} = x + \xi\beta_0 e^{-\gamma r^2}(x_g - x) + \alpha(rand - 0.5) \tag{3}$$

Where xnew is the new harmony, γ is the absorption factor of light intensity, ξ is the factor of maximum attraction, $\alpha(rand-0.5)$ is the random number in the interval 0–2, and d is the random disturbance term, whose purpose is to enhance the diversity of harmony, prevent premature convergence, and reduce the possibility of it falling into the local optimal. The process of improvisation is as follows (Fig. 1).

Where x_b, x_w are the optimal harmony and the worst harmony in the harmony memory bank, x_g are the harmony randomly selected from the high-quality harmony, TOPbest and TOPworst are the optimal harmony and the worst harmony in the high-quality harmony, T is the iteration number, Tmax is the maximum iteration number, and δ is the piecewise function.

$$\begin{cases} \delta = 2, T \leq 0.5T_{max} \\ \delta = 1, T > 0.5T_{max} \end{cases} \tag{4}$$

Harmonic new1 performs global optimal fast approximation and local search. Harmonic new2 has a larger search range and focuses on global search. In random selection, the positions of new1 and new2 on one dimension are reverse digits to each other, so that they can get more uniform distribution in solution space.

For HCMR, the algorithm needs a larger search range at the beginning of iteration, so the value is relatively low. At the end of iteration, the dependence on random selection is reduced and the value becomes higher. The value of the function is as follows

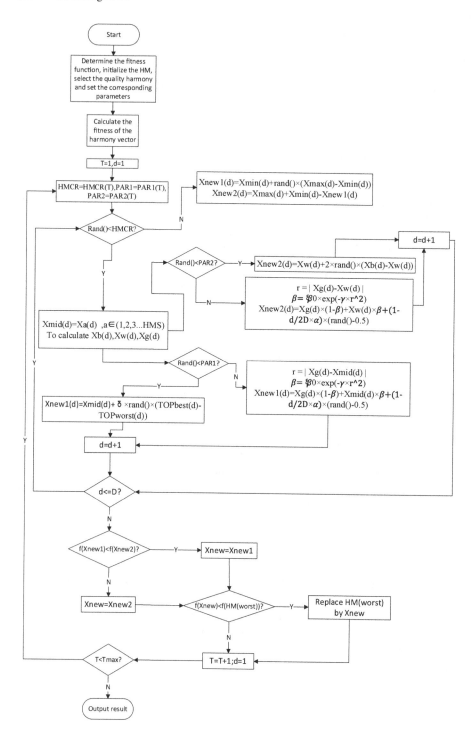

Fig. 1. RAGCHS flowchart

$$\begin{cases} c = \ln(HMCR_{max}/HMCR_{min})/T_{max} \\ HCMR = HMCR_{min} \times e^{c \times T} \end{cases} \tag{5}$$

Using this function will make HMCR smooth; For new1, if the PAR is large, the new harmony will be adjusted in the optimal harmony for many times, which is not conducive to the rapid convergence of the algorithm and is obviously not applicable at the initial stage of the algorithm. New2 focuses on global tuning, requiring a large search space at the beginning of the algorithm, which requires a relatively large PAR, and a relatively small PAR if the need for global search weakens later in the algorithm. The specific adaptive adjustment methods are as follows.

$$\begin{cases} PAR1 = PAR1_{min} + \frac{PAR1_{max} - PAR1_{min}}{T_{max}} \times 2T & (T < T_{max}/2) \\ PAR1 = PAR1_{max} & (T > T_{max}/2) \end{cases} \tag{6}$$

$$\begin{cases} PAR2 = PAR2_{min} + \frac{PAR2_{max} - PAR2_{min}}{T_{max}} \times 2T & (T < T_{max}/2) \\ PAR2 = PAR2_{max} & (T > T_{max}/2) \end{cases} \tag{7}$$

The RAGCHS algorithm has a simple flow, so it can well implement and solve other field-specific applications. HS series algorithm has certain advantages in practical application, because its population size is generally small, memory footprint is small and easier to implement, so it has better application value.

4   Experiment and Analysis

4.1   Experimental Scheme Design

In order to verify the performance of RAGCHS, HS, SGHS, IGHS, GCHS, HSFA and IDHS were selected for experimental comparison.

The HS algorithm is the most primitive algorithm, compared with which the improvement of RAGCHS can be clearly seen. Four of the remaining algorithms are variations of the HS algorithm. Among them, SGHS (Pan 2010) [11], IGHS (Xiang 2014) [12] are very classical HS variants, and GCHS (2016) [13] algorithm USES competitive selection mechanism, which is similar to the overall architecture of the algorithm in this paper, so it is compared as a reference. IDHS (Wang 2018) [15] is relatively novel and has a better optimization effect. These recently proposed HS variants have been shown to be effective and effectively solve optimization problems. HSFA [14] optimizes the firefly algorithm by using some theories of HS algorithm, which has certain reference value.

Six representative test functions are used in this paper, listed in Table 1.

Table 1. Experimental test functions

No	Functions	Space	$x^*(i = 1,...,D)$	$f(x^*)$
1	$f(x) = (x_1 - 1)^2 + \sum\limits_{i=2}^{D} i(2x_i^2 - x_{i-1})^2$	$(-10, 10)$	$2^{(2-2^i)/2^i}$	0
2	$f(x) = \max(|x_i|, 1 \le i \le D)$	$(-100, 100)$	0	0
3	$f(x) = 10D + \sum\limits_{i=1}^{D} (x_i^2 - 10\cos(2\pi x_i))$	$(-5.12, 5.12)$	0	0
4	$f(x) = \sum\limits_{i=1}^{D}(\sum\limits_{j=1}^{D}(j+10)^2(x_j^i - \frac{1}{j^i}))^2$	$(-D, D)$	$1/i$	0
5	$f(x) = \sum\limits_{i=1}^{D-1}(100(x_{i+1} - x_i^2)^2 + (x_i - 1)^2)^2$	$(-5, 10)$	1	0
6	$f(x) = g(x_1, x_2) + g(x_2, x_3) + \cdots g(x_D, x_1)$ $g(x, y) = 0.5 + \frac{\sin^2(\sqrt{x^2 + y^2}) - 0.5}{(1 + 0.001(x^2 + y^2))^2}$	$(-100, 100)$	0	0

In order to observe the performance of RSGCHS algorithm in the test more intuitively, the RSGCHS algorithm was compared with the above 6 algorithms in terms of the optimal fitness of the same scale and the number of iterations. The relevant experimental parameters are shown in Table 2 (where the values of PAR1 and PAR2 are the same, so they are combined; in the experiment, the high quality harmony number of algorithm parameters in this paper is 5).

Table 2. Parameters of the test algorithm

Algorithms	HMS	PAR	HMCR	BW	Other
HS	5	0.3	0.9	0.001	–
SGHS	5	$PAR_m = 0.9$	$HMCR_m = 0.98$	$BW_{min} = 0.0005$ $BW_{max} = (UB-LB)/10$	LP = 100
GCHS	5	0.5	0.99	–	–
IGHS	20	$PAR_{min} = 0.9$ $PAR_{max} = 0.95$	$HMCR_{min} = 0.9$ $HMCR_{max} = 0.99$	–	$\alpha = 0.25$ $\beta = 0.05$ $\xi = 0.1$
IDHS	20	$PAR_{min} = 0.1$ $PAR_{max} = 0.9$	$HMCR_{min} = 0.8$ $HMCR_{max} = 0.9$	–	F2 = 0.6
HSFA	–	0.1	0.9	0.001	$\gamma = 1$ $\alpha = 0.5$ $\beta_0 = 1$ NP = 10 KEEP = 2
RAGCHS	10 '5'	$PAR_{min} = 0.3$ $PAR_{max} = 0.7$	$HMCR_{min} = 0.8$ $HMCR_{max} = 0.95$	–	$\alpha = 0.5$ $\beta_0 = 1$ $\gamma = 1$

4.2   Experimental Results and Analysis

The first three functions in Table 1 are all single-mode functions, which are mainly used to measure and verify the convergence speed and local search ability of the algorithm in the calculation process. The last three functions are highly nonlinear multi-mode functions with multiple extremums, which can be used to test the algorithm's global search ability and the ability to jump out of local optima. In order to simplify the calculation, all the test functions were calculated by taking into account the 4-dimensional distribution. 30 experiments were conducted for each test, and the average value was recorded as the global optimal value. IGHS algorithm was iterated for 50000 times, and the maximum iteration times of other algorithms were all 30000, as shown in Table 3.

Table 3. Function test results

Functions		Algorithms						
		HS	SGHS	GCHS	IGHS	IDHS	HSFA	RAGCHS
f1	mean	8.89E − 02	4.82E − 02	1.56E − 01	6.27E − 11	2.59E − 01	5.55E − 01	**1.06E − 23**
	h	+	+	+	+	+	+	NONE
f2	mean	9.91E − 04	5.56E − 05	**0.00E + 00**	5.11E − 11	2.86E − 06	5.77E + 01	6.41E − 08
	h	+	+	−	−	+	+	NONE
f3	mean	1.28E + 00	1.25E + 00	**0.00E + 00**	**0.00E + 00**	3.63E − 09	8.27E + 00	**0.00E + 00**
	h	+	+	NA	NA	+	+	NONE
f4	mean	2.23E + 01	8.14E + 01	5.05E + 01	3.77E + 00	7.12E + 00	2.33E + 02	**2.05E − 01**
	h	+	+	+	NA	NA	NA	NONE
f5	mean	3.65E + 00	8.86E + 00	9.43E − 01	2.01E − 02	1.71E − 02	1.01E + 00	**3.45E − 03**
	h	+	+	+	NA	NA	+	NONE
f6	mean	1.36E + 00	1.97E + 00	1.39E − 01	1.22E − 01	6.86E − 02	1.49E + 00	**5.82E − 03**
	h	+	+	+	+	+	+	NONE

Table 3 shows the experimental results of six test functions. The best test results for each function are in bold font. The average value of the test error and the comparison between other algorithms and the algorithm in this paper are shown in the table. '+ ' means that the algorithm in this paper is more accurate than its algorithm, '−' means that the algorithm in this paper is less accurate than its algorithm, and 'NA' means that there is no significant difference between the algorithm in this paper and its algorithm. From the experimental results, the RAGCHS algorithm has a significant advantage in multi-extremum functions and a satisfactory precision in single-mode functions.

In order to further compare convergence speed of the algorithm, using the number of Function evaluation strategy (Function Evaluations, FEs), set the maximum number of Function evaluation for 25000. Figure 2 shows the convergence process of all test algorithms on six test functions (IGHS algorithm is not included).

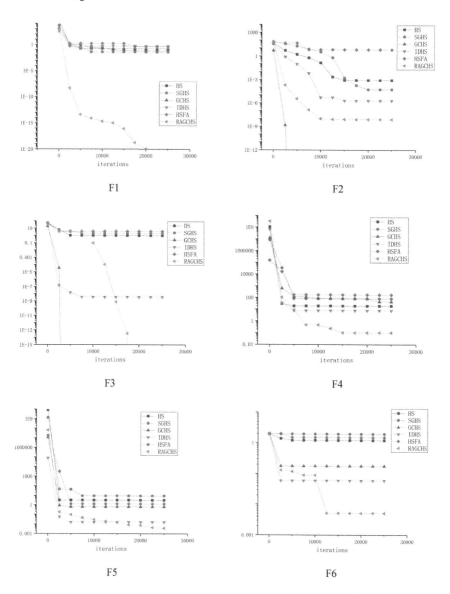

Fig. 2. Convergence velocity curves of different algorithms under six test functions

From the above experiments, it can be seen that for single-mode functions, all the seven algorithms can maintain good local exploration ability, among which GCHS, IGHS and RAGCHS have outstanding performance and excellent local exploration ability. On the whole, HS series algorithm has higher accuracy than HSFA algorithm. As for the multi-mode test function, it can be seen that the HS variant algorithm has a great improvement over the original HS algorithm, and the algorithm in this paper is the most prominent, with a better optimization effect. From the point of convergence speed, RAGCHS algorithm convergence speed faster, and not easy to fall into local optimal solution, in f6 test function, remove RAGCHS algorithm are outside the premature convergence problem, in 30 consecutive experiments, RAGCHS algorithm appear only once premature convergence condition, its precision compared with other algorithm improves the at least one order of magnitude. In the experiment of 250,000 times of function evaluation, it is not difficult to find that with the increase of iteration times, the accuracy of RAGCHS algorithm is significantly improved and stagnation is rarely generated. In summary, the RAGCHS algorithm can both rapidly converge in all single-mode and multi-mode functions and avoid algorithm stagnation in local optimization.

5 Conclusions

A global competitive harmonic search algorithm based on stochastic attraction is proposed in this paper. Adaptive global adjustment and harmony initialization strategies are designed to make the algorithm continuously converge in the whole iteration process on the basis of accelerating the convergence rate. On this basis, the algorithm in this paper combines global search and local search with the help of global competition strategy, which solves the contradiction between global search and local search of existing harmony search algorithms, and the contradiction between fast convergence and premature stagnation. Compared with the existing algorithms, the overall accuracy is improved by more than 50%, which has a good application prospect.

References

1. Valdez, F., Vazquez, J.C., Melin, P., et al.: Comparative study of the use of fuzzy logic in improving particle swarm optimization variants for mathematical functions using co-evolution. Appl. Soft Comput. **52**, 1070–1083 (2017)
2. Yu, W.J., Ji, J.Y., Gong, Y.J., et al.: A tri-objective differential evolution approach for multimodal optimization. Inf. Sci. **423**, 1–23 (2018)
3. Li, G., Cui, L., Fu, X., et al.: Artificial bee colony algorithm with gene recombination for numerical function optimization. Appl. Soft Comput. **52**, 146–159 (2017)
4. Chakri, A., Khelif, R., Benouaret, M., et al.: New directional bat algorithm for continuous optimization problems. Expert Syst. Appl. **69**, 159–175 (2017)
5. Wang, G., Guo, L.: A novel hybrid bat algorithm with harmony search for global numerical optimization. J. Appl. Math. **2013**, 1–21 (2013)
6. Wang, L., Yang, R., Xu, Y., et al.: An improved adaptive binary Harmony Search algorithm. Inf. Sci. **232**(Complete), 58–87 (2013)
7. Erdal, F., Do-An, E., Saka, M.P.: Optimum design of cellular beams using harmony search and particle swarm optimizers. J. Constr. Steel Res. **67**(2), 237–247 (2011)
8. Kaveh, A., Ahangaran, M.: Discrete cost optimization of composite floor system using social harmony search model. Appl. Soft Comput. **12**(1), 372–381 (2012)
9. Geem, W.Z.: Harmony search optimisation to the pump-included water distribution network design. Civil Eng. Environ. Syst. **26**(3), 211–221 (2009)

10. Landa-Torres, I., Manjarres, D., Salcedo-Sanz, S., et al.: A multi-objective grouping Harmony Search algorithm for the optimal distribution of 24-hour medical emergency units. Expert Syst. Appl. **40**(6), 2343–2349 (2013)
11. Pan, Q.K., Suganthan, P.N., Tasgetiren, M.F., et al.: A self-adaptive global best harmony search algorithm for continuous optimization problems. Appl. Math. Comput. **216**(3), 830–848 (2010)
12. Xiang, W., An, M., Li, Y., et al.: An improved global-best harmony search algorithm for faster optimization. Expert Syst. Appl. **41**(13), 5788–5803 (2014)
13. Xia, H., Ouyang, H., Gao, L., et al.: Global competitive harmonic search algorithm. Control Decis. **31**(2), 310–316 (2016)
14. Lihong, G., Gai-Ge, W., Heqi, W., et al.: An effective hybrid firefly algorithm with harmony search for global numerical optimization. Sci. World J. **2013**, 1–9 (2013)
15. Wang, L., Hu, H., Liu, R., et al.: An improved differential harmony search algorithm for function optimization problems. Soft. Comput. **23**(13), 4827–4852 (2019)
16. Zhai, J., Qin, Y.: Random cross global harmonic search algorithm. Comput. Eng. Appl. **54**(907)(12), 26–31 + 120 (2018)
17. Wang, H., Wang, W., Sun, H., et al.: Firefly algorithm with random attraction. Int. J. Bio-Inspired Comput. **8**(1), 33–41 (2016)
18. Zhao, J., Xie, Z., Lu, L., et al.: Deep learning firefly algorithm. Acta Electronica Sinica **46**(11), 75–83 (2018)
19. Tizhoosh, H.R.: Opposition-based reinforcement learning. J. Adv. Comput. Intell. Intell. Inf. **10**(4), 578–585 (2006)

Over-Smoothing Algorithm and Its Application to GCN Semi-supervised Classification

Mingzhi Dai[1,2], Weibin Guo[1(✉)], and Xiang Feng[1,2]

[1] Department of Information Science and Engineering,
East China University of Science and Technology, Shanghai, China
Y30180707@mail.ecust.edu.cn,
{gweibin,xfeng}@ecust.edu.cn
[2] Smart City Collaborative Innovation Center,
Shanghai Jiao Tong University, Shanghai, China

Abstract. The feature information of the local graph structure and the nodes may be over-smoothing due to the large number of encodings, which causes the node characterization to converge to one or several values. In other words, nodes from different clusters become difficult to distinguish, as two different classes of nodes with closer topological distance are more likely to belong to the same class and vice versa. To alleviate this problem, an over-smoothing algorithm is proposed, and a method of reweighted mechanism is applied to make the trade-off of the information representation of nodes and neighborhoods more reasonable. By improving several propagation models, including Chebyshev polynomial kernel model and Laplace linear 1st Chebyshev kernel model, a new model named RWGCN based on different propagation kernels was proposed logically. The experiments show that satisfactory results are achieved on the semi-supervised classification task of graph type data.

Keywords: GCN · Chebyshev polynomial kernel model · Laplace linear 1st Chebyshev kernel model · Over-smoothing · Reweighted mechanism

1 Introduction

Throughout the history of deep learning, neural networks constructed from simple nodes and edges have played a pivotal role. It's undeniable that convolutional neural network has witnessed great success in the research and application of artificial intelligence. Especially in recent years, almost all critical breakthroughs in the fields of image and speech recognition have been achieved by convolutional neural networks. It calculates the feature map by calculating the weight of the central pixel and the neighboring pixels to complete the extraction of spatial features. As the layers of the network continue to stack up, its ability to extract information continues to rise within a certain range. However, CNN also has limitations when processing some special data, that is, it can only process image or video data, which pixels is a neatly arranged matrix (Euclidean Structure), CNN will look stretched when dealing with social network, reference network and other topological graph data (Non Euclidean Structure) built with vertices and edges. Then GCN can use the theory of graph spectrum to implement

© Springer Nature Singapore Pte Ltd. 2020
P. Qin et al. (Eds.): ICPCSEE 2020, CCIS 1258, pp. 197–215, 2020.
https://doi.org/10.1007/978-981-15-7984-4_16

convolution operations on topological graphs, migrate traditional Fourier transforms and convolution operations to Graph, and use Chebyshev polynomials and other methods to fit the convolution kernel to reduce the calculation. Based on the above we can further get, owing to Laplacian matrix is a binary matrix, when convolution is performed using a spectral algorithm (Spectral domain) based on the Laplacian operator, the representation of discrepancy nodes will tend to converge to the same value. More precisely, the representation of nodes in the same connected component in the graph tends to converge to the same value, and there are only a few or one connected component in the graph structure of many tasks, which eventually leads to the node features of the Graph may become over-smoothing. Therefore, the article focuses on alleviating the above problems by incorporating a Reweighted mechanism and achieve better classification results on GCN semi-supervised classification.

The contributions of this paper are as follows:

1) We propose an improved graph convolution model based on the Chebyshev polynomial kernel. By analyzing its graph data classification performance in the GCN network, a new Re-w Chebyshev filter model is proposed;
2) The article also proposes an improved network based on a single parameter Laplacian hierarchical linear propagation model. By analyzing its semi-supervised classification performance in the GCN network, a new Re-w Renormalization trick model is also proposed;
3) We conducted a large number of semi-supervised classification experiments on commonly used graph data sets to verify the effectiveness of the models and methods proposed above in relieving the problem of excessive smoothing.

2　Related Work

Observe the development of graph convolutional networks in recent years, J. Bruna et al. [1] first proposed in the paper that CNN can be generalized to signals defined on a more general domain [2, 3] (i.e. non-Euclidean domain), the author defines two algorithms, one is a hierarchical clustering algorithm based on the spatial domain, and the other is an algorithm based on the spectral domain of the Laplace operator. The graph convolution method proposed by M. Niepert et al. [4] belongs to the typical spatial domain convolution, which proposes a framework for learning convolutional neural networks for arbitrary graphs, the author's opinion reveal that these graphs may be undirected, directed, and with both discrete and continuous node and edge attributes [5], they present a general approach to extracting locally connected regions from graphs, that is, find neighbors adjacent to each vertex, and then connect the nodes in the graph into a hierarchical structure in the spatial domain for convolution. The graph convolution method proposed by Thomas Kpif et al. [6] belongs to the algorithm of the spectral domain and uses the theory of graphs to perform convolution operations on topological graphs, the filters and graph signals of the convolutional network are moved to the Fourier domain at the same time for processing, and the Chebyshev polynomial is reduced to a parameter Laplacian-level linear model. In the end, Thomas Kpif et al. completed the semi-supervised classification task for typical graph datasets such as citation networks and knowledge graphs. This method is improved on a novel method proposed by David K. Hammond et al. [7] to construct a wavelet transform of a function defined on the vertices of an arbitrary finite weighted graph [8, 9], the author uses Chebyshev polynomials to recursively calculate the convolution kernel (Chebyshev kernel model), That is, the truncated expansion of Kth of Chebyshev polynomial $T_k(x)$ to approximate the convolution kernel g_θ. In addition, in the field of multi-label image recognition, some authors have proposed an end-to-end trainable multi-label image recognition framework (ML-GCN) [10], a simple method for multi-label recognition [11–13] is to train an independent binary classifier for each category, and explore "how to effectively model the collaborative relationship between labels", according to the dependencies between the labels For effective modeling, use the ResNet-101 network model [17] as the basic model for image feature extraction. The classification framework uses GCN to map label representations (such as word embeddings) to interdependent object classifiers, then image representation extracted by the convolutional network was input into the object classifier, and finally achieved good results in the field of multi-label image recognition. A large number of approaches for semi-supervised learning using graph representations have been proposed in recent years, most of which fall into two broad categories: methods that use some form of explicit graph Laplacian regularization and graph embedding-based approaches. Prominent examples for graph Laplacian regularization include label propagation [14], manifold regularization [15] and deep semi-supervised embedding [16].

3 Setup

3.1 Graph Theory

It is well known that in a data structure, for a graph G = (V, E), where V represents the nodes in the graph and E represents the edge connecting any two nodes in the graph. As shown in Fig. 1, the Laplacian matrix can be defined as L = D-A, where L is the Laplacian matrix, D represents the degree matrix of the vertices (the elements on the diagonal are the degrees of each vertex), and A represents the graph adjacency matrix.

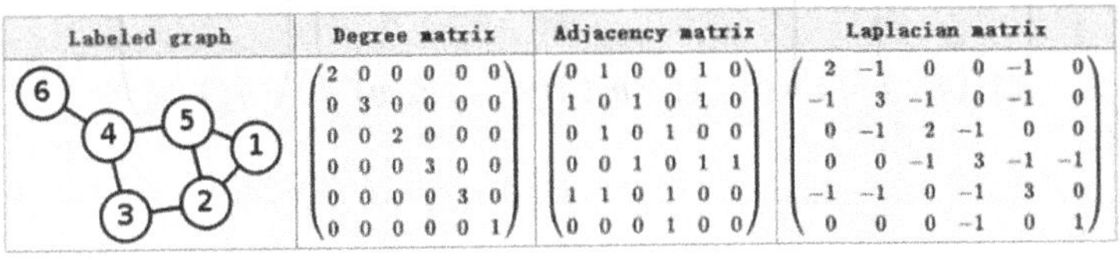

Fig. 1. Calculation method of Laplacian matrix.

As we know, the traditional Fourier transform is defined as Eq. 1.

$$F(\omega) = \mathcal{F}[f(t)] = \int f(t)e^{-iwt}dt \tag{1}$$

This formula represents the integral of the signal $f(t)$ and the basis function e^{-iwt}. In order to take the Fourier transform and convolution to Graph, the key work is to replace the Laplace operator eigenfunction e^{-iwt} with the eigenvector of the Laplacian matrix in Graph. Following the above definition, the Fourier transform on the graph can be obtained, as shown in Eq. 2.

$$F(\lambda_i) = \hat{f}(\lambda_i) = \sum_{p=1}^{n} f(p)u_i^*(p) \tag{2}$$

$$\begin{pmatrix} \hat{f}(\lambda_1) \\ \hat{f}(\lambda_2) \\ \vdots \\ \hat{f}(\lambda_n) \end{pmatrix} = \begin{pmatrix} u_1(1) & u_1(2) & \cdots & u_1(n) \\ u_2(1) & u_2(2) & \cdots & u_2(n) \\ \vdots & \vdots & \ddots & \vdots \\ u_n(1) & u_n(2) & \cdots & u_n(n) \end{pmatrix} \begin{pmatrix} f(1) \\ f(2) \\ \vdots \\ f(n) \end{pmatrix} \tag{3}$$

$$\hat{f} = U^T f \tag{4}$$

Where f represents an n-dimensional vector on Graph, $f(p)$ corresponds to the vertex of the graph one by one, and $u_i(p)$ represents the p-th component of the i-th feature vector. Then the Fourier transform of f under the feature value λ_i on the Graph is the inner product calculation with u_i^* (the conjugate form of the feature vector corresponding to λ_i). As shown in Eq. 3, the Fourier transform on the graph is generalized to the matrix multiplication form. And the matrix multiplication form of f on Graph can be simplified to Eq. 4.

$$\mathcal{F}^{-1}[F(\omega)] = \frac{1}{2\pi} \int F(\omega)e^{-i\omega t}d\omega \tag{5}$$

$$f(p) = \sum_{p=1}^{n} \hat{f}(\lambda_i)u_i^*(p) \tag{6}$$

$$\begin{pmatrix} f(1) \\ f(2) \\ \vdots \\ f(n) \end{pmatrix} = \begin{pmatrix} u_1(1) & u_1(2) & \cdots & u_1(n) \\ u_2(1) & u_2(2) & \cdots & u_2(n) \\ \vdots & \vdots & \ddots & \vdots \\ u_n(1) & u_n(2) & \cdots & u_n(n) \end{pmatrix} \begin{pmatrix} \hat{f}(\lambda_1) \\ \hat{f}(\lambda_2) \\ \vdots \\ \hat{f}(\lambda_n) \end{pmatrix} \tag{7}$$

$$\hat{f} = U^T f \tag{8}$$

As shown in Eq. 5, the traditional inverse Fourier transform is to integrate the frequencies. Similarly, Eq. 6 can be obtained. Then the Fourier transform of $\hat{f}$ under the feature value λ_i on the Graph is the inner product calculation with u_i (the feature vector corresponding to λ_i). As shown in Eq. 7, the inverse Fourier transform on the graph is generalized to the matrix multiplication form. And the matrix multiplication form of $\hat{f}$ on Graph can be simplified to Eq. 8.

$$\hat{h}(\lambda_l) = \sum\nolimits_{p=1}^{n} h(p)u_l^*(p) \tag{9}$$

$$\hat{h} = U^T h = \begin{pmatrix} \hat{h}(\lambda_1) & & \cdot \\ & \ddots & \\ \cdot & & \hat{h}(\lambda_n) \end{pmatrix} \tag{10}$$

$$(f * h)_G = \mathcal{F}^{-1}\big[\hat{f}(\varphi)\hat{h}(\varphi)\big] = \frac{1}{2\pi}\int \hat{f}(\varphi)\hat{h}(\varphi)e^{-i\varphi t}d\varphi \tag{11}$$

Theorem 1. The Fourier transform of the function convolution is the product of the function Fourier transform.

Promotion of Convolution. Suppose the convolution kernel function is $h(p)$, $\hat{h}(\lambda_l)$ is the Fourier transform of the convolution kernel h on Graph, as shown in Eq. 9. It can be known from Eq. 4 and Eq. 8 that it can be simplified and written as a diagonal matrix, as shown in Eq. 10. It can be inferred from Theorem 1 that the convolution of the function $f(p)$ and the convolution kernel $h(p)$ is the inverse transform of the product of its Fourier transform, as shown in Eq. 11. Therefore, the Fourier transform of the function convolution can be equivalent to the product of the Fourier transform of $f(p)$ and the Fourier transform of the convolution kernel $h(p)$, and then left multiplication by U (left multiplication U is the inverse of matrix multiplication transform).

So far, the convolution process on Graph has been completed, as shown in Eq. 12 below.

$$(f * h)_G = U(U^T h \cdot U^T f)$$

$$n = U \begin{pmatrix} \hat{h}(\lambda_1) & & \\ & \ddots & \\ & & \hat{h}(\lambda_n) \end{pmatrix} U^T f \tag{12}$$

Now assume that the input signal f is x, σ is the sigmoid activation function, and $g_\theta(\Lambda)$ represents the convolution kernel function. At this time, the convolution process is transformed into Eq. 13.

$$y_{output} = \sigma\big(U g_\theta(\Lambda) U^T x\big) \tag{13}$$

$$g_\theta(\Lambda) = \begin{pmatrix} \sum_{j=0}^{K} \alpha_j \lambda_1^j & & \\ & \ddots & \\ & & \sum_{j=0}^{K} \alpha_j \lambda_n^j \end{pmatrix}$$

$$= \sum\nolimits_{j=0}^{K} \alpha_j \Lambda^j \tag{14}$$

$$U \sum\nolimits_{j=0}^{K} \alpha_j \Lambda^j U^T = \sum\nolimits_{j=0}^{K} \alpha_j U \Lambda^j U^T = \sum\nolimits_{j=0}^{K} \alpha_j L^j \tag{15}$$

The convolution kernel $g_\theta(\Lambda)$ is designed as a diagonal matrix containing a polynomial $\sum\limits_{j=0}^{K} \alpha_j \lambda_i^j$, as shown in Eq. 14 above. Since the Laplacian matrix is a semi-definite positive symmetric matrix, it must be able to perform feature decomposition, and the feature vectors are orthogonal to each other. The matrices formed by the feature vectors are orthogonal matrices, so the feature matrix is simplified, as shown in Eq. 15.

When the input signal is x, according to Eq. 15, Eq. 13 can be reduced to Eq. 16 below,

$$y_{output} = \sigma\left(\sum\nolimits_{j=0}^{K} \alpha_j L^j x\right) \tag{16}$$

3.2   GCN Based on Chebyshev Polynomial Kernel

We can get from Sect. 3.1 that, supposing the input signal is x (x can be considered as a scalar of each input node), then the spectral convolution on Graph can be defined as the product of the convolution kernel ss and x, as shown in Eq. 17.

$$g_\theta \star x = U g_\theta U^T * x \tag{17}$$

Where $g_\theta(\Lambda) = \begin{pmatrix} \theta_1 & & \\ & \ddots & \\ & & \theta_n \end{pmatrix}$, $\theta \varepsilon \mathcal{R}^N$, $\star$ represents the convolution operation,

U is the eigenvector matrix of the second Laplace matrix L mentioned in Sect. 3.1 above, where $L^{sys} = D^{-\frac{1}{2}} L D^{-\frac{1}{2}} = I_N - D^{-\frac{1}{2}} A D^{-\frac{1}{2}}$, obviously, $U^T x$ is the Fourier transform of the input signal x on the Graph. In order to avoid the increase of matrix calculations caused by large graph networks, David K. Hammond et al. Used truncated expansion of Kth of Chebyshev polynomial $T_k(x)$ to approximate the convolution kernel $g_{\theta'}$. As shown in Eq. (18) below. The Chebyshev polynomial can be recursively expressed as Eq. 19.

$$g_{\theta'}(\Lambda) \approx \sum_{k=0}^{K} \theta_k' T_k(\tilde{\Lambda}) \tag{18}$$

Chebyshev Polynomial: $T_k(x) = 2x\, T_{k-1}(x) - T_{k-2}(x)$,

where $T_0(x) = 1$, $T_1(x) = x$ \hfill (19)

Where $\tilde{\Lambda} = \frac{2}{\lambda_{max}} \Lambda - I_N$, λ_{max} is the maximum eigenvalue of the Laplacian matrix L^{sys}, and $\theta' \in \mathcal{R}^N$ is a vector of Chebyshev coefficients. Bringing the convolution kernel $g_{\theta'}(\Lambda)$ in Eq. 18 into Eq. 17 can calculate the following Eq. 20,

$$g_{\theta'}(\Lambda) \star x \approx \sum_{k=0}^{K} \theta'_k T_k(\tilde{L}) * x \tag{20}$$

Where $\tilde{L} = \frac{2}{\lambda_{max}}L - I_N$, this formula is a K-order polynomial in Laplace operator, that is, it only depends on the nodes with a maximum of K steps from the central node (K-order neighborhood).

3.3 GCN Based on 1st Chebyshev Laplacian Linear Kernel

Based on the Chebyshev polynomial kernel in Sect. 3.2, Thomas Kpif et al. limited the order of the convolution kernel of the Chebyshev polynomial approximation to $K = 1$, then the above formula is simplified to the Laplace linear form on Graph. The linear form can alleviate the problem of overfitting the neighborhood structure of graphs with very wide node degrees (such as social networks, reference networks, knowledge graphs, and many other real-world graph data). Further assuming $\lambda_{max} = 2$, then Eq. 20 will be simplified to Eq. 21 to limit the number of parameters one more time to deal with overfitting and minimize the number of operations per layer (such as matrix multiplication). Then assume a single parameter $\theta = \theta'_0 = -\theta'_1$. Since $I_N + D^{-\frac{1}{2}}AD^{-\frac{1}{2}}$ has a eigenvalue range of [0, 2], utilize this operator in GCN will cause numerical instability and gradient explosion or disappear, so renormalization is used again. Assuming a single parameter $\theta = \theta'_0 = -\theta'_1$, it can be simplified as shown in the following Eq. 22,

$$g_\theta \star x \approx \theta'_0 x + \theta'_1 (L - I_N)x = \theta'_0 x - \theta'_1 D^{-\frac{1}{2}}AD^{-\frac{1}{2}} * x \tag{21}$$

$$g_\theta \star x \approx \theta\left(I_N + D^{-\frac{1}{2}}AD^{-\frac{1}{2}}\right) * x$$

$$\approx \theta\left(\tilde{D}^{-\frac{1}{2}}\tilde{A}\tilde{D}^{-\frac{1}{2}}\right) * x \tag{22}$$

where $\tilde{A} = A + I_N$ and $\tilde{D}_{ii} = \sum_j \tilde{A}_{ij}$.

3.4 Semi-supervised Node Classification

After introducing a simple and flexible model $f(X, A)$ to effectively propagate information on the graph, we return to the problem of semi-supervised node classification. Assume that the input signal $X (X \in \mathcal{R}^{N \times C})$ with matrix form $N \times C$, where N is the number of input nodes, C is the characteristic dimension of the input nodes (C input channels), and the parameter matrix of the convolution kernel is $\Gamma \in \mathcal{R}^{C \times F}$), then the output signal $Z(Z \in \mathcal{R}^{N \times F})$ can be written as Eq. 23,

$$Z = \tilde{D}^{-\frac{1}{2}}\tilde{A}\tilde{D}^{-\frac{1}{2}} * X * \Gamma \tag{23}$$

In the following, we consider a two-layer GCN for semi-supervised node classification on a graph with a symmetric adjacency matrix $\hat{A}$. First calculate $\tilde{D} = \tilde{A}^{-\frac{1}{2}}\tilde{A}\tilde{D}^{-\frac{1}{2}}$ in the pre-processing step.

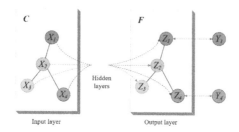

Fig. 2. Semi-supervised graph convolutional network.

As shown in Fig. 2, this is a semi-supervised learning GCN model with C input channels and F output channels. The structure of the graph (shown in black) is shared among the GCN layers, and the node labels are represented by Y_i. At this time, the two-layer GCN propagation model is Eq. 24 below,

$$Z = f(X, A) softmax\left(\hat{A}ReLU\left(\hat{A}XW^{(0)}\right)W^{(1)}\right) \tag{24}$$

Here, $W^{(0)} \in \mathcal{R}^{C \times H}$ is an input-to-hidden weight matrix for a hidden layer with H feature maps. $W^{(1)} \in \mathcal{R}^{H \times F}$ is a hidden-to-output weight matrix. The neural network weights $W^{(0)}$ and $W^{(1)}$ are trained using gradient descent. In this work, we perform batch gradient descent using the full dataset for every training iteration. The *softmax* activation function, defined as $softmax(x_i) = \exp(x_i)/\sum_i \exp(x_i)$, is applied row-wise. For semi-supervised multiclass classification, we then evaluate the cross-entropy error over all labeled examples, as shown in Eq. 25,

$$\mathcal{Q} = -\sum_{i \in Y_i} \sum_{f=1}^{F} Y_{if} \ln Z_{if} \tag{25}$$

where Y_i is the set of node which have labels.

4   Algorithm

4.1   Analysis of GCN Over-Smoothing

We can see that in the above two different nuclear propagation models, such as Eq. 20, the kernel model based on Chebyshev polynomial uses $\sum\limits_{k=0}^{K} \theta_k' T_k(\tilde{L}) * x$, it containing the Laplace matrix $L = I_N - D^{-\frac{1}{2}}AD^{-\frac{1}{2}}$, and in Eq. 22, the Laplace linear kernel containing $\tilde{D}^{-\frac{1}{2}}\tilde{A}\tilde{D}^{-\frac{1}{2}}(\tilde{A} = A + I_N)$. Observation shows that they all contains the adjacency matrix A (the index information of each paper in the citation network or the communication relationship of each object in the social network). Simply speaking, it's the information between the node in the graph and its neighbor. The information of these neighbor nodes is a binarization existence, means any two nodes in the figure are marked as 1 in the matrix when they are related, and are marked as 0 in the matrix when there is no connection. However, the directly problem caused by these binary relationship matrices is that it may lead to excessive smoothing. In other words, the characteristics of these nodes may be too smoothing, so that the nodes from different clusters may become indistinguishable, and two nodes with a closer topological distance (which can be reached after a few hops) are more likely to belong to the same class, otherwise the topological distance of two nodes far away are more likely to belong to different classes.

In addition, another key factor leading to excessive smoothing problems can be considered as excessive mixing of information and noise between nodes. Interactive messages from other nodes may be useful information or harmful noise. For example, in the vertex classification task, intra-class interaction can bring useful information, and information interaction between classes may cause indistinguishable noise between classes.

4.2 Reweighted Mechanism

Therefore, this paper focuses on alleviating the excessive smoothing problem of the above two kernel propagation models, and we propose a Re-weighted mechanism. The relevant algorithm is shown in Eq. 26 below,

$$A'_{ij} = \begin{cases} p/\sum_{j=1}^{C}, & \text{if } i \neq j \\ & i \neq j \\ 1 - p, \text{if } i = j \end{cases} \tag{26}$$

In theory, this mechanism can alleviate the problem of smoothing, where A'_{ij} is the relationship matrix after using the Re-weighted mechanism. p determines the weights assigned to the node itself and other related nodes. By doing this, when updating the node feature, we will have a fixed weight for the node itself and the weights for correlated nodes will be determined by the neighborhood distribution. On the one hand, when $p \to 1$, the characteristics of the node itself will not be considered, on the other hand, when $p \to 0$, the neighboring information will tend to be ignored.

After using the algorithm mechanism of Eq. 26, the above kernel model based on Chebyshev polynomial will be converted to Eq. 27,

Chebyshev Kernel Model (RW):

$$g_{\theta'} \star x \approx \sum_{k=0}^{K} \theta'_k T_k(\tilde{L}_{RW}) * x \tag{27}$$

Then this formula is a variant of the Laplacian K^{th} order polynomial, which means that the kernel model convolution based on the Chebyshev polynomial on the graph is performed under the new relationship matrix A'_{ij}. Among them, $\tilde{L}_{RW} = \frac{2}{\lambda_{max}}L_{RW} - I_N$, and $\tilde{L}_{RW} = \frac{2}{\lambda_{max}}L_{RW} - I_N$,

And the Laplace hierarchical linear model reduced to one parameter is converted to Eq. 28.

Renormalization Trick Kernel Model (RW):

$$g_\theta \bigstar x \approx \theta\left(I_N + D^{-\frac{1}{2}}A'_{ij}D^{-\frac{1}{2}}\right) * x$$

$$\approx \theta\left(\tilde{D}^{-\frac{1}{2}}\tilde{A}'_{ij}\tilde{D}^{-\frac{1}{2}}\right) * x \tag{28}$$

where $\tilde{A}'_{ij} = A'_{ij} + I_N$ and $\tilde{D}_{ii} = \sum_j \tilde{A}'_{ij}$.

We Also return to the problem of semi-supervised node classification. When the signal X in the form of an input matrix is assumed to $X(X \in \mathcal{R}^{N \times C})$, where N is the number of input nodes and C is the characteristic dimension of the input nodes (C input channels), the convolution kernel parameter matrix is defined as $\Gamma \in \mathcal{R}^{C \times F}$, consider a two-layer GCN for semi-supervised node classification of a Graph with a symmetric adjacency matrix A'_{ij}. At this time, the two-layer RWGCN propagation model is transformed into the following Eq. 29,

$$Z = f\left(X, A'_{ij}\right) = softmax\left(\hat{A}'_{ij}ReLU\left(\hat{A}'_{ij}XW^{(0)}\right)W^{(1)}\right) \tag{29}$$

Where $\hat{A}'_{ij} = \tilde{D}^{-\frac{1}{2}}\tilde{A}'_{ij}\tilde{D}^{-\frac{1}{2}}$, the experiment of semi-supervised learning also contains C input channels and F output channels Is carried out on the two-layer RWGCN model, the structural information of the graph is shared among the RWGCN layers, and the label is represented by Y_i.

5   Experiments

5.1   Preparation

We use a lot of experiments to verify the effect of RWGCN to improve the classification performance when mitigating excessive smoothness. The experiment runs in TensorFlow environment, the system reserves 16G memory space, and selects Nvidia graphics cards to assist in training and classification tasks. The graph datasets selected in the experiment include citation network graph datasets (Citeseer, Cora, and Pubmed). The nodes in the graph are documents, and the edges are reference links between documents. The article also uses the data set (NELL) extracted from the knowledge graph, which is a group of directed-connected information entities that are never-ending from the web page. As shown in Table 1 below, each data set has its own name, number of nodes, number of edges, data classification and number of features of the data set, and the labeling rate in the last column indicates the number of labeled nodes and data used for training The ratio of the total number of set nodes. For example, for the Citeseer dataset, the name belongs to the citation network, the number of nodes represents the number of papers it contains, the number of edges represents the citation relationship between any papers, the data classification of the data set represents which magazine the paper belongs to, and the number of features comes from all the word vector representation of the word training included in the paper (excluding a small number of commonly used vocabulary), and the mark rate indicates the ratio of the number of magazine-type papers marked in the data set to the total number of papers.

Table 1. Dataset statistics, as reported in Yang et al. [18].

Dataset	Type	Node	Edges	Classes	Features	Label rate
Citeseer	Citation network	3,327	4,732	6	3,703	0.036
Cora	Citation network	2,708	5,429	7	1,433	0.052
Pubmed	Citation network	19,717	44,338	3	500	0.003
NELL	Knowledge graph	65,755	266,144	210	5,414	0.001

The article is based on the RWGCN model described in the above Eq. 29, trained on multiple graph data sets, and evaluated the accuracy of the model prediction on a test set containing 1,000 labels. The article selects the same data set split as in the paper of Yang et al. [18], and uses 500 additional verification sets with labels for hyperparameter optimization (dropout layer is added to all layers. And several hidden nodes use L2 regularization), and the label of the verification set does not participate in training.

In the citation network graph data set, the experiment optimizes the hyperparameters on the Cora data set, and uses the same common parameter set on Citeseer and Pubmed. The article defines the maximum iteration maximum epoch as 200 when training two kernel models, both use the Adam optimizer [19] and set the learning rate to 0.01, and the early stop window is set to 10, that is, if the loss of the verification set is 10. If the iteration does not decrease, the training is stopped. The article uses the initialization method described in Glorot and Bengio [20] to initialize the weights and normalize the input feature vectors. In terms of hyperparameter settings, the experimental settings of the Citeseer, Cora and Pubmed datasets have a dropout rate of 0.5, the L2 regularization parameter is 5×10^{-4}, and the number of hidden layer nodes is set to 16; and the atlas dataset NELL The dropout rate is set to 0.1, the L2 regularization parameter is 1×10^{-5}, and the number of hidden layer nodes is set to 64.

5.2   Comparison of Methods

The article compares the method of Yang et al. [18], label propagation (LP) [14], semi-supervised embedding (SemiEmb) [16], under the same benchmark. Manifold regularization (ManiReg) [15] and skip-gram graph embedding (DeepWalk) [21]. At the same time, the method in TSVM is deleted [22], because it cannot extend one of our datasets which means a large number of classes. We further compare it with the iterative classification algorithm (ICA) proposed in Lu and Getoor [23], which combines two logistic regression classifiers, one for individual local node features, and the other is Sen et al. [24] uses local features and aggregation operators for classification. Similar to Thomas Kipf et al. [6], we first use all labeled nodes as the training set to train a local classifier, and use it to guide unlabeled nodes for the training of related classifiers. Local classifier guide) Iterate 10 times. L2 regularization parameters and aggregation operators are selected according to the performance of each data set on the validation set. Then, the article is compared with the best variant of the Planetoid model. Finally, on the basis of Thomas Kpif's RWGCN based on Chebyshev kernel and Laplace linear kernel model, a large number of experiments were performed on two improved kernel models with the same benchmark graph dataset. As shown in Table 2, it can be clearly seen that the method RWGCN (Renormalization trick) (in bold) has achieved unprecedented classification effects.

Table 2. Summary of results in terms of classification accuracy.

Method	Citeseer	Cora	Pubmed	NELL
ManiReg	60.1	59.5	70.7	21.8
SemiEmb	59.6	59.0	71.1	26.7
LP	45.3	68.0	63.0	26.5
Deep Walk	43.2	67.2	65.3	58.1
ICA	69.1	75.1	73.9	23.1
Planetoid	64.7	75.7	77.2	61.9
GCN	70.3	81.5	79.0	66.0
RWGCN	**72.3**	**83.1**	**79.5**	**68.0**

This paper also compares the effect of the proposed RWGCN based on two improved nuclear models on the citation network dataset. That is to verify the effectiveness of the Re-Weighted mechanism algorithm in the graph convolution network based on the Chebyshev polynomial kernel propagation model and the Renormalization trick kernel propagation model. The classification accuracy and comparison with previous methods are shown in Table 3.

Table 3. Comparison of propagation models.

Description	Propagation model	Citeseer	Cora	Pubmed
1st-order model	$X\Theta_0 + D^{-\frac{1}{2}}AD^{-\frac{1}{2}}X\Theta_1$	68.3	80.0	77.5
1st-order term only	$D^{-\frac{1}{2}}AD^{-\frac{1}{2}}X\Theta$	68.7	80.5	77.8
Single parameter	$(I_N + D^{-\frac{1}{2}}AD^{-\frac{1}{2}})X\Theta$	69.3	79.2	77.4
Chebyshev filter	$\sum_{k=0}^{K} T_k(\tilde{L})X\Theta_k$	69.6	81.2	73.8
Chebyshev filter kernel model (RW)	$\sum_{k=0}^{K} T_k(\tilde{L}_{Re-w})X\Theta_k$	**71.4**	**81.7**	**78.9**
Renormalization trick	$\tilde{D}^{-\frac{1}{2}}\tilde{A}\tilde{D}^{-\frac{1}{2}}X\Theta$	70.3	81.5	79.0
Renormalization trick kernel model (RW)	$\tilde{D}^{-\frac{1}{2}}\tilde{A}'_{ij}\tilde{D}^{-\frac{1}{2}}X\Theta$	**72.3**	**83.1**	**79.5**

In Table 3, the graph convolutional network based on the Chebyshev polynomial kernel propagation model is represented by Chebyshev filter (RW) (shown in bold), and the Chebyshev polynomial is reduced to a single parameter Laplacian linear propagation kernel model by Renormalization trick (RW) (Shown in bold) means that 'RW' means the nuclear propagation model with Re-Weighted mechanism added. In other cases, the propagation model of the neural network layer will be replaced with the specified propagation model (Description). It can be seen that the semi-supervised classification performance of commonly used graph data sets has been significantly improved on both nuclear propagation models, indicating that our Re-weighted mechanism has a contribution to alleviating the problem of over-smoothing of node features.

5.3 Parameter Discussion

For the RWGCN based on the Chebyshev filter (RW) kernel propagation model, in order to explore the effect of different values of the parameter p on the semi-supervised classification, the article iterates through the set $\{0, 0.1, 0.2, ..., 0.9, 1\}$ to inquiry.

Table 4. CA of Chebyshev filter kernel model (RW) with p-value interval 0–1.

CA	Cora	Citeseer	Pubmed
$p = 0.1$	80.6	70.4	78.3
$p = 0.2$	80.4	69.8	76.9
$p = 0.3$	79.1	69.3	76.8
$p = 0.4$	78.4	69.3	75.3
$p = 0.5$	78.9	69.1	74.1
$p = 0.6$	78.5	69.0	74.1
$p = 0.7$	78.7	69.0	74.2
$p = 0.8$	78.9	69.5	74.7
$p = 0.9$	78.6	69.4	75.2
$p = 1.0$	78.8	70.2	75.8
Average	**79.09**	**69.5**	**75.54**

As shown in Table 4, the figure shows the effect of the network balancing the weight of node information and neighbor node information on mitigating excessive smoothness when updating node characteristics. On the left side of Fig. 3 is the line chart, the abscissa axis represents parameter p of Re-weighted mechanism, and the ordinate axis inform the classification accuracy which represent by 'CA'. On the right side of the Fig. 3 can intuitively represent the classification accuracy on each dataset of Chebyshev filter kernel model (RW). We can observe from the trend of the line chart in Fig. 3 that when the p-value belongs to [0.01, 0.1], there may be a relatively high classification performance, and then we can further to explore the label recognition rate in this interval.

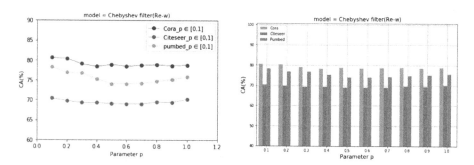

Fig. 3. Trend of CA of Chebyshev filter kernel model (RW) with p-value interval 0-1 on each data set.

Table 5. CA of Chebyshev filter kernel model (RW) with p-value interval 0.01–0.1.

p	0.01	0.02	0.03	0.04	0.05	0.06	0.07	0.08	0.09
Cora	77.8	80.2	81.3	81.6	81.5	81.7	81	80.6	80.9
Citeseer	68.8	69.2	70.3	70.7	71.2	71.4	71.1	70.7	70.6
Pubmed	71.5	75.2	76.5	76.7	78.8	78.9	78.8	78.8	78.2
AVE	**72.7**	**74.9**	**76.0**	**76.3**	**77.2**	**77.3**	**77.0**	**76.7**	**76.6**

As shown in Table 5 and Fig. 4, the abscissa axis of Fig. 4 represents Reweighted parameter p and the ordinate axis inform the 'CA' as mentioned above. Obviously, when $p = 0.06$, the model can reach the highest classification accuracy in the three citation network datasets. Therefore, it can be inferred that more consideration is given to the characteristics of the nodes to obtain a higher label recognition rate in Chebyshev polynomial kernel model, and the information of the edges has little effect on classification capability of model.

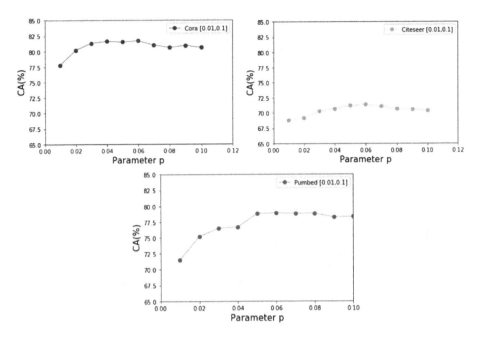

Fig. 4. Trend of CA of Chebyshev filter kernel model (RW) with p-value interval 0.01–0.1 on each data set.

For the RWGCN based on the Renormalization trick (RW) nuclear propagation model, in order to explore the effect of the parameter p on the classification performance, the article also iteratively explores in the set {0, 0.1, 0.2, …, 0.9, 1}.

Table 6. CA of Renormalization trick kernel model (RW) with p-value interval 0–1.

CA	Cora	Citeseer	Pubmed	NELL
$p = 0.1$	76.9	68.8	76.1	8.8
$p = 0.2$	79.9	70.6	76.2	29
$p = 0.3$	81.6	71.3	78.8	15
$p = 0.4$	81.8	71.3	79	24.1
$p = 0.5$	82.6	72	79.2	39.6
$p = 0.6$	82.6	71.1	79.2	55.1
$p = 0.7$	81.8	72.1	78.5	63.3
$p = 0.8$	81.2	71.9	77.8	66.2
$p = 0.9$	81.4	70.8	78	62.7
$p = 1.0$	81.2	69.8	78.1	61.7
AVERAGE	**81.1**	**70.97**	**78.09**	**42.55**

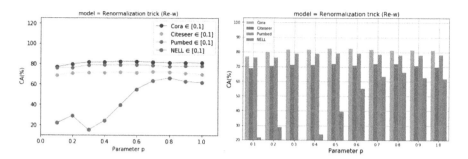

Fig. 5. Trend of CA of Renormalization trick kernel model (RW) with p-value interval 0–1 on each data set.

From the accuracy data in Fig. 5 and Table 6, we can observe that for the Cora dataset, when p-value belongs to [0.5, 0.6], Renormalization trick (RW) kernel model may have relatively high accuracy. For the Citeseer dataset, when p-value belongs to [0.6, 0.8], the model may have a relatively high accuracy. Next, when the p-value of the Pubmed dataset belongs to [0.4, 0.6] and the p-value of the NELL dataset belongs to [0.7, 0.9], the classification accuracy of the model may also be better. Then, we next explore the label recognition rate of these intervals in each data set,

Table 7. CA of Renormalization trick kernel model (RW) with p-value respectively.

p-Value of Cora	0.51	0.52	0.53	0.54	0.55
CA (%)	82.6	82.6	82.6	82.4	82.6
p-Value of Citeseer	0.62	0.64	0.66	0.68	0.70
CA (%)	71.6	71.7	71.9	71.9	72.1
p-Value of Pubmed	0.42	0.44	0.46	0.48	0.50
CA (%)	79.1	79.2	79.2	79.3	79.2
p-Value of NELL	0.72	0.74	0.76	0.78	0.80
CA (%)	65.0	66.5	68.0	67.9	66.2
p-Value of Cora	0.56	0.57	0.58	0.59	0.60
CA (%)	82.6	82.7	83.1	82.6	82.6
p-Value of Citeseer	0.72	0.74	0.76	0.78	0.80
CA (%)	72.2	72.0	72.1	72.3	71.9
p-Value of Pubmed	0.52	0.54	0.56	0.58	0.60
CA (%)	79.1	79.0	79.0	78.8	79.2
p-Value of NELL	0.82	0.84	0.86	0.88	0.90
CA (%)	64.6	63.7	63.2	63.4	62.7

We can get from Fig. 6 and Table 7 that, when $p = 0.57$, the Renormalization trick (RW) kernel model can obtain the highest accuracy rate in the Cora dataset. When $p = 0.78$, the model can obtain the highest accuracy on the Citeseer dataset; meanwhile, when $p = 0.49$, the model can obtain the best accuracy on the Pumped dataset; when $p = 0.76$, Renormalization trick (RW) kernel model can obtain the highest accuracy on the NELL dataset. Therefore, it can be inferred that the Renormalization trick (RW) kernel model more considers the edge information, node edge information has a certain effect on improving the accuracy of model recognition, thereby improving the classification accuracy of the Cora dataset; and the model is more sensitive to the information on the edges of the Citeseer dataset. When adding more edge information, it can use the information of the node edge to improve the accuracy of label classification; for Pumped, it can be seen that the model uses more information of the node itself, and the sensitivity of its own and edge More balanced, that is, when the highest label recognition rate is obtained, node and edge information are equally considered. At the same time, it is obvious that the sensitivity of the model to the information of nodes and edges of the NELL dataset is similar to that of the Citeseer dataset, that is, when more information about the edges is added, it can be used in combination with the information about the edges of the nodes to improve label recognition performance.

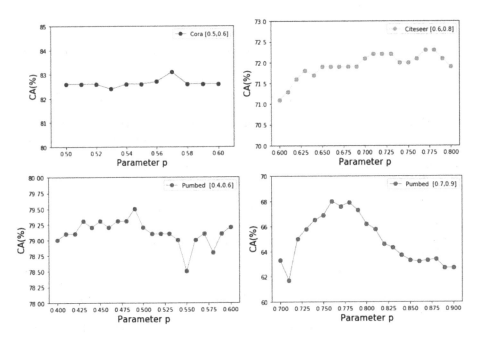

Fig. 6. Trend of CA on each dataset of the Renormalization trick kernel model (RW) with different p-values.

6 Conclusion

The article is guided by GCN semi-supervised classification. In order to alleviate the over-smoothing problem of the two propagation models of Chebyshev convolution kernel and Renormalization trick kernel in processing graph data set classification, an effective Re-weighted mechanism algorithm is applied. And a model named RWGCN was proposed based on two nuclear propagation models of Chebyshev filter (RW) and Renormalization trick (RW). In addition, a large number of experiments were carried out on a variety of common graph data sets (Cora, Citeseer, Pubmed, NELL, etc.), and finally label recognition performance of semi-supervised classification was improved. The influence of the parameter p on the model classification was explored, and the best parameter p of data set was obtained. This article provides research issues for similar over-smoothing problems, and plays a certain role in the future work.

References

1. Bruna, J., Zaremba, W., Szlam, A., LeCun, Y.: Spectral networks and locally connected networks on graphs. In: Proceedings of International Conference on Learning Representations, pp. 253–267 (2014)
2. Belkin, M., Niyogi, P.: Laplacian eigenmaps and spectral techniques for embedding and clustering. In: NIPS, vol. 14, pp. 585–591 (2001)
3. Coifman, R.R., Maggioni, M.: Diffusion wavelets. Appl. Comput. Harm. Anal. **21**(1), 53–94 (2006)
4. Niepert, M., Ahmed, M., Kutzkov, K.: Learning convolutional neural networks for graphs. In: Proceedings of the International Conference on Machine Learning, pp. 2014–2023 (2016)
5. Berkholz, C., Bonsma, P.S., Grohe, M.: Tight lower and upper bounds for the complexity of canonical colour refinement. In: Proceedings of the European Symposium on Algorithms, pp. 145–156 (2013)
6. Kipf, T.N., Welling, M.: Semi-supervised classification with graph convolutional networks. In: Proceedings of the International Conference on Learning Representations, pp. 23–37 (2017)
7. Hammond, D.K., Vandergheynst, P., Gribonval, R.: Wavelets on graphs via spectral graph theory. Appl. Comput. Harmonic Anal. **30**, 129–150 (2011)
8. Hilton, M.: Wavelet and wavelet packet compression of electrocardiograms. IEEE Trans. Biomed. Eng. **44**, 394–402 (1997)
9. Sendur, L., Selesnick, I.: Bivariate shrinkage functions for wavelet-based denoising exploiting interscale dependency. IEEE Trans. Signal Process. **50**, 2744–2756 (2002)
10. Chen, Z.M., Wei, X.S., Wang, P.: Multi-Label Image Recognition with Graph Convolutional Networks, pp. 23–45 (2019)
11. Ge, W., Yang, S., Yu, Y.: Multi-evidence filtering and fusion for multi-label classification, object detection and semantic segmentation based on weakly supervised learning. In: CVPR, pp. 1277–1286 (2018)
12. Lee, C.-W., Fang, W., Yeh, C.-K., Frank Wang, Y.C.: Multi-label zero-shot learning with structured knowledge graphs. In: CVPR, pp. 1576–1585 (2018)
13. Li, X., Zhao, F., Guo, Y.: Multi-label image classification with a probabilistic label enhancement model. In: UAI, pp. 1–10 (2014)

14. Zhu, X., Ghahramani, Z., Lafferty, J.: Semi-supervised learning using gaussian fields and harmonic functions. In: International Conference on Machine Learning (ICML), vol. 3, pp. 912–919 (2003)
15. Belkin, M., Niyogi, P., Sindhwani, V.: Manifold regularization: a geometric framework for learning from labeled and unlabeled examples. J. Mach. Learn. Res. (JMLR) 7, 2399–2434 (2006)
16. Weston, J., Ratle, F., Mobahi, H., Collobert, R.: Deep learning via semi-supervised embedding. In: Montavon, G., Orr, G.B., Müller, K.-R. (eds.) Neural Networks: Tricks of the Trade. LNCS, vol. 7700, pp. 639–655. Springer, Heidelberg (2012). https://doi.org/10.1007/978-3-642-35289-8_34
17. Carlson, A., Betteridge, J., Kisiel, B., Settles, B., Hruschka, E.R., Mitchell, T.M.: Toward an architecture for never-ending language learning. In: AAAI, vol. 5, pp. 3–12 (2010)
18. Yang, Z., Cohen, W., Salakhutdinov, R.: Revisiting semi-supervised learning with graph embeddings. In: International Conference on Machine Learning (ICML) (2016)
19. Kingma, D., Ba, J.: Adam: a method for stochastic optimization. In: Computer Science, pp. 256–278 (2014)
20. Glorot, X., Bengio, Y.: Understanding the difficulty of training deep feedforward neural networks. In: AISTATS, vol. 9, pp. 249–256 (2010)
21. Perozzi, B., Al-Rfou, R., Skiena, S.: Online learning of social representations. In: Proceedings of the 20th ACM SIGKDD International Conference on Knowledge Discovery and Data Mining, pp. 701–710 (2014)
22. Joachims, T.: Transductive inference for text classification using support vector machines. In: International Conference on Machine Learning (ICML), vol. 99, pp. 200–209 (1999)
23. Lu, Q., Getoor, L.: Link-based classification. In: International Conference on Machine Learning (ICML), vol. 3, pp. 496–503 (2003)
24. Sen, P., Namata, G., Bilgic, M., Galligher, B., Eliassi-Rad, T.: Collective classification in network data. AI Mag. 29, 93 (2008)

Improved Random Forest Algorithm Based on Adaptive Step Size Artificial Bee Colony Optimization

Jiuyuan Huo[1,2(✉)], Xuan Qin[1],
Hamzah Murad Mohammed Al-Neshmi[1], Lin Mu[1], and Tao Ju[1,2]

[1] School of Electronic and Information Engineering,
Lanzhou Jiaotong University, Lanzhou 730070, China
huojy@mail.lzjtu.cn
[2] CERNET Co., Ltd., Beijing 100084, China

Abstract. The traditional random forest algorithm works along with unbalanced data, cannot achieve satisfactory prediction results for minority class, and suffers from the parameter selection dilemma. In view of this problem, this paper proposes an unbalanced accuracy weighted random forest algorithm (UAW_RF) based on the adaptive step size artificial bee colony optimization. It combines the ideas of decision tree optimization, sampling selection, and weighted voting to improve the ability of stochastic forest algorithm when dealing with biased data classification. The adaptive step size and the optimal solution were introduced to improve the position updating formula of the artificial bee colony algorithm, and then the parameter combination of the random forest algorithm was iteratively optimized with the advantages of the algorithm. Experimental results show satisfactory accuracies and prove that the method can effectively improve the classification accuracy of the random forest algorithm.

Keywords: Random forest algorithm · Artificial bee colony algorithm · Unbalanced data · Classification problem

1 Introduction

Since Google's AlphaGo defeated Korean go player Li Shishi, machine learning has attracted widespread attention [1]. As an effective machine learning algorithm, the random forest algorithm (RF) can be used in regression and classification tasks. It has the advantages of a good prediction effect under default super parameters, not easy to overfit, strong anti-noise ability, and considerable classification effect in solving non-linear problems. In addition, few parameters affect the performance of the random forest algorithm. And it is easy to see the importance of data characteristics. Its accuracy and efficiency in processing high-dimensional data are not inferior to neural network and SVM algorithm [2]. In view of these advantages, random forest algorithm has been widely used in various industries at home and abroad in recent years, such as medicine, finance, biology, geographic information and agriculture [3]. Although the application of random forest algorithm is very extensive, but in the environment of increasing data volume, the defects of random forest algorithm gradually appear. So, it

© Springer Nature Singapore Pte Ltd. 2020
P. Qin et al. (Eds.): ICPCSEE 2020, CCIS 1258, pp. 216–233, 2020.
https://doi.org/10.1007/978-981-15-7984-4_17

is necessary to improve the random forest algorithm and improve its classification ability. There are three ways to improve the random forest algorithm for data imbalance: the first is to preprocess the data, that is, to combine the random forest algorithm with the data preprocessing. For example, Wu Qiong and others in literature [4] deal with the data with NCL (neighbor cleaning rule) technology and classify the data with the random forest algorithm; the second is to improve the sampling method and the method of weighting the decision tree. For example, Chang Yuqing and others in literature [5] put forward the improved bootstrap resampling method, and then combined with the unbalanced coefficient to weigh the decision tree; the third is to prune the generated random forest. For example, in literature [6], the decision tree in the random forest is selected and added to the optimal subset based on the weighting of the classification interval to improve the real-time classification.

2   Traditional Random Forest Algorithm

The machine forest algorithm is an integrated learning method whose base classifier is a decision tree, which refers to a classifier that uses multiple trees to train and predict samples. This classifier was first proposed by Leo Breiman and Adele Cutler [7]. The flow of the traditional random forest algorithm is shown in Fig. 1.

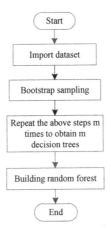

Fig. 1. Traditional random forest algorithm.

The main steps for building a random forest model are as follows:

(1) The data set is sampled by bootstrap sampling to construct a decision tree.
(2) In the decision tree, the attribute is selected as the splitting feature, and the appropriate feature is selected from the splitting feature for optimal splitting. The leaf node represents the prediction result of classification.
(3) The new data were classified and predicted by random forest.

3   Adaptive Step Size Artificial Bee Colony Algorithm

3.1   Artificial Bee Colony Algorithm

Artificial bee colony algorithm (ABC) is a global optimization algorithm based on swarm intelligence, which was proposed by the karaboga [8] in 2005 to optimize the algebraic problem. It is a bionic intelligent calculation method that simulates the behavior of the bee colony to collect honey. It consists of three parts: food source, Employed bee and non-Employed bee, and three control parameters: the number of honey sources N, the limited search *limit* and the maximum number of iterations *MaxCycle*.

3.2   Improved Artificial Bee Colony Algorithm

Although compared with the general swarm intelligence global optimization algorithm, the artificial bee colony algorithm is not easy to fall into the local optimization and has the advantages of simple operation, less setting parameters and fast convergence speed [9]. However, in the later stage of the algorithm, the convergence speed is slow, there is a premature phenomenon, and it is easy to fall into the local optimum. In this paper, the adaptive step size artificial bee colony (ASSABC) is proposed by adjusting the position search formula with the adaptive idea. Referring to the reference [10], the global optimal solution is introduced, and the step variable S and the global optimal solution are introduced based on the original position search formula. Equation (3) is the updating formula of the step variable. In order to reduce the useless search times of the artificial bee colony algorithm and improve the convergence accuracy and search efficiency of the ABC algorithm, the new honeybee source is explored and mined according to formula (2).

$$v_{ij} = x_{ij} + r_{ij}(x_{ij} - x_{kj}) \tag{1}$$

$$v_{ij} = x_{ij} + S_{ij}r_{ij}(x_{ij} - x_{kj}) \tag{2}$$

$$S_{ij} = S_{\min} + (S_{\max} - S_{\min}) \times \frac{|fit_i - fit_{\min}|}{fit_{\text{best}}} + R \tag{3}$$

In the formula, S_{ij} is the current adaptive step size of a honeybee, S_{max} and S_{min} are the historical maximum step size and minimum step size of honeybee respectively. The initial maximum step size and minimum step size of honeybee are both 0. Then, the step size is updated according to the change of the current global optimal solution of each iteration. Every iteration of the program, the step size variable determines whether the step size is updated according to the calculated fitness value and whether the global optimal solution is updated. fit_{best} is the fitness of the global optimal honey source, fit_i is the current honey source fitness, fit_{min} is the fitness of the global minimum honey source, and R is the random number of $[-1, 1]$. X_{best} term can push the new candidate solution to the global optimal solution, and the current step size is proportional to the distance between the current solution and the global optimal solution. The purpose of this design is to make the way to search the honey source more tractive and purposeful. When the distance between the bee and the global optimal solution is relatively long, the step size variable in the position updating formula will change correspondingly, so that the search step size will increase, and the convergence speed of the algorithm will be accelerated; when the distance between the bee and the global optimal solution is relatively close, the search step size will be smaller, and the search step size will always be close to the optimal solution to avoid missing the optimal solution due to too large step size.

4 Unbalanced Accuracy Weighted Random Forest Algorithm

Although the accuracy and efficiency of the traditional random forest algorithm are better than other classification algorithms in the processing of high-dimensional data, the random forest algorithm has its shortcomings in some data. (1) Random forest algorithm cannot predict a few classes well when dealing with unbalanced data; (2) The number of votes in decision tree of traditional random forest algorithm is 1, but the classification ability of each decision tree in random forest is good or bad. If the voting weight is the same, there will be errors in the final result, especially in unbalanced data set; (3) The default parameters of random forest algorithm usually produce a satisfactory prediction result. However, the default parameters are not necessarily the best choice. It is not efficient to find the best parameters by using the traditional grid search method.

In order to solve the appealing problem, this paper proposes an Unbalanced Accuracy Weighted Random Forest algorithm (UAW_RF) that tends to be applied to unbalanced data. This algorithm mainly improves the original algorithm from two parts: the first step is to restrict the bootstrap sampling in the random forest algorithm; the second step is to generate a decision tree group in advance, select a decision tree with strong classification ability in the decision tree group through AUC value to form a random forest, then vote and weight the decision tree in the random forest, and finally use the adaptive step size artificial bee colony algorithm optimizes the parameters of UAW_RF model and finally realizes the two classifications.

4.1 Bootstrap Sampling and Screening

Bootstrap is to generate many new samples with the same dimension from the original samples without using other sample data. For example, for a sample of size N, we want to extract m samples from it for training. Steps are as follows:

(1) Take a random sample from a sample data, write it down, put it back and take the sample again
(2) Repeat (1) m times to get a new sample.

There is random sampling for replacement, and the probability of each sample being drawn is the same. The basic idea of Bootstrap sampling is that when all samples are unknown, Bootstrap uses re-sampling to put the information in the samples that are estimated to be not dried out from the samples into the confidence interval.

Introduce the unbalanced coefficient E of unbalanced data. If the sample set is F, the minority class in the data set is F_{min}, and the majority class is F_{max}. The formula of unbalance coefficient E is as follows:

$$E = \frac{F_{min}}{F_{max}} \tag{4}$$

Using the unbalance coefficient E to select the results of bootstrap sampling and select the useful subset, that is, the samples with more minority classes to build the decision tree.

4.2   Weighted Decision Tree

How many decision trees to build in random forest can only generate as many decision trees for classification and prediction, and the classification ability of decision trees in random forest is uneven. Therefore, a decision tree group more extensive than the random forest for classification, and prediction is established in advance. Then the top decision tree is selected according to AUC value to build the desired decision tree size. For example, if we want to use 100 decision trees for classification, set up a decision tree group of 10 times the size of 100 decision trees in advance, and select the top 100 decision trees according to the AUC value from the 1000 decision trees. The AUC value is chosen as the index of selecting decision tree, which can well reflect the classification ability of decision tree. At the same time, a single accuracy rate can not fully show the strength of the classification ability of decision tree.

The classification ability of every decision tree in random forest is different. In traditional random forest algorithm, the right of every decision tree is the same, which is unreasonable. Decision tree weighting refers to giving weight to the decision tree in the voting process of the decision tree, so that the voting of the decision tree is not recorded as 1 vote, but $1 \times w$, i.e. w vote. The selection of the weight directly affects the final classification result, so it is relatively reasonable to assign the voting weight of the decision tree according to its classification ability. In order to improve the classification ability of random forest on unbalanced data, avoid the situation that the results of random forest classification are heavily biased to the majority of classes, and make the effect of random forest classification on the minority of unbalanced data set also perform well. The weight of decision tree voting is as follows:

$$w_i = E_i + \frac{min_i}{L} \tag{5}$$

Among them, E is the imbalance coefficient of the data sampled from the training set, min_i is the number of correct minority samples predicted by the decision tree, and L is the data sample length of the decision tree. The purpose is to improve the results of the final prediction model on minority data. The fewer minority samples the decision tree predicts correctly and the smaller the proportion of minority samples, the smaller the voting weight. The process is shown in the Fig. 2.

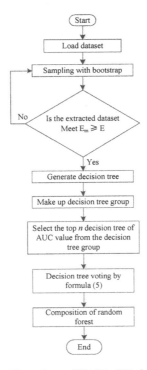

Fig. 2. Flow chart of UAW_ RF algorithm.

4.3 Optimization of Random Forest Model Based on Adaptive Step Size Artificial Bee Colony Algorithm

The default parameters of random forest usually produce a good classification result, but the default parameters are not necessarily the optimal parameter selection. The selection of parameters has a direct impact on the final classification result. The grid search method is to automatically adjust parameters, input parameters, divide the parameters to be optimized into grids in a specific space range, and search for the optimal parameters by traversing all intersections in the grid. However, the efficiency of this method is very low, which is suitable for small data sets [11].

The artificial bee colony algorithm can find the global optimal value through the local optimization behavior of individuals in the bee colony, which is suitable for various optimization problems. This paper introduces the algorithm of ASSABC into the optimization of random forest parameters (see Fig. 3), in which the number of parameters to be optimized corresponds to the dimension of ASSABC algorithm, the objective value of function corresponds to the AUC value of random forest algorithm. The position of the function corresponds to the optimal combination of parameters. The corresponding relationship is shown in Table 1.

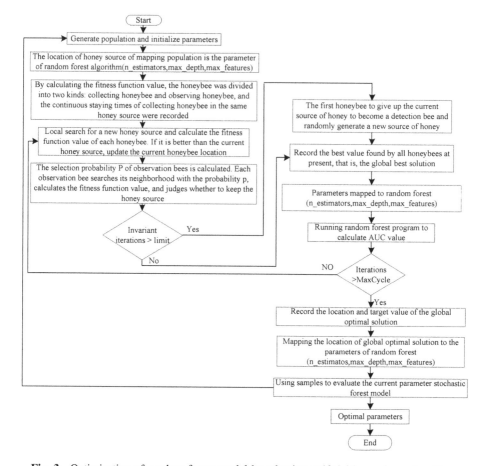

Fig. 3. Optimization of random forest model by adaptive artificial bee colony algorithm.

Table 1. The corresponding relationship between honey foraging behavior and the optimization of random forest parameters.

Honeybee's behavior in collecting honey	Optimization of random forest parameters
Nectar source location	Parameter combination of Stochastic Forest optimization
Source of nectar	AUC value of random forest
Speed of finding nectar	The process of finding random forest parameters
Maximum nectar	Maximum AUC value

In this paper, the main steps to optimize the parameter combination of UAW_ RF algorithm are as follows:

(1) The initial parameters of the algorithm are determined. The number of decision trees is randomly set as *n_estimators*; the depth parameter of decision trees is max_depth and the number of attributes to be considered for the optimal split point is max_features.

(2) Bootstrap sampling is used to sample data sets. N training sets are randomly generated, and M training sets with unbalanced coefficients $E \geqslant E_i$ are selected.

(3) M training sets establish M decision trees, and the decision trees with the highest AUC value are selected. According to the formula (3.4), the voting weight is carried out to form a random forest model to calculate AUC value.

(4) Set the population size, iterations, limited searches, parameters to be optimized, i.e. dimensions, and upper and lower bounds.

(5) The AUC value is taken as the objective function value, the population fitness after initialization is calculated, and the global optimal fitness value is found.

(6) The three parameters mentioned in step (1) are optimized, and the parameters are transferred to the random forest algorithm, which is used to build the model and calculate the AUC value.

(7) Iterative optimization is repeated (6) until the cycle condition is terminated.

5 Experimental Results and Analysis

5.1 Parameter Setting

(1) ABC and ASSABC algorithm parameter settings

The parameters of the comparative experiment between ABC algorithm and ASSABC algorithm are as follows: population size $N = 30$, *limit* = 100, number of iterations *MaxCycle* = 5000, dimension $D = 100$. In the experiment, the ASSABC and ABC algorithms run more than 10 times independently.

(2) RF and UAW_RF algorithm parameter settings

Three parameters of the random forest algorithm, which are the number of decision trees *n_estimators* (optimization range: 1–100), the depth of decision trees *max_depth* (optimization range: 1–45) and the number of attributes to be considered for the optimal split point *max_features* (optimization range includes int type: 2–20 and String type: "auto", "sqrt", "log2"). The default parameters of the traditional random forest algorithm and UAW_RF algorithm are (100, none, auto). The parameters of ASSABC algorithm are population size $N = 60$, iteration times *MaxCycle* = 200, limited search times *limit* = 20, dimension $D = 3$, which is the number of optimization parameters. In KEEL (https://sci2s.ugr.es/keel/imbalanced.php#subA) database, segment 0 data set is selected. At the same time, KDD cup 1999, a multi-classification data set, is divided into three data sets, U2R, R2L and PROBE, according to the three unusual cases of normal ratio U2R, R2L and PROBE. Taking 70% of the data in the data set as the training set, the remaining 30% as the test sample, and take the median value of the results after 5 experiments. KEEL database classifies and arranges each data set according to its characteristics, which is convenient for searching on-demand. However, the number of samples in the data set is small, the imbalance coefficient is not large, and users need to process the characteristic data themselves. WGRF in the figure refers to using the grid search method to find the optimal parameters of RF algorithm, and ASSABCRF refers to using the ASSABC algorithm to find the optimal parameters of RF, and so on.

5.2   Result Analysis

5.2.1   Optimization Performance Comparison of Bee Colony Algorithm

Five representative test functions were selected from the relevant literature [10, 12–16] and, of which F_1 and F_2 are continuous unimodal functions, F_3 and F_5 are nonlinear multi-modal functions, F_4 is a complex multi-modal function with strong correlation between variables [10, 12–16] (Table 2).

Table 2. Test functions.

Function expression	Range of values	Optimal solution
$F_1(x) = \sum_{i=1}^{n} x_i^2$	$[-100, 100]$	0
$F_2 = \sum_{i=1}^{n} \|x_i\|^{i+1}$	$[-1, 1]$	0
$F_3(x) = 1 + \sum_{i=1}^{n} \left(\frac{x_i^2}{4000}\right) - \prod_{i=1}^{n} \left(\cos\left(\frac{x_i}{\sqrt{i}}\right)\right)$	$[-600, 600]$	0
$F_4(x) = \sum_{i=1}^{n-1} \left[100\left(x_{i+1} - x_i^2\right)^2 + (x_i - 1)^2\right]$	$[-30, 30]$	0
$F_5(x) = 10n + \sum_{i=1}^{n} \left(x_i^2 - 10\cos(2\pi x_i)\right)$	$[-5.12, 5.12]$	0

Table 3. Optimization results of test function in 100 dimensions.

Function	ABC			ASSABC		
	Average value	Minimum value	Mean square deviation	Average value	Minimum value	Mean square deviation
F_1	2.08E−14	4.64E−15	3.40E−14	3.14E−15	2.87E−15	2.48E−16
F_2	3.11E−09	1.58E−10	2.90E−09	1.20E−15	8.98E−17	1.12E−15
F_3	4.73E−14	1.12E−14	6.14E−14	2.78E−15	9.99E−16	1.93E−15
F_4	1.90E+00	1.07E−01	2.02E+00	6.34E+01	2.56E+00	3.33E+01
F_5	8.30E−07	1.30E−10	2.04E−06	5.46E−13	2.27E−13	2.67E−13

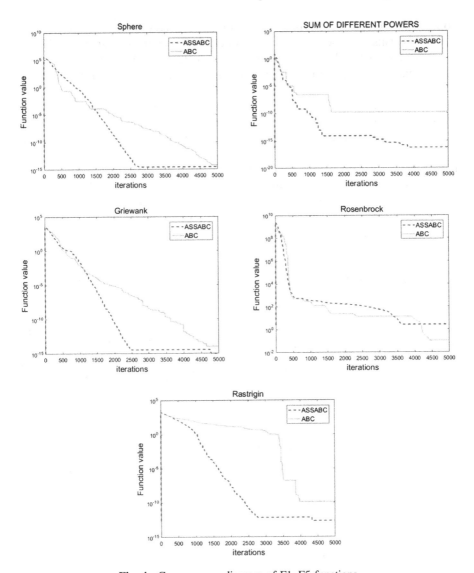

Fig. 4. Convergence diagram of F1–F5 functions.

Comparing the data of ABC and ASSABC algorithm in Table 3 and the convergence graph in Fig. 4, the following conclusions are drawn. The results of the two algorithms in function F_1 show that the convergence speed of ASSABC algorithm is improved. The result of function F_2 shows that the convergence accuracy of ASSABC is better than that of ABC. This shows that bee learning in the process of mining accelerates the convergence speed and accuracy of the algorithm. From the variance point of view, ASSABC algorithm has a certain stability, so we can see that ASSABC algorithm is much better than ABC algorithm in convergence accuracy and algorithm stability. From the operation results of functions F_3 and F_4, it can be seen that ASSABC algorithm fails to optimize the function and only accelerates the convergence speed of the function. This is because bees learn from each other in the mining process and do not balance the relationship between each item, which makes the mining ability of the algorithm insufficient. From the optimization results of F_5 test function, it can be seen that ASSABC algorithm enhances the ability to jump out of the local optimum and avoids falling into the local optimum while increasing the convergence speed and accuracy. From the results of the two algorithms in functions F_3, F_4 and F_5, it can be seen that although the ability of ASSABC algorithm to jump out of the local optimum is improved, the sensitivity of step variable in some multimodal functions is not enough, which leads to the instability of the program and easy to fall into the local optimum. It can be seen from this that although the convergence speed, accuracy and local optimization ability of ASSABC algorithm in some test functions have been improved, the results in complex multimodal functions are not very ideal, and it is easy to fall into local optimization due to insufficient mining capacity and insufficient sensitivity of step variable.

5.2.2   Performance Comparison of Improved Random Forest Algorithm

Traditional classifiers generally use classification accuracy to evaluate the classification performance of the model, but this index is not suitable for unbalanced data sets [17]. Therefore, AUC value and kappa coefficient based on the confusion matrix (see Table 4) are selected as the evaluation index of a classifier. AUC value is based on the confusion matrix, which is shown in Table 4. TP and TN respectively represent the number of positive and negative samples predicted correctly; FN represents the number of samples predicted as negative in fact; FP represents the number of samples predicted as positive in fact. Select from Table 5 (Figs. 5 and 6).

Table 4. Confusion matrix.

	Predictive positive class	Predictive negative class
Real positive class	TP	FN
Actual negative class	FP	TN

Table 5. Unbalanced data set information.

Data set	Sample size	Characteristic number	Unbalance coefficient
Segment0	2307	19	6.01
U2R	812867	41	813.63
R2L	812865	41	15631.02
PROBE	826664	41	58.68

Table 6. Algorithm performance.

Dataset	Name	Optimization	Parameter	Train set accuracy	Test set accuracy	Train set AUC	Test set AUC	Kappa
Segment0	RF	Default	100, None, auto	100%	99.20%	1	0.9705	0.9650
	RF	Grid	250, None, auto	100%	99.36%	1	0.9764	0.9722
	RF	ASSABC	168, 26, log2	100%	99.36%	1	0.9764	0.9722
	UAW_RF	Default	100, None, auto	100%	99.36%	1	0.9764	0.9722
	UAW_RF	Grid	107, None, auto	100%	99.36%	1	0.9863	0.9727
	UAW_RF	ASSABC	181, 25, 15	100%	99.52%	1	0.9823	0.9792
U2R	RF	Default	100, None, auto	99.99%	99.99%	0.9861	0.7812	0.6922
	RF	Grid	103, None, auto	100%	99.99%	1	0.8181	0.7368
	RF	ASSABC	2, 21, 23	99.99%	99.99%	0.9999	0.9210	0.6956
	UAW_RF	Default	100, None, auto	100%	99.99%	1	0.8529	0.8275
	UAW_RF	Grid	52, None, auto	100%	99.99%	1	0.8181	0.7368
	UAW_RF	ASSABC	17, 18, 12	100%	99.99%	1	0.8332	0.7389
R2L	RF	Default	100, None, auto	100%	99.99%	0.9998	0.9565	0.9476
	RF	Grid	162, None, auto	100%	99.99%	1	0.9904	0.9865
	RF	ASSABC	100, 29, auto	100%	99.99%	1	0.9674	0.9501
	UAW_RF	Default	100, None, auto	100%	99.99%	0.9999	0.9871	0.9817
	UAW_RF	Grid	167, None, auto	100%	99.99%	1	0.9873	0.9827
	UAW_RF	ASSABC	198, 27, log2	100%	99.99%	1	0.9898	0.9892
PROBE	RF	Default	100, None, auto	99.99%	99.99%	1	0.9981	0.9980
	RF	Grid	143, 21, auto	99.99%	99.99%	0.9999	0.9978	0.9972
	RF	ASSABC	142, 21, auto	99.99%	99.99%	1	0.9987	0.9983
	UAW_RF	Default	100, None, auto	100%	99.99%	1	0.9979	0.9978
	UAW_RF	Grid	134, None, auto	100%	99.99%	1	0.9983	0.9982
	UAW_RF	ASSABC	245, 25, auto	100%	99.99%	1	0.9988	0.9985

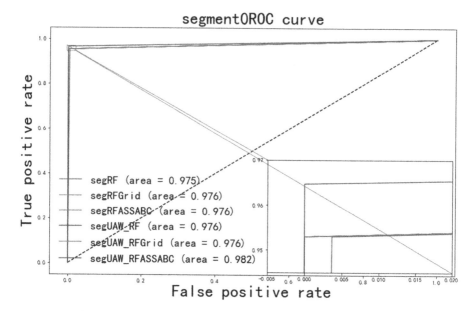

Fig. 5. ROC curve of segment0 data set.

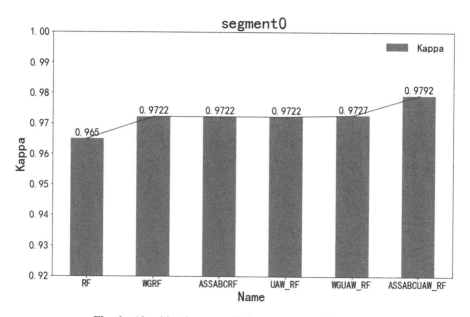

Fig. 6. Algorithm kappa coefficient on segment0 data set.

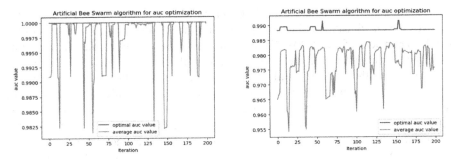

Fig. 7. Optimization iteration chart of segment 0 data set of ASSABC algorithm.

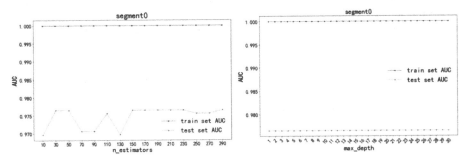

Fig. 8. Effect of n_estimators and max_depth on segment0 dataset.

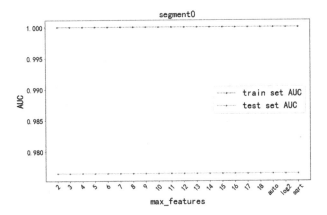

Fig. 9. Effect of max_features on segment0 dataset.

According to Table 6 and Fig. 5, it can be seen that the UAW_RF algorithm improves the classification ability compared with the RF algorithm. The results of the default parameters of the UAW_RF algorithm and the results of the RF algorithm optimization parameters are the same, indicating the effectiveness of the improved algorithm. From the comparison of parameters, it can be seen that the optimization of parameters in segment0 dataset is helpful to improve the classification ability of the algorithm. However, it is shown in Fig. 8 and Fig. 9 that in the segment0 dataset, parameter optimization only needs the number of decision trees, because in the other two parameters, it can be seen from the figure that any parameter can be selected. Random forest algorithm is to ensure the accuracy and AUC value of the algorithm. The smaller the scale of random forest is, the smaller the depth of tree is, and the higher the efficiency is, the better the algorithm is. Therefore, the correctness and effectiveness of the grid optimization and the ASABC algorithm on this data set. From the optimization results of the UAW_RF algorithm, it can be seen that the ASABC optimization results are better than the grid optimization results. From Fig. 7, it can be seen that the ASABC optimization results have one more calendar average AUC value comparison than the grid in performance analysis comparison. The ROC curve of the later data set is not pasted because the display of ROC curve is not easy to see the changes on the graph. WGRF in the figure refers to using grid search method to find the optimal parameters of RF algorithm, and ASSABCRF refers to using ASSABC algorithm to find the optimal parameters of RF, and so on (Fig. 10).

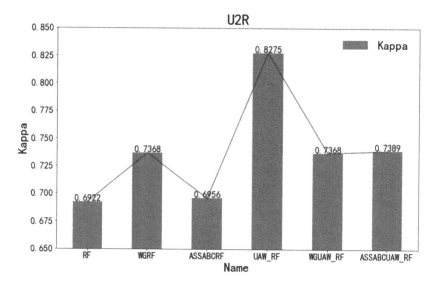

Fig. 10. Kappa coefficient of algorithm on U2R data set.

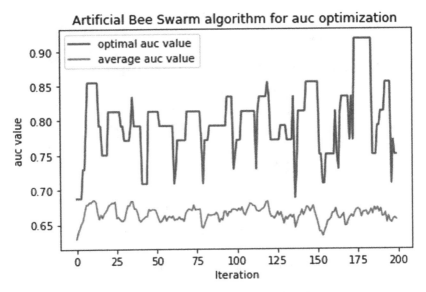

Fig. 11. ASSABC iterative optimization chart.

The performance of the improved algorithm is better than that of the traditional algorithm. From Fig. 10, it can be seen that the combination of parameters obtained by the two optimization methods is not better than the default parameters in U2R data set, which means that when we use the algorithm for optimization and model selection, we need to make an appropriate human judgment and do not rely on the algorithm excessively (Fig. 11).

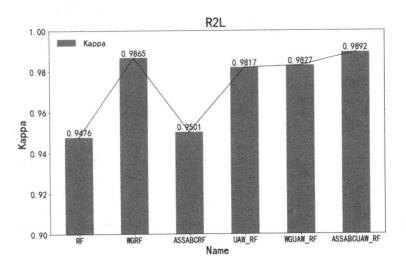

Fig. 12. Kappa coefficient of algorithm on R2L dataset.

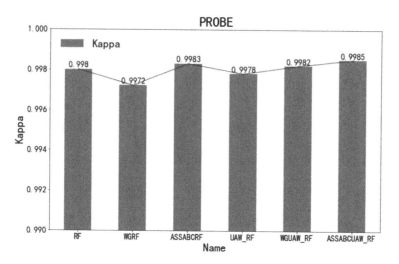

Fig. 13. Kappa coefficient of algorithm on PROBE dataset.

According to the results of R2L and probe data sets (Figs. 12 and 13), the UAW_RF algorithm improves the classification ability compared with RF algorithm. The results of the default parameters of UAW_RF algorithm and the optimization parameters of RF algorithm are the same, which shows the effectiveness of the improved algorithm. From the comparison of parameters, it can be seen that parameter optimization on R2L and probe data sets helps improve the classification ability of the algorithm. However, it is reflected in AUC diagram of parameter pair data set on R2L and PROBE data sets that parameter optimization only needs the number of decision trees, because in the other two parameters, it can be seen from the diagram that any parameter can be selected. According to Table 6 and the curve of AUC value and Kappa coefficient on each data set, we can see that the prediction ability of the improved random forest algorithm is better and it is necessary to find the appropriate combination of parameters for the random forest model in some data.

6   Conclusion

In view of the shortcomings of random forest algorithm in unbalanced data set, a weighted random forest algorithm with unbalanced accuracy UAW_RF algorithm is proposed. The adaptive artificial bee colony algorithm is used to optimize the three important parameters of random forest algorithm and UAW_RF. By comparing the performance of the four datasets, we can see that the classification ability of UAW_RF algorithm is better than that of RF algorithm. The results show that it is necessary to optimize the parameters in most datasets. At the same time, the parameters classification results obtained by the algorithm are better than those obtained by grid search in some datasets.

Acknowledgement. This work is supported by the CERNET Innovation Project (No. NGII20190315) and the Foundation of A Hundred Youth Talents Training Program of Lanzhou Jiaotong University.

References

1. Raska, S., Ming, G., Ying, X., Tao, H.: Python Machine Learning. China Machine Press, Beijing (2017)
2. Caruana, R., Karampatziakis, N., Yessenalina, A.: An empirical evaluation of supervised learning in high dimensions In: International Conference on Machine Learning, San Diego, California, USA, pp. 96–103. ACM (2008)
3. Ren, C.: Urban PM (2.5) concentration prediction based on parallel random forest. Taiyuan University of Technology (2018)
4. Wu, Q., Li, Y., Zheng, X.: Stochastic forest algorithm optimization for unbalanced training set classification. Ind. Control. Comput. **26**(7), 89–90 (2013)
5. Chang, Y., Sun, X., Zhong, L., Wang, F., Liu, Y.: State evaluation of industrial process operation based on Improved Stochastic Forest algorithm. Acta Autom. Sin., 1–10 (2019). https://doi.org/10.16383/j.aas.c190066
6. Xu, Y., Zhang, J., Gong, X., Jiang, K., Zhou, H., Yin, J.: Real time flow classification method of power business based on Improved Stochastic Forest algorithm. Power Syst. Prot. Control. **44**(24), 82–89 (2016)
7. Breiman, L.: Random forests. Mach. Learn. **45**(1), 5–32 (2001)
8. Karaboga, D., Akay, B.: A comparative study of artificial bee colony algorithm. Appl. Math. Comput. **214**(1), 108–132 (2009)
9. Zhang, Q., Li, P., Wang, M.: Optimization algorithm of artificial bee colony based on adaptive evolutionary strategy. J. Univ. Electron. Sci. Technol. **48**(04), 560–566 (2019)
10. Zhu, G., Kwong, S.: Gbest-guided artificial bee colony algorithm for numerical function optimization. Appl. Math. Comput. **217**(7), 3166–3173 (2010)
11. Zhou, T., Ming, D., Zhao, R.: Land cover classification by parameter optimization stochastic forest algorithm. Mapp. Sci. **42**(02), 88–94 (2017)
12. Liu, B., Jiang, M., Zhang, Z.: Artificial bee colony algorithm based on tabu search and its application. Comput. Appl. Res. **32**(07), 2005–2008 (2015)
13. Ma, A., Zhang, C., Zhang, B., Zhang, X.: An adaptive artificial swarm algorithm for classification problems. J. Jilin Univ. (Eng. Ed.) **46**(01), 252–258 (2016)
14. Du, Z., Liu, G., Han, D., Yu, X., Jia, J.: Elite artificial swarm algorithm based on global unbiased search strategy. J. Electron. **46**(02), 308–314 (2018)
15. Sankar, P., Vishwanath, N., Lang, H., et al.: An effective content based medical image retrieval by using ABC based artificial neural network (ANN). Curr. Med. Imaging Rev. **12**(999), 1 (2016)
16. Luo, J., Liu, Q., Yang, Y., et al.: An artificial bee colony algorithm for multi-objective optimisation. Appl. Soft Comput. **50**, 235–251 (2017)
17. Qian, X., Qin, J., Song, W.: Improved parallel stochastic forest algorithm and its out of package estimation. Comput. Appl. Res. **35**(06), 1651–1654 (2018)

Application

Visual Collaborative Maintenance Method for Urban Rail Vehicle Bogie Based on Operation State Prediction

Yi Liu[1,2,3(✉)], Qi Chang[3], Qinghai Gan[2], Guili Huang[2],
Dehong Chen[2,3], and Lin Li[1,2]

[1] CRRC Corporation Limited, Zhuzhou 412000, China
liuyi_hust@163.com
[2] National Innovation Center of Advanced Rail Transit Equipment,
Zhuzhou 412000, China
[3] Hunan University of Technology, Zhuzhou 412000, China

Abstract. The bogie is a crucial component of urban rail vehicles, and its performance plays a decisive role in the safe operation of vehicles. Aiming at the intelligent operation and maintenance requirements of rail transit equipment, in this paper, it takes several key parts of the urban rail vehicle bogie system as research objects, such as motor bearings, frames, fasteners, etc., and proposes a three-dimensional (3D) visual collaborative maintenance method. Firstly, a multi-sensor urban rail vehicle bogie running simulation experiment analysis platform was constructed, thereby establishing a database of running state and performance characteristics of the bogie in the whole life cycle. Then, the health status of key components of bogie was predicted by the state interval prediction model. Finally, the three-dimensional visual collaborative maintenance model proposed in this paper was integrated to realize the early warning of the bogie operation faults, 3D precise guidance of automatic location and maintenance operation information, and collaborative sharing of visual information among multiple users.

Keywords: Rail traffic equipment · State prediction · Intelligent operation and maintenance · 3D visualization

1 Introduction

As one of the key components of urban rail vehicles, the bogie plays an important role in moving guide, bearing and vibration reduction, and is also the final executor of traction and braking [1–3]. Because of the short start-stop interval, large passenger flow change and long running time, the urban rail vehicles are subjected to frequent random

This work was supported in part by the Key Project of Research and Development Plan of Hunan Province under Grant 2018GK2044, in part by the Natural Science Foundation of Hunan Province under Grant 2018JJ4084, in part by the National Natural Science Youth Fund Project under Grant 51805168, and in part by the Science and Technology Talent Project of Hunan Province – Huxiang Youth Talent under Grant 2019RS2062.

© Springer Nature Singapore Pte Ltd. 2020
P. Qin et al. (Eds.): ICPCSEE 2020, CCIS 1258, pp. 237–245, 2020.
https://doi.org/10.1007/978-981-15-7984-4_18

traffic change load in operation, which may lead to the failure of the bogie structure of the urban rail vehicles, and then lead to the deterioration of the running quality of the urban rail vehicles, and even lead to serious safety accidents such as fatigue cracking and derailment. At present, there are difficulties in the later maintenance of urban rail vehicle bogies such as high maintenance costs, low maintenance efficiency, and difficult maintenance. The reasons are as follows. On the one hand, due to the lack of reliable prediction data of operation status for urban rail vehicle bogies, the domestic maintenance mechanism of bogies usually adopts regular maintenance [4–10]. This maintenance method not only has high maintenance cost, but also tends to make trains run with diseases, which has hidden safety risks that cannot be ignored. On the other hand, the urban rail vehicle bogie is a complex electromechanical integration system with a large amount of professional knowledge, fault theory, and maintenance information [11, 12]. It is difficult for the maintenance personnel to quickly implement bogie fault diagnosis and effective maintenance through their experience and memory.

With the rapid development of Internet of things, big data and other technologies, rail traffic equipment production and operation industries are eager to know the health status of train equipment, components or systems at any time by studying intelligent maintenance modes for real-time sensing, monitoring and diagnosis. Finally, the generation of faults can be predicted and prevented, which can reduce the operation and maintenance costs of products and provide a safer and more reliable transportation operation environment for the society. At present, this intelligent maintenance mode is still in the exploration stage. Most of the domestic and foreign rail transportation fields still adopt two methods of planned maintenance and fault repair for equipment maintenance and repair, which brings heavy economic burden to enterprises [8]. The running data of urban rail vehicles have the characteristics of heterogeneous networks, multiple data types and large amount of data [13–15]. On the basis of fully understanding the operation and maintenance professional knowledge and fault diagnosis theory of urban rail vehicle bogies, we adopt methods such as IOT sensor, big data, machine learning, visual modeling, etc., to realize the transparency of operation data, the punctuality of fault alarm and the visualization of maintenance information for key components and parts of urban rail vehicles bogie. This is also a research direction in line with the intelligent development and broad application market of rail transit equipment industry.

This paper selects key components that have a decisive impact on the bogies of urban rail vehicles, such as wheels, axles, frames, motors, bearings, fasteners, etc., and conducts research on early fault feature identification, condition prediction, and visual maintenance method to establish a complete set of state prediction theory and visual collaborative maintenance platform for key components of the urban rail vehicle bogies. Finally, the transition of urban rail vehicle bogies from paper-based "planned maintenance" guidance to a new method of data-based "preventive maintenance" for visualizing collaborative maintenance is realized.

2   Technical Framework

This research combines the interdisciplinary integration of mechatronics, information science, computer graphics and physics, and uses a combination of theoretical analysis, simulation analysis and experimental verification on the technical route, mainly to realize data collection and analysis and visualization of collaborative maintenance modeling and other technical content. The overall technical architecture of this study is shown in Fig. 1.

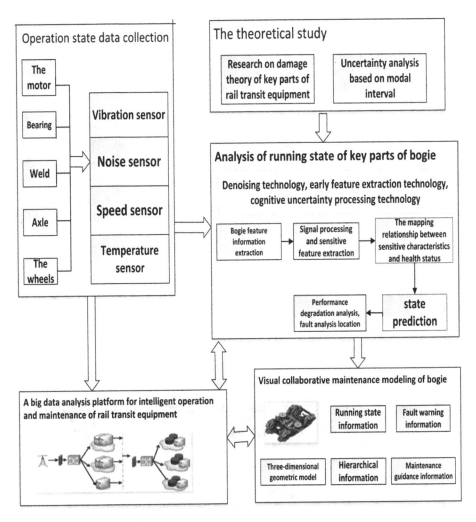

Fig. 1. The overall technical architecture.

3 Technical Proposal

3.1 Design of Multi-sensor Bogie Operation Simulation Experiment Analysis Platform

The operation simulation environment of bogie suitable for various operation environments is constructed by analyzing the components or structures that are easy to cause problems during the operation of urban rail vehicle bogies. This paper focuses on the motor operation simulation of the axle box or holding axle box, the state identification of the bogie frame weld, the state recognition of the tightness of the bogie bolt, and various sensors such as vibration, noise, temperature, and speed are installed at the key points of the experimental object. As shown in Fig. 2, the original data of bogie operation state is acquired through sensors and acquisition cards. Advanced wireless networking technology is applied to transmit data, which provides a real and reliable data source for the prediction of bogie operation conditions.

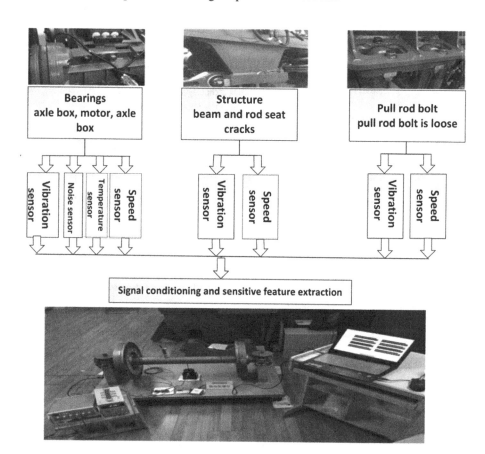

Fig. 2. The simulation platform for multi-sensor bogie operation.

3.2 State Prediction and Reliability Evaluation Method Based on Modal Interval

Based on the performance of state monitoring data and the experimental and theoretical analysis results of the urban rail vehicle bogie, the multi-layer (short, medium and long term) performance index set oriented to the operation state of the train bogie was constructed, and the technical standards and specifications for the evaluation and maintenance of the operation state of the urban rail vehicle bogie were established. Based on the theory of modal interval and hidden Markov model, this paper studies the dynamic performance and state identification algorithm of the bogie, analyzing the service performance and state evolution law of the bogie in the whole life cycle, and provides technical means for the evaluation and maintenance of the bogie. The implementation process is shown in Fig. 3.

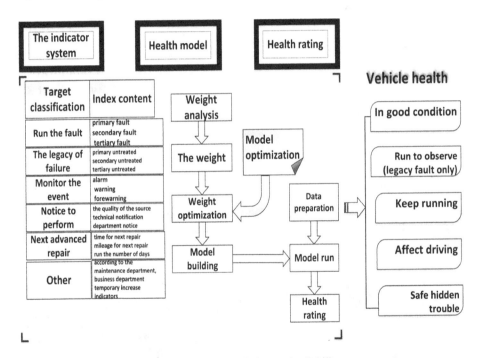

Fig. 3. The process of state prediction and reliability assessment.

3.3 Visual Cooperative Maintenance of Bogies with Real-Time Perception of Operating Status

The historical operation data, operation state prediction data, fault real-time diagnosis data, 3D geometric data and maintenance auxiliary guidance information of the key components of the bogie of urban rail vehicles are collected, and a 3D visual collaborative maintenance model of urban rail vehicles is established, so as to realize the integration of multi-dimensional maintenance assistance information (such as running state information, 3D models, 2D drawings, dynamic videos, etc.) in real scene. As shown in Fig. 4, in order to maintain the consistency of the data and visual scene of each collaborative work end, this paper uses the B/S network communication structure to achieve information communication between the maintenance personnel or between the maintenance personnel and the experts to complete collaborative operations or guidance behavior.

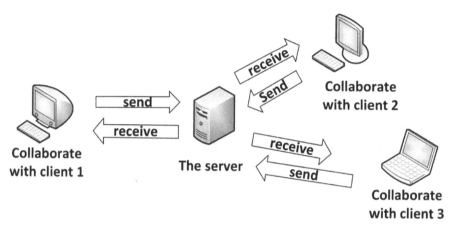

Fig. 4. The B/S network communication structure.

4   Applied Analysis

4.1   Data Acquisition

Aiming at the driving device, weld seam and fastener of bogie, this paper completed the design of multi-sensor simulation experimental platform, as well as the data acquisition and processing of weld, bolt tightening, motor bearing, which is shown in Fig. 5.

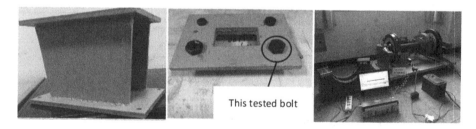

Fig. 5. Data acquisition simulation platform for traction seat, bolt and motor bearing.

4.2   Data Arrangement

This section takes the frame weld seam collection data as the research object, and the data with three states of the frame weld seam are selected respectively. Finally, ten groups of relatively ideal data are selected according to the specific experimental conditions. It is expected that six groups of experimental data will be used as training data, and the other four groups of data will be used as test data. The details are shown in Table 1.

Table 1. Experimental data.

States	Normal	Small crack	Big crack
Data labels	Normal	Small	Big
Single set of sample points	60000	60000	60000
Training data	6 groups	6 groups	6 groups
Test data	4 groups	4 groups	4 groups
All sample	10 groups	10 groups	10 groups

4.3   Data Analysis

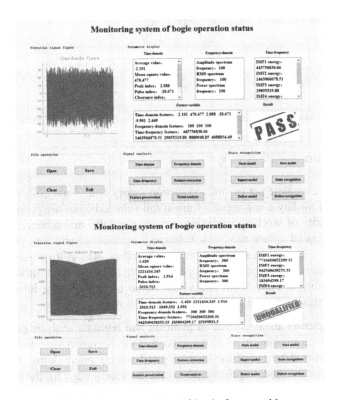

Fig. 6. State recognition of bogie frame welds.

The obtained experimental data are analyzed, and relevant experimental results are obtained, which can be clearly described in Fig. 6.

4.4   3D Visual Collaborative Maintenance Guidance

Based on the historical operation data, operation state prediction data, fault real-time diagnosis data, 3D geometry data and maintenance auxiliary guidance of the key parts of the bogie for urban rail vehicles, we establish a 3D visual coordination platform for the bogies of urban rail vehicles based on the B/S architecture to guide the maintenance personnel to complete the bogie maintenance tasks, which is shown in Fig. 7.

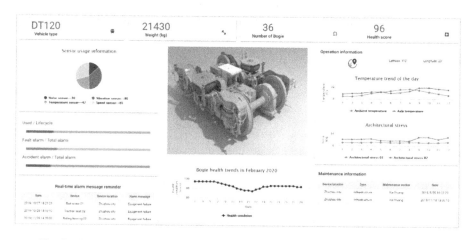

Fig. 7. 3D visual collaborative maintenance guidance for urban rail vehicle bogie.

5   Conclusion

In this paper, a complete set of state prediction theory and visual collaborative maintenance method for key components of urban rail vehicle bogies are proposed. In view of the particularity of working condition, structure, safety, maintenance requirements and other aspects of urban rail vehicle bogies, we take several key parts of bogie system, such as wheels, axles, frames, motors, bearings, etc., as the research object, to solve the problems of transparency of operation data of key parts of urban rail vehicle bogies, punctuality of fault alarm and visualization of maintenance information. The innovations of this study can be summarized as follows.

(1) The multi-sensor data acquisition platform and big data analysis center for real-time active monitoring were constructed, and the operation state and performance characteristics database of the inner city rail vehicle bogie during the whole life cycle was established, to provide advanced theoretical basis for the prediction and maintenance of the health state of key parts of the bogie.

(2) This paper revealed the performance laws of the key parts damage of the bogie in the monitoring signal under the dynamic load changing condition, and established the prediction model of the state interval of the key parts of the city rail bogie, and realized the controllability of making prediction of the key parts' performance of the bogie.

(3) This paper proposes a visual collaborative maintenance modeling method for urban rail vehicle bogie based on running state data, which realizes the transmission and sharing of 3D data model of urban rail vehicle and key parts from the design and production to the operation and maintenance, and innovates the guiding mode of rail transit equipment maintenance.

References

1. Zhang, Y., Li, J.: An overview of typical bogies of urban rail vehicles. Equip. Mach. (04), 14–18 (2017)
2. Hu, J.: Research and application of intelligent operation and maintenance system for Shanghai rail transit vehicles. Mod. Urban Transit (07), 5–9 (2019)
3. Gu, H.: Failure analysis and diagnostic treatment of bogie bearing in high speed EMU. Technol. Wind. (26), 177 (2019)
4. He, G.: Common fault analysis and handling measures of subway vehicle bogie. Eng. Technol. Res. 4(14), 144–145 (2019)
5. Xia L.: Simulation and virtual maintenance of failure monitoring and control of key equipment in subway stations. Guangdong University of Technology (2018)
6. Zhu, X.: Discussion on intelligent maintenance solution for urban rail transit vehicles. Mod. Urban Transit **07**, 16–21 (2019)
7. Huang, W.: Research on vehicle maintenance model and health evaluation of urban railway transit. Mod. Comput. (19), 73–75+100 (2019)
8. Li, X., He, D., Deng, J., Miao, J.: Overview on fault diagnosis methods of urban rain transit vehicle bogie. Equip. Manuf. Technol. **12**, 81–85 (2015)
9. Wang, H., Yang, J.: Overview on the development of railway van bogies in Europe. Locomot. Roll. Stock. Technol. (05), 12–14+20 (2012)
10. Li, F., An, Q., Fu, M., Huang, Y.: Survey on development of bogies for high speed multiple units and the dynamics features. Roll. Stock. (04), 5–9+47 (2008)
11. Liu, B., Zhu, J., Li, X.: Exploration and research on intelligent operation and maintenance system of urban rail transit vehicles. Mod. Urban Transit (06), 16–21 (2019)
12. Liu F.: Research on fault diagnosis method of urban rail transit vehicle bearing. Police Res. Explor. (Cent.) (06), 66–67 (2019)
13. Zhang, Ch., Wang, K., Li, B., Zhang, M., Li, X.: Random vibration fatigue analysis of urban rail transit vehicle ATP antenna beam. Urban Mass Transit **22**(06), 52–55 (2019)
14. Chang, Zh., Lu, X., Zhang, H.: Construction and implementation of big data management platform for rail transit vehicle industry. Urban Mass Transit **22**(02), 1–4 (2019)
15. Zhan, W., Xu, Y., Wang, Y.: Application research on intelligent operation and maintenance system of urban rail transit vehicle. Urban Public Transp. (12), 28–31+36 (2018)

Occlusion Based Discriminative Feature Mining for Vehicle Re-identification

Xianmin Lin[1], Shengwang Peng[1], Zhiqi Ma[1], Xiaoyi Zhou[2],
and Aihua Zheng[1(✉)]

[1] Anhui Provincial Key Laboratory of Multimodal Cognitive Computation,
School of Computer Science and Technology, Anhui University, Hefei, China
ahzheng214@foxmail.com
[2] School of Computer Science and Cyberspace Security, Hainan University,
Haikou, China

Abstract. Existing methods of vehicle re-identification (ReID) focus on training robust models on the fixed data while ignore the diversity in the training data, which limits generalization ability of the models. In this paper, it proposes an occlusion based discriminative feature mining (ODFM) method for vehicle re-identification, which increases the diversity of the training set by synthesizing occlusion samples, to simulate the occlusion problem in the real scene. To better train the ReID model on the data with large occlusions, an attention mechanism was introduced in the mainstream network to learn the discriminative features for vehicle images. Experimental results on two public ReID datasets, VeRi-776 and VehicleID verify the effectiveness of the proposed method comparing to the state-of-the-art methods.

Keywords: Vehicle re-identification · Occlusion · Attention

1 Introduction

With the hot research of person re-identification, vehicle re-identification has attracted more and more attention from researchers. Vehicle re-identification (ReID) is the task of retrieving particular vehicles across different cameras. It is very important task for many other fields such as intelligent video surveillance and smart transportation and social security. Despite of the great progress in recent years, it still faces challenges such as occlusion, light, and similar appearance of different vehicles.

Existing methods of vehicle ReID mainly reply on deep learning technology to learn robust features. They roughly fall into two categories, either constructing complex network models to extract discriminative features, or proposing various loss functions for metric learning. However, due to the relatively clean background with rare occlusion, the models are generally easy to fit on the training data while hard during testing especially with background clutters. Eg., the loss function easily converges during training with however unsatisfactory performance during testing.

On the one hand, due to the view differences across the non-overlapping cameras, one can consider that the vehicle occludes itself. For instance, as shown in Fig. 1(a), the front view vehicle images will occlude the rear parts. On the other hand, the

© Springer Nature Singapore Pte Ltd. 2020
P. Qin et al. (Eds.): ICPCSEE 2020, CCIS 1258, pp. 246–257, 2020.
https://doi.org/10.1007/978-981-15-7984-4_19

background occlusion is ubiquitous, such as the trees, other vehicles or pedestrians, as shown in Fig. 1(b).

(a)   (b)

Fig. 1. Occluded samples from VeRi-776. (a) Image pairs of vehicles with different viewpoints. (b) Some samples occluded by vehicles, trees, pedestrians and other objects.

Unfortunately, existing methods lack of consideration of the occlusion issue during vehicle ReID. Furthermore, training set in the existing datasets are generally with rare occlusion which results in the limited generalization ability of the deep networks while handling the complex scenarios in the real scenes.

Related work [6] evidences that increasing the diversity of the training set can effectively improve the generalization ability of the model. Therefore, we propose to increase the diversity of the training by synthesizing occlusion samples and propose an attention mechanism to mine the discriminative feature for vehicle ReID.

After obtaining the more diverse training data with synthesized occlusion samples, the key issue is how to learn a robust discriminative feature representation while relieve the influence of the occlusions. It is well known that attention plays a very important role in the human visual system [2, 5], which does not process the entire scene information when facing a scene. It instead selectively pay more attention on the prominent parts to better capture the useful information [7]. Inspired by this, we propose to utilize an attention model based feature mining framework for vehicle ReID in this paper. Specifically, we embed spatial attention module and channel attention module into ResNet-50 network and train the ReID model.

As summary, we propose a novel occlusion based discriminative feature mining method for vehicle ReID in this paper. The main contributions of this paper can be summarized as:

– We introduce a data augmentation scheme via synthesizing the occlusion samples to the training data, which can significantly increase the generality of the ReID model especially in the challenging scenarios with self or environmental occlusions.
– We utilize the spatial and channel attention module to emphasize the discriminative region for vehicle ReID by exploring the relationship between both the channel and spatial level on the features.
– Comprehensive experiments on the benchmark datasets evidence the promising performance of the proposed method comparing with the state-of-the-art methods on vehicle ReID.

2   Related Work

We briefly introduce the related vehicle ReID methods in two folds: i) the appearance based vehicle ReID, ii) the attributes and temporal information enhanced vehicle ReID.

2.1   Appearance Based Vehicle ReID

As the main information in computer vision, most of existing works rely on the appearance information for vehicle ReID. Zapletal et al. [18] and Sochor et al. [14] propose to using 3D-boxes to align different vehicle surfaces and use three visible surfaces feature extraction. Zhang et al. [20] design an improved triplet-wise training by classification-oriented loss for vehicle ReID. Li et al. [8] integrate the identification, attribute recognition, verification and triplet tasks into a unified CNN framework. Liu et al. [3] propose a coarse-to-fine ranking method for vehicle ReID, consisting of a vehicle model classification loss, a coarse-grained ranking loss, a fine-grained ranking loss and a pairwise loss. Zhouy et al. [22] propose a Viewpoint-aware Attentive Multi-view Inference (VAMI) model, by first exploiting conditional generative network to generate vehicle images in different views to solve the multi-view problem. Peng et al. [12] propose a cross-camera adaptation framework (CCA), which smooths the bias by exploiting the common space between cameras for all samples. He et al. [4] propose a two-module framework that combines appearance and corresponding license plate features for vehicle Re-ID. In the appearance module, they designed a Two-Branch Network to extract comprehensive global features.

2.2   Attribute and Spatio-Temporal Information Enhanced Vehicle ReID

To overcome the appearance limitation across the cameras, auxiliary information, including attributes and spatio-temporal information have been integrated in vehicle ReID task. Liu et al. [11] design a progressive searching scheme which employed the appearance attributes of the vehicle for coarse filtering. Li et al. [8] introduced the attribute recognition into the vehicle ReID framework together with the verification loss and triplet loss into a unified vehicle ReID framework. Shen et al. [13] combine the visual spatio-temporal path information for regularization. Wang et al. [16] introduce the spatial-temporal regularization into the proposed orientation invariant feature embedding module to boost the vehicle ReID. Liu et al. [10] propose a deep region-aware model (RAM) to extract features from a series of local regions and encourage the deep model to learn discriminative features in both global and local levels. Zhong et al. [21] propose to predict the spatio-temporal motion of the vehicle via the Gaussian distribution probability model, followed by the driving direction estimation via CNN embedded pose classifier, to boost the ReID task. Wang et al. [15] propose a novel Attribute-Guided Network (AGNet) to learn global representation with the abundant attribute features in an end-to-end manner.

3   Methods

This paper mainly propose an occlusion based discriminative feature mining (ODFM) method to consider the occlusion issue in vehicle ReID. The pipeline of the proposed method is shown in Fig. 2, which falls into three phases: First, we train the ReID model until convergence. Second, we find the discriminative regions according to some strategy (which we will describe in Sect. 3.2) to synthesize the occluded samples. Finally, we retrain the ReID model on both original and synthesized occlusion samples (selected according to a certain strategy introduced below) until convergence. It is worth noting that the labels of the synthesized occluded samples are consistent with their original ones.

3.1   Re-identification Network

During the training phase, each vehicle is treated as a class like the common re-identification model. The probability of the vehicle image belonging to each class can be obtained after passing through the FC layer as the classifier. During the test, we remove the classifier layer and utilize the remaining layers for feature extraction.

Given a training set T containing N images of M vehicles (i.e.: M classes), each training image is recorded as $(I_i, m_i), i \in 1, 2, \ldots, N$, where m_i represents the label of the image I_i. The ReID network can be regarded as a function that maps the image I_i to the classification score vector $S_i = g(I_i)$, which is further normalized into a probability distribution by a softmax function (y_{ij} is the predicted value):

$$p(y_{ij}|I_i) = \frac{\exp(s_{ij})}{\sum_{m=1}^{M} \exp(s_{im})}, j = 1, 2 \ldots M \tag{1}$$

We use the cross-entropy loss function to train the model:

$$l_i(\theta) = -\log\left(p(y_{im_i}|I_i)\right) \tag{2}$$

where θ represents the parameters of the network. Therefore, the loss function on the entire training set T can be calculated as:

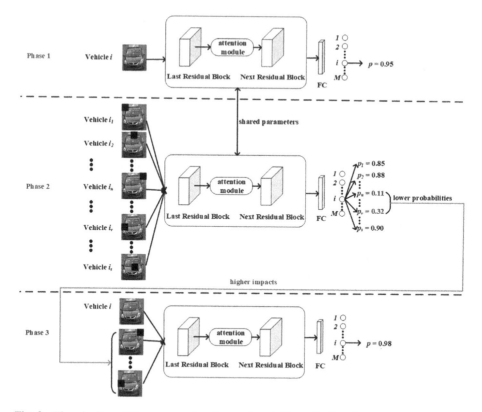

Fig. 2. The pipeline of the proposed occlusion based discriminative feature mining (ODFM) method. Phase 1 trains a ReID model to convergence with the original data. Phase 2 synthesizes the candidate occluded samples and simultaneously evaluates the impact values of each candidate via the pretrained ReID network. Phase 3 inputs both the original data and the selected occlusion samples with high impact values to retrain the ReID model. Note that, we use the attention mechanism all through the procedure.

$$l(\theta) = \frac{1}{N} \sum_{i=1}^{N} l_i(\theta) \tag{3}$$

Minimizing the loss function is equivalent to maximizing the posterior probability of the true value. We use stochastic gradient descent and mini-batch samples to optimize the loss function.

3.2   Occlusion Sample Synthesizing Strategy

To increase the diversity of the training set with more severe scenarios, we propose to synthesize the occlusion samples for data augmentation. Specifically, we use a sliding occlusion mask to mine the discriminative regions. For an image of size $H * W$ and an occlusion mask of size $d * d$, we move from left to right and top to bottom with steps

S_w and S_h, respectively. Therefore, a total of $S = \left(\left\lfloor \frac{W-d}{S_w} \right\rfloor + 1 \right) * \left(\left\lfloor \frac{H-d}{S_h} + 1 \right\rfloor \right)$ new samples are synthesized to the candidate sample pool.

Give a training image and its predicted true value probability p, the impact value of the l-th candidate can be defined as:

$$\widetilde{p}_l = \begin{cases} p - p_l, & p > p_l \\ 0, & p \leq p_l \end{cases} \tag{4}$$

where $p_l, l(l \in 1, 2, \ldots, S)$ indicates the probability value of the l-th candidate. To fairly consider the impact caused by different occlusion positions, we normalize the impact values of the S candidates of each training image into a distribution:

$$\overline{p}_l = \frac{\widetilde{p}_l}{\sum_{j=1}^{S} \widetilde{p}_j}, l = 1, 2, \ldots, S \tag{5}$$

according to which we sample the occluded images, which tend to have higher impact to the ReID task.

3.3   Spatial and Channel Attention Mechanism

To better explore the most discriminative regions in the vehicle images, we propose to introduce the channel and spatial attention mechanism. Given any feature map $X \in R^{C*H*W}$, one-dimensional channel attention feature $M_c \in R^{c*1*1}$ and two-dimensional spatial attention feature $M_s \in R^{1*H*W}$ can be obtained respectively through the channel attention module and the spatial attention module. The entire attention process is as follows:

$$X_1 = M_c \odot X, \; X_2 = M_s \odot X_1 \tag{6}$$

where $\odot$ represents the element-wise production. During the multiplication process, the attention values are sequentially propagated: the channel attention values are propagated along the spatial dimension, and vice versa. X_2 is the feature produced by the attention module. We shall elaborate each attention modules in the following.

Channel Attention Module. We use feature-to-channel relationship to produce channel attention maps. As mentioned in [19], each channel of a feature map can be considered as a feature detector, and channel attention will focus on "what information" of a given image is meaningful. In order to effectively obtain the channel attention map, we compress the input 3D features along the spatial dimension.

First, we use average pooling and maximum pooling to aggregate spatial information to respectively produce two different spatial context descriptors: F_{avg} and F_{max}, followed by the shared network to produce a channel attention map $M_c \in R^{C*1*1}$. The shared network consists of a multilayer perceptron (MLP) and a hidden layer. To reduce the parameters, we set the output size of the hidden layer to $R^{\frac{C}{r}*1*1}$, where r

represents the reduction ratio. Then the two features from the shared network are added together. In short, the channel attention feature is obtained as:

$$M_c = sigmoid(MLP(AvgPool(X)) + MLP(MaxPool(X)))$$
$$= sigmoid\left(W_1\left(W_0\left(F_{avg}\right)\right) + W_1\left(W_0\left(F_{max}\right)\right)\right) \quad (7)$$

where $sigmoid()$ is the activation function, $W_0 \in R^{\frac{C}{r}*C}$, $W_1 \in R^{C*\frac{C}{r}}$ are the shared weights of the multilayer perceptron. A ReLU activation function is added after W_0 (Fig. 3).

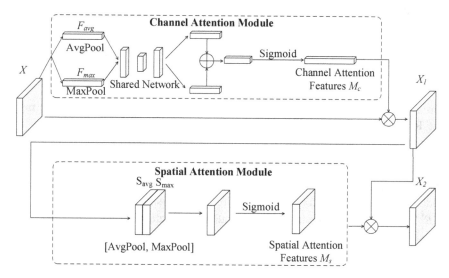

Fig. 3. Attention module diagram. The upper and lower parts demonstrate the channel and spatial attention modules respectively.

Spatial Attention Module. We use the spatial relationship of features to produce spatial attention maps. Unlike channel attention, spatial attention focuses meaningful regions. Note that applying the pooling operation along the channel dimension can effectively highlight the informative regions [17]. Herein, to obtain the spatial attention map, we first apply the average pooling and maximum pooling operations along the channel dimension, to aggregate the channel information of the input features as $S_{avg} \in R^{1*H*W}$ and $S_{max} \in R^{1*H*W}$. Then we concatenate them to synthesize an effective spatial feature descriptor. The concatenated these features are fed into a standard convolutional layer to obtain a spatial attention map $M_s \in R^{H*W}$. The process of obtaining the spatial attention map can be summarized by the following formula:

$$M_s = sigmoid\left(f^{7*7}([AvgPool(X); MaxPool(X)])\right)$$
$$= sigmoid\left(f^{7*7}\left([S_{avg}; S_{max}]\right)\right) \quad (8)$$

where f^{7*7} represents a convolution operation with a kernel size of 7 * 7.

3.4   Implement Details

We use ResNet-50 as our backbone network and employ Euclidean distance to measure similarity during the testing phase. We employ Gaussian weights with the bias as zero to initialize the last layer of ResNet-50, while preserving ImageNet pre-trained weights in all other layers. We use Stochastic Gradient Descent (SGD) to optimize the model with momentum as 0.9 and weight decay as 5e-4. We set the learning rate of the last layer to 0.02, while 0.01 for the other layers, followed by multiplication by 0.1 for every 25 iterations. During the training, we introduce a dropout layer before the classifier to regularize the network. We perform our experiment on Pytorch under Ubuntu16.04 system with a single TITAN xp. The input image is randomly flipped with a probability of 0.5. The batch size is set to 32.

4   Experiments

We evaluate our ODFM on two benchmark vehicle ReID datasets VeRi-776 [11] and VehicleID [9] comparing with the state-of-the-art methods. We use the commonly used metrics, mAP and rank-n to evaluate our experimental results. mAP indicates the mean average precision, while rank-n indicates the right hits in the first n rankings to the query image.

4.1   Experimental Results and Analysis

Results on VeRi-776. Table 1 reports the comparison results with other methods on the VeRi-776 dataset. It can be observed that our method significantly beats the prevalent methods especially on mAP, by 5.83% improvement than the second best method VFL [1]. It is worth noting that our method hasn't used any auxiliary information, like the license plate or spatio-temporal information in Siamese + Path-LSTM [13] and FACT + Plate-SNN + STR [11]. However, it still surpasses them on all metrics, which verifies the competitive performance of the proposed method.

Figure 4 demonstrates five examples of matching results on VeRi-776 dataset [11]. Our model generally achieves satisfactory results. Although some query vehicles are severely occluded by the environment, such as query (a) and (e), our method can still hit most of the right matchings. Furthermore, as shown in the results of query (b) to (d) in Fig. 4, it can correctly hit the gallery vehicles with occlusions or viewpoint changes even the query images are without occlusion or in totally different viewpoints.

Results on VehicleID. Table 2 demonstrates the comparison results with prevalent methods on VehicleID dataset. Generally speaking, our ODFM outperforms the state-of-the-art methods in a large margin. Our method consistently beats the prevalent methods on all the three subsets with promising performance, which yields to a new state-of-the-art for vehicle ReID.

Figure 5 shows several examples of the matching results on VehicleID dataset. Note that most of the vehicle images in VehicleID dataset is either front or back view with only one correct matching image in the gallery. From Fig. 5 we can see, our

Table 1. Comparison results on VeRi-776 [11]. The best three results are highlighted in red, blue and green respectively.

Method	mAP	rank1	rank5	reference
LOMO	9.64	25.33	46.48	CVPR2015
BOW-CN	12.20	33.91	53.69	ICCV2015
GoogLeNet	17.89	52.32	72.17	CVPR2015
FACT	18.49	50.95	73.48	ICME2016
FACT+Plate-SNN+STR	27.70	61.44	78.78	ECCV2016
Siamese-Visual	29.48	41.12	60.31	ICCV2017
Siamese+Path-LSTM	58.27	83.49	90.04	ICCV2017
NuFACT	48.47	76.76	91.42	TMM2018
VAMI	50.13	77.03	90.82	ICPR2018
VRSDNet	53.45	83.49	92.55	CVPR2018
AAVER	58.52	88.68	94.10	ICCV2019
VFL	59.18	88.08	94.63	ICIP2019
ODFM	65.01	91.00	97.85	**Ours**

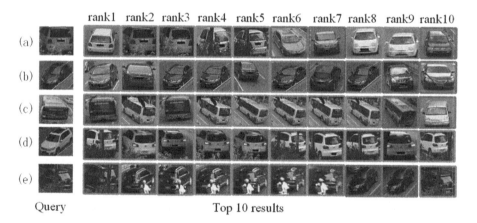

Query   Top 10 results

Fig. 4. Examples of matching result on VeRi-776 dataset [11]. The first column represents the query image, followed by the top 10 ranking results highlighted in green bounding boxes (indicating the right hits) and red bounding boxes (indicating the wrong hits), respectively. (Color figure online)

method can hit the unique right matching with different viewpoint (as shown in query (c)) and occlusion (as shown in query (d)). Meanwhile, it can handle the scenarios with severe illumination or scale changes across the cameras, such as query (a), (b) and (e).

4.2 Ablation Study

To verify the contribution of our method, we further evaluate the two key components in our model: occluded sample synthesizing strategy and attention module on VeRi-776 and VehicleID dataset on 800 test size. As shown in Table 3, both occluded sample synthesizing strategy and the spatial channel attention module play important roles in the ReID model, which verifies the contribution of the two components.

Table 2. Comparisons with state-of-the-art methods on VehicleID [9] (in %). The best three results are highlighted in red, blue and green respectively.

Methods	800		1600		2400		reference
	mAP	rank1	mAP	rank1	mAP	rank1	
LOMO	-	19.76	-	18.85	-	15.32	CVPR2015
BOW-CN	-	13.14	-	12.94	-	10.20	ICCV2015
GoogLeNet	46.20	47.88	44.00	43.40	38.10	38.27	CVPR2015
FACT	-	49.53	-	44.59	-	39.92	ICME2016
NuFACT	-	48.90	-	43.64	-	38.63	TMM2018
VAMI	-	63.12	-	52.87	-	47.34	CVPR2018
VRSDNet	63.52	56.98	57.07	50.57	49.68	42.92	ICPR2018
AAVER	-	72.47	-	66.85	-	63.54	ICCV2019
VFL	-	73.37	-	69.52	-	67.41	ICIP2019
ODFM	83.16	76.57	77.87	71.55	74.75	68.23	**Ours**

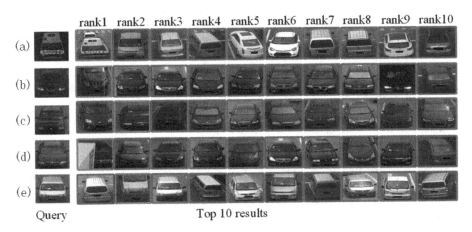

rank1 rank2 rank3 rank4 rank5 rank6 rank7 rank8 rank9 rank10

(a)

(b)

(c)

(d)

(e)

Query   Top 10 results

Fig. 5. Examples of matching result on VehicleID [9]. The first column represents the query image, followed by the top 10 ranking results highlighted in green bounding boxes (indicating the right hits) and red bounding boxes (indicating the wrong hits), respectively. (Color figure online)

Table 3. Ablation study on the occluded sample synthesizing strategy and the spatial channel attention module.

Methods	VeRi-776		VehicleID (800)	
	mAP	rank1	mAP	rank1
baseline	60.55	88.20	80.71	74.20
+attention	62.03	88.92	81.01	74.69
+sample synthesizing	64.62	90.41	82.29	76.29
+sample synthesizing + attention	65.01	91.00	83.16	76.57

5   Conclusion

In this paper, we propose an occlusion based discriminant feature mining (ODFM) method for vehicle re-identification. It first simulated the occlusion problem in real scenes by synthesizing occlusion samples, thereby increasing the diversity of the training set. In addition, the attention model was utilized in the mainstream network to learn the discriminative features of vehicle images. The experimental results on two public datasets VeRi-776 and VehicleID verify the superiority of the proposed method, which yields a new state-of-the-art for vehicle ReID.

Acknowledgement. This research is supported in part by the National Natural Science Foundation of China (Nos. 61976002), Hainan Provincial Natural Science Foundation (Grant No. 117063), the Natural Science Foundation of Anhui Higher Education Institutions of China (KJ2019A0033), and the National Laboratory of Pattern Recognition (NLPR) (201900046).

References

1. Alfasly, S.A.S., et al.: Variational representation learning for vehicle re-identification, pp. 3118–3122 (2019)
2. Corbetta, M., Shulman, G.L.: Control of goal-directed and stimulus-driven attention in the brain. Nat. Rev. Neurosci. **3**(3), 201–215 (2002)
3. Guo, H., et al.: Learning coarse-to-fine structured feature embedding for vehicle re-identification, pp. 6853–6860 (2018)
4. He, Y., Dong, C., Wei, Y.: Combination of appearance and license plate features for vehicle re-identification, pp. 3108–3112 (2019)
5. Itti, L., Koch, C., Niebur, E.: A model of saliency-based visual attention for rapid scene analysis. IEEE Trans. Pattern Anal. Mach. Intell. **20**(11), 1254–1259 (1998)
6. Krizhevsky, A., Sutskever, I., Hinton, G.E.: ImageNet classification with deep convolutional neural networks, pp. 1097–1105 (2012)
7. Larochelle, H., Hinton, G.E.: Learning to combine foveal glimpses with a third-order Boltzmann machine, pp. 1243–1251 (2010)
8. Li, Y., et al.: Deep joint discriminative learning for vehicle re-identification and retrieval, pp. 395–399 (2017)
9. Liu, H., et al.: Deep relative distance learning: tell the difference between similar vehicles, pp. 2167–2175 (2016)

10. Liu, X., et al.: RAM: a region-aware deep model for vehicle re-identification, pp. 1–6 (2018)
11. Liu, X., et al.: A deep learning-based approach to progressive vehicle re-identification for urban surveillance. In: European Conference on Computer Vision (2016)
12. Peng, J., et al.: Eliminating cross-camera bias for vehicle re-identification. In: arXiv: Computer Vision and Pattern Recognition (2019)
13. Shen, Y., et al.: Learning deep neural networks for vehicle re-ID with visual-spatio-temporal path proposals, pp. 1918–1927 (2017)
14. Sochor, J., Herout, A., Havel, J.: BoxCars: 3D boxes as CNN input for improved fine-grained vehicle recognition, pp. 3006–3015 (2016)
15. Wang, H., et al.: Attribute-guided feature learning network for vehicle re-identification. In: arXiv: Computer Vision and Pattern Recognition (2020)
16. Wang, Z., et al.: Orientation invariant feature embedding and spatial temporal regularization for vehicle re-identification. In: 2017 IEEE International Conference on Computer Vision (ICCV) 2017
17. Zagoruyko, S., Komodakis, N.: Paying more attention to attention: improving the performance of convolutional neural networks via attention transfer (2017)
18. Zapletal, D., Herout, A.: Vehicle re-identification for automatic video traffic surveillance, pp. 1568–1574 (2016)
19. Zeiler, M.D., Fergus, R.: Visualizing and understanding convolutional networks, pp. 818–833 (2014)
20. Zhang, Y., Liu, D., Zha, Z.: Improving triplet-wise training of convolutional neural network for vehicle re-identification, pp. 1386–1391 (2017)
21. Zhong, X., et al.: Poses guide spatiotemporal model for vehicle re-identification, pp. 426–439 (2019)
22. Zhouy, Y., Shao, L.: Viewpoint-aware attentive multi-view inference for vehicle re-identification, pp. 6489–6498 (2018)

StarIn: An Approach to Predict the Popularity of GitHub Repository

Leiming Ren[✉], Shimin Shan, Xiujuan Xu, and Yu Liu

School of Software, Dalian University of Technology,
Dalian 116620, Liaoning, China
martin.rlm@foxmail.com

Abstract. The popularity of repository in GitHub is an important indicator to evaluate its quality. Exploring the trend of popularity is a crucial guideline to study its development potential. Herein, StarIn, a stargazer-influence based approach is proposed to predict the popularity of GitHub repository. Using the followers in GitHub as a basic dataset, stargazer-following based network was established. The indicator, stargazer influence, was measured from three aspects of basic influence, network dynamic influence and network static influence. Experiments was conducted, and the correlation of StarIn was analyzed with the popularity of repository from the perspective of six characteristics. The experimental evaluation provides an interesting approach to predict the popularity of repositories in GitHub from a new perspective. StarIn achieves an excellent performance of predicting the popularity of repositories with a high accurate rate under two different classifiers.

Keywords: Stargazer influence · Popularity prediction · Network relationship · GitHub

1 Introduction

With the development of the Internet and cloud computing technologies, social coding becomes a hot research field [5, 20, 21]. Software developers gradually realized that distributed software coding can fully exploit the talents of each development entity, which enables the development of software projects break boundary in time and region. GitHub, the pioneer warehouse in the field of code hosting, has become the largest open source code social platform for learning, collaboration and communication among scholars and developers worldwide [12, 19]. In order to make it more convenient for users to learn and use, GitHub not only allows users to create, submit, modify, and comment on repositories [11], but also provides users to star, fork, and clone repositories [20]. Users could express their recognition or satisfaction of a repository through the behavior of starring or forking. Therefore, star or fork is regarded as a manifestation of popularity [5, 9, 17].

© Springer Nature Singapore Pte Ltd. 2020
P. Qin et al. (Eds.): ICPCSEE 2020, CCIS 1258, pp. 258–273, 2020.
https://doi.org/10.1007/978-981-15-7984-4_20

It is challenging for plagues developers to choose the code which are suitable for their task with a good prospect. On the other hand, it is also crucial for enterprises or companies to filter high-popularity code. Hence, an appropriate method or tool for effectively predicting its popularity is urgently needed.

So far, there are only few studies about predicting popularity of GitHub repository. For instance, Borges et al. [4] are the pioneers to focus on the study of this direction. They analyzed the main factors that influence the popularity of repository, including application domain and programming language. Meanwhile, they found and verified several questions about repository popularity. For example, whether the popularity is related to the number of forks, contributors and submissions, how early repository became popular, etc. For further research, they proposed a traditional linear regressions method, which is applied to two models to predict the popularity of GitHub repositories in different situations [3]. They found that the predicted rankings have a strong correlation with the real ones when newcomers are not considered. However, the study remained several defects that cannot be ignored. First, the experiment of estimating the repositories ranks was conducted under a precondition, that is the proposed models were pre-trained with the number of stars received in a period of time, usually about six months. As a result, it works not very well for the newly released repositories. Second, the experiment was conducted on the premise of a hypothesis of newcomers being not considered. Third, it paid more attention to the repositories themselves, ignoring the influence of GitHub users on them. Taken together, a more comprehensive and effective approach is warranted.

As the previous studies show [1, 3, 5, 8], the star is the most intuitive manifestation of popularity. Consequently a stargazer, who gives stars to a repository, plays an essential role in the development of repository. Stargazers may have different degrees of seniority, and the existing work has explained how the stargazers work on the popularity. For example, Blincoe et al. [2] provided a comprehensive analysis about what motivates people to follow others and how the popular users influence on their followers. Experiments indicated that followers are indeed influenced by popular users, and a major way is that popular users usually guide their followers to the projects they newly release or star. He mentioned that a new type of leadership is emerging through the behavior of following and popularity of users can be more influential than their contribution on the followers. Another important behavior in GitHub is unfollowing. Unfollowing is a kind of relationship dissolution actually. Jiang et al. [10] focused on studying the motivation behind unfollowing. They conducted researches and analyzed the potential impact factors that can cause unfollowing behavior. All these studies indicate that in addition to the quality of repository, the following relationship among users plays a very important role in repository popularity.

In this paper, we present a prediction method of repository popularity called StarIn, which is based on the stargazer influence of repository. A high-popularity repository will attract more influential stargazers. Similarly, influential stargazers paying attention to a repository, will bring more widespread attention to the repository and increase its popularity because they will guide their followers or other users to focus on the repository. Therefore, the aim of this paper is to assess the impact of repository from the perspective of stargazer, and explore the underlying relationship between stargazer influence and popularity in the development process of repository. Specifically, we first establish a stargazer-following based network, and introduce the concept of HFN, which is an evaluation method based on the number of followers of a user. Then we analyze the stargazer influence from three aspects of basic influence, network dynamic influence and network static influence. We carry out an experiment to demonstrate the rationality of our proposed approach. Finally, to evaluate the performance of StarIn, we extract randomly 200 repositories involving different fields and types that the total number of stargazers is over one million. Our experimental results show that the StarIn achieves a good performance of predicting the popularity of repositories with a high accurate rate under two different classification criteria.

Our study provides a number of insights into popularity analysis on GitHub repositories, and compared to previous works, the contribution of this paper is twofold:

– We are the first to evaluate the influence of GitHub repository from the view of stargazer.
– Our proposed StarIn has high prediction performance under different classification criteria.

The remaining of this paper is organized as follows. The workflow and approach we proposed are presented in Sect. 2. In Sect. 3, the rationality of approach is demonstrated. Section 3 also evaluates its performance on popularity prediction, followed by Sect. 4 that discusses our work. Finally, we summarize this paper in Sect. 5.

2   Definition and Methodology

In this section, we will introduce the analysis process firstly, and then present the necessary definitions. Finally, we will describe the approach we proposed to evaluate the stargazer influence. Our workflow is presented in Fig. 1, which consists of the following steps:

1. We collect the GitHub data source, and divide them into the follower, stargazer and repository dataset after preprocessing, and store them in the local database.
2. We establish stargazer-following and stargazer-starring relation based on the datasets.
3. We evaluate the stargazer influence from three aspects of basic influence, network dynamic influence and network static influence, and then demonstrate its rationality and effectiveness.
4. We conduct experiments to analyze the relation between the stargazer influence and popularity from multiple perspectives, and conclude the prediction criteria of repository popularity.

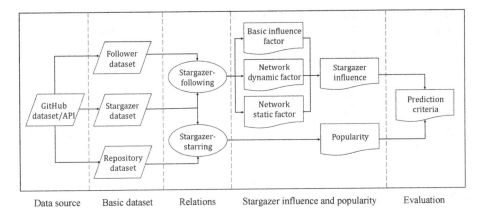

Fig. 1. The workflow of analysis.

2.1   Definition

HFN. In this paper, we define **HF** as the followers of a user, and **HFN** as the number of *HF*. Give a network of following relationship as shown in Fig. 2. *HFN* is the indegree of node in graph theory.

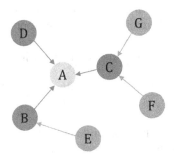

Fig. 2. The network of following relationship. Nodes of the same color are at the same follower level if we choose a target user A.

Each node in Fig. 2 represents a user, and the directed arrow among nodes represents a following relationship, that is, an arrow pointing from f to u represents that u is followed by f. In Fig. 2, suppose A is the target user we analyze, then *HFN* of A is 3, *HFN* of B is 1, and *HFN* of C is 2, etc.

HFN. The network dynamic factor (*NDF*), is an evaluation factor generated by the joint influence of all nodes and link relationships in the network. It has the following characteristics:

1. The weight calculation of the network node is based on the entire network.
2. Adding a new node or relationship affects the weight of all other nodes in the network.
3. Network computing overhead is large.

NSF. The network static factor (*NSF*), is an evaluation factor generated by the targeted nodes and their link relationships in the network. It has the following characteristics:

1. The weight calculation of the network node is based on the network formed by the targeted nodes and finite layer link relationships.
2. Adding a new node or relationship only affects the weight of nodes in the finite layer subnetwork, and nodes outside the finite layer subnetwork are not affected.
3. Network computing overhead is small, but it increases exponentially as the number of layers increases.

2.2   StarIn

The repository will increase a certain number of stargazers per time node (in days). We establish the network of stargazer-following relationship and calculate the influence of each stargazer. It must be pointed out that the stargazer with more followers does not always mean that his/her contribution to promoting the influence of repository is more valuable. This is because there is no direct correlationship between a stargazer's starring behavior and his/her follower number in the network. The behavior itself is the real reflection of high popularity of the repository, and it reflects the basic influence of repository to some degree. Therefore, it is of vital importance to consider the basic influence of repository when evaluating its stargazer influence. Based on this characteristic, our proposed approach to calculating stargazer influence can be formulated as:

$$SI_i = \frac{1 - \alpha - \beta}{N} + \alpha * \frac{NDF_i}{\sum NDF} + \beta * \frac{NSF_i}{\sum NSF} \tag{1}$$

where the first term represents the basic influence of each stargazer, NDF_i and NSF_i describe the global and local characteristics of the stargazer in the network respectively. α and β are coefficient, and $\alpha + \beta < 1$, N is the total number of stargazers of the repository. We calculate SI_i of each stargazer per day, and take the mean value of SI_i as the day stargazer influence, named as SI_t:

$$SI_t = \frac{\sum SI_i}{n} \tag{2}$$

where t represents the t_{th} day from the date of statistics, and n is the number of stargazers on t_{th} day.

From the date of release, the repository will be in two states: 'popular period' and 'flat period'. No repository will be in 'popular period' all the time. Usually, the repositories have a tendency to become popular right after their first release. After released, the growth rate of the repositories tends to stabilize [4]. When repository pass its 'popular period', it will enter 'flat period', and the maintenance time of 'flat period' is difficult to estimate. A new topic or an international conference will make the repository active again. Therefore, we need to pay attention to which state the repository is in when evaluating the stargazer influence. If the stargazer influence has been at a low level for a while, in other words, the maximum value of SI_t is below a certain threshold, it indicates the repository enters the 'flat period'. As a result, the influence change caused by the first increase of popularity after the 'flat period' will be discounted, so we introduce the time weight:

$$SI_t = \begin{cases} SI_t * Weight_t, & if\ in\ flat\ period, \\ SI_t, & else \end{cases} \tag{3}$$

where 'flat period' means that the maximum of SI_i is less than a threshold of stargazer influence, and $Weight_t$ is the time weight.

3   Experiment

In this section, we will present the specific calculation factors applied to our proposed approach in the previous section. Then we will demonstrate the rationality and validity of factors. Finally, we will analyze the relationship between stargazer influence and popularity of repository and conclude the prediction criteria of popularity.

3.1   Data Collection

The page[1] released GitHub's datasets of different versions, and we choose follower dataset as basic dataset after preprocessing [7]. The real-time information of repositories or users can also be accessed through GitHub APIs[2].

Follower dataset describes the following relation of users in GitHub. Since not all users follow others or are followed by others, the users in follower dataset are not all users of GitHub. Table 1 displays the overall information of the dataset. Users with followers account for less than 12% of all users.

Table 1. The overall information of follower dataset.

Being followed	Relationships	Following or being followed	Average	Median
3,833,652	29,809,738	5,375,944	7.7	2

We remove users having no followers and only display the users having followers with the scatter and proportion distribution, as shown in Fig. 3.

---

[1] http://www.ghtorrent.org/downloads.html.

[2] https://api.github.com/repos/torvalds/linux/stargazers for getting stargazer information of a repository, https://api.github.com/users/torvalds/followers for getting follower information of a user.

Figure 3(a) displays the scatter distribution of users with different follower number, where the horizontal axis represents the follower number, and the vertical axis is the user number. We find that when the logarithm of variables is taken, the following relationship is consistent with the power-law distribution in the social network, which can be formulated as $f(x) = cx^{-\alpha}$.

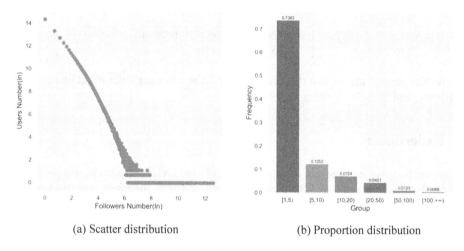

(a) Scatter distribution   (b) Proportion distribution

Fig. 3. The user distribution of different follower number.

The proportion of users with different follower number is shown in Fig. 3(b). In Fig. 3(b), the horizontal axis represents different ranges of follower number, and the vertical axis represents the proportion of users of different ranges in all users having followers.

3.2 Selection of Calculation Factors

Based on the characteristics of *NDF* and *NSF*, we choose two methods to calculate the stargazer influence: *PageRank* and *HFN*. *PageRank* is a classic network node link algorithm [15], and *HFN* is an evaluation method based on the follower number. In order to facilitate measurement and analysis, we quantify the influence factors, that is:

$$IF(index), index \in \{PageRank, HFN\} \tag{4}$$

where *IF* denotes the *InfluenceFactor* (hereafter called *IF*), which is the metric to measure the influence of *index*. The *PageRank* algorithm applied to the network of following relationship can be formulated as:

$$PageRank(u) = \frac{1-q}{N} + q \sum_{f\ follows\ u} \frac{PageRank(f)}{Followees(f)} \tag{5}$$

where u is the user to be analyzed the influence, f is a follower of u, $Followees(f)$ denotes the number of users that f follows, N is the total number of users in the network, q is a damping factor, and it is generally set to 0.85. $PageRank$ can directly derive the weight of nodes according to the node-link algorithm, and use the weight as the value of Eq. (4), that is:

$$IF(PageRank) = PageRank \tag{6}$$

The most intuitive and convenient way to evaluate whether a user has high impact is to determine the number of the followers for a user. Since HFN cannot directly describe the influence of each user through the weight of nodes like $PageRank$, we need to establish a mapping function to calculate the value of Eq. (4).

We first obtain the scatter distribution of HFN frequency through the follower dataset, and then use the Least Square methods to fit the scatter distribution to obtain the frequency function (see Eq. (7)). Then we map the frequency to weighted value. Since the user's distribution is a power-law distribution, we choose the function Eq. (8). Finally, considering that the user's frequency changes slightly when HFN is high, in order to make the high HFN users with differentiation, we perform power calculation on HFN, that is Eq. (9):

$$frequency(HFN) = fitting(HFN) \tag{7}$$

$$weight(HFN) = -\log(frequency(HFN)) \tag{8}$$

$$IF(HFN) = weight(HFN) * HFN^{\frac{1}{n}} \tag{9}$$

We group all users in follower dataset by HFN to get frequency distribution of different HFN, and use linear fitting, cubic fitting, inverse fitting, exponential fitting and exponential-inverse fitting to fit the scatter distribution. Experiments show that the exponential-inverse fitting can make the best fitting performance, and the fitting function is:

$$frequency(HFN) = a * e^{(b/(HFN+0.1))} + c * (1/(HFN+0.1)) + d \tag{10}$$

where a, b, c, and d are the fitting coefficients, which are generated during the fitting process, and different sample data generates different fitting coefficients. For the follower dataset, the coefficients are: $a = -1.61207626e{-}01$, $b = 7.53413079e{-}06$, $c = 2.80668098e{-}01$, $d = 1.55082975e{-}01$. Since the proportion of users with no followers is as high as 90% of all users, we set $weight(HFN)$ of them as 0, and the power exponent of Eq. (9) is set to 1/5.

3.3   Rationality Demonstration

We take the repository[3] as an example (hereafter called Repo_GAN) for demonstrating the rationality of influence factors. Repo_GAN describes a new training methodology for generative adversarial networks, which is currently an area with a high degree of attention in the field of deep learning. Therefore, it is of higher significance and representativeness to study its development.

We calculate the daily average of $IF(HFN)$ of Repo_GAN. As is shown in Fig. 4, the larger the curve slope of *Popularity* is, the higher (i.e. the red and yellow parts) the *Mean(IF)* is, which suggests that *IF* is consistent with the popularity closely. Then we calculate the stargazer influence of Repo_GAN for a total of 3,730 stargazers, and analyze the relationship between stargazer influence and popularity. Considering the factor of 'popular period' and 'flat period', we introduce the time weight. Figure 5 show the relationship between stargazer influence and popularity after introducing time weight.

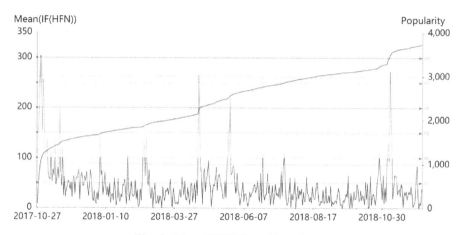

Fig. 4. Mean(IF(HFN)) and Popularity.

We observe that the stargazer influence with time weight achieves a reasonable and effective transition from 'popular period' to 'flat period', which makes the change of stargazer influence on different days more smooth. More interestingly, besides keeping consistent with the popularity on the overall trend, the stargazer influence has a distinct feature. It does not always synchronize with the change of popularity on the same day, but there is a chronological order. From Fig. 5, we can see that the peak of blue curve will be accompanied by red curve. It indicates that the stargazer influence rises earlier than popularity. The time interval is less than 20 days, which means that the popularity of repository will increase significantly in a certain period of time after the rise in stargazer influence.

---

[3] https://github.com/tkarras/progressive_growing_of_gans.

Fig. 5. Weighted stargazer influence and Popularity.

3.4   Evaluation and Results

In order to evaluate the performance of StarIn, a set of 200 repositories are manually selected by 14 software engineers who are working on different directions in software engineering. The 200 repositories involve different fields and types from GitHub. We evaluate the performance of StarIn from the perspective of following six characteristics:

1. The stars increase but the stargazer influence decreases, then the stars will decrease next day.
2. The stars increase but the stargazer influence decreases, then the daily average stars for the next 3 days are lower than the stars on that day.
3. The stars and stargazer influence increase, and the average incremental influence is less than the average influence of all stargazers, then the daily average stars for the next 3 days are lower than the stars on that day.
4. The stars decrease but the stargazer influence increases, then the daily average stars for the next 7 days are higher than the stars on that day.
5. The stars decrease but the stargazer influence increases suddenly, then the stars in the next 20 days increase suddenly (to 30) or keep high (6 stars per day) in consecutive days (5 days).
6. The stars and stargazer influence decrease, and the average reduced influence is greater than the average influence of all stargazers, then the daily average stars for the next 7 days are lower than the stars on that day.

We utilize two classifiers to study the performance of StarIn. One is author-type based classifier, and the other is project-type based classifier.

For the author-type based classifier, we classify the repositories according to whether the author is user or organization [6, 14], and record respectively the prediction results of six characteristics with different influence factor weights in Table 2.

Table 2. The prediction results of six characteristics with different factor weights.

Author type	Characteristic	Best factor weight $(1 - \alpha - \beta, \alpha, \beta)$	Prediction Acc
User	Characteristic 1	(0.8, 0.1, 0.1)	0.8066
	Characteristic 2	(0.8, 0.1, 0.1)	0.8275
	Characteristic 3	(0.3, 0.1, 0.6)	0.9378
	Characteristic 4	(0.6, 0.4, 0)	0.7417
	Characteristic 5	(0.7, 0.1, 0.2)	0.8169
	Characteristic 6	(0.6, 0.4, 0)	0.6949
Organization	Characteristic 1	(0.3, 0.1, 0.6)	0.7519
	Characteristic 2	(0.1, 0.1, 0.8)	0.7830
	Characteristic 3	(0.4, 0.1, 0.5)	0.9239
	Characteristic 4	(0.8, 0.1, 0.1)	0.7423
	Characteristic 5	(0.8, 0.1, 0.1)	0.7735
	Characteristic 6	(0.7, 0.1, 0.2)	0.6376

For project-type based classifier, we classify the repositories by languages (Python, Java, C, PHP, JavaScript, HTML, CSS, Objective-C, Go, Ruby) [1, 4, 16] and properties (tool, framework, algorithm, application, document) [13], and calculate the prediction accuracy of the six characteristics separately. The specific meanings of the five properties are as follows:

- Tools usually refer to application interfaces or third-party libraries.
- Frameworks are often the mature and complete components or packages to facilitate the design reusable and extensible.
- Algorithms usually refer to code sets that are used to solve specific problems and have a clear purpose.
- Applications usually refer to software or finished projects.
- Documents usually refer to tutorials for the use of certain tools and software or systematic documents summarized by predecessors on a topic.

Figure 6(a) and Fig. 6(b) show that the prediction accuracy of six characteristics obtained by classifying the repositories according to languages and properties. It can be concluded the following findings:

A. All languages have high and stable prediction performance on characteristic 1 to 4, with the prediction accuracy reaching 70%–80%.
B. Almost all languages have higher prediction accuracy on characteristic 2 than characteristic 1.
C. Java, PHP and Go have extremely high prediction performance on characteristic 5, and the prediction accuracy can reach over 90%.
D. Characteristic 4 and 5 have obvious symmetry characteristics, that is, the language on characteristic 4 with lower prediction accuracy has higher prediction accuracy on characteristic 5.
E. Except Objective-C, C, and JavaScript, the prediction performance on characteristic 6 is not obvious in other languages.

F. Algorithmic repositories have the highest prediction performance on characteristic 1 and 6, the prediction accuracy can reach 87% and 72%.

G. Except the algorithmic repositories, the prediction performance of repositories of other properties on characteristic 2 is improved relative to the characteristic 1.

H. The repositories of framework have a better prediction performance on characteristic 5 than others.

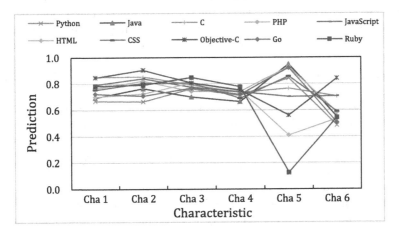

(a) The prediction performance with languages

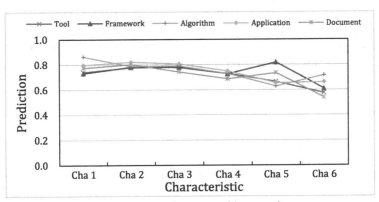

(b) The prediction performance with properties

Fig. 6. The prediction performance.

Repositories of different languages and properties have different prediction performance on different characteristics. Most repositories are sensitive to changes in stargazer influence, that is, the popularity of repositories will increase and decrease correspondingly in a short period of time after the stargazer influence increasing or decreasing. It can be seen from finding D that characteristic 4 and 5 are symmetrical. It indicates that after the stargazer influence rises or even surges, the stars of repository will either rise within a week, or surge or continuously keep high within 20 days. From finding F, G and H, repositories of algorithm and framework have a more positive response to the change of stargazer influence. In contrast, repositories of document, tool and application are insensitive. This is because repositories of algorithm and framework are more technical, and they are likely to be concerned by high-influence stargazers. When the stargazer influence of such repositories rises, it usually means that they begin to enter the public view. For the repositories of document, tool and application, they are usually paid attention to only when the stargazers have requirements. Therefore, the attention of high-influence stargazers to the repositories often does not have a distinct driving effect.

The quality of repositories determines whether it will attract high-influence stargazers. In order to discover the predictive characteristics of different quality repositories, we divide the repositories into 0–5k, 5k–10k, 10k–20k and 20k+ according to the number of stars, and describe the prediction results of six characteristics respectively in Fig. 7.

Figure 7 shows that the repositories have excellent prediction performance on others except characteristic 5 and 6, and prediction accuracy reaches 70%–80%. The number of stars reflects the quality of repository to a certain extent. Therefore, it can be seen from the variation curve of characteristic 5 that the higher the quality of repository, the more significant the prediction performance of characteristic 5. In other words, it is extremely possible that the popularity will become higher in the future after the stargazer influence of high-quality repositories surges. When the number of stars reaches 10 k+, the probability is close to 1. It is not difficult to explain this phenomenon. High-quality repositories will be more likely to be noticed by high-influence stargazers, and their attention will guide more people to notice the repositories. While people who star the low-quality repositories are mostly users with relatively low stargazer influence, the guiding effect they produce is more contingent and accidental, which is difficult to represent the trend of popularity.

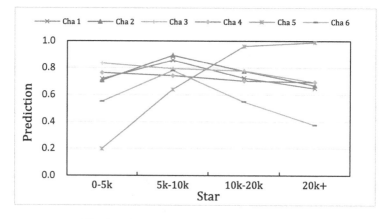

Fig. 7. The prediction performance with stars.

4   Discussion

4.1   Why Does StarIn Work?

StarIn achieves a great success in predicting the popularity of GitHub repositories, and the precision can reach an extremely high level. We conclude several points that can explain the effectiveness of StarIn.

Firstly, StarIn is designed to evaluate the influence of repositories based on the obtained stars, which is the most intuitive manifestation of popularity. It is more convincing than evaluating the influence by measuring the quality of project itself.

Secondly, stargazer is one who gives stars to a repository. Therefore, stagazers play an essential role in the development of repository. A change in the number of stargazers means an increase or decrease of people's attention to repository. In addition, stargazers have different degrees of seniority, the influence of stargazers determines whether the popularity of the project will develop in a good or bad direction.

At last, StarIn is highly universal because it not only considers the impact of stargazers on the repository, but also the basic influence of project itself. The design of StarIn will be suitable for predicting the popularity of repositories with different programming languages, usages and quality.

4.2   Threats to Validity

Our proposed approach is to predict the popularity of GitHub repository through measuring its stargazer influence. There is a threat of the causal relationship between the characteristics of influential stargazers and the construct of popularity. To mitigate this threat, we perform a verification experiment before conducting the evaluation experiment. In the verification experiment, we utilize a representative sample, Repo_GAN, which is one of the 'Top Ten Global Breakthrough Technologies' published by the MIT Science and Technology Review in 2018, to demonstrate that there exists a close relationship between influential stargazers and popularity. Based on this argument, we evaluated the effectiveness of approach on a dataset of 200 repositories.

In our evaluation experiment, we use two classification criteria, namely, author-type based classifier and project-type based classifier. They are widely used classification criteria in previous literature and contain most repository types. We consider that such classification criteria is reasonable. In real-world repositories search, developers usually focus on what the author type or project type of the repository is. In addition, each classification type contains dozens of samples in our experiment. Therefore, StarIn works well even if a new repository comes.

5   Conclusion

In summary, we analyzed the critical factor associated with the popularity of repository, the stargazer influence. We explored a variety of characteristics in the chronological order between the stargazer influence and popularity. We found that the stargazer influence can effectively predicted the popularity of repository. First, we propose the concept of HFN, and establish influence factor of *PageRank* and *HFN* when evaluating the stargazer influence. Second, we divided the influence factor into network dynamic factor and network static factor according to the characteristics of the factor in the social network, and established an evaluation criteria of stargazer influence. Third, we demonstrated the rationality of influence factor through the relationship between the influence factor and popularity. Finally, we found that the stargazer influence exhibits an excellent prediction performance to the popularity by analyzing the repositories with different languages, properties and stars.

Our results provide an approach to predict the popularity of repositories in GitHub from a new perspective. The approach we proposed is of high extensibility and flexibility because the influence factor is alternative and the parameters is also dynamic to meet different demands. In the future, we propose to develop a method of associating the repositories with papers. Furthermore, we will try to adopt the prediction method to evaluate academic papers, which will contribute to the evaluation of the scientific success with a more comprehensive perspective. Meanwhile, the prediction method will help developers to gain insights into how to choose a high-quality repository for their work [18], and help GitHub to develop a better recommendation scheme, which may promote the development of software or the coding society.

Acknowledgment. This work is supported in part by the Natural Science Foundation of 453 China grant 61502069, 61672128, and by the Fundamental Research Funds 454 for the Central Universities grant DUT18JC39, DUT18GF108.

References

1. Bissyande, T.F., Thung, F., Lo, D., Jiang, L., Reveillere, L.: Popularity, interoperability, and impact of programming languages in 100,000 open source projects. In: 2013 IEEE 37th Annual Computer Software and Applications Conference, pp. 303–312. IEEE (2013)
2. Blincoe, K., Sheoran, J., Goggins, S., Petakovic, E., Damian, D.: Understanding the popular users: following, affiliation influence and leadership on GitHub. Inf. Softw. Technol. **70**(1), 30–39 (2016)
3. Borges, H., Hora, A., Valente, M.T.: Predicting the popularity of GitHub repositories. In: Proceedings of the 12th International Conference on Predictive Models and Data Analytics in Software Engineering, p. 9. ACM (2016)
4. Borges, H., Hora, A., Valente, M.T.: Understanding the factors that impact the popularity of GitHub repositories. In: 2016 IEEE International Conference on Software Maintenance and Evolution (ICSME), pp. 334–344. IEEE (2016)
5. Borges, H., Valente, M.T.: What's in a GitHub star? Understanding repository starring practices in a social coding platform. J. Syst. Softw. **146**(10), 112–129 (2018)
6. Chatziasimidis, F., Stamelos, I.: Data collection and analysis of GitHub repositories and users. In: 2015 6th International Conference on Information, Intelligence, Systems and Applications (IISA), pp. 1–6. IEEE (2015)
7. Gousios, G.: The ghtorent dataset and tool suite. In: Proceedings of the 10th Working Conference on Mining Software Repositories, pp. 233–236. IEEE Press (2013)

8. Jarczyk, O., Gruszka, B., Jaroszewicz, S., Bukowski, L., Wierzbicki, A.: GitHub projects. quality analysis of open-source software. In: Aiello, L.M., McFarland, D. (eds.) SocInfo 2014. LNCS, vol. 8851, pp. 80–94. Springer, Cham (2014). https://doi.org/10.1007/978-3-319-13734-6_6

9. Jiang, J., Lo, D., He, J., Xia, X., Kochhar, P.S., Zhang, L.: Why and how developers fork what from whom in GitHub. Empir. Softw. Eng. **22**(1), 547–578 (2017)

10. Jiang, J., Lo, D., Yang, Y., Li, J., Zhang, L.: A first look at unfollowing behavior on GitHub. Inf. Softw. Technol. **105**(6), 150–160 (2019)

11. Kalliamvakou, E., Gousios, G., Blincoe, K., Singer, L., German, D.M., Damian, D.: An in-depth study of the promises and perils of mining GitHub. Empir. Softw. Eng. **21**(5), 2035–2071 (2016)

12. Long, Y., Siau, K.: Social network structures in open source software development teams. J. Database Manag. (JDM) **18**(2), 25–40 (2007)

13. Marques, O.: Tools, frameworks and applications for high performance computing: minisymposium abstract. In: Kågström, B., Elmroth, E., Dongarra, J., Waśniewski, J. (eds.) PARA 2006. LNCS, vol. 4699, p. 239. Springer, Heidelberg (2007). https://doi.org/10.1007/978-3-540-75755-9_29

14. Munaiah, N., Kroh, S., Cabrey, C., Nagappan, M.: Curating GitHub for engineered software projects. Empir. Softw. Eng. **22**(6), 3219–3253 (2017)

15. Page, L., Brin, S., Motwani, R., Winograd, T.: The PageRank citation ranking: bringing order to the web. Technical report, Stanford InfoLab (1999)

16. Ray, B., Posnett, D., Filkov, V., Devanbu, P.: A large scale study of programming languages and code quality in GitHub. In: Proceedings of the 22nd ACM SIGSOFT International Symposium on Foundations of Software Engineering, pp. 155–165. ACM (2014)

17. Robles, G., González-Barahona, Jesús M.: A comprehensive study of software forks: dates, reasons and outcomes. In: Hammouda, I., Lundell, B., Mikkonen, T., Scacchi, W. (eds.) OSS 2012. IAICT, vol. 378, pp. 1–14. Springer, Heidelberg (2012). https://doi.org/10.1007/978-3-642-33442-9_1

18. Sun, X., Xu, W., Xia, X., Chen, X., Li, B.: Personalized project recommendation on GitHub. Sci. China Inf. Sci. **61**(5), 1–14 (2018)

19. Treude, C., Leite, L., Aniche, M.: Unusual events in GitHub repositories. J. Syst. Softw. **142**(1), 237–247 (2018)

20. Tsay, J., Dabbish, L., Herbsleb, J.: Influence of social and technical factors for evaluating contribution in GitHub. In: Proceedings of the 36th international conference on Software engineering, pp. 356–366. ACM (2014)

21. Wang, T., Zhang, W., Ye, C., Wei, J., Zhong, H., Huang, T.: FD4C: automatic fault diagnosis framework for web applications in cloud computing. IEEE Trans. Syst. Man Cybern. Syst. **46**(1), 61–75 (2015)

MCA-TFP Model: A Short-Term Traffic Flow Prediction Model Based on Multi-characteristic Analysis

Xiujuan Xu[1,2], Lu Xu[1], Yulin Bai[1], Zhenzhen Xu[1,2], Xiaowei Zhao[1,2], and Yu Liu[1,2(✉)]

[1] School of Software, Dalian University of Technology, Dalian 116620, China
[2] Key Laboratory for Ubiquitous Network and Service Software of Liaoning Province, Dalian 116620, China
{xjxu,xzz,xiaowei.zhao,yuliu}@dlut.edu.cn

Abstract. With the urbanization, urban transportation has become a key factor restricting the development of a city. In a big city, it is important to improve the efficiency of urban transportation. The key to realize short-term traffic flow prediction is to learn its complex spatial correlation, temporal correlation and randomness of traffic flow. In this paper, the convolution neural network (CNN) is proposed to deal with spatial correlation among different regions, considering that the large urban areas leads to a relatively deep Network layer. First three gated recurrent unit (GRU) were used to deal with recent time dependence, daily period dependence and weekly period dependence. Considering that each historical period data to forecast the influence degree of the time period is different, three attention mechanism was taken into GRU. Second, a two-layer full connection network was applied to deal with the randomness of short-term flow combined with additional information such as weather data. Besides, the prediction model was established by combining these three modules. Furthermore, in order to verify the influence of spatial correlation on prediction model, an urban functional area identification model was introduced to identify different functional regions. Finally, the proposed model was validated based on the history of New York City taxi order data and reptiles for weather data. The experimental results show that the prediction precision of our model is obviously superior to the mainstream of the existing prediction methods.

Keywords: Urban transportation · Short-term traffic flow prediction · Multi-characteristic analysis · MCA-TFP model

1 Introduction

Intelligent transportation system (ITS) plays a huge role in urban traffic. Through ITS, drivers can select their travel routes in advance, avoiding

This work was supported in part by the Natural Science Foundation of China grant 61672128, 61702076; the Fundamental Research Funds for the Central Universities DUT18JC39.

congested road sections during the high peek period, so as to reduce their travel time and improve the travel efficiency. Eventually, ITS could adjust the reasonable distribution of traffic flow on each section of road network and alleviate the pressure on urban traffic. By analyzing the real-time traffic flow to predict traffic flow in the next short time, traffic resources could be reallocated reasonably. Therefore, short-term traffic flow forecasting is indispensable for ITS. Real-time and effective short-term traffic flow prediction can achieve path guidance, help people choose better travel routes, and alleviate traffic congestion [17]. The main work of this paper is as follows:

(1) The research background and significance of traffic flow forecasting are introduced. The classification of existing forecasting methods is summarized. The related technologies used in this model are introduced in detail.

(2) Through the visual analysis of historical traffic flow data, three characteristics that affect short-term traffic flow data, namely spatial correlation, time correlation and randomness, are obtained. The time correlation mainly includes proximity, daily periodicity and weekly periodicity.

(3) Based on the above-mentioned characteristic analysis, we define the prediction problem of this paper, divide the urban areas, and establish the prediction model. The model mainly consists of three modules. Firstly, the convolutional neural network with residual unit is used to deal with the spatial correlation; three gated loop unit networks are used to deal with three characteristics in time correlation, respectively, and then for each time characteristic, the attention mechanism is used to assign weights to different time periods; for the randomness processing module, this paper mainly considers the influence of weather, holidays, and the day of the week on the prediction model, and uses a two-layer fully connected network for the processing. Then, the output that is fused after the above-mentioned attention mechanism and the output of the randomness processing module are added by units and the final prediction result is obtained through a tanh function.

(4) Historical taxi order data of New York City is collected and weather data corresponding to the time is obtained through the crawler. The collected data is pre-processed. The taxi order data processing includes data screening and cleaning, extraction of traffic flow data for each time period, and conversion of traffic flow data into a matrix; weather data processing includes data screening and patching, One-hot encoding, data normalization and extraction of weather data for each time period.

(5) For the pre-processed data, the short-term traffic flow is predicted by using the established prediction model, and the influence of the multi-characteristics proposed in this paper on short-term traffic flow is analyzed by the prediction result combined with the identified functional area. Then, through comparison with some existing prediction methods, it is found that the prediction model proposed in this paper is better than the current mainstream method. Compared with the more advanced method of ST-ResNet proposed by researchers in the past two years, the RMSE is reduced by 1.95, and the MAE is reduced by 1.3. The prediction effect is also improved.

2   Related Work

Traffic flow forecast have crossed researchers' radar in the late 1960s. Foreign researchers began to study technology-related traffic flow prediction city since then. It has also been related to predictive models, and used them in actual traffic flow prediction. Based on previous studies, in accordance with the difference in traffic flow forecasting scholars in the field of the methodology used, we can divided the existing traffic flow forecasting model into three categories.

(1) The model based on mathematical logic is as the model based on the historical average prediction method (Historical Average), the model based on time series (time series model), the model based on Kalman filter (Kalman filter model), the model based on autoregressive integrated moving average (ARIMA) and so on.
(2) The model based on theoretical of artificial intelligence includes the model based on Neural network and the model based on support vector machine.
(3) The model is based on combined model approach, such as the combination of time series model and neural network model, the combination of Kalman filter algorithm model and time series model, the combination of nonparametric regression and moving average models.

2.1   Prediction Model of Mathematical Logic

Prediction model of mathematical logic is relatively early used for traffic flow prediction, Stephanedes et al. [6] first proposed a model based on the historical average method for traffic control. They considered the influence between a road segment to be predicted and a road segment in a surrounding area, and then established a prediction model by Kalman filter algorithm. At the same time, they used a large number of predictions to improve the model parameters error.

Williams et al. [12] predicted urban expressway traffic flow by the seasonal autoregressive integrated moving average model (ARIMA) which based on the analysis of traffic flow data. Z. Gu et al. [16] found out that the traffic flow can be decomposed into a basis series representing the change of traffic flow and a deviation series representing random interference and measurement noise. Partial deviation series can be appropriately filtered out, and only the base series is predicted. They proposed method called Basis Prediction method aimed at improving the accuracy of short-term traffic flow prediction. Based on the K-Nearest Neighbor (KNN) algorithm, L. Yang et al. [4] forecasted short-term traffic flow, and defined the factors affecting short-term traffic flow in the port, such as port working status and operation instruction characteristics, as state vector, and used K-Dimension Tree (KD tree) to reduce the time complexity of neighbor searching.

2.2   Prediction Model Based on Artificial Intelligence

In the prediction model based on artificial intelligence, generally speaking, the original data sets are divided into training sets and test sets. Then some existing artificial intelligence related methods are used to establish prediction models. The training sets are used to obtain some parameters of the model through training, and the test sets are used to evaluate the prediction of the model. Neural Network (NN) prediction model and Support Vector Machine (SVM) prediction model are common prediction models based on artificial intelligence. Polson N G et al. [7] proposed a deep learning model which is a linear model combined with L1 regularization and tanh layer sequence fitting. It is mainly used to predict the traffic volume in the case of acute non-linear problems caused by special events. Chen Y et al. [1] proposed a prediction algorithm based on deep learning (DeepTFP), which uses three deep residual neural networks to model the time proximity, period and trend characteristics of traffic flow. DeepTFP aggregates the output of three residual neural networks to optimize the parameters of time series prediction model. Ma X et al. [5] proposed a method based on convolution neural network to learn traffic as an image, so as to predict large-scale, network-wide traffic speed with high accuracy. This method applies convolution neural network to image through two steps: traffic feature extraction and traffic speed prediction within the network.

Koesdwiady A et al. [3] proposed a prediction model based on deep belief network. The traffic data and weather data are fused based on decision-level data to improve the accuracy of weather condition prediction. Wu Y et al. [13] proposed a traffic flow prediction model based on deep belief network, which takes full advantage of the cyclical and spatio-temporal characteristics of weekly/daily traffic flow, and introduced an attention-based model, which can automatically learn to determine the importance of traffic flow in different time periods in the past. Zheng Yu et al. [15] proposed a prediction model that takes the whole city as an image and constructs three identical residual convolution structures to capture trends, cycles and proximity information over time.

2.3 Combined Prediction Model

In recent years, with the deepening of related research, researchers have found that there are some unavoidable shortcomings in a single traffic flow forecasting model. Researchers began to combine two or more models for traffic flow forecasting to improve the accuracy of forecasting. Wang Jian et al. [10] combined ARIMA model with BP neural network, and used Bayesian theory to adjust the weight of model parameters, and finally found that the prediction ability is better than a single model. Li Y H et al. [14] combined the moving average model and non-parametric regression method to build the prediction model, and finally found that the prediction accuracy is higher. Guo Jiang S et al. [8] used time series model and Kalman filter method to construct prediction model, and the final prediction accuracy was significantly improved. G. Dai et al. [2] proposed a short-term traffic flow prediction model that combined the spatio-temporal analysis with a Gated Recurrent Unit (GRU), compared to the prediction results with CNN and GRU, the results show that the model is superior to CNN model and GRU model in accuracy and stability.

3   Problem Definition

We present the traffic flow prediction problem firstly. Second, we define the short-term traffic flow characteristic. In field of traffic, the close correlation in space is strong but the long correlation is weak, including recent time correlation, daily period correlation, weekly period correlation in time correlation. Meanwhile, it exists randomness caused by abnormal weather and holidays. We consider the three characteristics of the short-term traffic flow prediction model based on feature analysis.

In this paper, we adopt the grid method to divide cities into $I * J$ grids with equal size according to longitude and latitude of a city.

Definition 1: A grid represents a region, and a region can be represented as a tuple (i, j). The whole region is represented as $L = \{l_1, l_2, \ldots, l_i, \ldots, l_{i*j}\}$. In addition, time is divided into non-overlapping intervals which are defined as $T = \{t_1, t_2, \ldots, t_T\}$.

In order to understand the short-term traffic flow in a certain area of the city, we present Definition 2 as follows.

Definition 2: The traffic flow of a certain area (i, j) in a city in a period of time t is defined as $x_t^{i,j}$, which can be expressed by the following Eq. 1:

$$x_t^{i,j} = \sum_{m \in (i,j), n \in t} |l_m \in L \wedge t_n \in T| \tag{1}$$

where $|\bullet|$ represents the cardinality of a set, that is the set of all cases satisfying the conditions. m represents the set of all points in the region (i, j), and n denotes the set of all moments within the time interval t. Finally, we propose the short-term traffic flow prediction problem in this paper in Definition 3.

Definition 3: The short-term traffic flow prediction problem refers to the historical traffic flow data $x_n^{i,j} | n = 0, 1, \ldots, t - 1$ at time t and before a given area (i, j) to predict the traffic flow $x_t^{i,j}$ of this area at the moment t, where each moment is 15 min. For simplicity, we write $x_t^{i,j}$ as x_t.

4   Overview of Traffic Prediction Model

The short-term traffic flow forecasting can be considered as the sequence modeling task [11]. Considering the above three correlations including spatial correlation, temporal correlation and randomness, we propose a short-term traffic flow prediction model based on multi-characteristic analysis in order to predict the short-term traffic flow of a city more accurately.

As shown in Fig. 1, our model has three parts to stimulate the spatial correlation, temporal correlation and randomness. First, we deal with spatial correlation through the convolutional neural network with residuals, and then divide the time axis into three modules, representing recent time, daily period and

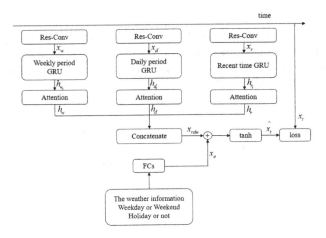

Fig. 1. Three parts of MCA-TFP model.

weekly period respectively. Second, the output of res-conv is used as the input of GRU network, and three GRU networks are used to process temporal correlation. Meanwhile, considering the different degrees of influence of different time periods on the traffic flow of the predicted time, we introduce attention mechanism in three GRU networks for weight allocation of different time periods, and the output of the three modules is fused into x_{rdw}. For the effects of the randomness, under the premise of considering more random factors as much as possible, we consider the weather information, holidays and the day of the week. We input information such as the weather data, holidays and the day of the week after processing to a two layer of neural network with connect fully, the output results can be represented as x_e. Further concentrate of x_{rdw} and x_e. Consequently, considering that the convergence rate of the function $tanh$ in the back-propagation learning process is faster than the standard logistic function, we fused results are mapped to $[-1, 1]$ by a function tanh. Finally, the loss function is used to calculate the difference between predicting traffic flow and real traffic flow, and then it is returned to the model to continuously adjust parameters and reduce errors to obtain the prediction model.

4.1 MCA-TFP Algorithm

We present the algorithm of our model as follows:

Algorithm: MCA-TFP algorithm

Input: historical data of short-term traffic flow, input of random processing module (including weather, holidays and other information), input length of three time characteristics;

Output: traffic flow at the next time t.

(1) Identify urban functional areas;
(2) Build the historical traffic flow data matrix of recent time, daily period time and weekly period time modules;
(3) Input the matrices corresponding to the three time modules into the convolutional neural network with residuals;
(4) Take the output of convolutional neural network as the input of GRU network;
(5) Attention mechanism is used to represent the distribution weight value of each time period in the GRU network;
(6) Output of the above three time modules is output by means of parameter fusion;
(7) External factor module adopts two-layer fully connected network processing to get an output;
(8) Add the above two outputs by bits and get the predicted result through a tanh function;
(9) Calculate losses and update model parameters to achieve the best prediction effect.

4.2 Identify Urban Functional Areas

We define short-term traffic flow prediction based on urban grid area division, and proposes an urban functional area recognition model based on the divided urban grid area, so as to verify the spatial correlation of short-term traffic flow on the predictive model.

4.3 Model Spatial Correlation Processing

In general, a city is so big that traffic flow of adjacent area in a city may influence each other. Considering that *CNN* is used to handle with the local structure of image/video data is a very powerful tool, we use *CNN* to deal with the space correlation. In addition, there is also a correlation between traffic flows in two distant parts of a city. A convolution layer only reflects the spatial correlation between regions close distance, but it could not grab the correlation between remote area. We design with multiple layers of *CNN* to capture spatial correlation among all areas in a city. Meanwhile, we do not use a pooling layer in our model in order to keep spatial correlation among remote regions from being lost due to pooling operation.

A very deep convolutional neural network will affect the effect of model training, but our model still need a very deep convolutional neural network to capture the dependencies of all different regions in a large city. For urban traffic flow data, assume the size of the input region is $16 * 7$, and the size of the convolution kernel is fixed at $3 * 3$. If the spatial correlation of all regions is to be established, that is, the final prediction result of each region depends on the input of all regions, at least 8 layers of convolution are required. In order to solve the problem of poor training effect caused by deep *CNN*, we use residual network in the model.

Compared with the ordinary *CNN*, residual network to deep convolution neural Network effect is very good.

In this paper, residual units are connected after the first layer of convolution, and another convolution layer is connected after the residual units. As shown in Fig. 2, the overall structure of Res-Conv is presented. In this figure, each residual unit includes two convolution layers with kernel size of (3, 3) and two relu activation functions. The operation of each residual unit can be expressed by Eq. 2 (Fig. 3):

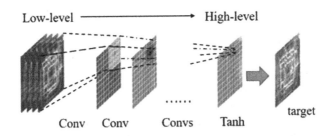

Fig. 2. Convolution operation.

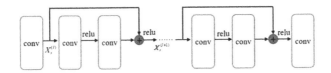

Fig. 3. Res-Conv structure.

$$x_r^{(l+1)} = x_r^{(l)} + \Gamma(x_r^{(l)}; \theta_r^{(l)}) \tag{2}$$

where Γ is the residual function, $l = 1, 2, \ldots, L$ represents the position of the current residual unit in Res-Conv, θ_r^l represents all the parameters to be learned in the Lth residual unit, x_r^{l+1} represents the input of the $(L+1)$th residual unit corresponding to the adjacent time segment, and x_r^l represents the input of the Lth residual unit corresponding to the adjacent time segment.

4.4   Model Time Correlation Processing

Because of the traffic flow dynamic change characteristics, we use *RNN* based on the correlation between time series of traffic flow. *RNN*'s parameters are trained by back propagation (*BP*) algorithm to learn the rules over time. The *BP* algorithm with time is in accordance with the time of the reverse transfer error information step by step forward. We use gated recurrent unit (*GRU*) to solve gradient explosion and disappearance problem. The GRU effect is equivalent to LSTM but has fewer parameters, which is suitable for constructing a large training model [2].

Three Modules of Time Correlation. For the traffic flow in a certain area of the city, it may be affected by the traffic volume in the first few hours in the same area. For example, if traffic jam occurs in a certain area at 8 a.m., this area may still be in the state of traffic jam at 9 a.m. Therefore, if we want to predict the traffic flow at a certain time in future, the traffic flow data of the first few hours before that time is essential for prediction. In addition, the traffic situation in the morning rush hour may be similar to that in a continuous working day and repeat every 24 h. That is to say, because the commuting time of urban population is similar, the daily travel peak time will be relatively concentrated, so the daily traffic flow will be in a state of periodic change. Besides, considering that the travel habits of urban population are not only different on weekdays and weekends, but also the traffic flow is not exactly the same on weekdays from Monday to Friday, this paper also considers the weekly periodicity of traffic flow.

In order to deal with recent time correlation, daily period correlation, weekly period correlation in time correlation, this paper divides time correlation into three modules according to the order of time axis, namely recent time module, daily period time module and weekly period time module. Each module is a *GRU* network. Meanwhile, the output of Res-Conv model corresponding to spatial correlation is taken as the input of *GRU*. For example, we use the recent time module as input to the *GRU*. Recent time *GRU* network structure is shown in Fig. 4 with Eq. 3 as follows:

$$h_{r_m} = GRU(h_{r_{m-1}}, x_{r_m}) \tag{3}$$

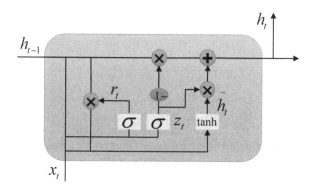

Fig. 4. Gated recurrent unit (*GRU*).

where h_{r_m} represents the output of a hidden layer of the region at time m. $h_{r_{m-1}}$ denotes the output of the hidden layer of the region at time $m-1$. x_{r_m} represents the input of the region (i, j) at time m, namely the output of Res-Conv module.

Considering the near time module, each time period for the current period to forecast the influence degree of different. For example, if we predict the traffic flow at 9:30 a.m., so before this time period will affect the current prediction

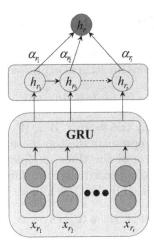

Fig. 5. Recent time module Att-GRU.

of traffic flow, but in this time before 9:00 traffic flow to traffic than 8:30 a.m. to 9:30 a.m. influence is greater. Similarly, in the daily period time module, the influence of the corresponding time period on the traffic flow in the predicted time period is also different. In the weekly period time module, the influence of the corresponding time period of each week on the predicted traffic flow is also different. Therefore, attention mechanism is introduced for three time processing modules respectively, and different weights are assigned for different time periods to represent different degrees of influence of different times on traffic flow in predicted time periods.

Attention Mechanism. In this paper, the weight measures for the three time modules can be represented by Fig. 5. Taking the adjacent time module as an example, the weighting of each time interval can be expressed by formula 4.

$$h_r = \sum_{m-1}^{t} a_{r_m} h_{r_m} \tag{4}$$

where h_r represents the output vector of the whole adjacent time module, and a_{r_m} represents the weight allocated for each time interval. This weight is obtained by calculating the proportion of the matching score of the output state of the hidden layer and the output of the whole adjacent time module in the total score. Such comparison can be expressed by Eq. 5.

$$\alpha_{r_m} = \frac{exp(score(h_{r_m}, \tilde{h}_r))}{\sum_{m-1}^{t} exp(score(h_{r_m}, \tilde{h}_r))} \tag{5}$$

where the score function represents the score value of the output of the mth hidden layer in the output of the whole adjacent time module. When the score value is larger, the output of the entire adjacent time module is more dependent on the input time period of the current moment. Meanwhile, the random initialization will be used as a parameter to update step by step during the training process. In this paper, the score function is expressed by Eq. 6.

$$score_{h_1,h_2} = V^T tanh(W_H h_1 + W_X h_2 + b_X) \tag{6}$$

where h_1 and h_2 represents any two vectors, W_H, W_X, b_X and V are the parameters to be learned, and V^T represents the transpose matrix of the matrix V.

After Attention mechanism of the three modules of the time axis, we use a fusion method based on parameter matrix to fuse the output of the three modules, which can be expressed by Eq. 7 as follows.

$$x_{rdw} = W_r \circ H_r + W_d \circ h_d + W_w \circ h_w \tag{7}$$

where $\circ$ is Hadamard, h_r, h_d, h_w represents the output of three modules. W_r, W_d, W_w are the parameters to be learned.

Model Randomness Processing. Short-term traffic flow will be affected by many complex external factors, such as abnormal weather and unexpected events, which are collectively referred to as random factors. In this paper, the influence of weather conditions, holidays and day of the week on short-term traffic flow is mainly considered on the premise of considering as many random factors as possible. On the network structure of the model, two-layer fully connected network is adopted for processing, where the first layer network can be regarded as the embedding layer of each sub-factor, followed by an activation function. The second layer network is used to map low-dimensional to high-dimensional information with the same shape as x_t. Finally, the output after the two-layer fully connected network is x_t.

Module Fusion. The final prediction result is obtained by adding the output of the three time modules fused with the output of the random modules by bits and then passing through a tanh function, which is expressed by Eq. 8 as follows.

$$\hat{x}_t = tanh(x_{rdw} \oplus x_e) \tag{8}$$

where tanh function is used to ensure that the output value is within $[-1, 1]$. Finally, the entire model updates all parameters by minimizing the loss function. The loss function of the model in this paper adopts mean square error, which can be expressed by Eq. 9.

$$\gamma(\theta) = \frac{1}{N} \sum_{i=1}^{N} (x_t - \hat{x}_t)^2 \tag{9}$$

where θ is the parameter to be learned in the model, x_t is real traffic flow, $\hat{x}_t$ is predicted traffic flow, and N is the total data.

5   Experiments and Results

This paper builds an experimental overall model by combining TensorFlow and Keras frameworks. The use of these two frameworks greatly improves the efficiency of the training model. In general, the performance of the analytical prediction model is actually the performance metric for the regression task. Currently, several common performance indicators have Mean Absolute Error (MAE) and Mean Square Error (MSE). Root Mean Square Error (RMSE).

5.1   Data Set

The historical traffic flow data used in the experiment is the yellow taxi historical order data published by the New York City Government public data platform for 6 months from June to September 2015. The data volume per month is about 1.8 G. The number of data in the month is about 11 million. These data completely record the driving information of the yellow taxi in New York City. Each of the data contains some basic parameters, such as getting off time, getting on and off the latitude and longitude coordinates, driving Distance and so on.

The original weather data in this experiment was derived from the Global Weather Accurate Forecast Network, which is the weather record data of the day, obtained by using the Beautiful soup reptile. The raw data set is data including time, temperature, wind speed, weather conditions and so on, in hours.

5.2   Experimental Parameters

In the spatial correlation Res-Conv module for processing the prediction model, for the convolution operation, the convolution kernel size of the convolutional layer Conv1 and all residual units is set to 3 * 3, the number is 64, and the convolution layer is set. Conv2 has a convolution kernel size of 3 * 3 and a number of 64. In the three GRU modules that process the temporal correlation of the prediction model, the input length of the GRU module set in the neighborhood is set to 12, that is, the traffic volume of the historical data of the first 3 h before the predicted time period is treated. The input length of the daily cycle GRU module is set to 7, that is, the historical data of the same time period of the first 7 days of the date corresponding to the time period to be predicted has an influence on the traffic flow of the predicted time period, and the input length of the weekly cycle GRU module is set. It is 4, that is, the historical data of the same time period corresponding to the date of the first four weeks corresponding to the date to be predicted has an influence on the traffic flow of the predicted time period. The dimensions of the hidden layer output of the three GRUs are all set to 128.

In this experiment, 80% of the data was used as a training set, and 20% of the data was used in the test set, that is, the taxi history order data from June 1 to September 10, 2015 was used for the training set, September 2015. Taxi historical order data from 11th to September 30th is used for the test set. Batchsize is set to 64, the learning rate is set to 0.0003, the maximum epoch is set to 200, and Early stopping is used to improve training efficiency.

5.3 Analysis of Experimental Results

Influence of Different Residual Unit Numbers on the Model. The effect of different residual unit numbers on the training effect of the model is different in the experiment. Table 1 is the result of three performance indicators obtained by training different residual unit numbers.

Table 1. Performance metrics corresponding to the number of different residual units

Number of residual units	RSME	MSE	MAE
1	21.64	468.29	11.18
2	20.07	402.80	10.42
3	18.88	356.45	9.83
4	18.14	329.06	9.01
5	17.56	308.35	8.92
6	17.41	303.11	8.75
7	17.45	304.50	8.80
8	17.43	303.80	8.77

From Table 1 we can see that when the number of residual units increases from 1, the three performance indicators gradually decrease as the number of residual units increases. It can be seen that when the residual network is used in the model, The deeper the network, the better the model training effect, and the better the prediction results will be. However, when the number of residual units reaches a certain number, the training effect of the network tends to be saturated, which means that the network has fully learned the spatial correlation, and then increasing the number of network layers can not make the network training better. At the same time, as the number of network layers deepens, the training model takes longer and longer. Therefore, in order to reduce the model training as much as possible while ensuring the training accuracy of the model, the number of residual units finally selected in the experiment is 6.

In order to verify the validity and advancement of the model, the short-term traffic flow prediction model proposed in this paper is compared with the current common prediction methods. The main methods used are:

(1) Historical average (HA)

In this paper, the average value of traffic flow at the same time in the same region of the data set is used as the prediction result of the historical average method. Considering that the traffic flow will be affected by various factors and the fluctuation is large, the fluctuation is large. The arithmetic mean of the corresponding time period of the first three days of the predicted time period is used as the predicted result of the time period to be predicted.

(2) Autoregressive moving average model (ARIMA)

This is a well-known model for predicting data for a certain time period in the time series. The ARIMA model can be represented by $ARIMA(p, d, q)$, where d represents the order of the difference and is used to transform the non-stationary sequence. For a stationary sequence, p represents the order of autoregression, and q represents the moving average order. In this paper, after multiple differential processing and autocorrelation plot and partial correlation graph test of the data used in the plotting experiment, the final value of p used in this paper is 5, the value of d is 1, and the value of q is 0.

(3) XGBoost

The idea of XGBoost is to constantly add trees to the original structure, and to grow a tree by constantly performing feature splitting [9]. The process of adding a tree each time is actually learning a new function to fit the residual of the last prediction. In this paper, after repeated comparison experiments, the eta value is set to 0.01, the max_depth value is 7, and the subsample value is 0.8.

(4) LSTM

As mentioned above, the cyclic neural network can be well used for prediction based on time series data, and LSTM, as a variant of the cyclic neural network, can solve the gradient explosion and gradient disappearance problems of the circulating neural network. In this paper, the Batch Size used for training is set to 64, the number of neurons in the hidden layer is 128, and the time step is 12.

(5) ST-ResNet

ST-ResNet is a *CNN*-based deep learning framework for traffic prediction. The model uses three CNNs with residual units to capture trend, period, and neighbor time information, respectively. In this paper, the lengths of the three CNN modules are set to 12, 4, and 4 respectively.

In order to quantitatively measure the prediction effect of different methods, this paper calculates the performance indicators of each method prediction results. As shown in Fig. 6 and 7, there are comparison charts of performance indicators of several methods. From the figure, it can be seen that the traditional statistical methods have poor prediction effects, while the current mainstream neural network related methods LSTM and XGBoost effects are relatively traditional. The prediction effect has been improved. For ST-ResNet, a complex network structure, the prediction effect is better than simple network structures such as LSTM and XGBoost, but because the model only divides short-term traffic data into trend, periodicity and proximity. The time is three modules, and the time correlation within the three modules is not considered, so the training effect is not as good as the prediction model proposed in this paper. Relative to this method, the RMSE is reduced by 1.95 and the *MAE* is reduced by 1.3. In summary, the short-term traffic flow prediction model proposed in this paper is significantly better than the existing prediction model and has certain practicability.

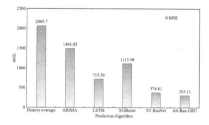 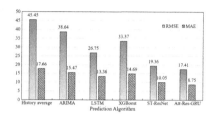

Fig. 6. MSE of predicted results by different methods.

Fig. 7. RMSE and MAE of predicted results by different methods.

6 Conclusion

The ITS plays a huge role in solving urban traffic problems. Through the ITS, drivers can plan their travel routes in advance, avoiding some congested road sections during this period, and thus reduce the travel time and improve the travel efficiency. It achieves a reasonable distribution of traffic flow in each section of the road network, alleviating urban traffic pressure. An important part of the realization of ITS is short-term traffic flow prediction. By analyzing the real-time traffic flow data and predicting the traffic flow information in the next short time period, traffic resources can be allocated reasonably. However, the current traffic prediction model cannot meet the needs of urban transportation system in terms of real-time, accuracy and response to emergencies. Therefore, this paper proposes a short-term traffic flow prediction model based on multi-characteristic analysis, which effectively solves the above problems.

References

1. Chen, Y., et al.: DeepTFP: mobile time series data analytics based traffic flow prediction (2017)
2. Guowen, D., Changxi, M., Xuecai, X.: Short-term traffic flow prediction method for urban road sections based on spacetime analysis and GRU. IEEE Access **7**, 143025–143035 (2019)
3. Koesdwiady, A., Soua, R., Karray, F.: Improving traffic flow prediction with weather information in connected cars: a deep learning approach. IEEE Trans. Veh. Technol. **65**(12), 9508–9517 (2016)
4. Lijin, Y., Qing, Y., Yonghua, L., Yuqing, F.: K-nearest neighbor model based short-term traffic flow prediction method. In: 18th International Symposium on Distributed Computing and Applications for Business Engineering and Science (DCABES), pp. 27–30 (2019)
5. Ma, X., Dai, Z., He, Z., Ma, J., Wang, Y., Wang, Y.: Learning traffic as images: a deep convolutional neural network for large-scale transportation network speed prediction. Sensors **17**(4), 818 (2017)
6. Okutani, I., Stephanedes, Y.J.: Dynamic prediction of traffic volume through kalman filtering theory. Transp. Res. Part B Methodol. **18**(1), 1–11 (1984)

7. Polson, N., Sokolov, V.: Deep learning predictors for traffic flows. Transp. Res. Part C Emerg. Technol. **79**, 1–17 (2016)
8. Shen, G.J., Wang, X.H., Kong, X.J.: Short-term traffic volume intelligent hybrid forecasting model and its application. Syst. Eng. Theory Pract. **31**, 562–568 (2011)
9. Tianqi, C., Carlos Ernesto, G.: XGBoost: a scalable tree boosting system. In: 22nd ACM SIGKDD International Conference on Knowledge Discovery and Data Mining, p. 785–794 (2016)
10. Wang, J.: Short-term freeway traffic flow prediction based on improved bayesian combined model. J. Southeast Univ. **42**(1), 162–167 (2012)
11. Wentian, Z., Yanyun, G., Tingxiang, J., Xili, W., Feng, Y., Guangwei, B.: Deep temporal convolutional networks for short-term traffic flow forecasting. IEEE Access **7**, 114496–114507 (2019)
12. Williams, B.M., Hoel, L.A.: Modeling and forecasting vehicular traffic flow as a seasonal ARIMA process: theoretical basis and empirical results. J. Transp. Eng. **129**(6), 664–672 (2003)
13. Wu, Y., Tan, H., Qin, L., Ran, B., Jiang, Z.: A hybrid deep learning based traffic flow prediction method and its understanding. Transp. Res. Part C Emerg. Technol. **90**, 166–180 (2018)
14. Ying-Hong, L.I., Liu, L.M., Wang, Y.Q.: Short-term traffic flow prediction based on combination of predictive models. J. Transp. Syst. Eng. Inf. Technol. **13**(2), 34–41 (2013)
15. Zhang, J., Yu, Z., Qi, D., Li, R., Yi, X., Li, T.: Predicting citywide crowd flows using deep spatio-temporal residual networks. Artif. Intell. **259**, S0004370218300973 (2017)
16. Zhiyang, G., Sun, Z.: Basis-prediction method for short-term traffic flow prediction and its application. In: IEEE International Conference on Service Operations and Logistics, and Informatics (SOLI), pp. 197–202 (2019)
17. Zhiyang, G., Sun, Z.: Short-term traffic flow prediction and its application based on the basis-prediction model and local weighted partial least squares method. In: 2019 14th International Conference on Computer Science & Education (ICCSE), pp. 992–997 (2019)

Research on Competition Game Algorithm Between High-Speed Railway and Ordinary Railway System Under the Urban Corridor Background

Lei Wang[(✉)] and Yao Lu

School of Economics and Management, Lanzhou Jiaotong University,
Lanzhou, Gansu, China
wanglei@mail.lzjtu.cn

Abstract. The coexistence of high-speed railway and ordinary railway in public transport corridors has led to competition and cooperation between the two transports systems. In this case study of Baolan high-speed railway, a game theory model is established, including three types of players of high-speed railway, ordinary railway and passenger, and involving three kinds of influencing factors of economy, fastness and comfort. The concept of "linear city" was used to simulate the position of the passengers and passenger's origin and destination into the two-dimensional O-D matrices. Based on the utility theory and heuristic algorithm, the Nash equilibrium problem was solved, and the reasonable pricing strategy of high-speed railway and ordinary railway was obtained.

Keywords: High-speed railway · Ordinary railway · Game theory · Heuristic algorithm · Utility theory

1 Research Background

Baolan high-speed railway is the west of XuLan high-speed rail in national medium and long-term railway network plan, is also an important part of the high-speed rail network with "four vertical and four horizontal" in China. After the completion of this high-speed railway, train running time will be reduced to 2 h from Baoji to Lanzhou, this will greatly shorten the time and space between the northwest with the eastern and central regions. That forming a rapid railway passage on the silk road, is an important railway transportation infrastructure of the "One Belt" And "One Road" strategy. The BaoLan high-speed railway, with a total length of more than 400 km and a design speed of 250 km per hour, is led from Baoji South Railway Station to Lanzhou West Railway Station, which was officially put into operation in July 2017.

At present, regardless of the railway or the highway, the direction to the Silk Road is basically the same, such a city corridor is not uncommon. For example, in Taiwan, Taipei and Kaohsiung high-speed railway, South Korea high-speed train (KTX), Japan Shinkansen and others are basically as such situation. With the development of the city along the urban corridor, it is always an important issue to meet the growing demand of

P. Qin et al. (Eds.): ICPCSEE 2020, CCIS 1258, pp. 290–308, 2020.
https://doi.org/10.1007/978-981-15-7984-4_22

inter city traffic. Due to the relatively short distance involved in BaoLan high-speed railway, the air transport system has never gained a competitive advantage while the road and railway system provide most of the inter-city transportation (Zhang, Luan and Zhao) [1].

As two kinds of mode of transportation for railway system, high-speed railway and ordinary railway both focus on passengers' rapid transit, after the opening of Baolan high-speed railway, the high-speed train will form competition and cooperation with the ordinary railway.

2 Related Research Review

The introduction of high-speed railway system can significantly affect the spatial structure and market share of existing modes of transportation. The existing research focuses on competition between high-speed railway systems and existing modes of transport. Chang proposed a static transportation distribution method to predict the market share of high-speed railway in the northwest-southeast corridor of Korea, and to evaluate the price structure and capacity constraints of all competing modes of transport (Chang and Chang) [2]. Roman, Espino, and Martin used the information provided by the reference database to analyze the potential competition between the Spanish-Martin-Barcelona corridor's high-speed railway and aviation through a non-agglomeration model, which is considered to be a long-distance travel Aspects of holding a smaller market share, and relative to the car and bus traffic, high-speed rail more competitive (Roman, Espino and Martin) [3]. Allport and Brown suggested that the relative impact of high-speed rail is more significant in smaller cities than in large metropolitan areas (Allport and Brown) [4]. An interpretation is that smaller cities are not as well served by airlines due to infrequent flights, whereas the travel time to high-speed railway stations is relatively low.

However, there are also studies that show there is also the possibility of cooperation between high-speed railway systems and other modes of transport. Givoni and Banister (2006) argue that airlines can use railway services as an additional service to airports hub in their networks to complement and replace existing aircraft services and assess their benefits and limitations (Givoni and Banister) [5]. Hsu and Chung (1997) evaluated the market distribution of high-speed railway and ordinary railway in the transport corridor, which showed better ability to serve half-way and long-distance travel markets, while ordinary railway showed better service short-distance travel market capacity (Hsu and Chung) [6].

In the case of urban corridor, high-speed rail and the existing ordinary railway system share a corridor interaction. Because of their relatively close location of the network configuration, parallel to each other and share some of the platform, is a competitive and cooperative relationship, but the competitive situation is the main trend. They cooperate with each other by attracting more passengers to use complementary railway services and to create value together while traveling. And they produce competition when the passengers in the choice of more than two railway systems it produces competition. Thus, competitive threats and cooperation complement co-exist. The research base on game theory is less in China, and high-speed railway and game are used as keywords to input CNKI. The literature is only a single digit and concentrated on the competition research of high-speed railway and air transport, and there is no system research on competition game and Cooperation between high-speed railway and ordinary railway. This article is based on the city corridor background and mainly studies the competition and cooperation between high-speed railway and ordinary railway in Baolan city corridor.

3   Game Model Establishment

3.1   The Basic Elements of the Game Model

The most important component of the game theory model is the player, the strategy set, and the gain function.

Player
Due to the relatively short distance involved in the Baolan high-speed railway, the air transport system has never won the competitive advantage, and the road passenger transport is limited to the short-distance transportation. Considering that the purpose of this paper is the competition between high-speed railway and the ordinary railway and cooperation; Thus the model consists of three factors: passengers, ordinary railway, high-speed railway.

Strategic Set
High-speed railway, ordinary railway and passengers will make choices to seek to maximize the benefits or utility during the game process, all of these choices constitute the strategic set of players.

This paper assumes that high-speed railway and ordinary railway are for the purpose of profit-oriented transportation system, so in the choice of strategy will choose to maximize the pursuit of benefits, and other difficult to change or the effect is not obvious factors (such as product quality, advertising efficiency), The price strategy is a relatively easy tool for the high-speed rail and ordinary railway to pursue the benefits, and the majority of the general travel costs of the passengers are directly or indirectly related to the price strategy adopted by the high-speed rail and the ordinary railway; Thus the price constitutes a strategic set of high-speed rail and ordinary railway.

Both the high-speed railway and the ordinary railway are constantly adjusting their prices in order to maximize their own profits. Each price is a strategy for both railroads. Each of them can choose a price range of strategies. This price range is a linear change strategy set. This can make the game process complicated because it can't list all the possible situations.

For passengers, they choose transportation in order to maximize their utility (generally travel cost minimization). Since this article mainly discusses the game of high-speed railway and ordinary railway, the passengers' strategy set for {high-speed railway, ordinary railway}.

Gain and Gain Function
It is clear that the game process belongs to the dynamic game, taking into account the passengers, high-speed railway, ordinary railway are aware of other stakeholders in the

game-related factors, and therefore a complete information game. Because the strategy of high-speed railway and ordinary railway is set as the price range, it can't list the gain function of high-speed railway and ordinary railway. This paper adopts formal definition; in addition, it can't enumerate the function of profit function of passengers.

3.2 Railway System Benefit Function Analysis

For high-speed railway and ordinary railway, their profit π can be obtained by revenue T_s minus the operating costs T_m.

$$Max\pi = T_s - T_m \tag{1}$$

The income T_s can be calculated by passenger flow volume, the unit cost and distance by the high-speed railway and ordinary railway.

$$T_s = vp_m \times q_m \times d_m \tag{2}$$

Type: vp_m indicates the unit distance of railway, q_m indicates the passenger volume flow of railway, d_m indicates the running distance of railway, m indicates the ordinary railway or high-speed railway.

Railway transport operating costs include two parts - fixed and variable costs. Among them, Fixed cost refers to the cost that does not change with the change of passenger volume for a certain period of time, which is related to the number of train compartments, Mainly refers to the cost of buying vehicles and others infrastructure, depreciation of fixed assets and so on (Yang and Wang) [7]. Variable cost refers to the cost that is proportional to the change in passenger traffic over a given period of time, which is related to the running distance of the train, mainly including the unit cost of train's running and the cost of stopping. The relationship between the total cost of the railway system transportation, the unit transportation cost and the passenger volume are as follows:

$$T_m = FP_m + VC_m q_m \tag{3}$$

Type: T_m refer to the total passenger cost of the railway, FP_m refer to the fixed cost of rail passenger transport, VC_m refer to the unit cost of the railway, q_m refer to the passenger volume of the railway, m refers to the ordinary railway and high-speed railway.

This kind of two part cost pricing strategy is realistic and validated in the marketplace. However, in order to analyze the fares of the railway system under the premise of maximizing the profit of the railway system, taking into account the invariance of the fixed cost of railway system and the corresponding capital investment and subsidy, this paper mainly analyzes the variable cost of railway passenger transport. The comprehensive solution model of the maximum profit of the railway system is as follows:

$$\text{Max } \pi_m = \sum_{r \in \text{ROUTE}} vp_m \times d_m^r \times q^r - \sum_{r \in \text{ROUTE}} VC_m \times d_m^r \times q^r$$

$$\text{s.t.} \begin{cases} VP_m^l \le vp_m \le VP_m^u \\ q^r = |\{g | U_r < U_{r'}g \in G, r' \ne r \in \text{ROUTE}\}|, \ \forall r \in \text{ROUTE} \end{cases} \quad (4)$$

Type: π_m refer to the max profit of railway system, vp_m refer to the unit cost of railway m, q^r refer to passenger volume of railway m, d_m refer to the distance of railway m, U_r refer to passengers' utility.

The decision variables of two kinds of mode of transportation in the price competition in the model are unit of distance fare vp_{m_1} and vp_{m_2}, m_1, m_2 respectively is ordinary railway and high-speed railway. The government's guidance plays a very important normative role in the development of transport corridors. The government allocates capital investment in railway and makes decisions based on political and environmental considerations and sets the maximum price for the railway system. The lowest price is determined by the cost of transporting the railway system, thus promoting high-speed railway, ordinary railway reasonable competition. These values should be based on the provisions of the government control in a reasonable range.

3.3 Analysis of Passenger Generalized Trip Utility Function

Analysis on the Factors Influencing Passenger Journey

In view of economy, quickness, convenience, safety, comfort on the choice of passengers have a greater impact on the way, these five factors can be used as a passenger utility function of the variables to start a preliminary analysis.

Economy

Economy factors mainly involving the fares of high-speed railway and ordinary railway. According to the above discussion, ticket price factor is the same in transportation distance, variable factors of it are single kilometers distance fares. The factors of the economy can be expressed as:

$$C_g^r = vp_m \times d_m^r \quad (5)$$

Type: C_g^r is economy factor, refer to ticket price that passenger g spend, vp_m refer to price for unit distance of railway m, d_m refer to distance of railway m, m refer to high-speed railway and ordinary railway.

Quickness

This game model is mainly used the time cost (time × per unit time value) of passengers in whole journey to measure quickness. Time including the time passengers spent on train, waiting time and the time in and out of the station, and passenger unit time value associated with different passenger categories. Quickness factor index expressed as following type:

$$T_g^r = t_r \times V_g \tag{6}$$

Type: T_g^r is quickness factor for passengers, t_r refer to the sum of the travel time spent on the route r, the waiting time of the station, and the time of entry and exit, V_g refer to unit value of passenger g.

Passengers in different parts have different time value. This paper calculate the passenger travel time value based on the production method (Zhang and Peng) [8].

$$V_a = \frac{GDP_a}{t_a \times q_a} \tag{7}$$

$$V_b = \frac{GDP_b}{t_b \times q_b} \tag{8}$$

$$V_g = \frac{V_a + V_b}{2} \tag{9}$$

Type: V_a and V_b are the two city passengers time value in transport channels at both ends city a and city b respectively, GDP_a and GDP_b respectively are gross national product of city a and city b, t_a and t_b respectively are the city per capita annual working hours in city a and city b, q_a and q_b respectively are population of city a and city b, V_g refer to passengers' time value between city a and city b.

Convenience
The time spend on buying ticket and waiting time are similar between high-speed railway and ordinary railway. so related research usually don't consider the different convenience caused by that, mainly considering the extra time and economic cost that passengers spend on transit transfer. but on the a short distance urban corridor like Baolan railway (Baolan high-speed railway's running time is only 3 h), under the condition of the same originating station transfer rarely occur, so this paper is temporarily not considered about convenience.

Comfort
In this game model, comfort index is mainly affected by the transport time factors, ignoring the influence of other factors, can be thought of time needed for fatigue and travel time curve connection, use the formula (Ai 2006) [9] to express as:

$$T_m = \frac{H}{1 + \alpha e^{-\beta t_m}} \tag{10}$$

$$F_g^r = T_m \times V_g \tag{11}$$

Type: T_m is passengers' fatigue recovery time by m, H refer to passenger's fatigue recovery limit time, α、β refer to coefficients of recovery fatigue by m, t_m is the total travel time of traveling by m, F_g^r is comfort factor, V_g is passenger's unit time value.

According to the relevant existing research literature (Dai) [10], H usually take 24 h, railway coefficients of recovery fatigue to be determined by coefficient of value (49,0.33).

Safety

Safety has a "one-vote veto" for passenger transport choice. If a transportation security is too low, no matter how significant is the utility, passengers won't choose this kind of way to travel, but the railway transport system in China country, regardless of the high-speed railway or ordinary railway has extremely high security features, public psychology rarely consider security features when choosing a mode of railway transportation, so temporary not consider security in this paper.

Passenger Group Classification

Here first, it is necessary to classify the target passenger groups according to the travel characteristics and traveler characteristics of the traffic (Tang, Xiong, Sun and Du) [11], and then analyze the weight of the above factors in each group so that the factors of the Baolan city corridor can be evaluated more accurately.

In the process of travel choice, passengers will show different preferences of the above services factors according to their respective travel demand with its own characteristics, follow the principle of benefit maximization, eventually choose ordinary railway or high-speed rail travel. According to the passengers' social and economic characteristics, they could be divided into different types, different types of passengers want different transportation service level without loss of generality. This article will divided passengers roughly into three categories: high level passengers, medium level passengers and low level passengers (Lu) [12].

Factors that affect passengers to select transportation mode include economy, fastness and comfort. Formula (12) said the generalized utility function of passengers when choosing a travel router.

$$U_{ir} = X_{i1}C_g^m + X_{i2}T_g^m + X_{i3}F_g^m \qquad (12)$$

Type: U_{ir} is utility of passengers, it refer to the utility of the passenger i by transport m, taking into account the common role of the passengers in the decision-making process, C_g^r, T_g^r, F_g^r refer to economic costs, fast cost, comfort costs respectively.

In the formula, X_{in} (n = 1, 2, 3) denote the weight value of the third kind of passenger, which indicates the degree of attention of the passenger to the factor, so as to obtain the different class passenger function as follows:

$$U_{1r} = x_{11}^{vp_m}C_1^r + x_{12}T_1^r + x_{13}^{vp_m}F_1^r \qquad (13)$$

$$U_{2r} = x_{21}^{vp_m}C_2^r + x_{22}T_2^r + x_{23}^{vp_m}F_2^r \qquad (14)$$

$$U_{3r} = x_{31}^{vp_m}C_3^r + x_{32}T_3^r + x_{33}^{vp_m}F_3^r \qquad (15)$$

Analysis of Passengers' Generalized Travel Cost Based on Utility Theory
In view of the utility theory can make a reasonable measure of subjective factors such as human preference, inclination and hobby, this paper introduces the utility theory into the game model as the basis for the definition of passenger generalized travel cost. In the ordinary railway and high-speed railway game process, different types of consumers have different tendencies, attitudes and choices. As mentioned above, the type of transportation chosen by passengers is affected by many factors, and the degree of subjective influence of each influencing factor is different. Therefore, in the calculation of passenger travel utility, the influence factors are different category passengers give different weight parameters.

According to the utility theory, the economy, fastness, and comfort of different types of passengers has different characteristics on the sensitivity of the change of the fare. Therefore, it is necessary to analyze the various utility cost values corresponding to the three passenger groups respectively.

Economy
The price of railway will affect the purchase intention of passengers obviously. According to the theory of utility, when other factors unchanged, when high-speed railway or ordinary railway ticket prices rise, this kind of mode of transportation of the economic costs to rise. But for different types of passengers, the sensitive degree of high level passengers, medium level passengers, low level passengers to economy change increases in turn, and utility reduced. This especially for low level passengers, showed no tolerance of the height of the rise in ticket prices, so low level passengers will usually show a preference for ordinary rail; while the high-level passenger showed the height of the fare increases tolerance, so the high level passenger usually will not give up due to higher ticket prices generally rice in fares and prefer for high-speed railway. Thus, for the passengers' utility function, the economy of the coefficient should be a reverse function related to the price.

Quickness
With the strengthening of social concept of time, passengers gradually strengthen the requirements of the quickness. When other factors constant, the length of travel time and cost of passengers present a tendency of positive correlation coefficients, for the same passenger perception of a certain mode of railway transportation, as the growth of the travel time, passenger utility decreases gradually. For different types of passengers, in the high level passenger, medium level passengers, low level passenger, in turn, have less attention to the quickness, sensitivity to the quickness changes in turn down.

Comfort
Along with the social development and people's consumption level is improving, the requirement of passengers to their comfort present a rising trend. Passengers' comfort including the size of the space, service quality, environment and so on. For the same passenger, in the case of other factors unchanged, passenger utility cost inversely proportional relationship with comfort. For different types of passengers, high level, medium level and low level of passengers' emphasis on comfort degree reduced in turn. At present, comfort and fare show a high correlation, the higher the fare is often the more comfortable ride, so the game model in the comfort weight and economic weight to take the same value.

4   Analysis of Passenger Quantity Based on Utility Function

The path selection model has a wide range of applications in the study of transportation problems. In this paper, the optimal traffic network OD matrix selection model is used to describe how the passengers choose the traffic mode according to their own tendency. Assuming that the traveler is a "rational economic man", that is, they always choose the lowest mode of transportation in the trip. This paper distributes the number of passengers of selecting various of transportation according to the utility ratio of various types of passenger to high-speed railway and ordinary railway.

The passenger's trip starts at the passenger station of the railway system and ends at its destination terminal. Visitors can choose their own departure station and terminal. Unit distance fares for high-speed railway and ordinary Railway affect the total fare between any two stations. The same method can be used to study the game between the high-speed railway and the ordinary railway between any two stations. This paper assumes that passengers are independent of each other and do not have a competitive relationship.

As mentioned above, in the case of this short-distance city corridor in Baolan high-speed railway (Baolan high-speed railway operation time is only 3 h), transfer rarely occur in the case of the same starting station, so passengers have two routes to choose from, (a) take only ordinary railway, (b) take only high-speed rail. On the one hand, due to the relatively small number of high-speed railway stations, there is only one high-speed railway station (H station in Fig. 1) in the neighboring areas of the city. On the other hand, the ordinary railway stations will be relatively high, and one of them is the main station (T station in Fig. 1), the other adjacent areas belong to the auxiliary station (t_1 station in Fig. 1).

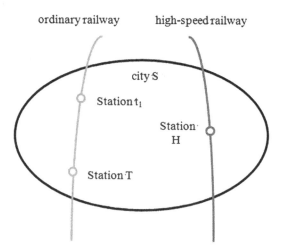

Fig. 1. Schematic diagram of urban rail distribution

Assuming that the passenger lives in the urban area near the city, he can choose to go directly to the main station T to take the intercity rail, and if the cost of a transit route is lower than the direct route, they can choose to ride the auxiliary line between the auxiliary lines And in the main station on the transfer to the railway between the city up. As can be seen in Fig. 1, regardless of whether the passengers traveling by ordinary railway use the auxiliary site on the extension, he needs to go through the master station T. Thus, all ordinary railway sites can be assumed to be a single site T.

Without losing the generality of the customer's choice, it is further assumed that the S city is a one-dimensional "linear city" between the T station and the H station, as shown in Fig. 2, where the passenger is located between the two stations, the passenger's personal utility function Determines its preference for selection, which is expressed as a degree of proximity of its distance T and H, that is, a point G on the T-line.

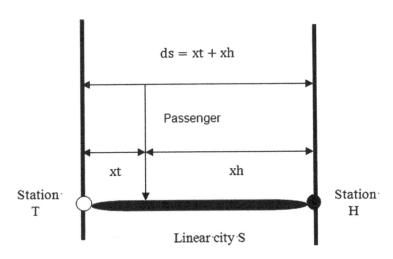

Fig. 2. Passenger g's originating site

Then the specific analysis of the starting point of the linear city. Different types of passengers choose ordinary railway or high-speed railway depends on the utility value of various modes of transport. G to the distance between the two end points respectively represent the tendency of the passenger to select the ordinary railway and the high-speed railway. The utility of the i-type passenger chooses the ordinary railway is $x^i_{s,t}$, and the utility of selecting the high-speed railway is $x^i_{s,h}$, the sum of the two is $x = x^i_{s,t} + x^i_{s,h}$. By combining the two linear cities of the departure and destination, an OD rectangle as shown in Fig. 3 is obtained, where both I and IV can represent two scenarios that the passenger described above may choose: (I) ordinary railway and (IV) high-speed railway. The position of the dividing line in the OD model depends, to a certain extent, on the utility function of the passenger, and determines the number of passengers selected at the starting point for high-speed railway and ordinary railway.

Among them, the subscript S and E represent the starting city and the city. The OD rectangle will play a central role in building the model and estimate the number of people choosing a variety of ride patterns based on the OD matrix. Since the number of segments of the population is related to the position of the dividing line, it is independent of the specific density of each point, it is assumed that the passengers in the OD matrix are evenly distributed and the number density of the passengers is equal, and this assumption does not interfere with the research of this paper. The number of the i passengers of the selected route r is determined by the Eq. (16).

$$q_r^i = \frac{X_r^i}{X_I^i + X_{IV}^i} \times Q_i \qquad (16)$$

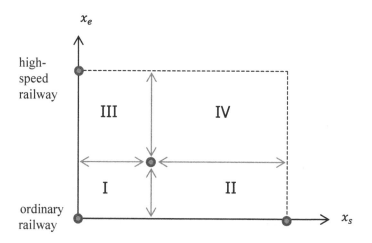

Fig. 3. Starting station passenger OD matrix

5   Heuristic Algorithm for Game Model

5.1   Nash Equilibrium

In this paper, there is a binary non-cooperative game between high-speed railway and ordinary railway. The Nash equilibrium expression for these two modes of transport is as follows:

Nash equilibrium expression for high-speed railway:

$$\left(\pi_H, \pi_T^*\right) \leq \left(\pi_H^*, \pi_T^*\right) \leq \left(\pi_H^*, \pi_T\right)$$

Nash equilibrium expression for ordinary railway:

$$\left(\pi_H^*, \pi_T\right) \leq \left(\pi_H^*, \pi_T^*\right) \leq \left(\pi_H, \pi_T^*\right)$$

In the above formula, * indicates that the item reaches the optimal strategy. According to the Nash theorem, there is at least one pure strategy or a mixed strategy of Nash equilibrium for each finite game, so the game must find an equilibrium solution. The government as a regulator will maintain the fair competition in the transport and safeguard the reasonable rights and interests of consumers to develop a reasonable price range of various modes of transport, this article will find a reasonable portfolio of high-speed railway and ordinary railway within this context.

5.2 Heuristic Algorithm Research

If the winning function is a continuous differentiable function, the Nash equilibrium can be solved by means of the partial differential standardization of the price. Though the winning function is continuous relative to price, it's not differentiable at every point, therefore, this paper proposes a heuristic algorithm to solve the Nash equilibrium problem of this game model.

The Nash equilibrium solution consists of two variable values, unit distance ticket price vp_m of the two players. The game process of ordinary railway and high-speed railway is a multi-stage full information dynamic game, the heuristic algorithm allows the two players to think that they have made the best response to the other player via an iterative optimization process and adjust unit distance ticket price constantly.

The Benefit Adjustment Function of Railway System

Adjust the benefit function of railway system listed in formula (4), add two subentries; In the process of iteration, when the market share of the participant m is close to 0 or 100%, one of these two constraints would impose heavy penalties to the objective function value(shown in formula 5), thus the search process is effectively guaranteed far from the boundary value.

$$\pi_m = \sum_{r \in ROUTE} vp_m \times d_m^r \times q^r - \sum_{r \in ROUTE} C_m \times q^r \times d_m^r$$
$$- \frac{100}{\sum_{r \in ROUTE} W_m^r q^r + 0.000001}$$
$$- \frac{100}{\sum_{r \in ROUTE} W_s^r q^r - \sum_{r \in ROUTE} W_m^r q^r + 0.000001}$$

In the formula, W_m^r is a 0–1 variable, if the route r includes the travel of the railway m, its value is 1, otherwise it is 0, the other variables have the same meaning as described above.

Initialization

Set the ordinary railway fare range is (vp_t^l, vp_t^u), high-speed railway fare range is (vp_h^l, vp_h^u), the fare range depends on the pricing range of the government and the cost of the railway to consider. After Baolan high-speed railway is opened, High-speed railway will refer to the traditional mode of transport of ordinary rail transport unit fare vp_t^0 and its own income function to develop their own unit fare vp_h^0. Passengers according to their own general travel costs to determine their own choice, high-speed rail and ordinary railway is to further calculate their own income.

Iteration

The calculation of each iteration process consists of two steps. The first step, ordinary railway finds the best fares within its range of fares (vp_t^l, vp_t^u) for the current fare of high-speed railway to form a new situation (vp_t^l, vp_h^0) in this situation, passengers determine their riding options according to their generalized travel costs, then ordinary railway gets the maximization profit; The second step, high-speed rail finds the best fares within its range of fares (vp_h^l, vp_h^u) for the current fare of ordinary railway, in this situation, passengers determine their riding options according to their generalized travel costs, then ordinary railway gets the maximization profit;

The Iteration Cycle and Termination

The Fig. 4 shows the flow chart proposed by the solution. Both of them converge gradually through the n times game, the judges in the paper is that the number of people

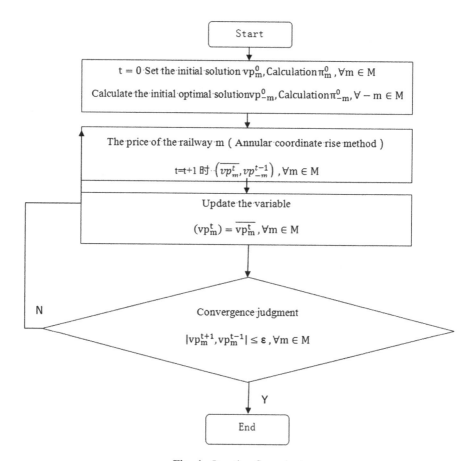

Fig. 4. Iterative flow chart

on the high-speed railway and the ordinary railway is basically stable, the price (vp_h^n, vp_t^n) of transport strategy is basically stable, at this point, both will not easily change their decision, reaching Nash equilibrium. From the above description, consider only one change of mode of transportation every time, labeled as "m", "-m" refers to another mode of transportation, when "m" changes its price strategies, the price of "-m" doesn't change.

The model is less restrictive, but the objective function is also a function of the distance between all the players in the system. Therefore, the model is relatively complex, and the image is not smooth at each place. The heuristic algorithm uses a ring-shaped coordinate rising method that is widely used to solve the optimization problem. The general idea of the method is to maximize the objective function value by optimizing the decision variables.

6 Game Calculation Between High-Speed Railway and Ordinary Railway of BaoLan City Corridor

Taking BaoLan transportation corridor as an example, the above game model is used to study the empirical study. According to the purpose of the study, the players include BaoLan high-speed railway, ordinary railway transport and passengers. Because the situation is similar, need only to consider the single line from Lanzhou to Baoji, and because the model mentioned in this paper has nothing to do with the starting station, we choose the most representative Lanzhou and Baoji two stations to do relevant research.

6.1 Game Model Parameters

The Initial Value
According to the current fare of Lanzhou Baoji section ordinary railway, it is estimated that the distance between Lanzhou and Baoji ordinary railway is about 0.12 yuan/km, and this value is the initial value of the distance from the ordinary railway unit.

Economy
China's current high-speed rail unit distance freight rates, the lowest is the price of Beijing-Guangzhou high-speed railway 0.376 yuan/km, the highest is the Guangzhou-Shenzhen high-speed railway 0.73 yuan/km. Reference to China's current situation of high-speed railway fares, the upper and lower bounds on the BaoLan high-speed railway fares, set BaoLan high-speed rail unit distance rates, floating range of 0.35 yuan/km to 1 yuan/km. With the emergence of high-speed railway, the ordinary railway from Lanzhou to Baoji gets a big competitive challenge of transport market, In the short term, Ordinary railway fares will not rise much on the basis of the original price, so the floating range of the ordinary railway freight unit distance is set to 0.1 yuan/km to 0.4 yuan/km.

Referring to the railway odometer, we can see that the distances of various routes between Lanzhou and Baoji are shown in Table 1. Find out the function relation between the income of the railway transportation mode and the number of passengers by substituting the unit distance fare and the distance of transportation into the formula (5).

Table 1. Distances of various routes between Lanzhou and Baoji

Distance (km)	Lanzhou - Dingxi	Lanzhou - Tianshui	Lanzhou - Baoji	Dingxi - Baoji	Tianshui - Baoji
Ordinary railway	118	432	587	469	155
High-speed railway	93.99	269.44	400.95	306.96	131.51

Quickness

According to the statistical data of Lanzhou City in 2016 and the relevant economic data of Baoji City Statistical Yearbook in 2016, figure out the average time cost of passengers is calculated as 0.5 yuan/person per minute. Without loss of generality, the average time cost of low-level passengers, middle-level passengers and high-level passengers is set to 0.2 yuan/person per minute, 0.5 yuan/person per minute, and 2 yuan/person per minute. According to the calculation, the rapidity indexes of various transportation modes are shown in Table 2, In order to verify the assumption that passengers do not usually choose to transfer, each factor index of the transfer is taken into account.

Table 2. Indicators for the rapidity of various modes of transportation

Time cost (yuan)	High-speed railway	Ordinary railway	High-speed to ordinary railway	Ordinary to high-speed railway
Transfer in Dingxi	76	142	75	90.2
Transfer in Tianshui			101.8	84.7

Comfort

According to the comfort measurement formula, by inputting the relevant data the comfort cost is shown in Table 3.

Table 3. Cost of comfort for various paths

Time cost (yuan)	High-speed railway	Ordinary railway	High-speed to ordinary railway	Ordinary to high-speed railway
Transfer in Dingxi	28.98	288.87	81.32	32.32
Transfer in Tianshui			39.41	45.47

The Scale of Low, Medium and High Level Passengers

According to the passenger flow data of the railway transportation corridor from Lanzhou to Baoji, the annual passenger flow between Lanzhou and Baoji is about 30000 people, under the premise that there is no significant impact on the conclusions of the study, in order not to lose generality, the scale of low-level passengers, middle-level passengers and high-level passengers is 6:3:1, low-level passengers are 18000 people, middle-level passengers are 9000 people, high-level passengers are 3000 people.

Service Attribute Parameter Weight

According to the analysis of the relationship between the high-level, mid-level and low-level of the three class passengers and the three transport factors (Comfort, Economy, Quickness), calibrate the service attribute parameter weight of ordinary railway and high-speed railway. To differentiate between three types of passengers, the weight may be considered as an exponential function with good differentiation. In the empirical analysis, the coefficient is setting as follows (Table 4):

Table 4. Factor setting of utility factor of passengers

	Coefficient of economy	Coefficient of rapidity	Coefficient of comfort
High-level passengers	100^{vP_m}	0.5	500^{vP_m}
Middle-level passengers	200^{vP_m}	0.3	200^{vP_m}
Low-level passengers	500^{vP_m}	0.2	10^{vP_m}

6.2 Passenger Travel Path Selection Analysis

In order to verify the foregoing conclusion that transfer rarely occurs in the short distance urban corridors such as the BaoLan high-speed rail and the same starting station case, the following calculations are carried out.

According to the linear city theory described above, Lanzhou City can be identified as the existence of a high-speed railway station and an ordinary railway station, if considering the middle of the transfer, the passengers near Lanzhou go to Baoji may exist the following six kinds of travel options.

Take the ordinary railway from Lanzhou to Baoji;

Take the high-speed railway from Lanzhou to Baoji;

Take the ordinary railway from Lanzhou, change to the high-speed railway in Dingxi, and then get to Baoji;

Take the high-speed railway from Lanzhou, change to the ordinary railway in Dingxi, and then get to Baoji;

Take the ordinary railway from Lanzhou, change to the high-speed railway in Tianshui, and then get to Baoji;

Take the high-speed railway from Lanzhou, change to the ordinary railway in Tianshui, and then get to Baoji;

Passengers will choose one of the six routes. In the case of (3) to (6), the transfer mode situation will be appeared for these four cases, if the passenger utility value is smaller than the passenger utility value in the case of direct access, the passenger chooses to discard the transfer route. The convenience here is usually a function of the sum for the passenger transfer time and the transfer cost, which can be defined as the transfer time * the time value cost of the passenger + the transfer cost.

The utility values of the four influencing factors and the weight value of the different influencing factors are taken into account. The utility values of the high-range, middle-range and low-range passengers for the six kinds of driving routes are shown in the following Table 5.

Table 5. Utility values for various travel routes

	Route 1	Route 2	Route 3	Route 4	Route 5	Route 6
High-level passengers	519.99	310.48	959.69	1107.67	1065.00	1002.16
Middle-level passengers	190.80	251.38	440.82	419.45	425.25	434.99
Low-level passengers	156.36	260.76	329.92	283.24	296.34	317.81

By comparing the results, it is found that the transfer utility cost of the high-level, middle-level and low-level passengers is much higher than the direct, thus validating the conclusions mentioned above.

6.3   Results and Analysis of Heuristic Algorithms

Through dynamic game, the unit distance Nash equilibrium fare of ordinary railway and high-speed railway from Lanzhou to Baoji is 0.35 RMB and 0.47 RMB; It can be obtained at this equilibrium price from the game results that there are 17876 people travel by ordinary railway, accounting for 59.6% of the total number, and the profit is 3447624.2 RMB; There are 12124 people travel by high-speed railway, accounting for 40.4% of the total number, and the profit is 961725.4 RMB. Under this circumstance of along the city corridor, this paper argues that the reasonable explanation of the balance between the high-speed railway and the ordinary railway as follows: High speed rail provides more efficient and comfortable transportation services by increasing the price of the ticket to get high-end and middle-end tourist source and more profit, while ordinary railway obtains middle-end and low-end tourist source by lower price to get more market share; Such a market division with Nash equilibrium makes both sides have improved profit. The results of this study provide a reference for the price adjustment of Baolan ordinary railway and high-speed railway.

In fact, the unit distance fare from Lanzhou to Baoji is 0.12 RMB, lower than the equilibrium price calculated in this paper, that's maybe caused by the government regulation or the semi-public nature of the railway.

7 Summary

This paper studies the cooperation and competition between the high-speed railway and the ordinary railway in the urban corridor. It constructs a competitive game model which involves three categories of player–ordinary railway, high-speed railway and passengers, involving economy, quickness and comfort factors. Then it introduces the utility theory, and solves the problem of the Nash equilibrium price using heuristic algorithm based utility theory. The reasonable pricing strategy of high-speed railway and ordinary railway is made, and successfully applied to the competition game between the high-speed railway and the ordinary railway in the BaoLan urban corridor. It proves that passengers usually do not choose the intermediate transition in the middle and short distance urban corridors.

There is still space for further expansion of this research, this paper is mainly considered the internal game railway system, without considering the highway transportation into the scope of the study, and also fails to conduct overall research in the entire transport field, which will become the focus of future research.

Acknowledgement. This work was partially funded by Social Science Planning Project of Gansu Province of China (No. YB060) and National Natural Science Foundation of China (No. 71461017).

References

1. Zhang, R., Luan, W., Zhao, B.: Pricing game analysis of high - speed railway and air ticket based on traveler choice. Railway Transp. Econ. **37**(1), 5–9 (2015)
2. Chang, I., Chang, G.L.: A network-based model for estimating the market share of a new high-speed rail system. Transp. Plann. Technol. **27**(2), 67–90 (2004)
3. Roman, C., Espino, R., Martin, J.C.: Competition of high-speed train with air transport: The case of Madrid-Barcelona. J. Air Transp. Manage. **13**(5), 277–284 (2007)
4. Allport, R.J., Brown, M.: Economic benefits of the European high-speed rail network. Transp. Res. Rec. **1381**, 1–11 (1993)
5. Givoni, M., Banister, D.: Airline and railway integration. Transp. Policy **13**(5), 386–397 (2006)
6. Hsu, C.I., Chung, W.M.: A model for market share distribution between high-speed and conventional rail services in a transportation corridor. Ann. Reg. Sci. **31**(2), 121–153 (1997)
7. Yang, Y., Wang, H.: Study on calculating the comprehensive cost of transportation for high-speed railway. J. Railway Eng. Soc. **1**, 102–106 (2009)
8. Zhang. Y., Peng, Q.: The influence of passenger dedicated line on the sharing rate of transportation corridor. Railway Transp. Econ. **28**(12), 16–19 (2006)
9. Ai, Y.: Research on competitive strategy of transportation mode in regional passenger transport channel based on game analysis. Lanzhou Jiaotong University, Lanzhou, vol. 4 (2016)

10. Dai, L.: A study on the competitive relationship between speed rail and other transportation modes-a case study of beijing-shanghai high-speed railway. Beijing Jiaotong University, Beijing, vol. 6 (2009)
11. Tang, Z., Xiong, F., Sun, J., Du, Xiang.: Quantitative evaluation of higher education audience's happiness index based on fuzzy analytic hierarchy process. J. Mudanjiang Univ. vol. **07**, 168–171 (2014)
12. Lu, W.: Research on competitive game between high - speed railway and civil aviation. Beijing Jiaotong University, Beijing, pp, 38–39 (2015)

Stock Price Forecasting and Rule Extraction Based on L1-Orthogonal Regularized GRU Decision Tree Interpretation Model

Wenjun Wu, Yuechen Zhao, Yue Wang[✉], and Xiuli Wang

School of Information, Central University of Finance and Economics,
Beijing, China
yuelwang@163.com

Abstract. Neural network is widely used in stock price forecasting, but it lacks interpretability because of its "black box" characteristics. In this paper, L1-orthogonal regularization method is used in the GRU model. A decision tree, GRU-DT, was conducted to represent the prediction process of a neural network, and some rule screening algorithms were proposed to find out significant rules in the prediction. In the empirical study, the data of 10 different industries in China's CSI 300 were selected for stock price trend prediction, and extracted rules were compared and analyzed. And the method of technical index discretization was used to make rules easy for decision-making. Empirical results show that the AUC of the model is stable between 0.72 and 0.74, and the value of F1 and Accuracy are stable between 0.68 and 0.70, indicating that discretized technical indicators can predict the short-term trend of stock price effectively. And the fidelity of GRU-DT to the GRU model reaches 0.99. The prediction rules of different industries have some commonness and individuality.

Keywords: Explainable artificial intelligence · Neural network interpretability · Rule extraction · Stock forecasting · L1-orthogonal regularization

1 Introduction

Forecasting stock prices is a hot topic in the fields of finance and computers. As the stock market is affected by many factors, traditional statistical prediction methods have certain limitations [1]. At the same time, the development of informatization makes deep learning an important tool for solving multi-factor, uncertain and non-linear time series prediction problems such as stock price prediction [2]. However, the specific process of deep learning is like a "black box", and the architecture of the end-to-end model and the working principle of optimizing model parameters are not familiar to most people. Therefore, it is particularly important to improve the interpretability of neural network models and make them more easily imitated and understood by human

This work is supported by National Defense Science and Technology Innovation Special Zone Project (No. 18-163-11-ZT-002-045-04).

P. Qin et al. (Eds.): ICPCSEE 2020, CCIS 1258, pp. 309–328, 2020.
https://doi.org/10.1007/978-981-15-7984-4_23

beings. Our key research questions are: Can we represent the decision-making process of neural networks through models that are easy to interpret? What are salient rules for forecasting stock prices?

In terms of stock price prediction, although Fama [3] put forward the "efficient market hypothesis", believing that stock prices are often unpredictable in the case of market efficiency, but, in reality, the market is not always so effective [4, 5]. Taking the Chinese stock market as an example, Chen and Sun [6] concluded in the empirical verification of CAPM model that neither A shares nor B shares in China are efficient markets and CAPM cannot pass the validity test. Neely and Christopher (2014) proved that technical indicators still play an important role in stock price forecasting, which has the same forecasting ability as macroeconomic variables and can better detect the typical decline (rise) in the equity risk premium near business-cycle peaks (troughs) [7].

Machine learning algorithms were introduced to stock prediction in the early 1980s and grew rapidly over the next few decades. White (1988) was the first to use neural network to predict IBM stock. Although the prediction effect was not very ideal, the results showed a new direction for subsequent researchers [8]. Kimoto et al. (1990) developed a variety of learning algorithms and prediction methods for the stock trading time prediction system of Tokyo stock exchange based on modular neural network. The prediction system achieves accurate prediction and the simulation of stock trading showed good profitability [9]. Yoon and Swales (1991) compared the prediction ability of neural networks with that of various discriminant analysis methods, and proposed that the neural network technology could predict the performance of stock prices more accurately than the traditional multivariate analysis technology of quantitative and qualitative variables, thus demonstrating the performance of neural networks [10].

In fact, there are many different model forms of neural networks, the most suitable form of stock prediction neural network is time recursive neural network (RNN). RNN effectively improves the problem of gradient explosion existing in CNN and greatly improves the accuracy of predicting stock price of neural network. As a kind of RNN, LSTM is suitable for stock market time series prediction, because it is good at exploring the non-linear relationship between time series data. Nelson et al. (2017) used the LSTM model to predict the future stock price trend and obtained a high accuracy rate of 55.9% based on the technical analysis index [11]. Gao et al. (2018) proposes a Convolutional Recurrent Neural Network (CRNN)-based architecture ConvLSTM, in which LSTM is used in RNN. And the results showed that LSTM improves the long-term dependence of traditional RNN and effectively improves the accuracy and stability of prediction [12]. In addition, Althelaya et al. (2018) noticed that a stacked LSTM architecture has demonstrated the highest forecasting performance for both short and long term [13].

However, as the neural network is a "black box" model, we cannot explore its internal decision-making principle. In order to explore the internal workflow of the neural network, the first method is to use visualization, such as heat map [14] and curve [15]. The other method is to extract human-understandable rules from the neural network through rule extraction technology to provide explanatory power. Zilke et al. (2016) introduced a new decompositional algorithm called "DeepRED" based on decision tree that is able to extract rules from deep neural networks. And the evaluation shows its ability to outperform a pedagogical baseline on several tasks [16]. Lakkaraju and Himabindu (2017) proposed Black Box Explanations through Transparent Approximations (BETA),which can learn (with optimality guarantees) a small number of compact decision sets, explaining the behavior of the black box model in unambiguous, well-defined regions of feature space [17]. Puri et al. (2018) presented an approach that learned if-then rules to globally explain the behavior of machine learning models that have been used to solve classification problems [18]. Wu et al. (2018) proposed to add tree regularization items to the GRU neural network and train the GRU model by using the average path length of the decision tree as a penalty term, so that the decision tree describing the prediction process has a small number of nodes for quick human tracking and prediction [19]. Schaaf et al. (2019) improved the tree regularization method proposed by Wu, and proposed to use the decision boundary of L1-orthogonal regularization simulation decision tree on the neural network model, so that the decision tree can better simulate the decision-making process of a neural network. This L1-orthogonal regularization method achieved high fidelity on the MLP model, and at the same time overcome the shortcomings that require a lot of training time and parameter adjustment in Wu's tree regularization [20].

For interpretable researches in the field of stock price forecast, Feuerriegel et al. (2019) proposed is a semantic path model that yields full interpretability when using financial news to predict macroeconomic indicators [21]. Rajab and Sharma (2019) proposed an efficient and interpretable neuro-fuzzy system for stock price prediction using multiple technical indicators with focus on interpretability–accuracy trade-off [22]. Wu (2019) applied Tree Regularization method to convert GRU to a decision tree to interpret its prediction rules. The discovered prediction rules actually reflect a general rule called mean reversion in stock market [23].

Contributions. In this work, we apply the newly proposed L1-orthogonal regularization method to optimize the GRU model, and demonstrate its effectiveness through the prediction of Chinese stock market. By using the L1-orthogonal regularization method, we not only construct the decision tree with low complexity to show the stock price prediction process of neural network, but also put forward some rule screening algorithms to select significant rules and present them in the form of decision table. In the empirical process, we choose 10 different industries' data for comparative analysis, and use the way of technical indicators discretization to make rules easy for people's thinking and decision-making process. Furthermore, we show the performance of L1-orthogonal regularization with some indicators such as precision, recall, F1, AUC, accuracy, APL (average path length of a decision tree), and fidelity, verifying that L1-orthogonal regularization can provide a good explanation ability for the GRU model without reducing the precision of the model in this scenario.

2   Background and Notation

2.1   GRU Neural Network

GRU is a variant of RNN and an important simplified form of LSTM. GRU simplifies the three-gating calculation of LSTM into two gating calculation, which makes the original complex gating structure more concise without losing the original calculation effect (Fig. 1).

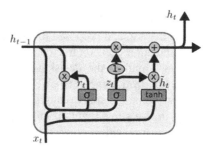

Fig. 1. Diagrammatic sketch of GRU

We can train GRU models via the following loss minimization objective:

$$\sum_{n=1}^{N} \sum_{t=1}^{T_n} loss(y_{nt}, \widehat{y_{nt}}(x_n, W)) + \lambda \cdot \Omega(W), \tag{1}$$

Where $\Omega(W)$ defines a regularization cost.

In the prediction using GRU neural network, we try to minimize the loss by adjusting the parameters.

2.2   Decision Tree

Decision tree is a common classification model, which can show its prediction process in a simple and intuitive way. In this paper, a decision tree is used for enhancing interpretability from a trained GRU (Fig. 2).

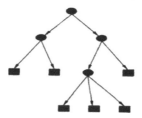

Fig. 2. Diagrammatic sketch of Decision tree

In the process of classification, a decision tree usually uses the best features to classify and predict the instance. It starts from the root node and selects the optimal feature splitting data each time.

In general, the deeper is the decision tree, the more difficult is it for people to understand. This paper focuses on the improvement of interpretability, so it adopts some pruning method to optimize it.

2.3 Model Interpretability and L1-Orthogonal Regularization

Interpretability is generally regarded as human imitability. We hope to know the specific prediction process of neural network to judge whether the prediction ideas are in line with the reality. For GRU, we don't know how it infers the prediction conclusion from the sample data set. For the decision tree, we can see clearly how it classifies and forecasts the data through the feature selection of each intermediate node. Therefore, this paper utilizes decision tree to simulate the decision-making patterns of GRU model and show the stock price prediction process.

In the method of extracting decision tree, Wu proposed a tree regularization method, which extracts decision tree by adding a penalty term of average path length (APL) in neural network [19]. However, APL is not differentiable, and it must be trained a substituted MLP model to estimate APL, which results in a lot of training time and complex parameter adjustment. In order to avoid the limitation of tree regularization, we use the latest L1-orthogonal regularization method proposed by Schaaf et al., which is easy to implement and differentiable, and force the decision boundary of GRU to be approximated by a decision tree [20].

Because the weight vector of the neural network represents the normal vector of the linear decision-making boundary, we use L1 regularization to make it forcibly sparse representation. It can make the linear decision-making boundary parallel to the axis. This representation is likely to harmonize well with decision trees [20]. The formula of L1 regularization term is as follows:

$$\Omega_1(W) = \sum_l \|W_l\|_1, \tag{2}$$

where W_l is the weight vector of the weight matrix of the neural network, and l is the layer number of total weight vectors.

In order to avoid too many decision boundaries being parallel to each other, that is to say, improving the interpretability should not reduce the prediction ability of the model, we add the orthogonalization term to make the weight vector close to the orthogonality. The formula of orthogonal regularization term is as follows:

$$\Omega_{0rth}(W) = \sum_l \|G_l - I_1\|, \tag{3}$$

where $G_l = W_l^T W_l$ and l is the layer number of total weight vectors.

Finally, we combine L1 regularization term (2) with orthogonal regularization term (3) to form L1-orthogonal regularization term, also known as sparse orthogonal regularization term:

$$\Omega(W) = \lambda_1 \cdot \Omega_1(W) + \lambda_1 \cdot \Omega_{Orth}(W), \tag{4}$$

where λ_1 and λ_{orth} are the regularization intensity of L1 regularization and orthogonal regularization, respectively. We can trade off the regularizer independently through these two parameters, which is very useful for empirical research (Fig. 3).

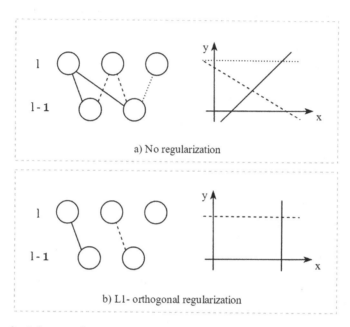

a) No regularization

b) L1- orthogonal regularization

Fig. 3. Influence of L1-orthogonal regularization on GRU weight vectors [20]

3 Research Data and Variables

3.1 Research Data

In order to explore the differences of stock price prediction rules in different industries, we take the constituent stocks of CSI 300 index as the stock pool, and divide the stock pool into 10 industries according to the industries classification standard of China Securities Index. The ten industries are materials, telecommunications, finance, industry, public, optional consumption, consumption, energy, medicine and information. We select the diurnal data of each industry from January 1, 2008 to December 31,

2019 as the basic data set. All the required data are downloaded from the official website of China Securities Index[1] and CSMAR database[2].

Here is a brief introduction to the 10 industry datasets.

1. Material Industry: 79,914 samples from 24 constituent stocks in the industry. The CSI 300 companies in this industry concentrate on copper, gold, steel, aluminum, etc.
2. Telecom Industry: 16,493 samples from 8 constituent stocks in the industry. The CSI 300 companies in this industry concentrate on mobile communications, data communications, etc.
3. Financial Industry: 134,427 samples from 77 constituent stocks in the industry. The CSI 300 companies in this industry are mainly banks, trusts, securities, etc.
4. Industrial sector: 129,736 samples from 53 constituent stocks of the industry. The CSI 300 companies in this industry concentrate on transportation, machinery, electric appliance, etc.
5. Public Utility Industry: 19,567 samples from 9 constituent stocks of the industry. The CSI 300 companies in this industry concentrate on electricity, hydropower, etc.
6. Optional Consumer Industry: 66,682 samples from 30 constituent stocks of the industry. The CSI 300 companies in this industry are mainly engaged in education, media, clothing and so on.
7. Consumer Industry: 36,341 samples from 16 constituent stocks of the industry. The CSI 300 companies in this industry are mainly engaged in retail, beverage, dairy, etc.
8. Energy Industry: 24,072 samples from 10 constituent stocks of the industry. The CSI 300 companies in this industry concentrate on oil, coal and so on.
9. Pharmaceutical Industry: 70,713 samples from 29 constituent stocks of the industry. The CSI 300 companies in this industry concentrate on medical treatment, medicine, health care products, etc.
10. Information Industry: 67,589 samples from 34 constituent stocks of the industry. The CSI 300 companies in this industry concentrate on internet, astronautical technology, entertainment, etc.

3.2   Variable Selection

In terms of variable selection, this paper focuses on the impact of technical indicators on the short-term up and down trend of stock price. Technical analysis is often used in stock market trading, which can reflect the past behavior of the market. However, the historical price data itself does not have a clear economic meaning, and the prediction patterns lack theoretical basis, so it is often not convinced. At the same time, continuous technical indicators also make the model kind of difficult to explain. Patel et al. (2015) used continuous technical indicators and their discretization forms to predict the stock price trend, which proved that the discretization technical indicators have better

---

[1] http://www.csindex.com.cn/.

[2] http://www.gtarsc.com/.

performance in the prediction of stock price trend [24]. In fact, continuous technical indicators can't reflect the trend information contained in the indicators.

Referring to the current popular technical indicators and common parameter settings, we use 13 commonly used technical indicators as independent variables. In order to explore the internal decision-making principle of the technical indicators and prove the effectiveness of its prediction trend, we discretize them (1: rising signal, 0: falling signal) according to the discretization method in [24]. All technical indicators are calculated by using the Talib library in Python.

Table 1. Selected technical indicators and its discretization method

Technical indicators	Time parameter setting	Ups and downs signals	Judgement condition
SMA5	$T = 5$	Up	$Adj\,Price_t > SMA5_t$
		Down	$Adj\,Price_t \leq SMA5_t$
SMA10	$T = 10$	Up	$Adj\,Price_t > SMA10_t$
		Down	$Adj\,Price_t \leq SMA10_t$
WMA5	$T = 5$	Up	$Adj\,Price_t > WMA5_t$
		Down	$Adj\,Price_t \leq WMA5_t$
WMA10	$T = 10$	Up	$Adj\,Price_t > WMA10_t$
		Down	$Adj\,Price_t \leq WMA10_t$
MOM	$T = 10$	Up	$MOM_t > 0$
		Down	$MOM_t \leq 0$
STCK	$T_{fast} = 9,$ $T_{slow} = 3$	Up	$STCK < 10; 10 \leq STCK \leq 90$ and $STCK_t > STCK_{t-1}$
		Down	$STCK > 90; 10 \leq STCK \leq 90$ and $STCK_t \leq STCK_{t-1}$
STCD	$T_{fast} = 9,$ $T_{slow} = 3$	Up	$STCD < 20; 20 \leq STCD \leq 80$ and $STCD_t > STCD_{t-1}$
		Down	$STCD > 80; 20 \leq STCD \leq 80$ and $STCD_t \leq STCD_{t-1}$
RSI6	$T = 6$	Up	$RSI6 < 20; 20 \leq RSI6 \leq 80$ and $RSI6_t > RSI6_{t-1}$
		Down	$RSI6 > 80; 20 \leq RSI6 \leq 80$ and $RSI6_t \leq RSI6_{t-1}$
RSI12	$T = 12$	Up	$RSI12 < 20; 20 \leq RSI12 \leq 80$ and $RSI12_t > RSI12_{t-1}$
		Down	$RSI12 > 80; 20 \leq RSI12 \leq 80$ and $RSI12_t \leq RSI12_{t-1}$
MACD	$T_{fast} = 12,$ $T_{slow} = 26$	Up	$MACD_t > MACD_{t-1}$
		Down	$MACD_t \leq MACD_{t-1}$
WILLR6	$T = 6$	Up	$WILLR6 < 20; 20 \leq WILLR6 \leq 80$ and $WILLR6_t > WILLR6_{t-1}$
		Down	$WILLR6 > 80; 20 \leq WILLR6 \leq 80$ and $WILLR6_t \leq WILLR6_{t-1}$

(continued)

Table 1. (*continued*)

Technical indicators	Time parameter setting	Ups and downs signals	Judgement condition
WILLR10	$T = 10$	Up	$WILLR10 < 20; 20 \leq WILLR10 \leq 80$ and $WILLR10_t > WILLR10_{t-1}$
		Down	$WILLR10 > 80; 20 \leq WILLR10 \leq 80$ and $WILLR10_t \leq WILLR10_{t-1}$
A/D	–	Up	$A/D_t > A/D_{t-1}$
		Down	$A/D_t \leq A/D_{t-1}$
CCI	$T = 14$	Up	$CCI < -100; -100 \leq CCI \leq 100$ and $CCI_t > CCI_{t-1}$
		Down	$CCI > 100; -100 \leq CCI \leq 100$ and $CCI_t \leq CCI_{t-1}$

As shown in Table 1, we convert continuous technical indicators into discrete up or down signals. Note that the signals here are not equal to the buying and selling signals definitely.

For the prediction target, in general, given the stock closing price P_t at period t and the stock closing price P_{t+n} at $t + n$ period, we can measure the trend of stock rising and falling at period $t + n$ by the return rate $R_{t,n} = (P_{t+n} - P_t)/P_t$.

However, the fluctuation of stock price is very random, which will bring a lot of noise. If only the stock price of period $t + n$ is used to judge the trend, label is likely to vary drastically in the short term and produce more invalid trades, which is especially significant when the stock price fluctuation in the period is obvious. Therefore, in order to reduce the high noise interference to the trend and restore the real trend of the stock price, we use the stock price smoothing method to define the classification label y [25]. The calculation steps are as follows:

1. Calculate the m day moving average MA_t of the complex closing price is to smooth the stock price:

$$MA_t = \sum_{i=1}^{m} \frac{AdjClose_{t-i+1}}{m}, \tag{5}$$

2. Calculate the rate of change of the moving average MA_{t+n} relative to the moving average MA_t in n period.

$$R_t^* = \left(\frac{MA_{t+n}}{MA_t} - 1\right) * 100\%, \tag{6}$$

3. Use R_t^* to divide the short-term rise and fall trendy:

$$y = \begin{cases} 1, R_t^* \geq 0 \\ 0, R_t^* < 0 \end{cases},$$ (7)

In this paper, the parameter $m = 5$ is used as the smoothing period and $n = 5$ as the trend calculation period. Considering the lag and foresight of technical indicators, in this way, the short-term noise can be filtered well and the effective trend can be maintained.

4 L1-Orthogonal Regularized GRU-DT Interpretation Model

Figure 4 shows the model framework. First, the price data is processed into discretized technical indicators as the features and short-term trends as prediction objectives (Sect. 3). Second, the optimal GRU model is trained based on time series splits and cross validation. The decision tree (called GRU-DT) is trained to fit the GRU model for interpretation. Third, rules are extracted along the paths of the GRU-DT.

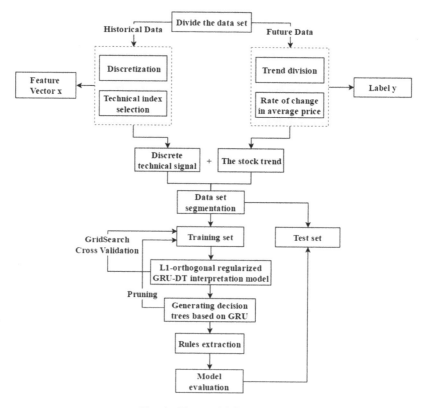

Fig. 4. The model framework

4.1   The Construction of GRU Model

In this paper, the GRU model is selected as the neural network. The loss function of the model is added with the L1-orthogonal regularization term, which makes the decision boundary easy to be approximated by the decision tree and enhances the interpretability of the model. During the experiment, we train the network strictly according to the steps of training, validation and testing.

The data sets are divided into training sets according to the first 80% of the time series and test sets according to the last 20%. The discretized technical indicators of stocks in various industries are taken into the input layer of the neural network, and the short-term trend of stock price y is taken into the output layer of the neural network, so as to predict the trend of rise and fall in the next short-term period.

In order to reduce the problem of overfitting and increasing the predictive performance, we use early stopping to realize the balance between training time and generalization errors, which makes the model stop when the validation loss is no longer significantly reduced.

We adjust the hyper-parameters of the model by means of Bayesian optimization and cross validation when training to get the optimal model. The hyper-parameters and adjustment ranges involved in the model are shown in Table 2. (Due to the use of early stop, we do not need to adjust the epoch parameter here and set its value to 50 by default).

Table 2. Model hyper-parameters and their adjustment ranges

Hyper-parameters	Adjustment ranges
Batch size	16, 32, 64, 128
Unit	128, 256, 512
λ_1	0–1
λ_{orth}	0–1
Learning rate	0.001, 0.003, 0.005, 0.1

Considering the order of the time series data, we further adopt the method of k-fold time series split cross-validation for tuning, whose principle is shown in Fig. 5. This method can better simulate the real environment for predicting the future [26], and at the same time provide an almost unbiased error estimate [27]. And we use the Hyperopt library in Python to implement Bayesian Optimization to adjust hyperparameters.

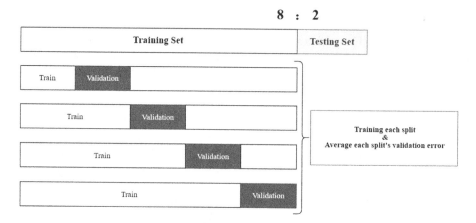

Fig. 5. An illustration of time series split cross validation

In the dichotomy of stock price trend prediction, we divide the results into four categories according to prediction and reality (Table 2). We regard a price uptrend as a positive example and a price downtrend as a negative example (parity price trend is very rare, which is ignored here temporarily) (Table 3).

Table 3. The four categories of prediction results.

TP (True Positive)	Prediction: up; Reality: up
FP (False Positive)	Prediction: up; Reality: down
TN (True Negative)	Prediction: down; Reality: down
FN (False Negative)	Prediction: down; Reality: up

Thus, we define the precision (P) and recall (R) of uptrend and downtrend:

$$P_{up} = \frac{TP}{TP + FP} \tag{8}$$

$$R_{up} = \frac{TP}{TP + FN} \tag{9}$$

$$P_{down} = \frac{TN}{TN + FN} \tag{10}$$

$$R_{down} = \frac{TN}{TN + FP} \tag{11}$$

Generally speaking, investors use "long equity" and "short equity" strategy to buy and sell stocks, that is, investors focus on one of the rising and falling situations. Then, in order to evaluate the investment opportunity, the precision and the recall should be calculated.

However, the precision and recall are contradictory. Therefore, we use F1 scores to represent the precision and recall, which is their harmonic mean and sensitive to the lowest value. Here, we evaluate F1 for both uptrend and downtrend that correspond to the long and short strategies, respectively, and calculate their mean for illustration.

$$F1_{up} = \frac{2}{\frac{1}{P_{up}} + \frac{1}{R_{up}}} \tag{12}$$

$$F1_{down} = \frac{2}{\frac{1}{P_{down}} + \frac{1}{R_{down}}} \tag{13}$$

$$F1 = \frac{F1_{up} + F1_{down}}{2} \tag{14}$$

In addition, we calculate the classification accuracy to measure the prediction accuracy of the model. If investors are concerned about both up and down, then classification accuracy is a good indicator.

$$Accuracy = \frac{TP + TN}{TP + FP + TN + FN} \tag{15}$$

Finally, we also use AUC to evaluate the model performance, which represents the area under ROC curve and can effectively face the problem of sample imbalance. AUC ranges from 0.5 to 1. The closer AUC is to 1.0, the higher the classifier performance is; the closer AUC is to 0.5, the worse the classifier performance is.

4.2 GRU-DT Construction and Rule Extraction

After the above tuning, the GRU model with the excellent performance has been obtained. In order to improve the interpretability of models and further obtain significant rules for stock price prediction of different industries, we take all features of the original training set as input, and the values predicted by the GRU model as output, so as to fit the decision tree (called GRU-DT). The maximum depth of the tree (max_depth) is one of the most important parameters to prune a decision tree. So, in this case, we find the optimal value of this parameter and the corresponding optimal decision tree for each industry data set by the combination of time series split cross validation and grid search. And the optimal GRU-DT is used to extract rules, so as to further improve the interpretability of the model and find an effective combination of technical indicators that is beneficial to the prediction of stock price trend. Fidelity and APL are used to measure GRU-DT.

Fidelity measures the extent to which a decision tree approximates a GRU network. In this paper, fidelity is defined as the probability that the prediction made by a GRU-DT is the same as that made by the original GRU model. Formula is as follows:

$$Fidelity = Prob\{f_{GRU-DT} = f_{GRU}|T\}, \tag{16}$$

where T is the tested data set (usually the test set), f_{GRU} is the prediction functions of GRU and f_{GRU-DT} is the prediction functions of the GRU-DT. The higher the fidelity, the better the decision tree reflects the prediction of the GRU model.

Model complexity is an important evaluation metric in interpretability research. Generally speaking, the lower the model complexity, the higher the interpretability of the model, that is, human beings understands the working principle of the model more easily for simpler models. In this paper, average path length (APL) is used to measure the complexity of a decision tree, which represents the average length from root node to leaf nodes for the data sample [19].

In the rule extraction part, we cut the original data set used for training by means of the k-fold time series cross-validation again (k = 4 in the experiments). The original data set for training is divided into the training set and the validation set. We use the optimal decision tree to extract rules and related metrics. In each fold of validation, the training set is used to extract rules, and the validation set is used to calculate evaluation metrics such as the precision of rules.

To evaluate a rule, we define $Support_{rule}$ to be the occurrence percentage of the rule in the data sample used, $Precision_{rule}$ to be the forecasting precision of the up or down trends (depending on the use of the rule for up or down forecasting), and $Utility_{rule}$ to be the investment utility of the rule, which is defined as follows:

$$Utility_{rule} = 2 \times (Precision_{rule} - 0.5) \times Support_{rule} \tag{17}$$

$Utility_{rule}$ compares $Precision_{rule}$ with 0.5 (the random forecasting precision) and is then multiplied by $Support_{rule}$, which reflects the utility of the rule used for investment. Generally, $Utility_{rule}$ should be larger than zero for a useful rule.

5   Model Evaluation and Result Analysis

5.1   Performance Evaluation of GRU-DT Model

Through Bayesian Optimization and k-fold time series split cross-validation, we find the optimal hyper-parameter of the L1-orthogonal regularized GRU model on different industry dataset which is shown in Table 4.

Table 4. Optimal hyper-parameters on different industry datasets

Industry data sets	Unit	L1 strength	Orth-strength	Learning rate	Batch size	Mean validation accuracy
Mat	64	0.0015	0.3348	0.003	128	69.69%
Tel	16	0.0340	0.4398	0.001	128	68.93%
Ind	32	0.0409	0.0082	0.003	128	69.78%
Fin	32	0.0206	0.0020	0.003	256	69.37%
Ut	16	0.0050	0.7777	0.003	128	69.88%
Opt	64	0.0110	0.4266	0.005	256	69.54%

(continued)

Table 4. (*continued*)

Industry data sets	Unit	L1 strength	Orth-strength	Learning rate	Batch size	Mean validation accuracy
Con	32	0.0036	0.8974	0.001	128	68.79%
En	128	0.0421	0.0243	0.003	256	70.08%
Med	64	0.0495	0.0176	0.001	128	69.40%
Info	64	0.0164	0.0601	0.001	128	69.46%

(Ma = Material, T = Telecommunication, In = Industry, F = Finance, U = Utility, O = Option, C = Consume, E = Energy, Me = Medicine, IT = Information)

Similarly, we find the optimal value of max_depth of the decision tree fitted with data from different industries which is shown in Table 5.

Table 5. The optimal parameter combinations

	Mat	Tel	Ind	Fin	Ut	Opt	Con	En	Med	Info
max_depth	7	6	6	7	7	6	8	6	5	2

(Ma = Material, T = Telecommunication, In = Industry, F = Finance, U = Utility, O = Option, C = Consume, E = Energy, Me = Medicine, IT = Information)

In the optimal model, we use AUC, accuracy, F1, APL and fidelity to compare the performance of the GRU model, GRU-DT and the decision tree model (DT) directly trained on the dataset for different industry datasets.

Table 6. The up and down predictive performance evaluation on test sets of different indus-tries

Data set	TP	TN	FP	FN	Total samples	P_{up}	P_{down}	R_{up}	R_{down}	$F1_{up}$	$F1_{down}$
Mat	5385	5631	2803	2180	15999	0.67	0.71	0.69	0.70	0.68	0.71
Tel	1163	1062	592	485	3302	0.66	0.69	0.71	0.64	0.68	0.66
Ind	8683	9348	3994	3994	25939	0.68	0.70	0.69	0.70	0.69	0.70
Fin	10277	10976	5018	4645	30916	0.67	0.70	0.69	0.69	0.68	0.69
Ut	1344	1389	630	630	3993	0.68	0.71	0.71	0.69	0.69	0.70
Opt	4540	4348	2345	2115	13308	0.66	0.67	0.68	0.65	0.67	0.66
Con	2800	2234	1186	1054	7274	0.70	0.68	0.73	0.65	0.71	0.67
En	1616	1729	739	734	4818	0.69	0.70	0.69	0.70	0.69	0.70
Med	5079	4753	2271	2051	14154	0.69	0.70	0.71	0.68	0.70	0.69
Info	4968	4353	2237	1975	13533	0.69	0.69	0.72	0.66	0.70	0.67

(Ma = Material, Tel = Telecommunication, Ind = Industry, Fin = Finance, Ut = Utility, Opt = Option, Con = Consume, En = Energy, Med = Medicine, Info = Information)

Table 7. Comparison of model performance evaluation of different industry test sets

Data set	GRU			GRU-DT				DT			
	AUC	Accuracy	F1	AUC	F1	APL	Fidelity	AUC	F1	APL	Fidelity
Mat	0.73	0.70	0.69	0.69	0.69	4.77	0.9986	0.69	0.69	7.91	0.9403
Tel	0.72	0.68	0.67	0.67	0.67	3.38	0.9961	0.68	0.68	5.94	0.8794
Ind	0.74	0.70	0.69	0.69	0.69	2.12	0.9999	0.69	0.69	5.99	0.9735
Fin	0.73	0.70	0.69	0.69	0.69	2.69	0.9999	0.69	0.69	6.00	0.9823
Ut	0.74	0.67	0.70	0.70	0.70	4.18	0.9982	0.70	0.70	6.99	0.9512
Opt	0.72	0.68	0.67	0.67	0.67	1.68	1.0000	0.70	0.70	5.99	0.8323
Con	0.74	0.67	0.69	0.69	0.69	3.88	0.9987	0.69	0.69	7.83	0.9175
En	0.74	0.70	0.69	0.69	0.69	2.11	0.9996	0.69	0.69	5.99	0.9732
Med	0.74	0.69	0.69	0.69	0.69	1.53	0.9999	0.69	0.69	5.00	0.9901
Info	0.73	0.69	0.69	0.69	0.69	1.47	1.0000	0.69	0.69	2.00	0.9735

(Ma = Material, Tel = Telecommunication, Ind = Industry, Fin = Finance, Ut = Utility, Opt = Option, Con = Consume, En = Energy, Med = Medicine, Info = Information)

As shown in Table 6, the performance of the model is similar in predicting the up and down labels, and the accuracy of predicting the down label is slightly higher than that of the up label, which indicates that our model is helpful in both "long equity" and "short equity" strategies. As shown in Table 7, for the GRU interpreting model, all industries have achieved high AUC, accuracy and F1 values (AUC value range is stable between 0.72–0.74, accuracy and F1 are stable between 0.67–0.70) considering randomness in the stock market, which proves that the method of using discrete technical indicators to predict the trend of stock price rise and fall is effective. the prediction performance difference between different industries is not significant, which reflects the universality of technical indicators, that is, all industries contain the same potential law hidden in historical data that will repeat.

By comparing GRU-DT and DT, we find that their prediction performance is similar, but GRU-DT often has a smaller APL (thus lower model complexity) and higher fidelity than DT, and therefore it has better practical usage.

5.2   Rule Extraction Results and Analysis

Applying the modeling steps and rule extraction methods mentioned above, we extract rules from the prediction processes of ten industries. Meanwhile, the support, precision and utility values of each rule are calculated. According to the above three values, the rules of each industry are sorted and filtered. And the rules are tested with the test set data. Based on these metrics, the effective rules are used to judge the trends of stock price in each industry are retained. We only show the rules that the support value is greater than 0.05 in the tables below (Table 8 and 9).

Table 8. Significant up rules in different industry dataset.

Data Set	Rule (Up)	Training set			Testing set		
		Support	Utility	Precision	Support	Utility	Precision
Mat	Rule1: SMA5 = up, WMA10 = up, A/D = up, WMA5 = up	0.414	0.160	0.694	0.399	0.159	0.699
Tel	Rule1: WMA10 = up, MACD = up, SMA10 = up	0.439	0.172	0.696	0.433	0.169	0.695
Ind	Rule1: SMA5 = up, WMA5 = up	0.471	0.189	0.701	0.457	0.180	0.697
Fin	Rule1: SMA5 = up, WMA5 = up	0.462	0.175	0.689	0.457	0.173	0.690
Ut	Rule1: SMA5 = up, WMA5 = up, WMA10 = up	0.431	0.174	0.701	0.412	0.177	0.715
Opt	Rule1: SMA5 = up	0.522	0.177	0.670	0.497	0.161	0.662
Con	Rule1: WMA10 = up, SMA5 = up, WMA5 = up	0.437	0.173	0.698	0.443	0.214	0.741
En	Rule1: SMA5 = up, WMA5 = up	0.466	0.182	0.696	0.464	0.182	0.696
Med	Rule1: WMA5 = up	0.517	0.202	0.696	0.501	0.198	0.698
Info	Rule1: SMA5 = up	0.513	0.189	0.684	0.506	0.203	0.702

(Ma = Material,   T = Telecommunication,   In = Industry,   F = Finance,   U = Utility,   O = Option,
C = Consume, E = Energy, Me = Medicine, IT = Information)

Table 9. Significant down rules in different industry dataset.

Data set	Rule (Down)	Training set			Testing set		
		Support	Utility	Precision	Support	Utility	Precision
Mat	Rule1: SMA5 = down, WMA10 = down, WMA5 = down, MACD = down, SMA10 = down	0.354	0.139	0.697	0.367	0.190	0.759
Tel	Rule 1: WMA10 = down, MACD = down, WMA5 = down	0.375	0.125	0.667	0.384	0.175	0.727
Ind	Rule 1: SMA5 = down, WMA5 = down	0.469	0.184	0.696	0.485	0.207	0.714
Fin	Rule1: SMA5 = down, WMA5 = down,WMA10 = down	0.431	0.160	0.686	0.441	0.193	0.719
Ut	Rule1: SMA5 = down, WMA5 = down, WMA10 = down, MACD = down, SMA10 = down	0.348	0.144	0.706	0.382	0.194	0.754
Opt	Rule 1: SMA10 = down, WMA5 = down	0.380	0.121	0.659	0.399	0.181	0.727
	Rule2: SMA10 = down, WMA5 = up, WILLR10 = down, CCI = down	0.070	0.024	0.6739	0.084	-0.014	0.415
Con	Rule1:WMA10 = down, WMA5 = down, MACD = down, WILLR6 = down	0.362	0.147	0.702	0.374	0.157	0.711
En	Rule1: SMA5 = down, WMA5 = down	0.476	0.191	0.701	0.474	0.210	0.721
Med	Rule 1: WMA5 = down, SMA5 = down	0.452	0.181	0.701	0.469	0.189	0.702

(Ma = Material,   T = Telecommunication,   In = Industry,   F = Finance,   U = Utility,   O = Option,
C = Consume, E = Energy, Me = Medicine, IT = Information)

To sum up, based on the above two tables, we can see that the rules extracted from GRU-DT have high precision in the testing set. Besides, SMA5 or WMA5 are the most important rules for predicting 'up' or 'down', followed by WMA10 in the downtrend rules, which are summarized in Table 10. Based on the above analysis, we further refine the general technical index combination, which can assist in the prediction of future stock price trend (regardless of index order):

Table 10. Common index combination applies to different industries

Trend	Common index combination	Applicable industries
Up	1: SMA5 = Up, WMA5 = Up	Ind, Fin, En
	2: SMA5 = Up	Opt, Info
	3: SMA5 = Up, WMA5 = Up, WMA10 = Up	Ut, Con
	4: WMA5 = Up	Med
	5: SMA5 = Up, WMA10 = Up, A/D = Up, WMA5 = Up	Mat
	6: WMA10 = Up, MACD = Up, SMA10 = Up	Tel
Down	1: WMA5 = Down, SMA5 = Down	Ind, En, Opt, Med, Info
	2: SMA5 = Down, WMA5 = Down, WMA10 = Down, MACD = Down, SMA10 = Down	Mat, Ut
	3: WMA10 = Down, MACD = Down, WMA5 = Down	Tel
	4: SMA5 = Down, WMA5 = Down, WMA10 = Down	Fin
	5: WMA10 = Down, WMA5 = Down, MACD = Down, WILLR6 = Down	Con

(Ma = Material, T = Telecommunication, In = Industry, F = Finance, U = Utility, O = Option, C = Consume, E = Energy, Me = Medicine, IT = Information)

6 Conclusion

In this paper, we use the latest L1-orthogonal regularization method to simulate the decision-making process of GRU neural network through decision tree, and put forward some rule screening algorithms to find significant rules for stock price trend prediction. In the experiment, we use 13 commonly used discretization technical indicators to predict the data sets of 10 industries in the CSI 300, and obtain the significant up and down rules of different industries. On this basis, we analyze the rules of each industry and obtain the combination rules. The results show that:

First, technical analysis is an important method of stock investment. The discrete technical indicators can effectively predict the short-term trend of stock price.

Second, the extraction results cater to the market characteristics of various industries, and the combination of SMA5 and WMA5 is the most frequent, which can be said to be the general short-term prediction rules.

Third, L1-orthogonal regularization can make a decision tree simulate a GRU neural network well. GRU-DT model has good predictive performance and strong interpretation ability, which is easy to be understood by human beings.

In the future work, we will compare the performance of various models, further verifying our model.

References

1. Li, C.: Prediction of stock index futures price based on BP neural network. Master Thesis of Qingdao University, Qingdao (2012)
2. Yu, Z., Qin, L., Zhao Z., Wen, W.: Stock price prediction based on principal component analysis and generalized regression neural network. Stat. Decis. Making **34**(18), 168–171 (2008)
3. Malkiel, B.G., Fama, E.F.: Efficient capital markets: a review of theory and empirical work. J. Financ. **25**, 383–417 (1970)
4. Bodie, Z., Kane, A., Marcus, A.J.: Investments, 10th edn. McGraw-Hill Education, New York (2014)
5. Liu, Z., Wang, Y.: An empirical study on the forecasting effectiveness of price-based technical indicators in bull and bear cycles of China's Shanghai stock market. In: Proceeding of 12th International Conference on Management of e-Commerce and e-Government (ICMECG 2018), pp. 412–417 (2018)
6. Chen, X., Sun A.: Effectiveness test of CAPM in Chinese stock market. J. Peking Univ. (philosophy and social sciences), 28–37 (2000)
7. Wang, J.-H., Leu, J.-Y.: Stock market trend prediction using arima-based neural networks. In: IEEE International Conference on Neural Networks, vol. 4, pp. 2160–2165. IEEE (1996)
8. White H.F.: Economic prediction using neural networks: the case of IBM daily stock returns. Earth Surf. Process. Land. **2**, 451–458 (1988)
9. Kimoto, T., Asakawa, K., Yoda, M., Takeoka, M.: Stock market prediction system with modular neural networks. In: 1990 IJCNN International Joint Conference on Neural Networks, vol. 1, pp. 1–6 (1990)
10. Yoon, Y., Swales, G.: Predicting stock price performance: a neural network approach. In: Proceedings of the Twenty-Fourth Annual Hawaii International Conference on System Sciences, Kauai, HI, USA, vol. 4, pp. 156–162 (1991)
11. Nelson, D.M.Q., Pereira, A.C.M., de Oliveira, R.A.: Stock market's price movement prediction with LSTM neural networks. In: 2017 International Joint Conference on Neural Networks (IJCNN), Anchorage, AK, pp. 1419–1426 (2017)
12. Gao, S.E., Lin, B.S., Wang, C.: Share price trend prediction using CRNN with LSTM structure. In: 2018 International Symposium on Computer, Consumer and Control (IS3C), Taichung, Taiwan, pp. 10–13 (2018)
13. Althelaya, K.A., El-Alfy, E.M., Mohammed, S.: Stock market forecast using multivariate analysis with bidirectional and stacked (LSTM, GRU). In: 2018 21st Saudi Computer Society National Computer Conference (NCC), Riyadh, pp. 1–7 (2018)
14. Bach, S., Binder, A., Montavon, G., Klauschen, F., Muller, K.-R., Samek, W.: On pixel-wise explanations for non-linear classifier decisions by layer-wise relevance propagation, PloS ONE, **10**(7), e0130140 (2015)
15. Friedman, J.H.: Greedy function approximation: a gradient boosting machine. Ann. Statist. **29**(5), 1189–1232 (2001)
16. Zilke, J.R., Loza Mencía, E., Janssen, F.: DeepRED – rule extraction from deep neural networks. In: Calders, T., Ceci, M., Malerba, D. (eds.) DS 2016. LNCS (LNAI), vol. 9956, pp. 457–473. Springer, Cham (2016). https://doi.org/10.1007/978-3-319-46307-0_29

17. Lakkaraju, H., et al.: Interpretable & explorable approximations of black box models. arXiv preprint, arXiv:1707.01154 (2017)
18. Puri, N., et al.: Magix: model agnostic globally interpretable explanations. arXiv preprint, arXiv:1706.07160 (2017)
19. Wu, M., Hughes, M.C., Parbhoo, S., et al.: Beyond sparsity: tree regularization of deep models for interpretability. In: Proceeding of the Thirty-Second AAAI Conference on Artificial Intelligence (AAAI 2018), pp. 1670–1678 (2018)
20. Schaaf, N., Huber, M.F.: Enhancing decision tree based interpretation of deep neural networks through L1-orthogonal regularization. arXiv preprint, arXiv:1904.05394 (2019)
21. Feuerriegel, S., Gordon, J.: News-based forecasts of macroeconomic indicators: a semantic path model for interpretable predictions. Eur. J. Oper. Res. 272(1), 162–175 (2019)
22. Rajab, S., Sharma, V.: An interpretable neuro-fuzzy approach to stock price forecasting. Soft Comput. J. 23(3), 921–936 (2019). https://doi.org/10.1007/s00500-017-2800-7
23. Wu, W., et al.: Preliminary study on interpreting stock price forecasting based on tree regularization of GRU. In: Mao, R., Wang, H., Xie, X., Lu, Z. (eds.) ICPCSEE 2019. CCIS, vol. 1059, pp. 476–487. Springer, Singapore (2019). https://doi.org/10.1007/978-981-15-0121-0_37
24. Patel, J., Shah, S., Thakkar, P., et al.: Predicting stock and stock price index movement using trend deterministic data preparation and machine learning techniques. Expert Syst. Appl. J. 42(1), 259–268 (2015)
25. Hong, J.H.: Research on stock price trend prediction based on GBDT model. Jinan university (2017)
26. Granger, C.W.J., Pesaran, M.H.: Economic and statistical measures of forecast accuracy. J. Forecast. 19(7), 537–560 (1999). Cambridge Working Papers in Economics
27. Varma, S., Simon, R.: Bias in error estimation when using cross-validation for model selection. BMC Bioinform. 7(1), 91–100 (2006). https://doi.org/10.1186/1471-2105-7-91

Short-Term Predictions and LIME-Based Rule Extraction for Standard and Poor's Index

Chunqi Qi, Yue Wang[✉], Wenjun Wu, and Xiuli Wang

School of Information, Central University of Finance and Economics,
Beijing, China
yuelwang@163.com

Abstract. In this paper, neural networks is proposed to predict the trend of the S.P. 500 index, comparing the effects of different data inputs and model types on the prediction results. Then the model was selected with good interpretation results to use the LIME interpretability algorithm to interpret the prediction results, and extract the prediction rules of the neural network model. It firstly compared three neural networks prediction models, including multi-layer perceptron (MLP), one-dimensional convolutional neural network (1D CNN) and long short-term memory (LSTM), with the price data, continuous technical indicators and discrete technical indicators of different time durations as inputs, and found the model and parameter configuration with good predicting effect for each type of inputs. In another aspect, the data of the first experiment was used, the rules were extracted in the model that predict the rise of the index by the LIME interpretability algorithm, and the rules with high investment utility were finally selected. The experimental result show high precision in predicting the trend of the stock index and high frequency of occurrence, with certain reference value for predicting the short-term index trend.

Keywords: Neural network · Stock prediction · LIME · Interpretability

1 Introduction

With the promotion of machine learning application, the research of neural network algorithms used in stock market prediction are more and more abundant. Neural network is one of the effective tools for the time series prediction. This paper tackles two problems that are not sufficiently addressed in the literature. The first one is the feature and model selection, and the second one is the interpretability.

The feature and model selection are vital to the stock prediction performance. The stock market price trend is often affected by many factors and its price sequence data is dynamic and nonlinear, so the difficulty of stock market trend prediction is that the data contains too much noise [1]. How to reduce the noise in the input sequence data and avoid neural network overfitting the noise is very important to improve the accuracy of the model for stock market forecasting. At present, the feature selection and processing

This work is supported by National Defense Science and Technology Innovation Special Zone Project (No. 18-163-11-ZT-002-045-04).

P. Qin et al. (Eds.): ICPCSEE 2020, CCIS 1258, pp. 329–343, 2020.
https://doi.org/10.1007/978-981-15-7984-4_24

of input data, neural network parameter selection and network structure optimization are the directions of many researches. In this paper, the pre-processing of input data is mainly converting the price data of the S&P index into Sample Moving Average (SMA), Moving Average Convergence Divergence (MACD), Commodity Channel Index (CCI) and other common technical indicators. Further, we discretized these technical indicators into buy or sell signals. We compare the prediction effects of the original price data with continuous and discrete technical indicators. We also test a variety of network structures, including Multi-Layer Perceptron, One-dimensional Convolutional Neural Network and Long Short-Term Memory. Then the good-performing models are selected.

The purpose of the interpretability analysis is to discover the reasons behind the neural network predictions. And we can determine whether the model has really learned the reasons we expected by interpreting the sample. This can also help users determine the rationality and credibility of the model and select the appropriate model based on it or improve the existing model. In this paper, we select the LIME (Local Interpretable Model-Agnostic Explanations) algorithm as the interpretability algorithm [2], which is a model-agnostic explanatory algorithm that we can apply to the interpretability analysis of any classification model. And the neural network model with good prediction performance of the S.P. index is selected, and the internal judgment rules are extracted by LIME. Then we check the rules against the actual situation and judge the rationality of the neural network model, and the actual reference value of the rules is further determined by testing.

2   Related Work

At present, the application of neural network models in stock market prediction mainly improves the prediction effect in two aspects. One approach is the processing and selection of input data, such as using the wavelet transform to reduce the noise in the data sequence [3, 4], optimizing the input data series by principal component analysis (PCA) to reduce data redundancy [5, 6]. Another approach is the selection and optimization of neural network model, the common neural network models include MLP, CNN and RNN [7–10].

In the data processing we converted the price data into technical indicators as input data by using the common methods of stock market technical analysis. The technical indicators mine the information in the price data, which is promisingly more reasonable and effective to reduce the noise in the price data. Complementing previous research, we compare the prediction performance based on the price data, continuous technical indicators and discrete technical indicators of different time durations as inputs.

In terms of interpretability, the neural network interpretability algorithms develop rapidly, especially in the field of image recognition and natural language processing. The commonly used interpretability models are sensitivity analysis, layer-wise relevance propagation (LRP) [11], model interpretation based on Shapley value [12], Local Interpretable Model-Agnostic Explanations (LIME) and so on. Through the interpretability model, we can judge the rationality of the original model, and unearth the judge rules from the model, so that the user can apply these rules. In the interpretation of stock market forecasting, there is very limited work using neural fuzzy system to study the interpretation of stock price forecast results [13], and using the decision tree to conduct interpretability analysis of stock market forecast results [14]. In this paper, we use the LIME algorithm on the stock market forecast and extract rules that predicted stock market trend from the neural network, so as to help people make a better judgment on the future trend of the stock market.

3 Comparing Input Features and Neural Network Models

3.1 Input Features and Preprocessing

We select daily price data for 5,030 days from January 1, 2000 to January 1, 2020, including the daily opening, closing, high, and low price indicators, and take the data of the first 3000 trading days as the training set, the data of the following 1000 trading days as validation set, and the remaining data as the test set. The whole closing price series is generally on the uptrend (Fig. 1.a), interspersed with cyclical fluctuations which is influenced by many volatile factors, such as the 2000 Southeast Asian financial crisis and the collapse of the dotcom bubble, the global financial crisis in 2008, the European debt crisis in 2011, and the election of Brexit in 2016. After a sustained rise, there are often more obvious local shocks. From the index daily return (Fig. 1.b) in the nearly 20 years, it's easy to see that the stock market volatility cycle is about 4 years, and at the end of the period, the trend will have a relatively large change, it has certain periodicity and regularity in the time series distribution.

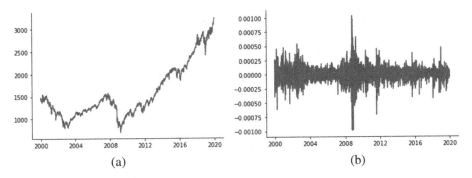

Fig. 1. The trend of S&P 500 from 2000 to 2020 a. (left) Daily closing price trend; b. (right) Trend of daily return rate

We chose three types of input features. And the durations of 1, 5 and 10 days of input features are considered in the experiments.

(1) The first is directly using the series of price data and daily return (calculated by the price data) for the model to make predictions.
(2) The price information is processed into continuous technical indicators, and then as input.
(3) The continuous technical indicators are discretized, and then as input.

For the standardization of the price data later used in the machine learning process, we standardize the price data of each duration separately. For example, for each 5-day price data, we standardize it by the standard z-score separately. Because of the long-term bull market of the S&P index, it is not reasonable to standardize all the price data at once, which leads to a low-price index in the early sequence and a high price index in the later sequence. However, our treatment can avoid the impact of this problem on the short-term forecasting.

The price sequence contains information about potential supply and demand. We process the daily price data and turn them into 9 commonly used indicators for stock market technical analysis: Simple Moving Average (SMA), Weighted Moving Average (WMA), Moving Average Convergence Divergence (MACD), Momentum indicator (MOM), Stochastic Oscillator K% (SLOWK), Stochastic Oscillator D% (SLOWD), Relative Strength Index (RSI), Larry William's R% (WILLR), and Commodity Channel Index(CCI). The calculation method is shown in Table 1 [15]. Through the technical indicators, we can show information of the stock market obviously, such as short-term fluctuations, overbought and oversold information, strong and weak market information, etc. These technical indicators may contain information that is difficult for the model to mine through the raw price data.

Table 1. Selected technical indicators and their formulas (C_t is the closing price, L_t the low price, H_t the high price at time t, $\text{DIFF} = \text{EMA}(12)_t - \text{EMA}(26)_t$, EMA exponential moving average, $EMA(k)_t$: $EMA(k)_{t-1} + \alpha \times \left(C_t - EMA(k)_{t-1}\right)$, α smoothing factor: $2/1 + k$, k is time period of k day exponential moving average, HH_{t-n} and LL_{t-n} mean lowest low and highest high in the last t days, respectively, M_t: $H_t + L_t + C_t/3$; SM_t: $\left(\sum_{i=0}^{n} M_{t-i+1}\right)/n$, D_t: $\left(\sum_{i=0}^{n} |M_{t-i+1} - SM_t|\right)/n$, Up_t means the upward price change, Dw_t means the downward price change at time t.)

Indicators	Time period	Formulas
SMA	10 days	$\frac{C_t + C_{t-1} + \cdots + C_{t-10}}{10}$
WMA	10 days	$\frac{((n) \times C_t + (n-1) \times C_{t-1} + \cdots + C_{t-10})}{(n + (n-1) + \cdots + 1)}$
MOM	9 days	$C_t - C_{t-n}$
SLOWK	5 days	$\frac{C_t - LL_{q-n}}{HH_{t-n} - LL_{t-n}} \times 100$
SLOWD	3 days	$\frac{\sum_{i=0}^{n-1} K_{t-i}\%}{n}$
RSI	14 days	$100 - \frac{100}{1 + \left(\sum_{i=0}^{n-1} Up_{t-i}/n\right)/\left(\sum_{i=0}^{n-1} Dw_{t-i}/n\right)}$
MACD	Fast period: 12 days Slow period: 26 days Signal period: 9 days	$MACD(n)_{t-1} + 2/n+1 \times \left(DIFF_t \times MACD(n)_{t-1}\right)$
WILLR	14 days	$\frac{H_n - C_t}{H_n - L_n}$
CCI	14 days	$\frac{M_t - SM_t}{0.015 D_t}$

It is worth noting that some MA indicators such as SMA and WMA are very related to stock prices and are often used by comparing with stock prices while RSI, CCI and other indicators can be directly used. As we show before, price goes up from 2000 to 2020, so we need to process the price-related technical indicators. The $BIAS_{SMA}$ and $BIAS_{WMA}$ are used as input, which are defined as:

$$BIAS_{SMA} = (C_t - SMA_t)/SMA_t \qquad (1)$$

$$BIAS_{WMA} = (C_t - WMA_t)/WMA_t \qquad (2)$$

Similarly, for MACD and MOM indicators, we define their diurnal relative rate of change as $\Delta MADC\%$ and $\Delta MOM\%$:

$$\Delta MACD_t\% = (MACD_t - MACD_{t-1})/MACD_{t-1} \qquad (3)$$

$$\Delta MOM_t\% = (MOM_t - MOM_{t-1})/MOM_{t-1} \qquad (4)$$

After that, we standardize each continuous technical indicator in the whole time period (2000–2020) for the machine learning process.

We then convert the continuous values of the technical indicators into discrete values 0 or 1 to represent the down (0) or up (1) signals. For example, RSI is often used to identify overbought and oversold. In general, if a stock has an RSI value of more than 70 in one day, it means that the stock is overbought, and its price is likely to fall in the near future (at this time the RSI signal is 0); If the value of RSI is less than 30, this means that the stock is oversold and its price may rise in the near future (at this point the RSI signal is 1). For the RSI values between 30 and 70, compare the value of the current-day indicator with the previous day's value, and if the RSI of the day is greater than the RSI of the previous day, the signal is 1, and 0 on the contrary. And the discretization method is shown in Table 2 referring to the method of Jigar Patel et al. [16]. In order to distinguish from continuous indicators, we add D () to denote discrete processing.

Table 2. Technical indicators discrete processing rules

Discrete technical indicators	Value	Processing rules
D(SMA)	1	$C_t \geq SMA_t$
	0	$C_t < SMA_t$
D(WMA)	1	$C_t \geq WMA_t$
	0	$C_t < WMA_t$
D(MOM)	1	$MOM_t \geq 0$
	0	$MOM_t < 0$
D(SLOWK)	1	$SLOWK_t \geq SLOWK_{t-1}$
	0	$SLOWK_t < SLOWK_{t-1}$
D(SLOWD)	1	$SLOWD_t \geq SLOWD_{t-1}$
	0	$SLOWD_t < SLOWD_{t-1}$

(continued)

Table 2. (*continued*)

Discrete technical indicators	Value	Processing rules
D(WILLR)	1	$WILLR_t \geq WILLR_{t-1}$
	0	$WILLR_t < WILLR_{t-1}$
D(MACD)	1	$MACD_t \geq MADC_{t-1}$
	0	$MACD_t < MADC_{t-1}$
D(RSI)	1	$RSI_t < 30$
		$30 \leq RSI_t \leq 70 \ and \ RSI_t > RSI_{t-1}$
	0	$RSI_t > 70$
		$30 \leq RSI_t \leq 70 \ and \ RSI_t < RSI_{t-1}$
D(CCI)	1	$CCI_t < -200$
		$-200 \leq CCI_t \leq 200 \ and \ CCI_t > CCI_{t-1}$
	0	$CCI_t > 200$
		$-200 \leq CCI_t \leq 200 \ and \ CCI_t < CCI_{t-1}$

Finally, in order to reduce the impact of randomness on stock price forecast, we do not use the daily up and down labels, but use a smoothed trend label. That is, if the SMA value of the next 3 days is greater than the SMA value of the previous 3 days, the trend is upward (1), and downward (0) conversely. The definition of the classification labels is as follows:

$$Label = \begin{cases} 1, SMA_{t-3} < SMA_{t+3} \\ 0, SMA_{t-3} \geq SMA_{t+3} \end{cases} \tag{5}$$

For the evaluation of the model, we train the model on the training set, validate it the on the validation set, and finally test it on the test set. We select the model accuracy on the test set, prediction precision, recall rate and F1 score of the upward trend as the evaluation criteria for the model prediction, defined as follows (Table 3).

Table 3. The definition of a confusing matrix

		Reality	
		UP (1)	DOWN (0)
Prediction	UP (1)	*TP (True Positive)*	*FP (False Positive)*
	DOWN (0)	*FN (False Negative)*	*TN (True Negative)*

$$Precision = \frac{TP}{TP + FP} \tag{6}$$

$$Recall = \frac{TP}{TP + FN} \tag{7}$$

$$F1 = \frac{2 \times Precision \times Recall}{Precision + Recall} \tag{8}$$

$$Accuracy = \frac{TP + TN}{TP + FP + TN + FN} \tag{9}$$

3.2   Neural Network Models

In this paper, three neural network models are compared: MLP, 1D CNN and LSTM. And we try the hyperparameter of each model within a reasonable range.

For the MLP model, we compare three-layer and four-layer feedforward neural networks. When the input data's duration is 5 or 10 days, the dimension is reduced using the flatten layer. The output layer is a double neuron, and SoftMax is selected as the activation function. The hidden layers use Tanh or ReLu as the activation function. The optimizer is the gradient descent with momentum or RMSprop. We have carried out a comprehensive parameter setting experiment to determine the parameters of various inputs. The hyperparameters of the MLP model are the number of neurons (n), the learning rate (lr), and the momentum constant (mo).

The principle of 1D CNN is the same as that of 2D CNN, and the main difference is the dimension of tensor data processed. The local features of data information are extracted by the convolution layer, so that they can handle the long feature sequence well. We use a convolution layer as the input layer in the experiment, and then it can be connected to next convolution layer or global pooling layer, and so on. The prediction results are output using the dense layer. In the experiments, the activation function of a convolution layer uses ReLu, and the final output layer uses the Sigmoid function. The optimizer is RMSprop. We experiment the predictive effect of one and two convolution layers, and also try different sizes of convolution kernels (ck) for feature extraction. The hyperparameters such as the size of neurons and the learning rate are also compared.

For the RNN model, we select LSTM as the input layer, then build a recurrent neural network from one to two layers, and the final output layer is a dense layer using the sigmoid activation function. The optimizer used is RMSprop. The hyperparameters are mainly the number of LSTM layers, the neuron size of each LSTM layer and the learning rate.

In order to improve the generalization capability of the model, we use the early stopping strategy to select the best training epochs. We also appropriately added L1 regularization (L1), L2 regularization (L2) and a certain dropout rate (dr) in the model to reduce overfitting and enhance generalization. The range of these hyperparameters are shown in Table 4.

Table 4. Hyperparameters settings and their ranges in neural network model

Hyperparameters (abbreviations)	Values
Neuron size of each layer (n)	$2, 8, 16, \ldots, 128$
Momentum constant (mo)	$0.1, 0.2, 0.3, \ldots, 0.7$
Learning rate (lr)	0.001, 0.01, 0.1
L1 regularization (L2)	0.0001, 0.001, 0.1
L2 regularization (L1)	0.0001, 0.001, 0.1
Dropout rate (dr)	$0.1, 0.2, \ldots, 0.5$
Size of convolution kernel (ck)	2, 3, 4, 5, 6

3.3   Evaluation

We take the price and daily return sequence, continuous technical indicator sequence and discrete technical indicator sequence as input sequence to each model, respectively, and try different hyperparameter combinations in each case. In the experiment, the training of the model adopts early stopping, and the number of training epochs with high validation accuracy or low loss function value is chosen as the final parameter in the course of training.

For a model of the same set of hyperparameters, the model's final F1 score and test set accuracy are slightly different due to different random initialization conditions. Therefore, we train a model multiple times, and the evaluation metrics are averaged to reduce the impact of incidental factors. The optimal hyperparameters and model metrics are shown in Table 5 to Table 7 below. And the best metrics are highlighted in bold.

Table 5. The predicting result of each model when the price and daily return sequence series are input features

Price (opening, closing, high and low price) and daily return sequence								
Model	Time	Optimal hyperparameters			Precision	Recall	F1	Accuracy
		Layers; n	L1; L2; dr	lr; mo				
MLP	1 day	3; 64-32-2	$10^{-4}; 10^{-4}; 0.1$	0.1; 0.3	0.6589	0.5114	0.578	0.5442
		4; 64-32-16-2	$10^{-4}; 10^{-4}; 0.1$	0.1; 0.5	0.6595	0.5051	0.5690	0.5430
	5 days	3; 64-32-2	$10^{-4}; 10^{-2}; 0.2$	0.1; 0.5	**0.6663**	0.5791	0.6163	0.5697
		4; 64-32-32-2	$10^{-4}; 10^{-3}; 0.2$	0.1; 0.4	0.6605	0.5890	0.6203	0.5678
	10 days	3; 64-32-2	$10^{-3}; 10^{-3}; 0.1$	0.1; 0.2	0.6592	**0.6132**	**0.6342**	**0.5724**
		4; 64-32-16-2	$10^{-4}; 10^{-4}; 0.2$	0.1; 0.5	0.6602	0.5710	0.6002	0.5602
		CNN layers; ck; n	L1; L2; dr	lr				
1D CNN	5 days	1; 3; 32	$10^{-4}; 10^{-3}; 0.2$	0.1	0.6535	0.5438	0.5871	0.5480
		2; 3; 64-32	$10^{-4}; 10^{-3}; 0.2$	0.1	**0.6709**	**0.5937**	**0.6281**	**0.5784**
	10 days	1; 5; 64	$10^{-4}; 10^{-4}; 0.2$	0.1	0.6607	0.4787	0.5498	0.5220
		2; 5; 64-32	$10^{-4}; 10^{-4}; 0.2$	0.1	0.6597	0.5831	0.6141	0.5642
		LSTM layers; n	L1; L2; dr	lr				
LSTM	5 days	1; 32	$10^{-3}; 10^{-3}; 0.2$	0.01	0.6443	0.4901	**0.6431**	**0.5854**
		2; 32-16	$10^{-4}; 10^{-3}; 0.1$	0.01	**0.6710**	**0.6185**	0.5481	0.5407
	10 days	1; 32	$10^{-3}; 10^{-3}; 0.2$	0.01	0.6545	0.5538	0.6285	0.5730
		2; 32-16	$10^{-4}; 10^{-3}; 0.1$	0.01	0.6658	0.5957	0.5984	0.5516

Table 6. The predicting result of each model when continuous technical indicators are input features

Continuous technical indicators sequence

Model	Time	Optimal hyperparameters			Precision	Recall	F1	Accuracy
		Layers; n	L1; L2; dr	lr; mo				
MLP	1 day	3; 32-16-2	10^{-4}; 10^{-4}; 0.1	0.1; 0.3	0.6326	0.5910	0.6104	0.5342
		4; 32-16-8-2	10^{-3}; 10^{-3}; 0.1	0.1; 0.5	0.6242	0.5724	0.5959	0.5159
	5 days	3; 64-32-2	10^{-4}; 10^{-2}; 0.2	0.1; 0.5	**0.6476**	0.6322	0.6357	**0.5631**
		4; 32-16-16-2	10^{-4}; 10^{-4}; 0.2	0.1; 0.5	0.6347	0.5366	0.5799	0.5258
	10 days	3; 64-32-2	10^{-4}; 10^{-3}; 0.1	0.1; 0.2	0.6382	0.6406	0.6386	0.5576
		4; 64-32-16-2	10^{-4}; 10^{-3}; 0.1	0.1; 0.2	0.6390	**0.6439**	**0.6391**	0.5591
		CNN layers; ck; n	L1; L2; dr	lr				
1D CNN	5 days	1; 2; 64	10^{-4}; 10^{-4}; 0.1	0.01	0.6329	0.6268	0.6087	0.5304
		2; 3; 64-32	10^{-4}; 10^{-3}; 0.2	0.01	**0.6407**	0.6638	0.6512	**0.5627**
	10 days	1; 5; 32	10^{-4}; 10^{-4}; 0.5	0.01	0.6114	**0.7032**	**0.6526**	0.5433
		2; 3; 64-32	10^{-4}; 10^{-3}; 0.3	0.01	0.6276	0.6025	0.6136	0.5344
		LSTM layers; n	L1; L2; dr	lr				
LSTM	5 days	1; 32	10^{-4}; 10^{-3}; 0.1	0.01	**0.6520**	0.6335	0.6418	**0.5649**
		2; 32-16	10^{-4}; 10^{-3}; 0.1	0.01	0.6129	0.6364	0.6212	0.5273
	10 days	1; 64	10^{-4}; 10^{-3}; 0.1	0.01	0.6367	**0.6668**	**0.6491**	0.5607
		2; 32-16	10^{-4}; 10^{-4}; 0.1	0.01	0.6341	0.6282	0.6301	0.5478

Table 7. The predicting results of each model when discrete technical indicators are input features

Discrete technical indicators sequence

Model	Time	Optimal hyperparameters			Precision	Recall	F1	Accuracy
		Layers; n	L1; L2; dr	lr; mo				
MLP	1 day	3; 32-16-2	10^{-4}; 10^{-4}; 0.1	0.1; 0.3	**0.6612**	0.5141	0.5779	0.5362
		4; 32-16-8-2	10^{-3}; 10^{-3}; 0.1	0.1; 0.3	0.6568	0.5378	0.5906	0.5398
	5 days	3; 64-32-2	10^{-4}; 10^{-2}; 0.2	0.1; 0.5	0.6418	**0.6472**	**0.6393**	0.5576
		4; 64-32-16-2	10^{-4}; 10^{-3}; 0.2	0.1; 0.5	0.6321	0.5571	0.5920	0.5269
	10 days	3; 32-32-2	10^{-4}; 10^{-4}; 0.1	0.1; 0.3	0.6505	0.5906	0.6172	0.5517
		4; 32-32-16-2	10^{-4}; 10^{-3}; 0.1	0.1; 0.2	0.6558	0.6137	0.6299	**0.5622**
		CNN layers; ck; n	L1; L2; dr	lr				
1D CNN	5 days	1; 2; 32	10^{-4}; 10^{-3}; 0.1	0.1	0.6211	0.6583	0.6321	0.5422
		2; 3; 64-32	10^{-4}; 10^{-3}; 0.2	0.01	**0.6496**	0.6306	0.6240	0.5611
	10 days	1; 5; 32	10^{-4}; 10^{-4}; 0.5	0.01	0.6325	0.6128	0.6068	0.5400
		2; 3; 64-32	10^{-4}; 10^{-4}; 0.1	0.1	0.6193	**0.7638**	**0.6934**	**0.5665**
		LSTM layers; n	L1; L2; dr	lr				
LSTM	5 days	1; 16	10^{-3}; 10^{-3}; 0.1	0.1	0.6317	0.6201	**0.6224**	0.5831
		2; 32-16	10^{-4}; 10^{-4}; 0.1	0.01	0.6026	**0.6221**	0.6030	0.5503
	10 days	1; 32	10^{-4}; 10^{-3}; 0.1	0.01	**0.6487**	0.5905	0.6095	**0.5914**
		2; 32-16	10^{-4}; 10^{-4}; 0.1	0.01	0.6485	0.5684	0.6039	0.5877

When price data and daily return sequence data are as input features (Table 5), it is easy to find that the F1 score and accuracy of the MLP model increase to a certain extent when the duration of the input features increases. However, for 1D CNN and LSTM models, the growth of duration length is not significant for enhancing the predictive effect of model, and may even lead to a decrease in accuracy. For the MLP model, the effect of the 3-layer neural network and the 4-layer neural network model is similar, and the good effect can also be obtained by using 3-layer network. The effect of using 2-layer convolution layer for the 1D CNN model is better than using only 1-layer convolution. The LSTM model using one recurrent layer works better than two.

For continuous technical indicators as input features (Table 6), the F1 score of various models has a very small increase compared to that of the price and daily return sequence data as input features, and the accuracy of the model prediction even decreases in some cases. The F1 score of the model is better with the increase of the time duration.

For the discrete technical indicators (Table 7), the improvement of the MLP model is not obvious. For 1D CNN and LSTM, some of their parameter combinations perform well. When the duration is 10 days, the F1 score of the 1D CNN model with two convolution layers can reach 0.6934, and the prediction accuracy of the one-layer LSTM model can reach 0.5914.

Combining the above experimental results, increasing the duration of the input features is helpful to improve the F1 score and accuracy. In addition, it can be found that proper pre-processing of the input features can also be helpful for the improvement in the predictive effect in some cases, but the improvement is limited on the average.

4 LIME-Based Rule Extraction

4.1 Introduction to LIME Algorithm

The core idea of the LIME interpretability algorithm is to disturb the characteristics of the input features, observe the change of the model prediction result caused by the disturbance, and find out the local prediction rules in the neural network model through multiple perturbation observations. Specifically, in the neural network to predict stock market rise and fall, the algorithm requires a well-trained neural network classifier model before some samples to be explained.

When using LIME algorithm to interpret the prediction results of a sample, first, a characteristic value of the sample is randomly disturbed, that is, a random feature value of the sample to be changed or erased. Then the perturbation sample is re-entered into the model, we observe the change of the model prediction results, select the most obvious result change, and measure the weight of this characteristic through the similarity of disturbance sample and the original sample, so as to find out the most important characteristics in the prediction. And the LIME algorithm may use a simple interpretable model (such as decision trees) to locally fit the original model prediction results on these features, and this model is used to analyze the local interpretability of the complex model. Therefore, the LIME algorithm is local interpretable and model agnostic: the interpretation result obtained by the LIME algorithm reflects the prediction of a model for a certain sample, it is the local rules of the model obtained when the model predicts the sample; The LIME algorithm only needs to process the input features, so there is no requirement for the prediction model. In that case, the algorithm can do interpretability analysis for any model, because it doesn't depend on it [17].

4.2 Evaluation

In the experiment of LIME-based rule extraction, in order to facilitate the demonstration of the explanatory effect of the interpretable algorithm, we select the 1-day data of continuous and discrete technical indicators as input, and choose the model with higher F1 for experiments.

Using the LIME interpretability algorithm, we make rule extraction on two MLP models, and do sample interpretation of the original data set (4000 data in training and validation set) one by one. Because there are different interpretation rules and different rule weights for different samples, we count the weight and occurrence of each rule in all the interpretation results, and then get the average weight of each rule (the sum of weights divided by the frequency of the rule). The importance of each rule is mainly reflected by the average weight of the rules, as shown in Table 8.

Table 8. Rule extraction from 4000 samples in the training and validation set

Rules for predicting uptrends							
Continuous technical indicator rules				Discrete technical indicator rules			
Rule[a]	Average weight	Sum of the weights	Frequency	Rule	Average weight	Sum of the weights	Frequency
WILLR > −11.56	0.1442	144.15	1000	D(CCI) = 1	0.0538	107.13	1992
SLOWD ≤ 35.40	0.1291	129.06	1000	D(MOM) = 1	0.0239	54.48	2278
−35.17 < WILLR ≤ −11.56	0.0761	76.06	999	D(WILLR) = 1	0.0159	31.51	1987
BIAS$_{SMA}$ > 0.0083	0.0557	55.65	1000	D(RSI) = 1	0.0146	29.69	2030
RSI > 61.04	0.0382	38.16	1000	D(SMA) = 1	0.0110	24.98	2277
CCI ≤ −75.45	0.0362	36.15	1000	D(MACD) = 1	0.0063	12.60	2010
35.40 < SLOWD ≤ 59.13	0.0178	17.76	1000	D(WMA) = 1	0.0011	2.41	2256
ΔMACD% > 0.0521	0.0164	16.35	1000	D(SLOWK) = 1	0.0007	1.38	1994

[a]We used the standardization of z-scores for continuous technical indicators in training and testing, and here they are restored.

After that, we do a back testing of the rules by applying the rules to the data and measure whether the rules are effective or not. Then we define the occurrence ratio of a rule as support rate ($Support_{rule}$), and $Precision_{rule}$ which is the precision of this rule to predict the uptrend. Based on the above two variables, the investment utility ($Utility_{rule}$) of the rule is defined as follows:

$$Utility_{rule} = 2 \times (Precision_{rule} - 0.5) \times Support_{rule} \qquad (10)$$

In the calculation of $Utility_{rule}$, $Precision_{rule}$ is compared with 0.5 (the precision of random prediction), and then multiply by $Support_{rule}$, so $Utility_{rule}$ can comprehensively reflect the practicability of the rule in the investment. The range of $Utility_{rule}$ is $[-1, 1]$. In practice, if the $Utility_{rule}$ of a rule is greater than 0, the rule is of practical use. In the experiment, we select the important rules whose prediction results is up from the two models, test them on the training and validation set, and further test the effectiveness of the rules on the test set. The experimental results are shown in Table 9 and Table 10.

Table 9. The results of rules when input features are continuous technical indicators in the 1-day MLP model

Rules for predicting uptrends	Training set & validation set			Test set		
	$Support_{rule}$	$Precision_{rule}$	$Utility_{rule}$	$Support_{rule}$	$Precision_{rule}$	$Utility_{rule}$
WILLR > −11.56	0.2528	0.7240	0.1132	0.3300	0.7848	0.1879
RSI > 61.04	0.2437	0.6625	0.0792	0.4190	0.6754	0.1470
$BIAS_{WMA}$ > 0.0083	0.2537	0.7448	0.1242	0.2030	0.8374	0.1370
ΔMACD% > 0.0521	0.3485	0.5753	0.0524	0.3190	0.7084	0.1330
−35.17 < WILLR ≤ −11.56	0.2457	0.5687	0.0337	0.2760	0.5797	0.0440
SLOWD ≤ 35.40	0.2505	0.5209	0.0104	0.1620	0.6296	0.0420

Table 10. The results of rules when input features are discrete technical indicators in the 1-day MLP model

Rules for predicting uptrends	Training set & validation set			Test set		
	$Support_{rule}$	$Precision_{rule}$	$Utility_{rule}$	$Support_{rule}$	$Precision_{rule}$	$Utility_{rule}$
D(MACD) = 1	0.5097	0.6753	0.1787	0.5226	0.7630	0.2749
D(SMA) = 1	0.5702	0.6492	0.1702	0.6656	0.6974	0.2628
D(WILLR) = 1	0.4990	0.6553	0.1550	0.4894	0.7427	0.2376
D(RSI) = 1	0.5120	0.6645	0.1685	0.4904	0.7412	0.2366
D(MOM) = 1	0.5702	0.6106	0.1262	0.6676	0.6636	0.2185
D(CCI) = 1	0.5017	0.6661	0.1667	0.4843	0.7234	0.2165

In addition, we combine the rules with higher investment utility into a new rule and experiment with them. However, due to the increased filtering criteria of multiple combinations of rules, the support of the combined rules may be reduced, and the precision may be increased. We find that the decrease of support rate is more obvious, and in general the investment utilities of rule combinations are less than single rules, so we only make a combination of 2 or 3 rules. Then the experiment is also carried on the training and validation set, and tested on the test set. The rule combinations are shown in Table 11 and Table 12:

Table 11. The results of the rule combinations when input features are continuous technical indicators in the 1-day MLP model

Rule combinations for predicting uptrends	Training set & validation set			Test set		
	$Support_{rule}$	$Precision_{rule}$	$Utility_{rule}$	$Support_{rule}$	$Precision_{rule}$	$Utility_{rule}$
WILLR > −11.56 RSI > 61.04	0.1690	0.7263	0.0765	0.2668	0.7698	0.1440
WILLR > −11.56 $BIAS_{WMA}$ > 0.0083	0.1580	0.7674	0.0844	0.1591	0.8291	0.1047
$BIAS_{WMA}$ > 0.0083 RSI > 61.04	0.1040	0.7644	0.0549	0.1158	0.8260	0.0755
WILLR > −11.56 $BIAS_{WMA}$ > 0.0083 RSI > 61.04	0.0985	0.7969	0.0585	0.1030	0.8543	0.0729
WILLR > −11.56 $BIAS_{WMA}$ > 0.0083 ΔMACD% > 0.0521	0.0947	0.7836	0.0537	0.1127	0.8214	0.0725

Table 12. The results of the rule combinations when input features are discrete technical indicators in the 1-day MLP model

Rule combinations for predicting uptrends	Training set & validation set			Test set		
	$Support_{rule}$	$Precision_{rule}$	$Utility_{rule}$	$Support_{rule}$	$Precision_{rule}$	$Utility_{rule}$
D(MACD) = 1 D(SMA) = 1	0.4720	0.6806	0.1704	0.4994	0.7621	0.2618
D(WILLR) = 1 D(RSI) = 1	0.4440	0.6712	0.1520	0.4139	0.7542	0.2104
D(SMA) = 1 D(WILLR) = 1	0.3335	0.7113	0.1410	0.3625	0.7667	0.1933
D(MACD) = 1 D(SMA) = 1 D(RSI) = 1	0.3130	0.7348	0.1470	0.2960	0.8061	0.1812
D(MACD) = 1 D(WILLR) = 1 D(RSI) = 1	0.2947	0.7260	0.1332	0.2608	0.8147	0.1641

In the experiment of extracting rules using the LIME interpretability algorithm, continuous input features obtain more rules than discrete ones, because a discrete value is only 0 or 1 and thus the number of extracted rules is limited.

After selecting the important rules, we respectively test the investment utility of each single rule (Table 9 and Table 10). It is easy to find that the investment utility of these two kinds of important rules are greater than 0 and effective. The prediction precision of the rules extracted from the continuous data is between 0.52 and 0.83, and that of the discrete data is between 0.61 and 0.76. Due to the binary value of a discrete technical indicator, the support of a discrete rule is generally higher than that of a continuous rule, which leads to the more stable utility of the final discrete rule. We also see that the precision of a rule with high average LIME weight does not necessarily have high precision. Therefore, LIME can only preliminarily screen important rules, and a valid rule needs to be tested on the data set.

In the rule combination experiment (Table 11 and Table 12), more rules in the combination, more decrease in support, but more helpful be the improvement of precision. For example, the precision of continuous rule combinations is between 0.72 and 0.85, and that of discrete rule combinations is between 0.67 and 0.81.

5   Conclusion

In this paper, we select the daily price data of S&P 500 as the original data, and then we compare the influence of different pre-processing data methods and various models on the prediction results of stock price trend. It is found that the pre-processing and feature screening of the data, which converts the original data into technical indicators and discretizes technical indicators, are helpful to improve the prediction effect of the model, but the improvement is limited. In the meantime, choosing the appropriate duration and model can improve the prediction.

In the aspect of rule extraction, this paper uses the LIME interpretability algorithm to preliminarily extract prediction rules inside the model, and tests these rules on the training and validation set to find out effective rules with high investment utility. Then, the valid rules are further tested on the test set.

There are still some directions of improvement, and the discrete technical indicators are only treated as up or down. In the future, more detailed discretization can be carried out and more discrete values can be added to distinguish the different levels of rise and fall. Besides, this paper only extracts the rules within MLP model based on the LIME algorithm. In the following research, we will extract the rules from other kinds of models such as LSTM, and filter rules from multiple models together, so as to get more reliable rules.

References

1. Fama, E.F.: Random walks in stock market prices. Financ. Anal. J. **51**(1), 75–80 (1995)
2. Ribeiro, M.T., Singh, S., Guestrin, C.: "Why should i trust you?" Explaining the predictions of any classifier. In: Proceedings of the 22nd ACM SIGKDD International Conference on Knowledge Discovery and Data Mining, pp. 1135–1144 (2016)
3. Xiaodan, L., Zhaodi, G., Liling, S., Maowei, H., Hanning, C.: LSTM with wavelet transform based data preprocessing for stock price prediction. Math. Probl. Eng. **2019** (2019)
4. Kumar, C.S.: Fusion model of wavelet transform and adaptive neuro fuzzy inference system for stock market prediction. J. Ambient Intell. Hum. Comput. (2019). https://doi.org/10.1007/s12652-019-01224-2
5. Hong, C., Rongyao, C.: Research on stock price prediction based on PCA-BP neural network. Comput. Simul. **28**(03), 365–368 (2011)

6. Fangzhong, Q., Shaoqian, L., Tingting, Y.: Stock price prediction model based on PCA and IFOA-BP neural network. Comput. Appl. Softw. **37**(01), 116–121+156 (2020)
7. Sayavong, L.: Research on stock price prediction method based on convolutional neural network. In: Proceedings of 2019 International Conference on Virtual Reality and Intelligent Systems (ICVRIS 2019), vol. I, pp. 187–190 (2019)
8. Yu, P., Yan, X.: Stock price prediction based on deep neural networks. Neural Comput. Appl. **32**(6), 1609–1628 (2019). https://doi.org/10.1007/s00521-019-04212-x
9. Ruoyu, Q.: Stock forecasting model based on neural network. Oper. Manag. **28**(10), 132–140 (2019)
10. Nelson, D.M.Q., Pereira, A.C.M., Oliveira, R.A.D.: Stock market's price movement prediction with LSTM neural networks. In: 2017 International Joint Conference on Neural Networks (IJCNN), pp. 1419–1426. IEEE (2017)
11. Li, J., Chen, X., Hovy, E., et al.: Visualizing and understanding neural models in NLP. Comput. Sci. (2015)
12. Lundberg, S., Lee, S.I.: A unified approach to interpreting model predictions. In: 31st Conference on Neural Information Processing Systems (NIPS 2017), Long Beach, CA, USA (2017)
13. Rajab, S., Sharma, V.: An interpretable neuro-fuzzy approach to stock price forecasting. Soft. Comput. **23**(3), 921–936 (2017). https://doi.org/10.1007/s00500-017-2800-7
14. Wu, W., et al.: Preliminary study on interpreting stock price forecasting based on tree regularization of GRU. In: Mao, R., Wang, H., Xie, X., Lu, Z. (eds.) ICPCSEE 2019. CCIS, vol. 1059, pp. 476–487. Springer, Singapore (2019). https://doi.org/10.1007/978-981-15-0121-0_37
15. Yakup, K., Melek, A.B., Ömer, K.B.: Predicting direction of stock price index movement using artificial neural networks and support vector machines: The sample of the Istanbul stock exchange. Expert Syst. Appl. **38**(5), 5311–5319 (2011)
16. Patel, J., Shah, S., Thakkar, P., et al.: Predicting stock and stock price index movement using trend deterministic data preparation and machine learning techniques. Expert Syst. Appl. **42**(1), 259–268 (2015)
17. Fei, W., Binbing, L., Yahong, H.: Interpretability of deep learning. Weapons Aviat. **26**(01), 39–46 (2019)

A Hybrid Data-Driven Approach for Predicting Remaining Useful Life of Industrial Equipment

Zheng Tan[1,2], Yiping Wen[1,2(✉)], and TianCai Li[1,2]

[1] School of Computer Science and Engineering, Hunan University of Science and Technology, Xiangtan, China
ypwen81@gmail.com
[2] Key Laboratory of Knowledge Processing & Networked Manufacturing, Hunan University of Science and Technology, Xiangtan, China

Abstract. Guaranteeing the safety of equipment is extremely important in industry. To improve reliability and availability of equipment, various methods for prognostics and health management (PHM) have been proposed. Predicting remaining useful life (RUL) of industrial equipment is a key aspect of PHM and it is always one of the most challenging issues. With the rapid development of industrial equipment and sensing technology, an increasing amount of data on the health level of equipment can be obtained for RUL prediction. This paper proposes a hybrid data-driven approach based on stacked denoising autoencode (SDAE) and similarity theory for estimating remaining useful life of industrial equipment, which is named RULESS. Our work is making the most of stacked SDAE and similarity theory to improve the accuracy of RUL prediction. The effectiveness of the proposed approach was evaluated by using aircraft engine health data simulated by commercial modular Aero-Propulsion system simulation (C-MAPSS).

Keywords: Remaining useful life · Prediction · Industrial equipment · Stacked Denoising AutoEncoder · Similarity theory · RULESS

1 Introduction

Industrial equipment have complex mechanical systems, different equipment systems have their own operating mechanisms, and may malfunction or fail to complete normal tasks and functions, resulting in irreparable economic losses and waste of resources. Therefore, there is an urgent need to accurately grasp the health level of industrial equipment. Industrial equipment generates a large amount of data every day, by analyzing data, we can grasp the future change trend or operating status of these equipments, so as to achieve fault Prognostics and health management (PHM). At present, PHM as a key technology to ensure the safety and reliability of equipment has achieved a wealth of theoretical results and has been widely used in the past few decades [1, 2].

PHM technology mainly includes two parts: remaining useful life (RUL) prediction and health management. RUL predicting has the guiding value for maintenance decisions, spare parts ordering, and has long been used as the foundation and core of

© Springer Nature Singapore Pte Ltd. 2020
P. Qin et al. (Eds.): ICPCSEE 2020, CCIS 1258, pp. 344–353, 2020.
https://doi.org/10.1007/978-981-15-7984-4_25

PHM technology. In recent years, with the explosive growth of equipment monitoring data volume and the development of storage technology and computing capabilities, the data-driven approach based on machine learning have become the mainstream technology in the field of RUL prediction.

For example, a relevance vector machine based high-speed train traction system RUL prediction method was proposed in [3] to improve the accuracy of predictions in uncertain scenarios. Recently, a principal component analysis (PCA) and deep learning based feature extraction method was utilized in [4] for the early diagnosis and RUL prediction of slowly changing faults. To account for time dependency between sequential data, a recurrent neural network method was proposed in [5] for industrial equipment data. After that, in order to overcome the problem of large long-term errors in traditional recurrent neural networks, a long short-term memory (LSTM) network was utilized in [6] for the RUL prediction. Further, a generative adversarial network based model was established in [7] to cope with the issue of long-term degradation progress for industrial equipment.

On the other hand, some method based similarity theory is proposed in recent years. The construction of health indicator (HI) for equipment is the basis of similarity-based RUL methods. For example, a linear regression model based method to construct HI has been proposed in [8, 9] for the RUL prediction of aircraft engine. However, the linear regression model based feature extraction method ignored the initial wear of equipment, to overcome this shortcoming, an equipment degradation assessment model has been built in [10] by PCA.

However, literature shows that a single model cannot provide the most accurate RUL prediction result. Hybrid methods that utilize the advantages of several prediction models may be better. For example, considering that the neural network has strong non-linear function fitting capabilities, a Back Propagation (BP) neural network and similarity theory based method was utilized in [11] for the RUL prediction of aircraft engine.

With these observations, this paper proposes a RUL estimation method based on stacked denoising autoencode (SDAE) and similarity theory, which is named RULESS. RULESS first sent the equipment historical monitoring data to the SDAE model as the input data, the HI of equipment is represented by the output of the model and trained the test model. The RUL was weight predicted by similarity measurement based on HI trajectory. We use the aircraft engine health data simulated by Commercial Modular Aero-Propulsion System Simulation (C-MAPSS) to verify our proposed method. This method is compared with the four primitive methods in mean absolute error (MAE), mean square error (MSE).

The rest of the paper is organized as follows: Sect. 2 gives a brief introduction of the preliminaries of this work. Section 3 describes the proposed prediction method. Section 4 presents the experimental results of the proposed approach and others primitive method for contradistinction. Conclusions are drawn in Sect. 5.

2 Preliminaries

2.1 Problem Description

In our problem, we have multi-dimensional run-to-failure equipment data from sensor measurements. In the first, we defined that $x^{(t,i)}$ denote the measurement value of sensor i at time t, and let $x^{(t)}$ denote a vector of multidimensional sensor measurement such that $x^{(t)} = [x^{(t,1)}, x^{(t,2)}, \ldots x^{(t,m)}]$, m is the number of sensors. After that, we denote a matrix of sensor measurement with $X_{(1,T)} = \{x^{(1)}, x^{(2)} \ldots x^{(T)}\}$, T is the total time step of data. Then the remaining useful life of the equipment at the time index t can be defined as:

$$RUL = T - t \tag{1}$$

The historical data at the time index t can be defined as $X_{(1,t)} = \{x^{(1)}, x^{(2)}, \ldots x^{(t)}\}$, the essence of remaining useful life prediction is to build a model from historical data $X_{(1,t)}$, which is equivalent to finding a non-linear mapping function $f(x)$, so that as much as possible:

$$RUL \approx f\left(X_{(1,t)}\right) \tag{2}$$

The goal of this article is to learn this non-linear mapping function through prediction methods based on deep learning and similarity theory.

2.2 Stacked Denoising Autoencoder

AutoEncoder (AE) is a branch of neural network. A typical AE has an input layer representing the raw data, a hidden layer representing the feature transformation and an output layer that matches the input layer and is used for information reconstruction, and hidden layer include encode layer and decode layer.

If the number of input layer nodes m is more than the number of hidden layer nodes p, the dimension of data is reduced from m to p by AE. When the input data is subjected to noise processing, we label it as Denoising AutoEncoder (DAE). When the number of hidden layers is greater than 1, we can obtain a SDAE that can be applied in industrial equipment [12].

3 Proposed Method

Our proposed RULESS method is mainly constructed based on deep learning and similarity theory. As the Fig. 1 shows, the RULESS method mainly includes three parts.

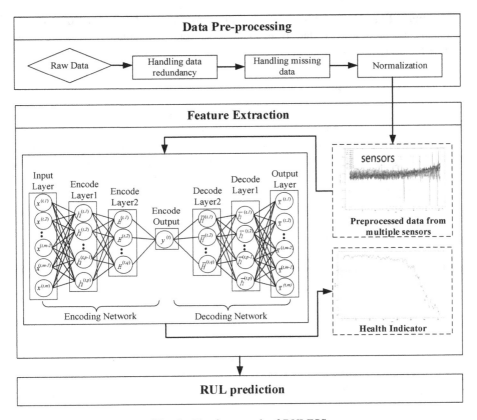

Fig. 1. The framework of RULESS

1. **Data Pre-processing:** Due to the complexity of operating environment in industrial equipment, there is generally exists redundancy and missing or noise in equipment data from sensor measurements, thus raw data needs to be preprocessed for RUL prediction.
2. **Features Extraction:** Extracting features from the preprocessed data by SDAE and take the output of the SDAE as HI.
3. **RUL Prediction:** HI based similarity is measured by a point-wise difference computation method, finally, assign the weight by similarity measurement to predict the RUL.

3.1   Data Pre-processing

Handling Data Redundancy. We remove the repeated data and remove the data which is significantly different with large offsets from neighboring time step.
Handling Missing Data. We remove the data missing some sensor measurements.

Normalization. Data needs to be normalized to remove dimensional differences. Finally the preprocessed data from multiple sensors is obtained. This is the input data for SDAE.

3.2   Feature Extraction

The preprocessed data from multiple sensors can be fused to produce HI which represents the health level of the equipment, the process to achieve this is called features extraction. In [11], multidimensional features are transformed into HI by BP neural network fitting. However, the BP neural network is a supervised algorithm, so this method still needs to define the health level of the equipment in advance as an output label. This type of label in real life is often difficult to get, also the BP neural network has limited ability to extract features.

We use a unsupervised SDAE with better feature extraction capabilities for health assessment, as the Fig. 1 shows:

$$h = f_g(x) \tag{3}$$

$$H = f_z(h) \tag{4}$$

$$y = f_h(H) \tag{5}$$

Where x is input data of SDAE, h is output of encode layer 1, H is output of encode layer 2, y is output of hidden layer 3, the number of nodes in hidden layer 3 is 1, $f_g(x)$ is transform function of hidden layer 1, and $f_z(h)$ is transform function of hidden layer 2, $f_h(H)$ is transform function of hidden layer 3.

The multidimensional sensor monitoring data $x^{(t)} = [x^{(t,1)}, x^{(t,2)}, \dots x^{(t,m)}]$ is used as input data for SDAE, and the output data of last hidden layer in model defined as $y^{(t)}$, that the $y^{(t)}$ can be seen as an effective single-dimensional representation of the multi-sensor monitoring data $x^{(t)}$ after features extraction. Let $y^{(t)}$ denote the HI of equipment at time index t. A HI trajectory is constructed by historical data $X_{(1,t)} = \{x^{(1)}, x^{(2)}, \dots x^{(t)}\}$ and used for the RUL prediction.

3.3   RUL Prediction

When the HI is constructed, the difference between the test HI trajectory and reference HI trajectory is computed by a point-wise difference computation method for similarity measure [8]. This similarity can be expressed as a distance between two vectors and is calculated as:

$$M(l) = \sum_{i=1}^{j} \left(y_1^{(i)} - y_2^{(i+l)} \right)^2, l = 1, 2. \dots, r - j \tag{6}$$

Where y_1 is the test HI trajectory, and y_2 is the corresponding reference HI trajectory, and j is the length of test trajectory, r is the length of referenced trajectory. M(l)

records the whole process that the test trajectory is moved to the end of reference trajectory.

And the minimum relative distance in M(l) is:

$$d = \min(M(1), M(2), \ldots M(l)) \qquad (7)$$

The location of the best-matching part at the reference trajectory, L, is calculated by:

$$L = \arg find(d = M(l)) \qquad (8)$$

$$l = 1, 2. \ldots, r - j \qquad (9)$$

Since similar HI trajectory represent similar health level of equipment, the RUL of referenced trajectory at time index $L + j$ is regarded as the RUL of a test trajectory at time index j. The RUL of the test trajectory at time index j can be expressed as:

$$RUL = \sum_i w_i \times RUL_i, \sum_i w_i = 1 \qquad (10)$$

$$RUL_i = r - (L + j) \qquad (11)$$

The weight w_i is assigned based on the minimum relative distance d. The smaller relative distance it will have, the greater the similarity and a larger weight is assigned.

4 Experiments

4.1 Experimental Settings

Dataset. The dataset we used in our experiments comes from C-MAPSS, it was developed by NASA to simulate realistic large commercial turbofan engines, and the data-set is public for experiment.

The features of the data-set are: It is divided into training data and testing data, the training data is used to build a prediction model, and the testing data is used to verify the prediction ability of model. Each data sample represents a run-to-failure turbine engine, so each data sample have a serious time series relationship. Each sample has unknown different initial wear and fabrication or installation errors. It is multidimensional data, including 26 sensor data and 3 operating condition setting data. The sensor data is seriously polluted by noise and exists obvious deviation.

Benchmarks. We use *MAE* and *MSE* to evaluate the performance of each method.

MAE is the average of the absolute errors. It can better reflect the actual situation of the prediction error, which is defined as follow:

$$MAE = \frac{1}{n}\sum_{i=1}^{n}|y_i - \hat{y}_i| \tag{12}$$

MSE is the average of the square errors. It can evaluate the degree of change of data, which is defined as follow:

$$MSE = \frac{1}{n}\sum_{i=1}^{n}(y_i - \hat{y}_i)^2 \tag{13}$$

Settings of SDAE. The parameters of SDAE network used for RULESS in our experiments are shown in Table 1.

Table 1. Settings of SDAE network for RULESS

Parameter	Value or setting
Number of nodes in the input layer (m)	21
Number of nodes in the encode layer 1 (p)	10
Number of nodes in encode layer 2 (q)	5
Number of samples used for training the network per	256
Noise rate	0.1
Loss function	Mse
Optimizer	Gradient descent algorithm
Activation function	Relu

Methods for Comparison. Four methods are used for comparison in our experiments.

- SADE+LSTM: The HI is constructed by SDAE is used as input data for the long short term memory (LSTM) neural network. The output of LSTM is predicted RUL.
- SADE+BP: The HI is constructed by SDAE is used as input data for the BP neural network. The output of BP is predicted RUL.
- PCA: The HI is constructed by PCA [10], then the RUL was predicted by similarity measurement based on HI trajectory by a pointwise difference computation method (Fig. 2).
- BP: The HI is constructed by BP [11], then the RUL was predicted by similarity measurement based on HI trajectory by a pointwise difference computation method.

4.2   Results and Discussion

In order to test the model's ability to predict the RUL of the equipment at the end of its life, and considering that the service cycle of most engine samples is between 150 and 200, so we randomly took the last 60 service cycles of the No. 9 engine as the prediction object. Tables 2 demonstrate the results of five methods on the metrics of *MAE* and *MSE* in details.

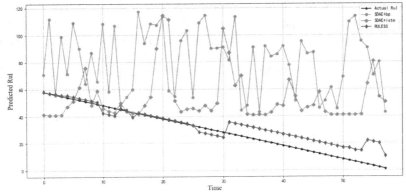

(a) Comparison of SADE+LSTM, SADE+BP and RULESS

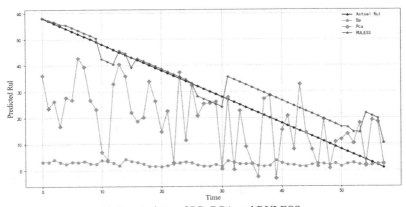

(b) Comparison of BP, PCA and RULESS

Fig. 2. Comparison results of five methods

Table 2. Detailed results of five methods on the metrics of MAE and MSE

Index	RULESS	SDAE+LSTM	SDAE+BP	BP	PCA
MAE	**9.4**	27.18	52.70	27.08	20.56
MSE	**51.13**	1185.77	3477.43	1000.57	738.27

The reason for the best prediction result of RULEES method may be: compared with the conventional method based on neural network, the RULESS method has the highest prediction accuracy, and with the increase of time for equipment operating, the prediction of the RUL for equipment is more accurate. This is due to the data from the early period of equipment has little effect for prediction, only when the equipment is in the late period of operation, and the components undergo severe wear, that the raw data from the equipment can obviously extract features to express the equipment health level at this time. As the amount of data increases at late period, other conventional methods based on the neural network model to form a mapping between the data and RUL would cause larger prediction errors at this time due to more data noise and insufficient optimization of the model. The RULESS method is more accurate in similarity measurement due to the larger amount of data, which leads to better prediction performance. the RULESS method is similar with BP network and PCA method in the similarity measurement part. The difference from the method referred above is the construction of the HI and this step is the most important for prediction method based on similarity. Appropriate HI construction methods should highly summarize the characteristics of the equipment degradation process by the data. Compared with other methods for constructing HI, the RULESS method has a higher ability of data generalization and feature extraction to construct HI, it is also unnecessary to manually define the fault of equipment by expert knowledge, which is more convenient, and also has the highest prediction accuracy,

5 Conclusion

In the RUL predicting process, the predicting method is usually established based on historical monitoring data, but the features implied in the historical monitoring data are often not fully analyzed and mined, the degradation progress of equipment cannot be accurately extracted, so that the accuracy of the predicting method fails to meet expectations. The RULESS method can use the SDAE's good feature extraction ability and the ability to fully exploit the historical data features to construct HI, and the RUL was weighted predicted by similarity measurement. Among them, deep learning and similarity theory are validly combined. Experiments on C-MAPSS data of aircraft engine verified the effectiveness of the proposed method. The experiment proves that RULESS can effectively enhance the precision of the RUL prediction method and provide a valid reference for PHM. However, the experiments in this article only consider the accuracy of RUL prediction, and ignore the uncertainty affects the RUL prediction in reality. The maintenance plans of equipment cannot be based only on the result of RUL prediction, and a reliable indicator of its uncertainty will be considered. Therefore, in the future we will plan to make reliability assessment in RUL predictions.

Acknowledgment. This work was supported by the National Key Research and Development Projectof China (No. 2018YFB1702600, 2018YFB1702602), National Natural Science Foundation of China (No. 61402167, 61772193, 61872139), Hunan Provincial Natural Science Foundation of China (No. 2017JJ4036, 2018JJ2139), and Research Foundation of Hunan Provincial Education Department of China (No.17K033, 19A174).

References

1. Pecht, M., Jaai, R.: A prognostics and health management roadmap for information and electronics-rich systems. Microelectron. Reliab. **50**(3), 317–323 (2010)
2. Shimada, J., Sakajo, S.: A statistical approach to reduce failure facilities based on predictive maintenance. In: International Joint Conference on Neural Networks, pp. 5156–5160 (2016)

3. Wang, X.L., Jiang, B., Lu, N.Y.: Relevance vector machine based remaining useful life prediction for traction systems of high-speed trains. Acta Automatica Sin. **45**(12), 2303–2311 (2019)
4. Zhou, F.N., Gao, Y.L., Wang, J.Y., et al.: Early diagnosis and life prognosis for slowly varying fault based on deep learning. J. Shandong Univ. **47**(5), 30–37 (2017)
5. Heimes, F.O.: Recurrent neural networks for remaining useful life estimation. In: International Conference on Prognostics and Health Management, pp. 1–6 (2008)
6. Wu, Y., Yuan, M., Dong, S.P., et al.: Remaining useful life estimation of engineered systems using vanilla LSTM neural networks. Neurocomputing **275**, 167–179 (2017)
7. Huang, Y., Tang, Y.F., Van Zwieten, J., Liu, J.X., Xiao, X.C.: An adversarial learning approach for machine prognostic health management. In: International Conference on High Performance Big Data and Intelligent Systems, pp. 163–168 (2019)
8. Wang, T., Yu, J., Siegel, D., Lee, J.: A similarity-based prognostics approach for engineered systems. In: International conference on prognostics and health management, pp. 4–9 (2008)
9. Zhao, Z., Liang, B., Wang, X., Lu, W.: Remaining useful life prediction of aircraft engine based on degradation pattern learning. Reliab. Eng. Syst. Saf. **164**, 74–83 (2017)
10. Jia, X., Jin, C., Buzza, M., Wang, W., Lee, J.: Wind turbine performance degradation assessment based on a novel similarity metric for machine performance curves. Renew. Energy **99**, 1191–1201 (2016)
11. Bektas, O., Jones, J.A., Sankararaman, S., Roychoudhury, I., Goebel, K.: A neural network filtering approach for similarity-based remaining useful life estimation. Int. J. Adv. Manuf. Technol. **101**, 87–103 (2018). https://doi.org/10.1007/s00170-018-2874-0
12. Lei, Y.G., Jia, F., Zhou, X., et al.: A deep learning-based method for machinery health monitoring with big data. J. Mech. Eng. **51**(21), 49–56 (2015)

Research on Block Storage and Analysis Model for Belt and Road Initiative Ecological Carrying Capacity Evaluation System

Lihu Pan[1,2(✉)], Yunkai Li[1,2], Yu Dong[2,3] (iD), and Huimin Yan[2]

[1] School of Computer Science and Technology, Taiyuan University of Science and Technology, Taiyuan 030024, China
panlh@tyust.edu.cn
[2] Institute of Geographic Science and Natural Resource Research, Chinese Academy of Science, Beijing 100101, China
[3] Jilin Provincial Department of Natural Resources, Changchun 130000, China

Abstract. Assessment and analysis of ecological carrying capacity are significant issues in regional sustainable development. Large-scale ecological carrying capacity research consume a lot of time and labor costs. In this paper, considering redundant storage, on-demand computing, and multi-scale representation, a block storage and analysis model (BSAM) is proposed to make computation faster. Taking "The Belt and Road Initiative" as an example, the ecological carrying capacity evaluation system (ECCES) was developed. The results show that the problems of low calculation efficiency and high time cost were effectively solved with improved data storage, analysis, and expression approaches used by the ECCES. And the calculation speed is improved with the BSAM. Moreover, in the BSAM, the complexity of the original data is under a more significant impact on computing speed. In contrast, the complexity of an algorithm has a smaller influence on computing speed. Furthermore, the early warning push function and automated report function can generate the ecological carrying capacity states or changes automatically, which can improve the real-time, usability, and convenience of the ecological carrying capacity assessment.

Keywords: Sustainable development · Multi-scale expression · Tile storage · Ecological carrying state · Level of detail

1 Introduction

The Ecological Carrying Capacity (ECC) refers to the coordination degree between human society and ecosystems with regional development. Some indicators in the ECC study can represent the occupation of natural resources by humans. Thus, ECC is an essential concept in the study of regional sustainable development [1–3]. In recent years, with the rapid progress of society and the rapid expansion of population, the ECC analysis has become a hot issue in many research fields such as ecology, resource science, and social science [4, 5]. Because the countries involved in the Belt and Road Initiative are mainly developing countries, most of them are in the window of economic

© Springer Nature Singapore Pte Ltd. 2020
P. Qin et al. (Eds.): ICPCSEE 2020, CCIS 1258, pp. 354–368, 2020.
https://doi.org/10.1007/978-981-15-7984-4_26

development. So, the intensification of human activities and the rapid development of the economy have caused a lot of regional ecological, environmental, and resource problems [6, 7]. Therefore, the design and implementation of the ECC Evaluation System (ECCES) for these areas have particular practical values for the development of the Belt and Road Initiative, such as maintaining ecological security, realizing regional sustainable development and formulating development plans by the government.

At present, the experimental system designs for ECC evaluation are mostly based on the analysis of a single geographical element, and there are few studies on comprehensive large-scale ECC evaluation systems [8, 9]. Moreover, the ECC research on the Belt and Road involves a wide area and requires the support of multiple data sources such as remote sensing, field research, and statistics [10, 11]. Fortunately, scale-matching ideas can be used to solve a variety of issues in data conversion, model parameter selection, and cartographic synthesis on different scales in the ECC analysis [12–14]. In the field of traditional GIS research, scale-matching research is mostly focused on data expression and synthesis. For example, the concept of Level of Detail (LOD) [15], image pyramid [16], and other concepts provide effective solutions of imagery fast display at different geographic scales [17]. In the field of Cloud GIS, scale matching based on the concept of LOD is always used to analyze various remote sensing indicators and geographic elements [18]. For example, Google Earth Engine realizes an on-demand calculation method to display a real-time map in the range visible to users and corresponding scales [19, 20]. In the study of ECC, multi-source, and multi-scale data analysis can use the scale matching and on-demand calculation approaches mentioned above to reduce the calculation load and increase the corresponding calculation speed.

As far as the purpose and result expression of ECC analysis are concerned, the focus of scientific research and business personnel is on the state of ECC overloaded areas and their changing trends. On the one hand, the temporal-spatial ECC variation analysis, which is used to reflect the intensity of human activities and the response of ecosystems, is one of the core issues of ECC research [21–23]. On the other hand, in the face of regularly updated data sources, early warning of ECC overloaded areas is also an important research topic [24, 25]. Therefore, using maps, trend charts, texts, and other expressions to display the ECC report is an essential task for researchers. Therefore, automatic report generation and early overloaded warning have become important directions and critical issues in the design of the ECCES.

By taking redundant storage, on-demand computing, and multi-scale representation as to the primary means, a Block Storage and Analysis Model (BSAM) was proposed to make computation faster. After that, by taking "The Belt and Road Initiative" as an example, we have developed the Ecological Carrying Capacity Evaluation System (ECCES). Based on the above-mentioned technical optimization and system implementation, an automated report function was developed based on templates and data filling. Moreover, an early warning function for ECC overloaded areas based on status thresholds and trend analysis can be used to analyze and express potential ECC overloaded areas for scientific researchers and business personnel. So, these functions can not only provide efficient and rapid data and technical support for more than the ECC assessment for countries but also help to forecast the ECC status in the future, recommend to shape policy.

2 System Hierarchy

This study has refined the calculation and application processes of relevant indicators for the ECC of various countries involved in the Belt and Road. Also, the users of the ECCES was divided into data producers, scientific researchers, the public, and government based on the demand analysis. Then, the virtual collaborative working relationship between these roles was modeled. For example, there is a mutual algorithm and data support relationship between data producers and scientists, and there is a mutual guidance relationship between government and scientists. According to the ECC calculation process and the role relationship above, the ECCES hierarchy includes the following four parts: cloud servers, data manage client, data analysis client, and the public portal. Notably, the primary users of the data management client and the data analysis client are data producers and scientists, and the public portal is open to the Internet (see Fig. 1).

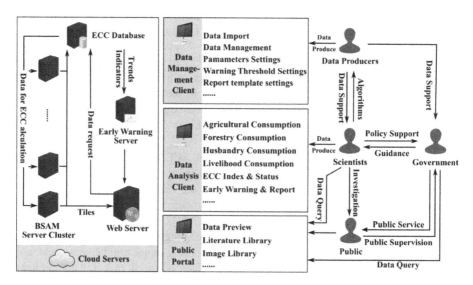

Fig. 1. Hierarchy of the Ecological Carrying Capacity Evaluation System (ECCES) for the Belt and Road Initiative

The cloud servers are the core parts of the entire system. The primary function of these servers is to store data related to ECC evaluation centrally and to open related business function modules for developers. Furthermore, the cloud servers include an ECC database server, a BSAM server cluster (includes a least one BSAM server), an early warning server, and a Web server.

(1) The database server contains the following two data persistence types: a relational database and an object storage system. The relational database is MariaDB, which is mainly used to store the system's necessary data tables and metadata

information of tiles and some origin image data. The object storage system is used to contain the original image data and its tiles.

(2) The BSAM server cluster provides algorithm interfaces through the HTTP protocol and the CGI interface. Also, this cluster allocates all the tile-calculate jobs into many servers to adapt BSAM. Because each tile of a raster data is relatively independent in ECC analysis, it can directly perform distributed calculations with the tile as the smallest unit. Therefore, the BSAM server cluster has strong adaptability in distributed computing.

(3) The early warning server extracts multiple indicators and time series trend data in the database, performs early warning according to the threshold set by the user, and pushes it to the webserver for facilitating inspection and browsing by researchers and the public.

(4) The Web server is the direct user interface of the cloud servers, including a client website, some map services, and image services, which is published through the REST interface protocol.

The data management client is usually used by the data producers for data import and management, alarm threshold setting, and report template setting. When the source data is uploaded to the server through this data management client, the system can tile the images automatically.

The data analysis client can help scientific researchers to set parameters for ECC evaluation, analysis, and result mapping. Moreover, the current server status can be monitored by administrators in the data analysis client.

The public portal is built on the Web server for public access and includes the three following main modules. (1) the data preview module, which is mainly used to browse the results of the ECC assessment. (2) an image library, which is mainly used to display and manage the collected field photos in ECC study. (3) a literature library, which is mainly used for preserving the literature and materials referenced or published in ECC research.

3   System Functional Design

According to the functional requirements of users in terms of rapid ECC assessment, automatic early warning, and automatic reporting, the core of ECCES for the Belt and Road is ECC evaluation, and BSAM makes the evaluation results real-time. Moreover, early warning function can push the evaluation results, which might be focused on, and automated reporting function ensure the rapid text generation of overall evaluation results, respectively. The data management function provides data support for the evaluation of the evaluation results.

3.1   Multi-scale Data Storage and Data Management

Multi-scale data storage and data management functions provide a data storage module for ECCES architecture. Multi-scale data storage is used to redundantly store the tiled data at different scales, which facilitates real-time scale matching and participation in the ECC evaluation. Also, it has the effect of improving the calculation and response speed. For different data source types, the system also has built-in several significant scale conversion tools as following. For vector data, the system integrates map synthesis tools such as Maplex and Subset. For raster data, the system integrates compositing tools such as average value compositing and max value compositing.

The data management function includes an image library and a literature library. Specifically, the field photos collected in the image library have GPS information and can be inserted into any electronic map. Moreover, the papers in the literature library can be entered into the area and scope involved as needed and used to verify and analyzing ECC results in regions.

3.2   Ecological Carrying Capacity Evaluation

Ecological carrying capacity evaluation is the core business function of the ECCES. This paper compares the ecological consumption aspect and the ecological supply aspect in a particular area and then evaluate the ECC result.

Ecological Consumption Evaluation

Ecological consumption can be calculated from the total living consumption [26], and its calculation formula is:

$$CNPPL = \left[\frac{CNPPP + (CNPPI_a + CNPPI_f + CNPPI_l)}{-CNPPE_a + CNPPE_f + CNPPE_l} \right] * (1 + LSP) \qquad (1)$$

In the formula: CNPPL is the total living consumption (unit: gC); $CNPPI_a$, $CNPPI_f$, $CNPPI_l$, $CNPPE_a$, $CNPPE_f$ and $CNPPE_l$ are agricultural import consumption, forestry import consumption, livestock import consumption, agricultural export consumption, forestry export consumption and livestock export consumption (unit: gC); LSP is the improvement rate of living standards, the initial value of the model is set to 0; CNPPP is the total production consumption (unit: gC), and its calculation formula as follows:

$$CNPPP = CNPPP_a + CNPPP_f + CNPPP_l \qquad (2)$$

In the formula: $CNPPP_a$, $CNPPP_f$, $CNPPP_l$ are the production consumption (unit: gC) of agriculture, forestry, and livestock, respectively. All the above-mentioned consumption data are derived from national-level statistics.

Ecological Supply Evaluation

Ecological supply is estimated by available ecological supply. The calculation formula of available ecological supply is as follows:

$$SNPP = \sum_{j=1}^{m} \sum_{i=1}^{n} \frac{(ANPP_{i,j} * \gamma)}{n} * ESP \qquad (3)$$

In the formula: SNPP is the Supplied Net Primary Productivity, which indicates the available ecological supply (unit: gC); γ is the grid pixel resolution; n is the year span of the data, i is the year; m is the number of regional grid pixels, and j is the pixel number; *ESP* is the ecological protection scenario parameter, and the initial value of the model is set to 0.7 (upper limit) or 0.5 (suitable amount); *ANPP* is the Available Net Primary Productivity of the ecosystem, which is calculated by the ecosystem Net Primary Productivity (*NPP*) and coefficient β:

$$ANPP = NPP \times \beta \tag{4}$$

β varies according to different vegetation types, and the default setting is 0.6; NPP data can be retrieved by remote sensing satellite data such as MODIS and Landsat according to the required scale.

Ecological Carrying Capacity Evaluation

The evaluation of ECC includes two leading indicators: ecological carrying capacity index (ECCI), and Ecological Carrying Capacity Status (ECCS). The *ECCI* can determine *ECCS* by the classification standard table (Table 1).

$$ECC = SNPP \div (CNPPL \div P) \tag{5}$$

$$ECCI = P \div ECC \tag{6}$$

In the formula: *P* is the population, which can be obtained by national statistics.

Table 1. State classification standard table for Ecological Carrying Capacity (ECC)

Ecological carrying index (EEI)	Ecological carrying status (EES)
<0.6	Wealthy
0.6–0.8	Surplus
0.8–1.0	More than balanced
1.0–1.2	Less than balanced
1.2–1.4	Overload
>1.4	Severe overload

3.3 Early Warning and Automated Reporting

To efficiently obtain the overload areas and their changes that are of interest to scientific researchers and business personnel, this study uses the state threshold (state warning) and the trend threshold (trend warning) to determine the early warning area. The status threshold that triggers the status alert can be set to More Than Balanced, Overload, or Severe Overload in ECCS. When the ECCS of a region transitions from a more surplus state to the set status threshold, the ECCES will push the status with related region information to the Web server, which provides an early warning function of ECC. The threshold of the changing trend is to analyze the change of the ECCI of a specific area in recent years by linear regression method and extract its slope: when the slope exceeds a certain threshold, the changes in ECC will be posted to the Web server. The status warning and the trend warning can reflect the current and historical perspectives of ECC in a region, respectively.

The automated reports of ECC are achieved by embedding real-time generated ECC assessment data into the monthly, quarterly, or annual report templates set by users in advance, to quickly and accurately generate corresponding reports.

4 Block Storage and Analysis Model (BSAM)

The Block Storage and Analysis Model (BSAM) provides on-demand calculation and fast response support for the ECCES in the base level, which includes two parts: block storage technology and block analysis strategy.

4.1 Block Storage Technology

All raster data in ECCES is divided into 512 * 512-pixel tiles to support data storage and analysis. Since these tiles are not only used for map display, but also for ECC analysis, the storage format is set to float point GeoTIFF data to ensure accuracy. Between adjacent LODs, the tiles of the higher level LOD correspond to the four tiles of the lower level (see Fig. 2a).

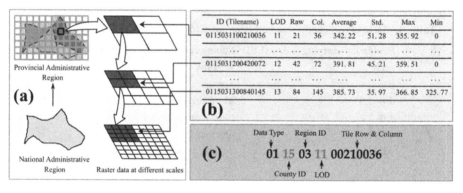

Fig. 2. The design of the block storage technology: (a) data storage at multiple scales; (b) tiled data tables in a relational database; (c) ID (filename) of tiles

Because this BSAM is designed for ECCES for the Belt and Road Initiative, we have chosen the Albers' for map projection. Thus, it is necessary to redesign the map range and tiles. Specifically, the longitude range of the Belt and Road countries ranges from 0 to 180° east longitude and 0 to 24.19° west longitude, and the latitude ranges from 10° south latitude to the Arctic. In the Asia North Albers Equal Area Conic, the

X-coordinate range of the Belt and Road envelope polygon is between −6,391,479–5,878,284, and the Y-coordinate range is between −3,940,112–6,412,663. So, the coordinate spans are 12,269,763 and 10,352,774, respectively. To facilitate the calculation of tiles, 12,800,000 spans are selected in the X direction and the Y direction to cover the entire tile area coverage in this study. Thus, the tile range coordinates on lower LODs do not produce decimals for increasing their readability.

Because the maximum value of the Y coordinate of the Belt and Road countries in Northern Asia is more than one-half the span, the study has modified the Albers projected coordinate system as follows: Move it along the negative Y-axis to 50000 Map units, that is, the "+y_0" parameter of its Proj4 string is set to −50000. The definition of the defined Albers projection is "+proj = aea +lat_1 = 15 +lat_2 = 65 +lat_0 = 30 +lon_0 = 95 +x_0 = 0 +y_0 = −50000 + datum = WGS84 +units = m +no_defs". At this time, by defining the tile range X coordinate and Y coordinate domains as [−6400000, 6400000], the entire Belt and Road countries can be included, as shown in Fig. 3.

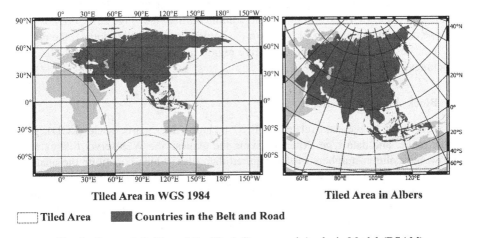

Tiled Area in WGS 1984   **Tiled Area in Albers**

☐ Tiled Area   ■ Countries in the Belt and Road

Fig. 3. Range definition of the Block Storage and Analysis Model (BSAM)

Regarding the definition method of the tile row and column numbers, the LOD is divided into a total of 20 levels from 1 to 20 under the Albers projection and coordinate range defined above. When LOD = 20, the ground resolution of any tile reaches 0.48 cm, which might store low-altitude remote sensing data lossless. In each LOD level, its row and column numbers start from the tile in the upper left corner and count from 0 (see Fig. 4). Each tile is relatively independent and has a corresponding geographic attribute, which can be one-to-one get any tile through LOD, row number, and column number.

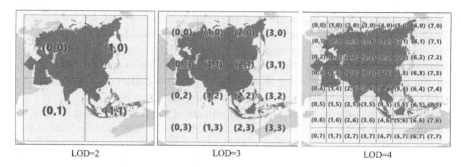

Fig. 4. Definition of row and column numbers for Block Storage and Analysis Model (BSAM)

To perform fast calculations and statistics with user data requests in the Web server, the relevant statistical information (such as average, standard deviation) of tiles is stored in a relational database (see Fig. 2b). To facilitate the system to read the necessary information of the tiles, the name (also the primary key of the tile in the relational database) of each tile has a uniform ID rule, including data type, country code, region code, LOD, row number, and column number (see Fig. 2c). Correctly, the data type is incremented from 01, indicating a particular type of data source; the region code indicates the provincial administrative region under a country to which the data tile belongs.

4.2   Block Analysis Strategy

Any request for image data related to ECC includes three parts: scale (LOD), geographic range, and data type. After receiving the request, the Web server first extracts the row and column numbers of the required data tiles, which constitute the analysis area, according to the request parameters. Usually, the range requested by the user is a rectangle (see Fig. 5), and the requested range can be determined by the maximum (xmax) and minimum (xmin) of the X coordinate of this rectangle, and the maximum (ymax) and minimum (ymin) of the Y coordinate of the rectangle.

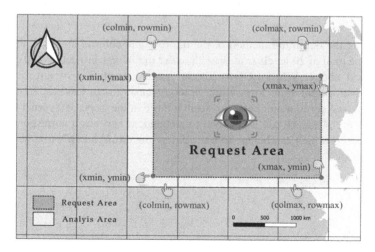

Fig. 5. Relationship between user request area and data analysis area

In order to obtain the complete data of the user request range, the analysis area range is generally bigger than the user request range (see Fig. 5). The most critical step is to use the coordinate range (xmin, xmax, ymin, and ymax). Four to calculate the minimum row number (rowmin), maximum row number (rowmax), minimum column number (colmin), and maximum column number (colmax) in the analysis area under a specific LOD. The calculation formula is as follows:

$$rowmin = floor\left(\frac{xmin}{L_{LOD}}\right) \tag{7}$$

$$rowmax = min\left[floor\left(\frac{xmax}{L_{LOD}}\right), N_{LOD} - 1\right] \tag{8}$$

$$colmin = N_{LOD} - floor\left(\frac{ymin}{L_{LOD}}\right) - 1 \tag{9}$$

$$colmax = min\left[N_{LOD} - 1 - floor\left(\frac{ymax}{L_{LOD}}\right), N_{LOD} - 1\right] \tag{10}$$

Among them, the floor function represents rounding down a float value; the max function and the min function represent finding the maximum and minimum of many values respectively; N_{LOD} represents the tile number of rows (or columns) under a particular LOD; L_{LOD} represents the distance along the X coordinate (or Y coordinate) in a single tile with a specific LOD represented by the projected map, which is calculated as:

$$N_{LOD} = 2^{LOD-1} \tag{11}$$

$$L_{LOD} = \frac{12800000}{2^{LOD-1}} \tag{12}$$

Every tile was obtained from the BSAM by following three steps (see Fig. 6). (1) determine whether the tile data exists at the requested scale, and return the data directly if it exists, otherwise proceed to the next step. (2) determine whether the complete tiles exist in a smaller-scale data. If they exist, ask the BSAM server cluster to perform the upscaling operation and return the result, and persist the tile in the database. Otherwise, proceed to the next step. (3) determine whether the original image data of the tile exists. If it exists, request the BSAM server cluster to calculate and return it, and persist the tile in the database, otherwise return an error. The above steps can perform data calculation on-demand at the scale required by the user, and respond quickly to the user's request.

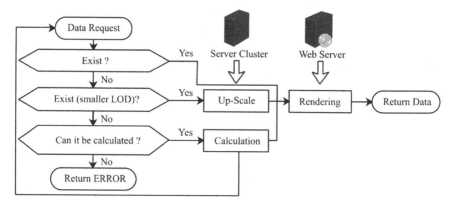

Fig. 6. The process of the block analysis strategy

5 Performance Analysis Between BSAM and Traditional Strategy

5.1 Performance Differences at Different Scales

This study uses the NDVI algorithm as an example to analyze the performance advantages of BSAM and compares the performance between traditional strategy and BSAM between LOD 8 and 20. Using the Landsat7 satellite Near-Infrared band and Red band data with a LOD of 20 as the data source, under the same analysis server, the NDVI of Taiyuan City (around 6,999 km^2) with a LOD of 8–20 is calculated. The traditional strategy calculates the NDVI where the LOD is 20 firstly and then up-scaling the NDVI image to the certain LOD we need. Oppositely, the BSAM generates the Near-Infrared band and Red band data in the particular LOD we need first and then calculates the NDVI. Besides, the study also compared the operation efficiency between the source data with a complete tile pyramid in BSAM (BSAM + redundant in Fig. 7) and no pyramid in BSAM (BSAM in Fig. 7). The results show that under the traditional strategy, the NDVI calculation time under any LOD is about 3 s, and it slightly increases with the LOD increase. This phenomenon is because no matter which LOD is used to calculate the NDVI, the NDVI with a LOD of 20 must be calculated first, so this strategy always takes up much time. Under the BSAM strategy, with the LOD increased from 8 to 20, the calculation time increased from 1.47 s to 3.12 s, which is a significant improvement over the traditional strategy. In the BSAM + redundant storage test, the NDVI calculation time between 8 and 20 LOD is between 0.51 and 3.11 s because the time to create a tiled pyramid is reduced. Therefore, compared with the traditional strategy, the operation efficiency of the BSAM strategy is significantly improved.

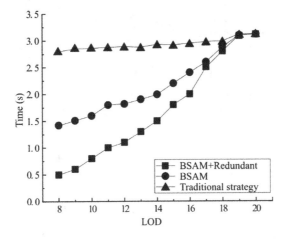

Fig. 7. Comparative analysis of performance between Block Storage and Analysis Model (BSAM) and traditional strategy in different scales

5.2 Performance Differences Under Different Index Complexity

In this study, the Normalized Difference Vegetation Index (NDVI), Modified Vegetation Index (MVI), and Enhanced Vegetation Index (EVI) were used to analyze the performance differences between BSAM strategy and traditional strategy under different index complexity [27]. The calculation formulas of the above three crucial underlying vegetation indices are:

$$NDVI = (NIR - R) \div (NIR + R) \tag{13}$$

$$MVI = \sqrt{(NIR - R) \div (NIR + R) + 0.5} \tag{14}$$

$$EVI = 2.5 \times \frac{NIR - R}{NIR + 6.0 \times R - 7.5 \times B + 1} \tag{15}$$

In the formula: NIR, R, and B are the surface reflectance in the Near-Infrared, Red, and Blue bands, respectively. It can be seen that the algorithm complexity of MVI is higher than that of NDVI and EVI, and EVI uses more original data bands than MVI and NDVI, which means the original data complexity of EVI is higher than the others.

To compare the complexity impact of the original data and the algorithm performance of BSAM and the traditional strategy, this paper uses the Landsat7 Surface Reflectance data with a LOD of 7 to calculate the NDVI, MVI, and EVI of the Taiyuan test area, and conduct performance analysis research (see Fig. 8). The results show that: in the BSAM strategy and BSAM + redundant strategy, the calculation time of each indicator is relatively short, and there is not much difference. While the calculation time of NDVI, MVI, and EVI is 0.92 s, 0.98 s, and 1.02 s in BSAM, the calculation time in BSAM + redundant storage mode is 0.44 s, 0.45 s, and 0.52 s. In the traditional strategy, the calculation time of each NDVI, MVI, and EVI is 3.02 s, 3.21 s, and

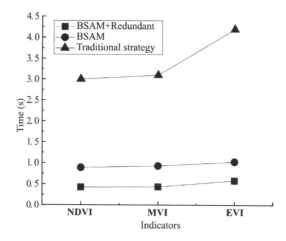

Fig. 8. Comparative analysis of performance between Block Storage and Analysis Model (BSAM) and traditional strategy with different indexes with different complexity

4.29 s, which shows that the performance of the BSAM strategy has been significantly improved compared to the traditional strategy.

From the perspective of indicator types, the EVI calculation speed in the BSAM + redundant strategy and the traditional strategy is slow. At the same time, the difference between NDVI and MVI in the BSAM + redundant strategy is not significant. This phenomenon shows that the original data complexity in BSAM has a more significant effect on the operation speed, while the algorithm complexity has a smaller effect on the operation speed of BSAM.

6   Conclusion and Discussion

The main goal of our study is to solve the issues of low computational efficiency and high time cost in the assessment of large-scale Ecological Carrying Capacity (ECC). By taking the Belt and Road region as an example, scale-matching, redundant storage, and on-demand calculation were used as the primary means in the Ecological Carrying Capacity Evaluation System (ECCES), which is oriented to multi-source, multi-scale, and continuously updated remote sensing data and socio-economic data. Furthermore, a Block Storage and Analysis Model (BSAM) was put forward where multi-scale representation as the basis for different storage layers. The results show that:

(1) The ECCES architecture combines ecological technology with GIS technology. It applies the block storage, block analysis, and expression methods of data related to ECC, which effectively solves a large amount of data and computing efficiency problems. The ECCES has broken through the technical bottleneck of large-scale ECC rapid analysis. Moreover, the image library and literature library of the ECCES also provide a reliable data source for verification and comparative analysis of the ECC evaluation results.

(2) The BSAM effectively improve the calculation speed of the raster data in ECC calculation, and can quickly display the evaluation results at an absolute scale required by the user, which achieving on-demand calculation and fast mapping. Also, BSAM provides an algorithm basis for distributed computing. Additionally, the original data complexity in BSAM has a more significant effect on the operation speed, while the algorithm complexity has a smaller effect on the operation speed.

(3) Early warning push and automated reports provide scientific researchers and end-users with automatically ECC state generating function of interest and its changing results, which significantly improves the real-time, availability, and convenience of ECCES.

The BSAM and ECCES for the Belt and Road Initiative proposed by this research innovatively combine raster tile technology and ecological carrying capacity algorithm, effectively improving the ECC calculation speed. The ECCES can quickly show the results of ECC evaluation on a certain scale, and achieve the main goal of on-demand calculation and rapid mapping. However, the ECCES is still in the experimental and testing stage, and there are still many things to be improved. For example, in the object storage aspect, tiling raster data will generate a large number of data fragments, which will reduce the efficiency of data reading, migration, and maintenance. In the future, a compact tile storage solution may solve this problem. Besides, the ECCES only supports the GeoTIFF tile format of 512 pixels * 512 pixels. The strict data format will cause insufficient tile compression and processing capabilities. Consequently, new compression algorithms can be combined later to reduce the space occupation problem effectively.

References

1. Arrow, K., et al.: Economic growth, carrying capacity, and the environment. Ecol. Econ. **15** (2), 91–95 (1995)
2. Rees, W.E.: Ecological footprints and appropriated carrying capacity: what urban economics leaves out. Environ. Urban. **4**(2), 121–130 (1992)
3. Feng, Z., et al.: Research on the carrying capacity of resources and environment in the past 100 years: from theory to practic. Resour. Sci. **39**(03), 379–395 (2017). (in Chinese)
4. Yan, H., et al.: Status of land use intensity in China and its impacts on land carrying capacity. J. Geog. Sci. **27**(4), 387–402 (2017). https://doi.org/10.1007/s11442-017-1383-7
5. Du, W., et al.: Evaluation methods and research trends of ecological carrying capacity. J. Resour. Ecol. **9**(02), 115–124 (2018). (in Chinese)
6. Rönnbäck, P., et al.: Ecosystem goods and services from Swedish coastal habitats: identification, valuation, and implications of ecosystem shifts. AMBIO J. Hum. Environ. **36** (7), 534–544 (2007)
7. Ouyang, Z., et al.: Improvements in ecosystem services from investments in natural capital. Science **352**(6292), 1455–1459 (2016)
8. Cuadra, M., Björklund, J.: Assessment of economic and ecological carrying capacity of agricultural crops in Nicaragua. Ecol. Ind. **7**(1), 133–149 (2007)

9. Byron, C., et al.: Calculating ecological carrying capacity of shellfish aquaculture using mass-balance modeling: Narragansett Bay, Rhode Island. Ecol. Modell. **222**(10), 1743–1755 (2011)
10. Xin, Z., Xu, Z.: Analysis of water resources carrying capacity of the "Belt and Road" initiative countries based on virtual water theory. J. Resour. Ecol. **10**(6), 574–583 (2019)
11. Teo, H.C., et al.: Environmental impacts of infrastructure development under the belt and road initiative. Environments **6**(6), 72 (2019)
12. Sun, P., Liu, S.: Large scale eco-hydrological model and its integration with GIS. Acta Ecol. Sin. **23**(10), 2115–2124 (2003)
13. Wang, Y., Huang, M.: Research on hierarchical inter-connectivity among multi-scale representations from GIS feature. J. Hunan Univ. Sci. Technol. (Nat. Sci. Edn.) **1**, 20 (2006)
14. Hu, Y., et al.: Design and implementation of agricultural product quality and safety traceability system based on grid management. Chin. Eng. Sci. **20**(02), 63–71 (2018). (in Chinese)
15. Wulder, M.: Optical remote-sensing techniques for the assessment of forest inventory and biophysical parameters. Prog. Phys. Geogr. **22**(4), 449–476 (1998)
16. Cheng, C., et al.: Research on remote sensing image subdivision pyramid. Geogr. Geo Inf. Sci. **26**(1), 19–23 (2010)
17. Li, L., et al.: An algorithm for fast topological consistent simplification of face features. J. Comput. Inf. Syst. **9**, 791–803 (2013)
18. Dean, J., Ghemawat, S.: MapReduce: simplified data processing on large clusters. In: Proceedings of Sixth Symposium on Operating System Design and Implementation (OSD 2004) (2004)
19. Gorelick, N., et al.: Google earth engine: planetary-scale geospatial analysis for everyone. Remote Sens. Environ. **202**, 18–27 (2017)
20. Hu, Y., Dong, Y.: An automatic approach for land-change detection and land updates based on integrated NDVI timing analysis and the CVAPS method with GEE support. ISPRS J. Photogramm. Remote Sens. **146**, 347–359 (2018)
21. Shi, Y., et al.: Research progress and prospect on urban comprehensive carrying capacity. Geogr. Res. **32**(1), 133–145 (2013)
22. Kang, P., Xu, L.: The urban ecological regulation based on ecological carrying capacity. Procedia Environ. Sci. **2**, 1692–1700 (2010)
23. Dong, Y., et al.: Analysis of spatiotemporal changes of grassland carrying capacity in Mongolian Plateau based on supply-consumption relationship. J. Nat. Resour. **34**(05), 1093–1107 (2019). (in Chinese)
24. Jie, F., Kan, Z., Wang, Y.: Basic points and progress in technical methods of early-warning of the national resource and environmental carrying capacity. Progress Geogr. **36**(3), 266–276 (2017)
25. Xu, W., et al.: Evaluation methods and case study of regional ecological carrying capacity for early-warning. Progress Geogr. **36**(3), 306–312 (2017)
26. Du, W.: Research on Ecological Carrying Capacity Based on Supply and Consumption. Changan University (2018)
27. Mcdaniel, K., Haas, R.: Assessing mesquite-grass vegetation condition from landsat. Photogramm. Eng. Remote Sens. **48**, 441–450 (1982)

Hybrid Optimization-Based GRU Neural Network for Software Reliability Prediction

Maochuan Wu[1], Junyu Lin[2(✉)], Shouchuang Shi[1], Long Ren[3], and Zhiwen Wang[3]

[1] China Shipbuilding Corporation 716 Research Institute, CSIC Information Technology Co., Ltd, Lianyungang, Jiangsu, China
wumc@jari.cn, 884040776@qq.com
[2] Institute of Information Engineering, Chinese Academy of Sciences, Beijing, China
linjunyu@iie.ac.cn
[3] Harbin Engineering University, Harbin, China
1262684570@qq.com

Abstract. Aiming at the problems of low prediction accuracy and weak generalization ability of current reliability prediction models, this paper proposes a hybrid multi-layer heterogeneous particle swarm optimization algorithm (HMHPSO) that can simultaneously optimize the structure and parameters of the GRU neural network. It first introduced a multi-layer heteromass particle swarm optimization (MHPSO) algorithm, which sets the population topology as a hierarchical structure and introduces the concept of attractors, so as to improve the update formula of particle speed, and enhance the information interaction ability between particles, increase the diversity of the groups, thereby improving the optimization ability of the algorithm. Then the HMHPSO used the quantum particle swarm optimization (QPSO) algorithm to determine the structure of the GRU, that is, the number of hidden nodes. Experimental results show that the algorithm can generate GRU neural networks with high generalization performance and low architecture complexity, and has better prediction accuracy in software reliability prediction.

Keywords: Software reliability · PSO · GRU · Prediction accuracy · Generalization performance

1 Introduction

Software reliability, as an important factor of software credibility, has become the focus of scholars' attention and research [1, 2]. Method for establishing software reliability model can be divided into two directions: based on statistical analysis and data-driven approach [3]. The model established by the former needs to be based on some specific assumptions, which are often difficult to meet in practice. The model established by the latter is a method based on data, and because it does not require assumptions and high applicability and prediction accuracy, it has received extensive attention and research, and has become a current hotspot of software reliability modeling research.

© Springer Nature Singapore Pte Ltd. 2020
P. Qin et al. (Eds.): ICPCSEE 2020, CCIS 1258, pp. 369–383, 2020.
https://doi.org/10.1007/978-981-15-7984-4_27

At present, data-based software reliability models mainly include support vector machine [4], gray theoretical [5], and hybrid prediction models based on multiple prediction methods [6–8], especially using artificial neural networks to implement software reliability prediction has become a research [9]. Literature [10] proposed a software reliability prediction model based on wavelet neural network, which has better prediction effect. Literature [11] applied the dynamic weighted combination model of rough feedforward neural network to software reliability prediction, and verified the effectiveness of the model. Literature [12] improved the traditional single neural network model, and connected and combined multi-layer neural networks to improve the performance. Literature [13] proposed an adaptive genetic algorithm-learning vector quantization neural network reliability model, and improved the reliability prediction accuracy in short-term prediction. Some studies [14, 15] proposed the use of intelligent optimization algorithms combined with dynamic fuzzy neural networks for reliability prediction, which improved prediction accuracy. The accuracy of the prediction model of the above prediction model has been improved to a certain extent, but the accuracy is still difficult to meet the actual application, resulting in its low popularity in practical applications.

A gated recurrent unit (GRU) neural network is an ideal method for processing time series data such as software defects and software failure data [16–18]. Compared with other neural networks, the existence of feedback connections makes the training of GRU networks more difficult. Therefore, the design of a new automated algorithm to determine the appropriate network configuration and connection parameters is essential. But current research on neural networks GRU focused on the improvement of the network connection structure, methods and theories have not been studied for the GRU network structure and parameters simultaneously.

This paper uses the PSO algorithm to automate the training process of the network, so as to realize the simultaneous learning of GRU network structure and parameters. The PSO algorithm may lead to a lack of population diversity in the later stages of the search process, so that the results are only locally optimal [19]. Based on this, this paper proposes a multi-layer heteromass particle swarm optimization algorithm (MHPSO). In this algorithm, the topological structure of the population is a hierarchical structure. By introducing the concept of an attractor, the attractor's ability to attract the particles themselves is considered. Then the particle's speed update formula is improved, which enhances the information interaction ability between particles, increases the diversity of the group, and further improves the algorithm's optimization ability.

In addition, this paper proposes a hybrid optimization algorithm that simultaneously optimizes the structure of the GRU neural network and its parameters. When training the GRU, the QPSO algorithm is used to determine the number of hidden nodes in the GRU, and the parameters are adjusted using the MHPSO algorithm. It can automatically generate a neural network with simple structure and high versatility.

2 Improved Particle Swarm Algorithm

The GRU is a kind of recurrent neural network [20, 21], in this paper, HMHPSO hybrid algorithm is used to find the optimal structure and parameters of GRU neural network. The MHPSO and hybrid algorithms proposed below will be introduced in detail.

2.1 GRU Neural Network

Recurrent Neural Networks (RNN) is an important branch of artificial neural networks. It introduces a feedback mechanism in the hidden layer to achieve effective processing of sequence data. Recurrent neural networks have the powerful ability to store and process contextual information, and have become one of the research hotspots in speech recognition, natural language processing, computer vision and other fields. The unique gate structure of Long Short-Term Memory (LSTM) solves the problem of the disappearance of the gradient of the time dimension of the traditional recurrent neural network and becomes an important part of the RNN structure.

Gated Neural Network (GRU) neural network is an improvement of the LSTM structure. It is a network model that has made major breakthroughs in recent years. It uses update gates and forget gates to solve the gradient disappearance problem of traditional recurrent neural networks. The two gating vectors, the update gate and the forget gate, determine which information can ultimately be used as the output of the gating loop unit, store and filter the information, and save the information in the long-term sequence. The update gate helps the model determine how much past information is to be passed to the future, or how much information from the previous time step and the current time step needs to be passed on. The forget gate mainly determines how much past information needs to be forgotten. However, GRU will lose important information during the filtering process, and the existence of the feedback connection makes the training of the GRU network more difficult, which will reduce the prediction ability of the model.

2.2 PSO Algorithm

In GRU, particles are generally distributed near the global best point, and the speed decreases in the later stage of searching for the point. When the speed decay is zero, the particles cannot continue to optimize, and the particles fall into a stagnation state, so that the result is only a local optimum. To solve this problem, literature [22] improved the particle movement rules to make the particles move towards particles with better fitness values in their vicinity to prevent premature convergence. Literature [23] set new parameters in the PSO structure to improve its performance. In order to avoid local optimal results, literature [24] introduces a new particle search strategy and adds it to the PSO algorithm, which improves the accuracy of the search results. In response to these problems, this paper optimizes the PSO algorithm by reconstructing the particle distribution and redefining its velocity decay formula.

2.3   MHPSO Algorithm

As the particles in PSO tend to move to their own local optimal position and global optimal position, a rapid convergence effect of the particle population is formed, and local extreme values or premature convergence are prone to occur. To solve these problem, this paper first proposes the MHPSO algorithm, which defines the distribution structure of particles and clarifies the interaction between multiple layers of particles. The algorithm structure is shown in Fig. 1, Each layer contains the same number of particles.

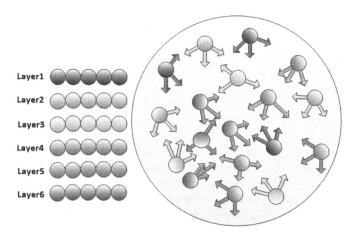

Fig. 1. Structure of particle swarm population

At each iteration, the algorithm will assign particles to different levels according to their fitness values, and assign levels from low to high according to the fitness value from large to small. In a hierarchical structure, particles are attracted by particles in the layer above them. The particles in this layer are its attractors, and they are themselves attractors of the particles in the lower layers. For example, the level of the particles is the third layer, then the particles in the fourth layer are the attractors of the particles, and they are themselves the attractors of the particles in the second layer.

The MHPSO algorithm adds an additional term from the attractor to the speed update formula, as shown in Eq. (1):

$$V_{i,j}^{t+1} = \omega V_{i,j}^t + c_1 r_{1,i,j}^t \left(\hat{y}_j^t - x_{i,j}^t \right) + c_2 r_{2,i,j}^t \left(y_{i,j}^t - x_{i,j}^t \right) + \sum_{a=1}^{A_j^i} c_3 r_{3,i,j}^t \left(x(i)_{a,j}^t - x_{i,j}^t \right) \quad (1)$$

Among them, $V_{i,j}^t$ is the velocity of particle i at time t; $x_{i,j}^t$ is the position of particle i; ω is the inertia weight, c_1, c_2 are acceleration coefficients or called learning factors; $y_{i,j}^t$ is the individual extreme point of the whole particle swarm; $\hat{y}_j^t$ is the global extreme point of the whole particle swarm; $r_{1,i,j}^t$, $r_{2,i,j}^t$ are independent uniformly distributed random numbers in the interval $[0,1]$; $V_{i,j}^t \in [-V_{max}, V_{max}]$, V_{max} is a constant, which is set artificially according to the specific problem. $x(i)_{a,j}^t$ is the position where the particle i attracts the particle a, A_j^i is the total number of the particles i attracted, c_3 is the constant acceleration coefficient.

In this paper, $r(i)_{a,j}^t$ is used to denote the attraction coefficient corresponding to the attraction particle a of the particle i. To ensure that particles are balanced by their attractors and improve the robustness of the algorithm, the particle's attraction coefficient is divided into the following two cases:

When $S(i)_{a,j}^t \leq \bar{S}(i)_j^t$,

$$r(i)_{a,j}^t = r_{min}^t + \frac{\left(r_{max}^t - r_{min}^t\right)\left(S(i)_{a,j}^t - S(i)_{min,j}^t\right)}{\bar{S}(i)_j^t - S(i)_{min,j}^t} \tag{2}$$

When $S(i)_{a,j}^t > \bar{S}(i)_j^t$,

$$r(i)_{a,j}^t = r_{min}^t + \frac{\left(r_{max}^t - r_{min}^t\right)\left(S(i)_{max,j}^t - S(i)_{a,j}^t\right)}{S(i)_{min,j}^t - \bar{S}(i)_j^t} \tag{3}$$

Among them, r_{min}^t and r_{max}^t respectively represent the minimum and maximum coefficients of suction. $S(i)_{min,j}^t$ and $S(i)_{max,j}^t$ respectively represent the minimum and maximum distances from all attractors of particle i to particle i. $S(i)_{a,j}^t$ represents the distance from the attractor a of the particle i to it. $\bar{S}(i)_j^t$ is the average distance from all attractors of particle i to it. When $S(i)_{a,j}^t \leq \bar{S}(i)_j^t$, the attractor a will correspond to a larger attraction coefficient; when $S(i)_{a,j}^t > \bar{S}(i)_j^t$, the attractor a will correspond to a smaller attraction coefficient.

3   MHPSO Algorithm

In order to optimize the performance of GRU neural network, this paper designs a QPSO and MHPSO interleaved execution algorithm (HMHPSO). In which, the QPSO is responsible for setting the structure of the GRU hidden layer, and MHPSO is responsible for optimizing its parameters.

3.1   Particle Coding Strategy

Since HMHPSO simultaneous optimization parameters and structure of the GRU, structure particles and parameter particles are defined separately, as shown in Fig. 2. In QPSO, the algorithm encodes structural particles to generate a binary string. If the value of each item in the binary string is 1, it indicates that there are hidden nodes. Otherwise, it means that the hidden node does not exist, and the length of the string depends on the maximum number of hidden nodes allowed by the GRU(*Max_SNode*). In MHPSO, parametric particles are encoded as real-coded vectors.

A structure particle:

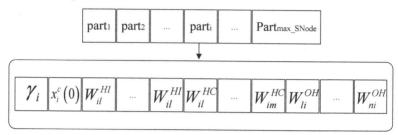

A structure particle:

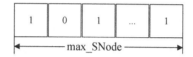

Fig. 2. Particle coding strategy

For the GRU neural network shown in Fig. 2 (where there are l node in the input layer, m nodes in the hidden layer, and n nodes in the output layer), the parametric particles represent the connection weights between nodes and the self-feedback coefficients. Therefore, the number of elements in the parameter particles is $m + m + ml + mm + mn$. Through such specific processing, the structure and parameter optimization can be conveniently performed by a hybrid optimization algorithm.

3.2 Fitness Function

In order to analyze the properties of the particles in HMHPSO, a fitness function must be set. For the GRU neural network, the fitness function is generally expressed by the average variance of the actual value and the predicted value, and the value of the function is proportional to the optimal performance of the particle, which is usually expressed by *RMSE*:

$$RMSE = \sqrt{\frac{1}{n * N_t} \sum_{t=1}^{N} \sum_{i=1}^{n} (y_{ti}(k) - y_{ti}(k))^2} \qquad (4)$$

Where N is the number of training samples, n is the number of network outputs, and $y_{ti}(k)$ and $y_{ti}(k)$ are the expected and actual output of the *t-th* sample at time k, respectively. Considering the size of the network and the convergence accuracy of the network, the paper designs a new fitness function to calculate the fitness value. The specific form is as follows:

$$fitness = RMSE + \beta((m - A)/(max\_SNode - A)) \tag{5}$$

Where β is the control coefficient that penalizes the network size, m is the number of hidden nodes in the network, and A is a fixed constant. If any two particles have the same fitness value, then the network with fewer hidden nodes will be considered better. Therefore, in the case that the network is considered effective (that is, there is at least one node in the hidden layer), the HMHPSO algorithm can make the network have better generalization performance and minimal complexity.

Where β is the control coefficient and is used to adjust the network size, m is the number of hidden nodes in the network, and A is a fixed constant. If any two particles have the same fitness value, then the network with fewer hidden nodes will be considered better. Therefore, in the case that the network is considered effective, the HMHPSO algorithm can make the network have better generalization performance and minimal complexity.

3.3   Implementation Process

The following are the main steps of the HMHPSO algorithm:

Step 1: Initialize structural population A;

Randomly generate particles of the initial structure population and structure particles (the number is between 1 and *Max_SNode*).

Step 2: For each *Ai* in *A*:

(A) Initialize the parameter population *P*. Each particle in *P* represents various parameters with the same network structure. The initial parameter particles are randomly generated to generate the parameter population, and the weights of each particle obey a uniform distribution in a certain range.

(B) The MHSPO algorithm is executed on the training set and ends when the specified number of iterations is reached or the optimal parameter particle is solved.

(C) Obtain the best particle. This particle records the best network found in the structure *Ai*, which is represented by *Ai.net*.

Step 3: Update the fitness value of *Ai.net* for each network through the validation set. When the algorithm does not reach the set number of iterations and the optimal network found is updated, skip to step 4; otherwise, skip to step 7.

Step 4: Iteratively update the structure particles using the QPSO algorithm to update the network structure, where each structure particle represents the number of hidden nodes of a network.

Step 5: If the network structure represented by the particle *Ai* changes, perform one of the following two operations:

(A) Add hidden layer nodes. The parameters of the new nodes in the hidden layer are optimized by MHPSO, and the original hidden layer nodes will not be updated during the search for the optimal newly added nodes. At this time, the fitness function of each step of the algorithm is the weighted mean square error after adding nodes, which reaches a minimum.

(B) Delete hidden nodes. The hidden layer nodes are trimmed in the reverse order when they are added to the network. The trimmed network can obtain better generalization ability and simpler connection structure. Then the parameters of the entire network are optimized by the MHPSO.

Step 6: Go to step 3.

Step 7: outputs the global optimal value Pg, and the algorithm terminates.

4 Experiment and Analysis

This paper tests the performance of the MHPSO algorithm by introducing four classic functions, which are shown in Table 1.

Table 1. Test function description

Function	Dimension	Search interval		Optimal value
Rosenbrock	30	[− 10,10]	(0, 0, …,0)	0
Griewank	10	[− 600,600]	(0, 0, …,0)	0
Rastrigin	30	[− 5.12,5.12]	(0, 0, …,0)	0
Ackley	10	[− 32,32]	(0, 0, …,0)	0

4.1 Particle Swarm Improvement Algorithm Comparison Experiment

In order to evaluate the performance of MHPSO, the paper uses four classic functions to test the performance of the algorithm, as shown in Table 1. Among these 4 benchmark functions, Rosenbrock is a unimodal function, whose function is to measure the ability of the algorithm under test to overcome and prevent premature convergence. Rastrigin, Ackley, and Griewank are multimodal functions, which have many local best features. Since Griewank's dimensionality is more than 15-dimensional when the function characteristics tend to be unimodal, 10-dimensional is chosen here. The experimental test parameters are listed in Table 2.

This experiment analyzes the performance of the algorithm by comparing the MHPSO algorithm with the PSO algorithm and the QPSO algorithm. The following evaluation criteria are set:

(1) Mean optimal value (MeanBst), the expectation of the optimal value obtained after running the algorithm 30 times, used to measure the average quality of particle optimization;

(2) Optimal value (Optv), after running 30 times, the best optimization result found by the algorithm;

(3) Mean Iterations (MeanC). When the algorithm optimization results are successful (that is, the accuracy requirements are met), the average number of iterations is needed to show the performance of the algorithm's convergence speed.

The data of the evaluation criteria shown in Table 3 were obtained through experiments. Experimental results show that the convergence speed of the MHPSO algorithm is significantly faster than that of QPSO. The MeanC index in Table 3 shows

Table 2. Experimental parameter settings

Parameter	Algorithm		
	PSO	QPSO	MHPSO
Population size	30	30	30
The maximum number of iterations	1000	1000	1000
Algorithm runs	50	50	50
ω	Decreases linearly from 0.7 to 0.2	Decreases linearly from 0.7 to 0.2	Decreases linearly from 0.7 to 0.2
c_1	Decreases linearly from 2.5 to 0.5	Decreases linearly from 2.5 to 0.5	Decreases linearly from 2.5 to 0.5
c_2	Increases linearly from 0.5 to 2.5	Increases linearly from 0.5 to 2.5	Increases linearly from 0.5 to 2.5
c_3	–	–	2.5 divided by the number of attractors
r_1	U (0, 1)	U (0, 1)	U (0, 1)
r_2	U (0, 1)	U (0, 1)	U (0, 1)
r_{min}^t	–	–	0.3
r_{max}^t	–	–	1.2

that for the Griewank and Rastrigin functions, both the QPSO and MHPSO algorithms have obtained or approached the theoretical optimal value, but the convergence speed of the MHPSO algorithm has increased by about 25%. This is due to the addition of the concept of multiple layers, and the use of memory particle swarm update features, which accelerates the convergence of the algorithm.

For Rosenbrock and Ackley functions, MHPSO and other algorithms have not reached the theoretical optimal value, but MHPSO algorithm will get better optimization results. Specifically, for the Rosenbrock function, MHPSO has a 20% improvement in convergence accuracy over QPSO. For the Ackley function, the average optimal value obtained by the MHPSO algorithm is 14 orders of magnitude higher than that of the QPSO algorithm, and the degree of convergence accuracy is greatly improved. This is because the improved adaptive operation of the MHPSO algorithm fully increases the diversity of the population and can effectively avoid the premature convergence of the algorithm. In summary, the MHSPO algorithm not only guarantees a good optimization effect through memory particle swarm update and adaptive operation, but also greatly improves the convergence speed of the algorithm.

4.2 Software Reliability Prediction Experiment

This experiment analyzes the optimization effect of HMHPSO algorithm on GRU neural network by comparing PSO-GRU (used to adjust the weight parameters of GRU network by modifying the structure), HQPSO-GRU (a hybrid algorithm of PSO and QPSO to optimize the structure and parameters of GRU neural network) and HMHPSO-GRU. The population number, iteration times and search space range of the three algorithms are given in Table 4, and other parameters are consistent with those in Table 3.

Table 3. Experimental parameter settings

Function	Algorithm	MeanBst	Optv	MeanC
RosenBrock	PSO	349.579	7.0226	–
	QPSO	25.6407	24.1649	–
	MHPSO	19.4867	17.9503	–
Griewank	PSO	0.0273	1.645e-010	546
	QPSO	0.8145	0.4681	428
	MHPSO	0.2748	0.3716	315
Rastrigin	PSO	44.3826	21.5742	–
	QPSO	0	0	–
	MHPSO	0	0	246.2

Table 4. Algorithm parameter settings

Parameter	Algorithm		
	QPSO	MHPSO	PSO
Population number	10	30	30
Number of iterations	1000	1000	1000
Search space	[−1,1]	[−1,1]	[−1,1]

In the experiment, the software reliability data set published by NASA and the software failure data set published by Bell Labs were used to predict software reliability.

(1) Bell Labs Software Fault Data Prediction

The software failure data of Bell Labs Network Security Information System Analysis Center (CSIAC) includes software failure data of five projects, which are Dataset1-Dataset5. The paper selects Dataset1 as the experimental data of the paper, which includes a total of 136 sets of data, of which the training set accounts for 50%, the validation set accounts for 25%, and the test set accounts for 25%.

In the experiment, mean absolute percentage error (MAPE) and root mean square error (RMSE) were used as evaluation indicators to evaluate the performance of each optimized neural network. The optimization of GRU neural network through PSO needs to determine the number of hidden layer nodes first. Because there is no unified node setting rule, this paper determines 6 nodes in the hidden layer through trial and error simulation experiments. The final experimental results are shown in Table 5, which are the prediction results obtained after the three algorithms run independently for 20 times.

Table 5. Running results of the three algorithms

Algorithm	Training set		Validation set		Validation set		Hidden
	MAPE (%)	RMSE	MAPE (%)	RMSE	MAPE (%)	RMSE	
PSO-GRU	1.5671	1.0651	0.4576	0.9276	0.6802	2.6481	6
HQPSO-GRU	0.6832	0.5468	0.2564	0.8616	0.2054	1.3526	4.90
HMHPSO-GRU	0.3849	0.2749	0.1537	0.6236	0.1946	0.5842	4.20

Table 5 shows the three algorithms of MAPE, RMSE, and the number of nodes in the hidden layer. It can be seen that the HMHPSO-GRU model has a good effect in the training and testing process. It has the lowest number of hidden layer nodes in the GRU network. This is because the HMHPSO-GRU algorithm simplifies the number of GRU hidden nodes by using the QPSO algorithm, and uses the MHPSO algorithm to optimize and adjust the particle parameters, and automatically generates a network with low architecture complexity and high generalization performance, thereby improving the accuracy of prediction.

(2) MDP Software Defect Data Prediction

The MDP database contains the number of software defects collected by NASA in 13 different software defect module measurement projects. The training set, validation set, and test set in each data set account for 50%, 25%, and 25% respectively.

This experiment uses the accuracy, precision, recall and F-measure values as evaluation criteria to test the performance of PSO-GRU, HQPSO-GRU, and HMHPSO-GRU. In the experimental process, the optimal number of hidden layer nodes in PSO-GRU is determined to be 7 through trial and error simulation experiments. Figure 3 shows the final prediction results of three models on seven data sets, and Fig. 4 shows the average prediction results of three models on seven data sets.

Table 6 shows that the HMHPSO-GRU has the fewest number of hidden layer nodes, which generates a more compact network model. The experimental results in Fig. 3 and Fig. 4 show that the overall prediction performance of the PSO-GRU and HQPSO-GRU algorithms is significantly lower than the HMHPSO-GRU algorithm.

The experimental results in Figs. 3 and 4 show that the overall prediction performance of the PSO-GRU and HQPSO-GRU algorithms is significantly lower than the HMHPSO-GRU algorithm. Since the PSO algorithm may fall into extreme values, the generalization performance of the PSO-GRU algorithm is the worst.

Compared with HQPSO-GRU, the average accuracy rate, average precision rate, average recall rate, and average F-measure value of HMHPSO-GRU on seven data sets have increased by 4.44%, 3.2%, 2.46%, and 0.99, respectively, compared with PSO-GRU, the prediction result is more obvious. This is because the HMHPSO-GRU algorithm prunes the hidden layer nodes. The pruned network can obtain better generalization ability and simpler connection structure, and optimize the parameters of the newly added nodes, so that the prediction results are better.

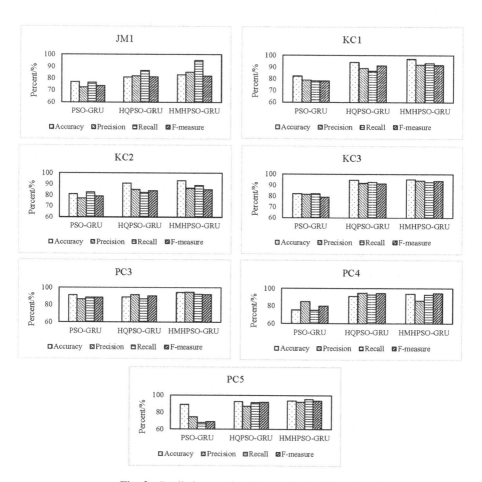

Fig. 3. Prediction results of every single data set

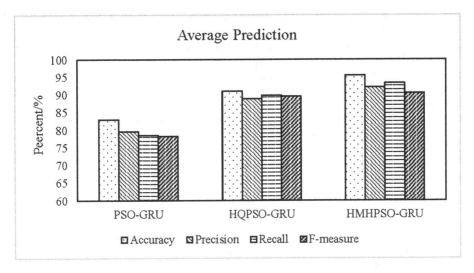

Fig. 4. Average prediction

5 Conclusion

Aiming at the problems of low prediction accuracy and weak generalization ability of the existing software reliability prediction models, this paper first proposes an improved particle swarm optimization algorithm MHPSO, and then combines the QPSO algorithm to propose the HMHPSO algorithm.

It is used to optimize the structure and parameters of GRU neural network, and the optimized model is used in software reliability prediction. Comparative experiments show that the MHSPO algorithm proposed in this paper not only guarantees a good optimization effect, but also greatly improves the convergence speed of the algorithm; HMHPSO-GRU algorithm can automatically determine the number of nodes in the hidden layer of the GRU network and generate its optimal parameters. The optimized GRU neural network has lower structural complexity and better generalization performance. It can process time series data such as software defects and software failure data well, and has better prediction accuracy in software reliability prediction. However, the particle swarm parameter setting of this algorithm is more complicated, and the adaptive convergence setting is more cumbersome, which is the future optimization direction.

References

1. Dinghui, R.: Current status of domestic software quality evaluation services. Software **37** (10), 51–54 (2016)
2. Guohua, S.: A review of software trustworthy evaluation research: standards, models and tools. J. Softw. **27**(04), 955–968 (2016)

3. Luo, L.: Application of software reliability design technology. Graduate School of Chinese Academy of Sciences (2013)
4. Nanshuai, W.: Software defect prediction model based on genetic optimization support vector machine. China Sci. Technol. Paper **10**(02), 159–163 (2015)
5. Jin, C., Jin, S.W.: Software reliability prediction model based on support vector regression with improved estimation of distribution algorithms. Appl. Soft Comput. **15**(2), 113–120 (2014)
6. Weidong, C.: Software reliability prediction model based on grey Elman neural network. J. Comput. Appl. **36**(12), 3481–3485 (2016)
7. Lo, J.H.: A data-driven model for software reliability prediction. In: IEEE International Conference on Granular Computing, pp. 326–331. IEEE Computer Society (2012)
8. Xiao, X., Dohi, T.: Wavelet shrinkage estimation for non-homogeneous poisson process based software reliability models. IEEE Trans. Reliab. **62**(1), 211–225 (2013)
9. Xing, Q.S., Liang, H.J., Hua, D.Z.: Software fault prediction and experimental analysis based on neural network integration. CEA **50**(10), 44–47 (2014)
10. Kiran, N.R., Ravi, V.: Software reliability prediction using wavelet neural networks. In: Conference on Computational Intelligence and Multimedia Applications, vol. 1, pp. 195–199. IEEE (2017)
11. Chatterjee, S., Nigam, S., Roy, A.: Software fault prediction using neuro-fuzzy network and evolutionary learning approach. Neural Comput. Appl. **28**(1), 1221–1231 (2016). https://doi.org/10.1007/s00521-016-2437-y
12. Lakshmanan, I., Ramasamy, S.: Improving software reliability estimation using multi-layer neural-network combination model. Int. J. Innov. Comput. Appl. **8**(2), 113–121 (2017)
13. Qiao, H.: Research of software reliability prediction model based on AGA-LVQ. Comput. Sci. **40**(1), 179–182 (2013)
14. Luo, L.: Software reliability growth model based on dynamic fuzzy neural network with parameter dynamic adjustment. CEA **40**(2), 186–190 (2013)
15. Luo, L.: Software reliability growth model based on fuzzy neural network combined with adaptive step size cuckoo search algorithm. J. Comput. Appl. **34**(10), 2908–2912 (2014)
16. Yang, Z., Duan, Q., Zhong, J.: Analysis of improved PSO and perturb & observe global MPPT algorithm for PV array under partial shading condition. In: Control And Decision Conference (CCDC), 2017 29th Chinese, pp. 549–553. IEEE (2017)
17. Delice, Y., Kızılkaya Aydoğan, E., Özcan, U., İlkay, M.S.: A modified particle swarm optimization algorithm to mixed-model two-sided assembly line balancing. J. Intell. Manuf. **28**(1), 23–36 (2014). https://doi.org/10.1007/s10845-014-0959-7
18. Chatterjee, S., Sarkar, S., Hore, S., Dey, N., Ashour, A.S., Balas, V.E.: Particle swarm optimization trained neural network for structural failure prediction of multistoried RC buildings. Neural Comput. Appl. **28**(8), 2005–2016 (2016). https://doi.org/10.1007/s00521-016-2190-2
19. Wachowiak, M.P., Timson, M.C., DuVal, D.J.: Adaptive particle swarm optimization with heterogeneous multicore parallelism and GPU acceleration. IEEE Trans. Parallel Distrib. Syst. **28**(10), 2784–2793 (2017)
20. Chung, J., Gulcehre, C., Cho, K.: Gated feedback recurrent neural networks. In: International Conference on Machine Learning, pp. 2067–2075 (2015)

21. Eberhart, R., Kennedy, J.: A new optimizer using particle swarm theory. In: Proceedings of the Sixth International Symposium on Micro Machine and Human Science, MHS 1995, pp. 39–43. IEEE (1995)
22. Mohan, V.J., Albert, T.A.D.: Optimal sizing and sitting of distributed generation using particle swarm optimization guided genetic algorithm. Adv. Comput. Scie. Technol. **10**(5), 709–720 (2017)
23. Foster, N., McCray, B., McWhorter, S.: A hybrid particle swarm optimization algorithm for maximum power point tracking of solar photovoltaic systems. In: 2017 NCUR (2017)
24. Wang, H., Cui, Z., Sun, H., Rahnamayan, S., Yang, X.-S.: Randomly attracted firefly algorithm with neighborhood search and dynamic parameter adjustment mechanism. Soft. Comput. **21**(18), 5325–5339 (2016). https://doi.org/10.1007/s00500-016-2116-z

Space Modeling Method for Ship Pipe Route Design

Zongran Dong[1,3](✉) ⓘ, Xuanyi Bian[2] ⓘ, and Song Yang[1,3]

[1] School of Software, Dalian University of Foreign Languages,
Dalian 116044, China
dongzongran@163.com
[2] School of Naval Architecture and Ocean Engineering,
Dalian University of Technology, Dalian 116024, China
[3] Research Center for Language Intelligence,
Dalian University of Foreign Languages, Dalian 116044, China

Abstract. Ship pipe route design (*SPRD*) is one of the most complex and time-consuming processes in ship detail design. Currently, there are many researches on the optimization of ship pipe routes, but there is still a lack of effective and convenient methods to build the pipe routing space. In order to solve this problem, a piping space modeling method for *SPRD* is proposed. This method is based on stereo lithographic (STL) file which is commonly used in data exchange, and it can convert the initial space model built in 3D-CAD software into the data model required by the pipe routing algorithms. For the application purpose, a piping space modeling utility (*PSMU*) is developed with Python and OpenGL, promoting the development of practical pipe routing system. Finally, the feasibility and practicability of the proposed method are verified by the experiment on the piping space of an actual ship fuel system.

Keywords: Ship pipes · Space modeling method · Ship pipe route design · Modeling utility

1 Introduction

1.1 Background of the Research

The ship design includes conceptual design, basic design, detailed design and production design, in which the detailed design is the one that costs the most manpower. In the detailed design stage, the most complicated and time-consuming task is the ship pipe route design [1]. Ships are not mass-produced products like automobiles. It means that ships with different specifications should have new pipe systems individually. Generally, the design of ship piping system is executed by experienced engineers. They need to determine the positions and connections of pipes according to the cost of materials, the availability of space and the installation of accessories. However, the manual work increases the time cost of ship production and brings challenges for the piping engineers.

To solve this problem, researchers have carried out lots of studies on ship pipe route design through automatic algorithms, which mainly used the improved deterministic algorithm, heuristic algorithm or the combination of both [2–5]. And in the aspect of intelligent pipe routing system, many studies focus on the design and implementation

P. Qin et al. (Eds.): ICPCSEE 2020, CCIS 1258, pp. 384–392, 2020.
https://doi.org/10.1007/978-981-15-7984-4_28

of functional modules [6, 7]. In general, an intelligent pipe routing system consists of the following modules:

- Interface Module

It converts the initial model produced by CAD software to the piping space model which contains information of equipment, working areas, restricted areas, and pipe nozzles.

- Optimizer Module

It uses optimization algorithms to automatically generate the optimal piping routes which meet the production demands.

- Display Module

It visualizes the result of pipe layout.

Therefore, the establishment of piping space model is the foundation for intelligent routing algorithms, and the investigation on efficient piping space modeling method is the precondition of pipe routing system from theoretical research to practical application.

Currently, the mainstream 3D-CAD software, such as Solidworks, CATIA, Napus, FORAN, *etc.*, are widely adopted by piping engineers to design the initial model. However, the initial model cannot be directly used for solving ship pipe route design (*SPRD*) problem. In addition, the initial model contains some unnecessary information, such as color, material, and script, which are not required by the pipe routing algorithms. Therefore, it is necessary to investigate new methods and tools that can extract the location and boundary information from the initial model, and then use the information to rebuild the simplified model for *SPRD*.

1.2 Related Works

Ship pipe route design has been a hot research topic for a long time resulting in various approaches. The research starts from 2D piping space with several simple obstacles to 3D piping space with complex multi-objective optimization and multiple constraints. There are four main approaches that are generally used to generate ship pipes: mathematical programming method, potential field method, skeleton search [8], and cell decomposition approach [9]. All these methods need to be carried out in the gridding space model which is the mainstream at present.

Previous works on piping space modeling can be grouped into the following two categories:

(1) Build the piping space in Optimizer Module. Xiaoning Fan [10] studied the automatic pipe routing based on the improved ant colony optimization (ACO) algorithm, which directly used MATLAB to build the layout space for piping. Hai Nguyen [11] adopted improvements for Dijkstra algorithm to design the branch and elbow pipes, and the piping space in his study was also built in MATLAB. This method is suitable for the algorithms to route pipes. However, it is difficult to create complicated space models for practical pipe routing.

(2) Build the piping space in Interface Module. The space model information will be transferred to Optimizer Module through Interface Module. Sang-Seob Kang [12] proposed an expert piping system that implemented an Interface Module between the expert system and the general-purpose CAD system. Asmara [13] also built an Interface Module in his automatic piping system (DelftPipe) which can read 3DDump files. Shin-Hyung Kim [14] designed CAD APIs in the Interface Module of his piping system to parse space model information, which can adjust the model by parameters.

As can be seen from their researches, building model in Interface Model is better than doing that in Optimizer Module. However, their methods are not very efficient and are difficult to verify the correctness of the extracted model. Consequently, there is still a lack of efficient methods and utilities for piping engineers or pipe routing systems to build the space model of *SPRD* algorithms. To deal with it, this paper proposes a novel pipe space modeling method and developed a utility based on the STL file.

The principle and functions of this method are introduced in Sect. 2, and then the practical piping space of a ship fuel system is used to verify the feasibility of this method in Sect. 3. Finally, Sect. 4 concludes the paper.

2   Piping Space Modeling Method

2.1   Method Overview

In order to meet the habit of piping engineers and their actual work needs, this paper proposes a piping space modeling method for auto-routing of ship pipes. The process of this method has two steps: first, built the initial model in 3D-CAD software, and then rebuild the space model for *SPRD* in the piping space modeling utility (*PSMU*), which is developed with Python. The flow chart of the method is shown in Fig. 1.

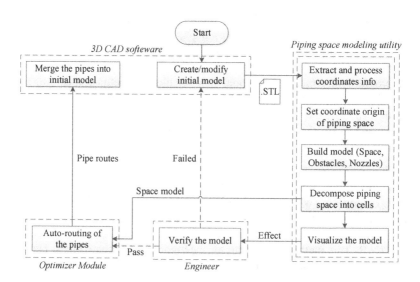

Fig. 1. Flow chart of piping space modeling method in pipe routing system

2.2   Modeling Method Description

Initial Model Creation. The first stage of piping space modeling method is to initialize a model in 3D-CAD software. Generally, the most frequently used method for simplifying 3D model is to build its axis-aligned boundary box (*AABB*). But referring to the research of Asmara [15], if the model is complex, the *AABB* method will be inaccurate, and then he proposed optimized subdivision boundary box (*OSBB*) to improve it. The irregular equipment is simulated by stacking cuboids of different sizes. An example of generator model established by *OSBB* is shown in Fig. 2.

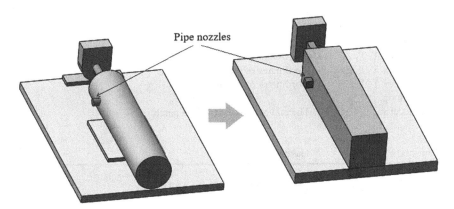

Fig. 2. An example for simplifying a generator model

The advantage of using the low precision model is that the original equipment in the engine room has complex geometry shapes, if the high precision model of the equipment is pursued, both the modeling complexity and computing time will be increased a lot. Moreover, ship pipes need to be arranged orthogonally, so *OSBB* is good enough to meet the requirement of *SPRD*.

It should be noted that the equipment, working areas and restricted areas are all regarded as obstacles, and simulated by cuboids or its combinations. The cylinder pipe nozzles are also simulated with small-size cuboids, which can clearly represent the positions and sizes of the nozzles.

The mainstream 3D-CAD software can output interface files for data exchange. In addition, it is very easy to set the unit, datum plane and center point of the model in 3D-CAD software. In this paper, the unit of the model is *mm* which is in line with the requirement of engineering design, and the center of the space is set at the center of the model to facilitate the model observation in the Display Module.

Coordinate Information Extraction. The STL file of the initial model can be created after the initial model has been built. The features of STL file include: (1) Widely used, i.e., the file can be created by majority 3D-CAD software; (2) Small size, i.e., only the necessary boundary information of the model are stored; (3) Easy to parse, i.e., the file format is structured and clear [16]. Therefore, the STL file can be employed as the interface to transfer location and volume information of the initial model between CAD software and pipe routing algorithms.

STL file is a kind of drawing interchange format file that approximates 3D solid model with triangular patches. It is a collection of several small unordered triangular patches. Each triangle patch is represented by the coordinates of its three vertices (X_i, Y_i, Z_i) and a normal vector (n_1, n_2, n_3) pointing to the outside of the model. The normal vector and the coordinates of the three vertices satisfy the right-hand spiral rule. STL file has two formats: ASCII and binary. The ASCII file format is adopted in this paper because the format is easy for parsing. Figure 3 shows the ASCII format of STL file.

```
solid filename              // File path and name
    facet normal n₁ n₂ n₃   // Component values of the normal vector
        outer loop
            vertex X₁ Y₁ Z₁  // Coordinates of the first vertex
            vertex X₂ Y₂ Z₂  // Coordinates of the second vertex
            vertex X₃ Y₃ Z₃  // Coordinates of the third vertex
        endloop
    endfacet                // The end of the first triangle patch
    ......
    ......
Endsolid                    // The end of the STL file
```

Fig. 3. The ASCII format of STL file

After generating the STL file of the initial model, as shown in Fig. 1, the STL file will be parsed by *PSMU*, which can reconstruct and visualize the piping space. The parsing process of STL files is as follows:

1. Open the STL file of the initial model, extract the vertex coordinates of all triangles, and put them into *pt_lst1*.
2. Divide the points in *pt_lst1* into groups (every 36 adjacent vertices will be put into a group), remove the duplicate points in each group, and put the left points of each group into *pt_lst2*.
3. Write the coordinates of the points in *pt_lst2* to the file 'model.txt', in which every 8 adjacent points stand for the 8 vertices of a cuboid.

Piping Space Modelling. As described in the previous section, 'model.txt' contains the vertex coordinates of each cuboid, which are extracted from the STL file. And then, a grid piping space needs to be built based on 'model.txt' for automatic piping algorithms. The modeling process of grid piping space is shown in Fig. 4.

1. Set the coordinate origin of the piping space, and then perform translation transform on the model coordinates in 'model.txt' according to the coordinate origin.
2. Classify model coordinates by categories of obstacle or pipe nozzle.
3. Transform the vertex coordinates of each cuboid belonging to obstacle into diagonal coordinates and store them into 'model.ini'.
4. Build obstacle models according to 'model.ini'.

5. Transform the vertex coordinates of each cuboid belonging to pipe nozzle into center coordinates, extract the diameter of pipe nozzle, and store them in the 'model_nozzles.ini'.
6. Build pipe nozzle models according to 'model_nozzles.ini'.
7. Build piping space model according to the size of practical layout space, and then put obstacle models and pipe nozzle models into it.
8. Decompose the piping space model into grid cells.

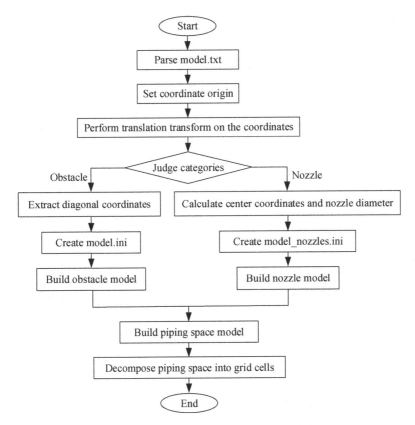

Fig. 4. Flow chart of the piping space modeling method

Visualization. The piping space model is visualized by OpenGL. It makes the model can be rotated, scaled, translated as well as be viewed from different perspectives.

The visualization of piping space model can make automatic routing algorithms use correct routing environment and reduce the workload of engineers. As shown in Fig. 1, engineers will decide whether to modify the initial model or use automatic algorithms to route pipes after observing the visual effect in *PSMU*.

3   Result and Discussion

3.1   Test Case

To verify the feasibility and practicability of the piping space modeling method, a piping space of the ship fuel system is created as the test case. Figure 5 shows the schematic diagram of this fuel piping system which includes fuel oil tank, fuel transfer pumps, fuel oil storages, marine main engine, diesel generators and boilers.

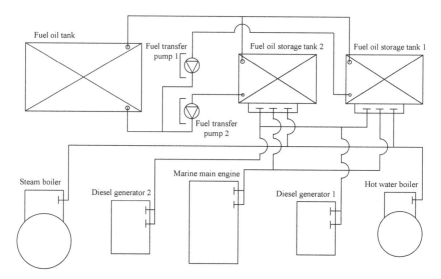

Fig. 5. A schematic diagram of the fuel piping system in a ship engine room

As mentioned in Sect. 2, the first step of the piping space modelling method is to build the initial model in 3D-CAD software. In this case, we build the initial model in SolidWorks, as shown in Fig. 6.

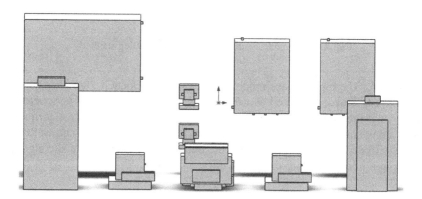

Fig. 6. The initial space model of the fuel piping system in SolidWorks

3.2   Result

PSMU parses the STL file of the initial model created by SolidWorks, and then generates the piping space model required by pipe routing algorithms. The visual effect of the space model is shown in Fig. 7.

The space model can be displayed as wireframe mode or solid mode in *PSMU* for checking. The feasibility and validity of the piping space modeling method can be verified by the comparison between Fig. 6 and Fig. 7.

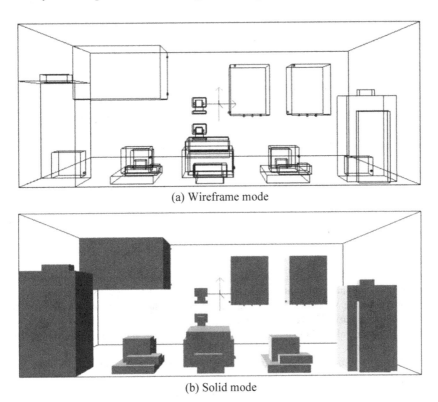

(a) Wireframe mode

(b) Solid mode

Fig. 7. The visualization of fuel piping system in *PSMU*

4   Conclusion

In this paper, a piping space modeling method is proposed, and a piping space modeling utility (*PSMU*) is implemented based on the method with Python and OpenGL.

With this method, a model space was built for pipe routing algorithms in only two steps. First, an initial model in 3D-CAD software was created, and then *PSMU* was used to generate the piping space model. By extracting and transforming the coordinates in STL file, *PSMU* can provide a simulated environment just required by the pipe routing algorithms. Finally, the proposed method and utility were tested with an actual case of the piping space for ship fuel system.

Further ongoing developments will focus on some practical aspects of piping space and the routing result modeling, *i.e.*, building the *OSBB* model intelligently and importing the result generated by routing algorithm into CAD systems for further modification.

Acknowledgements. This work is supported by the Doctoral Scientific Research Foundation of Liaoning Province (Grant No. 2019-BS-061), the Basic Research Foundation of Education Department of Liaoning Province (Grant No. 2019-JYT-07).

References

1. Shao, X.Y., Chu, X.Z., Qiu, H.B.: An expert system using rough sets theory for aided conceptual design of ship's engine room automation. Expert Syst. Appl. **36**(2), 3223–3233 (2009)
2. Kniat, A., Gdansk, T.U.: Optimization of three-dimensional pipe routing. Ship Technol. Res. **47**, 111–114 (2000)
3. Jiang, W.Y., Lin, Y., Chen, M., et al.: An ant colony optimization-genetic algorithm approach for ship pipe route design. Int. Shipbuild. Prog. **61**(3–4), 163–183 (2014)
4. Sui, H.T., Niu, W.T.: Branch-pipe-routing approach for ships using improved genetic algorithm. Front. Mech. Eng. **11**(3), 316–323 (2016). https://doi.org/10.1007/s11465-016-0384-z
5. Shiono, N., Suzuki, H.: A dynamic programming approach for the pipe network layout problem. Eur. J. Oper. Res. **277**, 52–61 (2019)
6. Asmara, A.: Pipe routing framework for detailed ship design. Delft University of Technology, Delft (2013)
7. Wang, Y.L., Yu, Y.Y., Li, K., et al.: A human-computer cooperation improved ant colony optimization for ship pipe route design. Ocean Eng. **150**, 12–20 (2018)
8. Aurenhammer, F.: Voronoi diagrams - a survey of a fundamental geometric data structure. ACM Comput. Surv. **23**, 345–405 (1991)
9. Ito, T.: A genetic algorithm approach to piping route path planning. J. Intell. Manuf. **10**(1), 103–114 (1999)
10. Fan, X., Lin, Y.: Ship pipe routing design using the ACO with iterative pheromone updating. J. Ship Prod. **23**(1), 36–45 (2007)
11. Nguyen, H., Kim, D.J., Gao, J.: 3D piping route design including branch and elbow using improvements for Dijkstra's algorithm. In: International Conference on Artificial Intelligence: Technologies & Applications, Bangkok (2016)
12. Kang, S., Myung, S., Han, S.: A design expert system for auto-routing of ship pipes. J. Ship Prod. **15**(1), 1–9 (1999)
13. Asmara, A., Nienhuis, U.: Automatic piping system in ship. International Conference on Computer and IT Application, pp. 269–280. Sieca Repro (2006)
14. Kim, S.H., Ruy, W.S., Jang, B.S.: The development of a practical pipe auto-routing system in a shipbuilding CAD environment using network optimization. Int. J. Naval Archit. Ocean Eng. **5**(3), 468–477 (2013)
15. Asmara, A., Nienhuis, U., Hekkenberg, R.: Approximate orthogonal simplification of 3D model. In: IEEE Congress on Evolutionary Computation, Barcelona. IEEE (2010)
16. Hallmann, M., Goetz, S., Schleich, B.: Mapping of GD&T information and PMI between 3D product models in the STEP and STL format. Comput. Aided Des. **115**, 293–306 (2019)

Simulation of Passenger Behavior in Virtual Subway Station Based on Multi-Agent

Yan Li, Zequn Li, Fengting Yan$^{(\boxtimes)}$, Yu Wei, Linghong Kong,
Zhicai Shi, and Lei Qiao

School of Electrical and Electronic, Shanghai University of Engineering Science,
Shanghai 2016204, China
yanfengting2008@163.com

Abstract. According to architectural structure of underground station and condition of passengers, the emergency evacuations are extremely important during emergency accidents caused in the stations. Based on the architectural features and situation of passengers in the station, Multi-Agent system is proposed to analyze passengers' behavior and elements which affect passengers' behavior. Base on the crowd mentality, the model of selective behavior of passengers' route was established to simulate passengers' actions during emergency accidents. It is feasible to find out the influence brought by crowd mentality affect the emergency evacuations, and the influence of remitting the situation from external factor through WebVR. At last, it provides references for emergency evacuation strategy.

Keywords: Underground station · Emergency accidents · Passengers' behavior · Multi-Agent

1 Introduction

As a means of transportation in many big cities, the subway is very busy. During holidays, the morning and evening shift travel periods, there are crowds peak. For example, Shanghai Xujiahui subway station is very heavy in terms of passenger flow and density. Once an emergency occurs, many passengers will be killed or injured [1]. The subway station is a large-scale closed space with large and concentrated passenger flow, so it takes a long time to evacuate the crowd [2]. In addition, passengers have complex identities and different cognitive abilities, which brings many difficulties to the emergency evacuation of people in subway stations [3]. In recent years, people have also paid more and more attention to formulating diversified emergency evacuation strategies in advance as a pre-plan for emergency evacuation of the crowd [4].

The establishment of emergency evacuation plan first needs to make a targeted emergency evacuation decision based on the environmental conditions of the subway station, combined with the behavioral characteristics of passengers, the number of passengers, and the composition of passengers [5]. Research on the environment of subway stations often needs to consider three main factors: 1) the characteristics of the topology of the subway station; 2) the size of the subway station and the number of passengers; 3) the number, width and location of the subway exits [6]. Research on

© Springer Nature Singapore Pte Ltd. 2020
P. Qin et al. (Eds.): ICPCSEE 2020, CCIS 1258, pp. 393–408, 2020.
https://doi.org/10.1007/978-981-15-7984-4_29

passenger behavior needs to take into account differences in gender and age of passengers, individual interactions and the interaction between individuals and the environment, as well as complex aggregation manifestations of conformity behavior under the influence of external factors [7, 8].

Based on the environmental factors of subway stations and the attributes of passengers, many researches use the method of MAS (Multi-Agent System) [9–11] to conduct crowd behavior research. MAS can use the group intelligence calculation of multiple single agents to complete the intelligent decision of a group. C.M. Henein [12] proposed the concept of Autonomous Agent based on a single agent, that is, an object in a specific environment perceives information in the environment, and then makes decision calculations based on its own purpose. This kind of decision calculation based on specific environment can form a decision model, which can be improved continuously according to historical data, and can predict the future situation. Based on the autonomous Agent technology [13], through the distributed decision calculation of each independent agent, one or more group intelligent decisions are formed, which improves the calculation efficiency, shortens the calculation time and improves the calculation accuracy.

Based on the above characteristics of MAS, Agent can be used to simulate the autonomous behavior of passengers and their synergistic effect, so as to simulate the system. Yongqiang [14] used MAS to calculate and simulate the phenomenon of passenger group in emergencies. The experimental results showed that MAS has advantages in complex system modeling. Fengting [15, 16] used MAS technology to conduct research in microscopic traffic simulation. The experimental results showed that the study of the mass passenger traffic simulation using this MAS technology builds a microscopic traffic model, and proves the advantages and necessity of the MAS-based method. In summary, in the mass passenger evacuation behavior, which is based on single passenger behavior using MAS, the individual personalization, the interaction between the individual and the environment, and the interaction between the groups can be considered, so that it can better simulate the mass passenger evacuation behavior in an emergency at a subway station.

This paper will model the behavior of passengers based on the characteristics of subway station construction scene, the individual characteristics of the passenger group, the interaction between the individual, the mutual influence between the individual and the environment, and the relationship mapped by the MAS technology between the individual passenger and the agent. Finally, the characteristics of passenger behaviors were described, which provide theoretical support for the establishment of contingency plans for subway station emergencies.

2 Establishment of Crowd Evacuation Model

The establishment of the crowd evacuation model includes the following: 1) the construction of the model environment; 2) the definition of the attributes of the Agent; 3) the rules of the agent behavior; 4) the definition of the external environmental impact; 5) the path decision algorithm of the Agent.

2.1   Model of a Subway Station Architectural Scene

To meet the needs of modeling the subway station building scene, this paper uses Xujiahui subway station as the research object. The scene of the entire subway station is shown in Fig. 1 below. This paper builds a 3D scene based on the actual scene map data of Xujiahui subway station. The vertical top view and the 45-degree angle top view are shown in Figs. 2 and 3 below (Fig. 4):

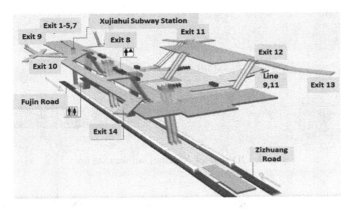

Fig. 1. Map of Xujiahui subway station (Shanghai City, China)

Based on the real building scene data of Xujiahui subway station, the simulation of the platform inside the subway station is performed, and the actual layout structure of the rail transit platform is simulated with the grid as a unit. The size of the grid usually corresponds to the size of the plane space occupied by the individual. Previous studies have found that the size of the plane space occupied by an individual is a square between 0.4 m × 0.4 m and 0.5 m × 0.5 m. Combined with the congestion caused by passengers trying to evacuate the station as soon as possible in an emergency, this paper defines the model environment as follows:

Definition 1: The environment is a platform divided by a grid of 0.4 m × 0.4 m, and finally forms a rectangular space of n × m grid. Each grid has its own attributes: Patch = {status, pass, no}. status indicates the attributes of the grid, including: obstacles such as walls, billboards, and pillars, evacuation exits or aisles; pass indicates whether passengers can pass this grid, and passengers will be distributed in a grid with a pass attribute of 1; no indicates the number of passengers in the grid, up to 1.

2.2   Definition of Individual Passenger Attributes

In the MAS system, the Agent should have the following properties: it has its own behavior library and can make autonomous decisions; it can send its own behavior information to the outside world; it can perceive the behavior information sent by other individuals and the surrounding environmental information, and modify self-behavior according to this information.

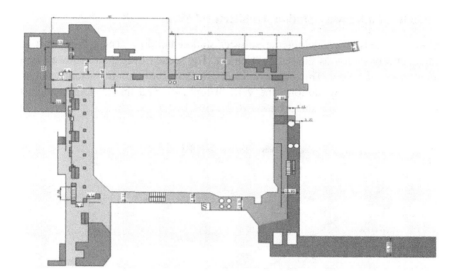

Fig. 2. 2D map of Xujiahui subway station

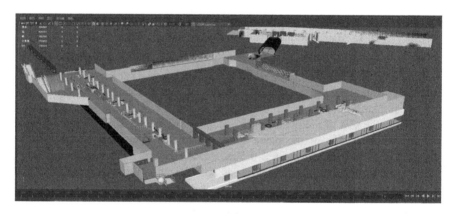

Fig. 3. 5-degree angle top view of scene model of Xujiahui subway station

Based on the above properties, this paper will define the attributes of individual passenger Agents as follows:

Definition 2: The individual passenger agent is written as P_Agent. P_Agent = {St, Vt, act, I}. St is the state set of the Agent at time t, Vt is the speed of the Agent at time t, act is the set of decision-making actions that the Agent may take, and I refers to the influence on Agent from the information of surrounding environment and other individual behavior.

Each sub-attribute set has the following definitions:

Definition 3: The state set St of the Agent at time t includes: $St = \{Pt, type, C, \rho\}$. Pt represents the position coordinates of the Agent at time t, and *type* is a collection of

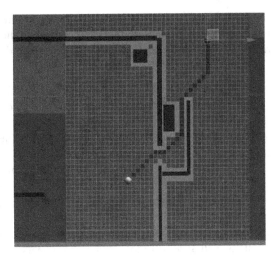

Fig. 4. Partial 2D grid map of subway station

passenger types. In this model, the passenger types are divided into children, young men, young women, and old people. Different types of passengers will have significant differences in walking speed, reaction speed, and strength (competitive ability to reach the same grid at the same time). C represents the individual's herd mentality index, or the dependence on herd behavior; ρ represents the density of the surrounding people at time t, that is, the average number of individuals around the Agent, the unit is person/m^2, this density will affect the walking speed of the person.

Definition 4: The specific speed Vt of the Agent at time t includes: $Vt = \{vt, dir, Pa\}$. Among them, the instantaneous displacement velocity vt of the Agent at time t will change with the change of the density of the surrounding people. The specific calculation formula is as follows:

$$v_t = \left(\begin{array}{ll} v_0, & \rho < 0.54 \\ v_0 \times \{0.3 \times [1.3 - 0.8\ln(\rho)] + 0.01 \times (3 - 0.76 \times \rho) + 0,2\}, & 0.54 < \rho < 3.8 \\ 0.37, & \rho > 3.8 \end{array} \right)$$

$$(1)$$

v_0 is the normal walking speed of passengers, which is determined by the type of passengers. According to the research, the normal speed of middle-aged and young men and women is 1.55 m/s and 1.45 m/s respectively, the speed of children is 1.41 m/s, and that of old people is 1.21 m/s. It can be seen that the moving speed of individual passengers will be affected by the surrounding environment, and the changes in the displacement of the individual passengers will also affect the surrounding passengers, so the Agent simulation can better describe the interaction and cooperation between passengers. *dir* represents the moving direction of the Agent. Passengers tend to move in the direction of the export during emergencies. In the model, combined with the eight-neighborhood principle, the passenger's target grid is more likely to be one of the five directions of the front, left front, right front, left, and right directions facing the exit direction. There is also a small probability that it will be one of the left or right rear grid, but it will never be the rear grid. *Pa* represents the selectable destination grid of the Agent.

Definition 5: There are three typical types of actions that may be taken by the Agent: the behavior of pursuing the shortest path (act_1), the behavior guided by experience (act_2), the completely herd behavior (act_3). The behavior of pursuing the shortest path is that the Agent is familiar with the environment, can quickly find the location of the exit, and is getting closer to the exit in this direction. The behavior guided by experience is that the Agent is not very familiar with the environment and leaves from the direction of entrances based on existing experience. The completely herd behavior means that the Agent is completely unfamiliar with the environment and can only move blindly with the crowd.

When a rail transit emergency occurs, due to the consistent characteristics of the station entrance and exit, as well as less complicated environments, people can quickly find the exit location. Therefore, in the model in this paper, the Agent adopts the single decision behavior of act_1, but this does not mean that the Agent will only constantly move to the nearest exit. The Agent will also be affected by the surrounding environment information and other Agent information.

Definition 6: The set of impacts on the Agent by external information: $I = \{g, j, f...\}$. The g indicates the impact of external information sources such as guidance personnel, broadcast, and indicator lights on passengers, and j indicates the impact of the target grid being occupied by other Agents, that is, the impact of crowding and blocking. f indicates the impact of the fire or other events that may cause casualties on the Agent. There are also some other events that may affect passengers, due to their low probability of occurrence, they are not directly defined in the model. When specific issues are involved, they are defined according to actual conditions. I can also be an empty set, which means that passengers rely entirely on self-organization for behavioral decisions.

2.3 Passenger Behavior Rules with External Environmental Impact Definition

2.3.1 Passenger Self-organizing Behavior Rules

In an emergency of the rail transit system, passengers' behavioral decisions mainly follow the shortest path principle after ruling out the influence of external guidance:

Definition 7: The shortest path principle means that passengers will preferentially choose the route to the exit. Therefore, the Agent will preferentially choose the grid facing the exit direction as the target grid for next time step. The next best is the grid on the front left and front right. In the model, the attractiveness of each grid to the Agent needs to be quantified, so the attractiveness index A is introduced.

$$A(i,j) = \frac{D_{\min}}{D(i,j)} \tag{2}$$

D_{min} represents the distance from the next grid in the shortest path to the exit. $D(i,j)$ represents the distance from the target grid to the exit. The bigger A is, the more attractive the grid is to the Agent.

In emergencies, due to the critical situation, under the influence and pressure of most individuals, passengers individual will give up their subjective thinking and intuition, and blindly adopt decision-making behaviors consistent with most people. This is a common herd phenomenon. This phenomenon can be said to be a double-edged sword, which has a dual effect on the implementation of emergency evacuation strategies. Its positive effect is reflected in that passengers who are unfamiliar with the structure of rail transit stations can follow the group of passengers who are familiar with the situation, which can reduce the potential loss of property and life. Its negative effect is reflected in the fact that many evacuation routes that could have played a role cannot be effectively used because passengers follow the group action blindly. In addition, the emergence of the herd phenomenon will greatly change the behavior of the group in the emergency evacuation exercise of rail transit. This phenomenon is a major obstacle to the formulation of emergency evacuation strategies for rail transit.

Therefore, in the model of this paper, the attractiveness of the element grid will also be related to the density of the target grid and the surrounding grids. So the grid attraction formula is improved from formula (2) to:

$$A'(i,j) = C \times \frac{D_{min}}{D(i,j)} \times \frac{\rho(i,j)}{\rho_{max}} \ldots\ldots \tag{2'}$$

C is the herd index of the individual in Definition 3. Due to the unpredictability of individual psychology, C of each Agent will be randomly selected from 1 to X. X depends on the criticality of the emergency. $\rho(i,j)$ represents the actual crowd density of the target grid and surrounding grids. ρ_max is the maximum density that the target grid and surrounding grids can accommodate. When selecting the optimal route, the Agent of passenger will consider the maximum attractive grids calculated by formulas (2) and (2'), and take the larger attractive grid. By defining the wayfinding mode of the Agent in the above manner, the Herd behavior of the Agent can be well simulated, and the model will be closer to reality.

2.3.2 The Rules of Conduct Under External Influence

In a rail transit emergency, passenger behavior decisions will inevitably be subject to external interference. According to Definition 6, the external environment interference in this model mainly includes the following three aspects: the guiding factors issued by rail transit companies and the government, such as broadcasting guidance. The environmental factors that may cause loss of life and property, such as closed passages caused by fire; the influencing factors caused by other passenger behaviors, such as changes in route selection caused by crowding. Its impacts are as follows:

(1) Under the influence of $I(g)$, that is, under the influence of external guidance factors, according to the survey, 90% of passengers choose to follow the guidance and 10% of passengers choose to stick to their own judgment. The grid of Agents who choose to follow the guidance will be completely dominated by the external

environment. And because the role of external guiding factors is mainly to guide passengers to make the right choice, it is basically similar to the rules of passenger's self-organized behavior.

(2) Under the influence of $I(j)$, that is, under the influence of congestion, the target grid of the next time step selected by multiple agents will be the same. In this case, the strength of each Agent will be compared. The strength of the Agent is determined by the type attribute of the Agent, middle-aged man > middle-aged woman > children > old people. Agents of the same type will be randomly selected to enter the target grid, and the other Agents will choose the sub-optimal grid as the target grid for the next time step. In this case, the Agent not only needs to select the target grid according to its own situation, but also considers the properties of the next grid. If there are more empty grids can be selected at the next grid position, the higher the probability that it will be selected as the target grid, that is, the higher its attractiveness index to the Agent.

(3) Under the influence of $I(f)$, that is, under the influence of disaster events that may cause passenger casualties, individuals will automatically make decisions to stay away from disaster event points. Therefore, when Agents determine the target grid, only the grids farther from the disaster event point will be selected.

2.4   Path Selection Algorithm for Passenger Agent

When a rail transit emergency occurs, the main objective of the passenger agent will immediately change, it will change to moving to the grid representing the exit at the fastest speed and avoiding the danger that may arise in the middle. In this process, path selection will be performed after each grid is reached according to the situation on site. This selection will integrate the rules of passenger self-organizing behavior with the external environment. After the main target is changed, the Agent will choose the path through the following algorithm:

> **Step 0:** Determine whether the status attribute of the grid where the Agent is located is an exit. If so, the main target of the Agent is reached, the Agent is cleared, and the number of people evacuated successfully increases by one. Otherwise, go to *Step 1*.
> **Step 1:** determine the exit direction, calculate the moving speed according to formula (1), screen the reachable grid in the front, left front, right front, left and right of the exit direction, and remove the grid with *pass* attribute of 0 (the grid is obstacles or cannot pass) from the *Pa* set in Definition 3.
> **Step 2:** Calculate the attractiveness of the remaining grids according to formulas (2) and (2′), and sort them according to the attractiveness. Pick the most attractive grid.
> **Step 3:** Select the most attractive grid as the candidate grid. If no other agent selects this gird as the candidate grid, go to Step 4. If other agents select this grid as an alternative grid at the same time, go to *Step 5*.
> **Step 4:** The Agent moves to the alternative grid, enters the next time step, and goes to Step 0.
> **Step 5:** compare the strength (Pt in Definition 3) between itself and the Agents with the same target grid. If its strength is stronger than other Agents, go to Step 4; otherwise remove the grid from *Pa* and go to Step 2.

Each Agent uses the above algorithm for path selection. After all Agents are cleared, the evacuation process ends, and the simulation terminates. The flow chart of the Agent path selection algorithm is shown in Fig. 5:

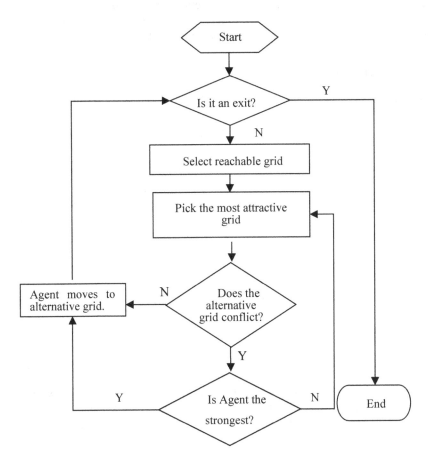

Fig. 5. Flow chart of path selection algorithm for passenger

2.5   Summary of the MAS Model

In this paper, a 0.4 m × 0.4 m grid is used as a unit to simulate the layout of a rail transit station and build a model environment. Define each passenger individual as an Agent, and define the attributes of the individual Agent according to Definitions 1 to 6. According to the rules of the passenger individual behavior in Sect. 2.3.1, the agent's movement rules are defined; according to the external influence factors described in Sect. 2.3.2, Define the rules that affect the Agent's movement; according to Sect. 2.4, the agent routing algorithm is established; finally A MAS model capable of describing passenger behavior characteristics on emergency rail transit stations is established.

3 Establishment of Passenger Group Model

The multi-index statistics of the number of people at this large-scale subway station are conducted. Among them, the number of people at subway exits 1–20 at different peak periods is counted, and the statistics of different age stages and gender of each subway station exit are classified. And the group modeling of these data is carried out.

3.1 Peak Period Crowd Data of Xujiahui Subway Station

As the crowd peak at Xujiahui subway station mainly occurs during the morning, evening, noon and afternoon peak hours, the number of passengers at each of the 20 entrances and exits is counted, as shown in Table 1 below:

Table 1. Population peak data table of Xujiahui subway station (traffic at each exit).

Exit number	Morning peak hours (8:00–9:00)	Noon peak hour (11:00–12:00)	Afternoon peak hour (15:00–16:00)	Evening peak hour (18:00–19:00)
10	3120	1023	1256	2766
15	1765	569	644	1534
16	3524	983	1312	2844
17	1683	503	588	1385
18	1157	261	342	755
19	930	168	186	497
20	1275	249	324	803

Based on the statistics of the passengers at the 20 entrances and exits in 4 time periods, the passengers are further classified according to their age and gender, and are divided into 4 categories: old people, young men, young women, and children. The real data information obtained in this paper is shown in Table 2:

Table 2. Every 10 min of traffic flow at each station (afternoon time).

Exit number	Old people	Young man	Young woman	Children	Sum
10	19	172	244	26	461
15	11	49	96	5	161
16	15	143	152	27	328
17	9	50	83	5	147
18	6	16	32	3	57
19	0	12	18	1	31
20	5	22	2	3	54

3.2 Data Modeling

Based on the differences in passenger age and gender, this paper tracks the walking speed of the above four types of passengers, and obtains the median, mean, and standard deviation of the walking speed of these four types of passengers.

This paper selects the channel of Xujiahui Subway Station Line 1 transfer to Line 9 and Line 11 as the main data collection point. During the half hour between 16:30 and 17:00 during the evening peak hours, the traffic volume of this channel is as high as 4,495. Among them:

Number of people going to Lines 9 and 11: 1921 (person); including the aged and children: 64 (person);

Number of people going to Line 1: 2574 (person), including the aged and children: 81 (person).

The details are shown in Table 3 below:

Table 3. Walking speed V of four types of passengers.

Type	Sample size (person)	Median (m/s)	Mean (m/s)	Standard deviation (m/s)
Young man	50	1.38	1.38	0.37
Young woman	50	1.25	1.28	0.18
Children and old people	100	1.14	1.17	0.37

4 Experimental Results and Analysis

This paper uses Unity3D and Maya modeling tools, and uses C# language to carry out scene modeling and simulation of crowd evacuation in Xujiahui subway station, and realizes the crowd evacuation based on MAS technology and the mentioned algorithm. Through the guidance calculation based on the environmental grid, the waiting platform of xujiahui subway station of Shanghai rail transit is selected as an example. After field investigation, it is found that there are 20 entrances and exits on the upper and lower four floors of the station, and there are five intersections of subway lines. This subway station can basically be regarded as a rectangular connected space or two subway station spaces connected by aisle. This station has a total of 20 exits, the exit at both ends and the middle exit allow the traffic flow of 2 people/second and 1.5 people/second respectively. This paper chooses the off-peak platform situation for analysis. After actual investigation and simplification, it is assumed that there are 100 people in the station, of which 30% are elderly and children. A security checkpoint is selected to conduct a survey on the flow of people.

1) After statistical analysis, the average speed of passing the security inspection machine is 13.61S.
2) The average speed of passing the ticket gate is 4.68 s.

3) Every 10 min, 78 people enter the subway station and 49 people leave the subway station.

After simplification, the environment model is established in this paper, and the specific layout is shown in Fig. 6. In this simulation model, *turtle* is used to simulate passengers, and its color represents different types of passengers, where gray represents young adults and black represents elderly and children. *patch* is used to simulate the grid of the floor plan of the station. Its color represents different grid attributes. The gray indicates the subway exit, and the black indicates the impassable area. In addition, some fire areas with Pass attribute of 0 will be randomly defined to simulate emergency events. The rules of passenger behavior are simulated by certain instructions, and the *direction* switch is used to control whether they are affected by $I(g)$. This example model will compare the passenger's completely self-organized behavior under the influence of herd mentality and the behavioral decision making under the influence of the guiding factor $I(g)$.

Fig. 6. Simulation scene of Xujiahui subway station platform

In the case of only self-organization by passengers, the situation inside the station is shown in Fig. 7 after 20 s. It can be seen that due to the influence of herd mentality, the crowd is more likely to gather, the leftmost exit is more crowded than the other two exits, and the elderly and children are crowded behind the crowd, making the evacuation effect less desirable. Under the influence of $I(g)$, the situation inside the station is shown in Fig. 8 after 20 s. The personnel distribution at each exit is relatively average. The evacuation effect has been greatly improved compared to the completely self-organization by passenger.

Based on the time when passengers leave from the exit in the simulation, a line chart of the number of passengers in the station over time is drawn, as shown in Fig. 9. The black lines and the gray lines represent the time it took the passenger to leave when there was no guidance factor and when there was a guidance factor, respectively.

Fig. 7. No guidance factors

Fig. 8. With guiding factors

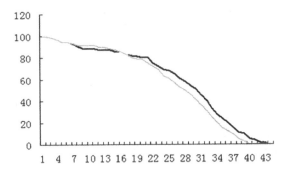

Fig. 9. Schematic of passenger numbers over time

Based on the above simulation results, during the rail transit emergency, passengers will be in a chaotic state for the first 5 s, but within 20 s, individual passengers will choose their own target exits and gradually gather in groups. Under the influence of herd mentality, there will be a state of overcrowding in a certain exit, resulting in a slow overall evacuation efficiency. Therefore, in the process of emergency evacuation, herd mentality is the most important problem to be solved. From the final simulation results, when the passengers are completely self-organized, it takes 43 s for all passengers to leave. Under the influence of $I(g)$ (outside guidance), it takes 38 s for all passengers to leave. This shows that external guidance can make passengers more evenly distributed at each exit, thereby reducing the time required for emergency evacuation.

The passage from area 1 to area 2 in the subway station is shown in Fig. 10 below:

Fig. 10. Channels connecting Area 1 and Area 2

The queuing of passengers in the subway station was also simulated. The following Fig. 11 is the scene of the subway entering the station.

The queuing scene of people in the subway station is shown in Fig. 12 below:

5   Prospection

After a rail transit emergency, due to the concentrated passenger flow and closed space, the behavior of passengers is different from the group behavior in other public places. It can be seen from the model in this paper that passengers will choose the appropriate path to leave according to the actual situation, but due to factors such as physical strength and their own experience, the elderly and children are easily left behind the crowd. At the same time, due to the influence of herd mentality, the crowd will tend to gather together, so one or a part of the exits will be crowded. This phenomenon will affect the evacuation efficiency to a large extent, and will have a great impact on

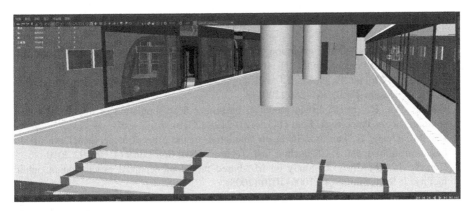

Fig. 11. The scene of the subway entering the subway station

Fig. 12. Scenes of queuing people in a subway station

emergency evacuation. Therefore, more consideration should be given to this factor in emergency evacuation. After adding guidance factors such as broadcast information, this situation will be greatly alleviated, and the speed of passenger evacuation will also be significantly accelerated.

In subsequent studies, practical research will be used to investigate specific data and add more influential factors that are in line with reality, such as the influence of passage blockage caused by fire or earthquake, and the influence of obstacles such as billboards. It will further improve the passenger behavior model and provide a theoretical basis for emergency plan formulation.

References

1. Liu, D.: Analysis on the evolution of group emergencies under different government emergency management modes. Theory Pract. Syst. Eng. (11), 1968–1976 (2010)
2. Helbing, D., Molnar, P., Farkas, I.J.: Self-organizing pedestrian movement. Environ. Plann. B: Plann. Design **28**(3), 361–383 (2001)
3. Gao. P., Xu, R.:. Event driven model in passenger flow simulation of urban rail transit station. Theory Pract. Syst. Eng. (11), 2121–2128 (2010)
4. Liu, M., Zhang, S., Pan, X.: Study on simulation to the process of personnel evacuation based on multi-agent. J. Changchun Inst. Technol. (Nat. Sci. Edn.) (4), 90–92 2010
5. Cui, X., Li, Q., Chen, J.: Study on MA-based model of occupant evacuation in public facility. J. Syst. Simul. (4), 1006–1010 (2008)
6. Franklin, S., Graesser, A.: Is it an agent, or just a program?: A taxonomy for autonomous agents. In: Müller, J.P., Wooldridge, M.J., Jennings, N.R. (eds.) ATAL 1996. LNCS, vol. 1193, pp. 21–35. Springer, Heidelberg (1997). https://doi.org/10.1007/BFb0013570
7. Chen, Y., Wang, B.: A study on modeling of human spatial behavior using multi-agent technique. Expert Syst. Appl. **39**, 3048–3060 (2012)
8. Henein, C.M., White, T.: Microscopic information processing and communication in crowd dynamics. Phys. A **389**, 4636–4653 (2010)
9. Sun, S., Zhao, D., Li, Y.: Micro traffic simulation method based on multi-agent and cellular automata. J. Dalian Maritime Univ. **30**(4), 38–42 (2004)
10. Wang, J.: Research of cruise terminal emergency evacuation strategy base on cellular automaton. Wuhan University of Technology (2011)
11. Wang, W., Wu, S., Cheng, J.: Research on mathematical model of evacuation plan of building. J. Wuhan Univ. Sci. Technol. (11), 155–158 (2010)
12. Zhang, Q., Zhao, G.: Performance-based design for large crowd venue control using a multi-agent model. Tsinghua Sci. Technol. **14**(3), 352–359 (2009)
13. Tan, W., Hu, X., Lei, W.: The implementation of Unity3D based A-star algorithm in the game, 15 November 2018
14. Ji, Y., Sun, Y., Jia, B.: Pedestrian walking speed survey and research in Xi'an, 15 December 2018
15. Yan, F., Jia, J., Hu, Y., Guo, Q., Zhu, H.: Smart fire evacuation service based on internet of things computing for Web3D. J. Internet Technol. **20**(2), 521–532 (2019)
16. Yan, F., Hu, Y., Jia, J., Guo, Q., Zhu, H., Pan, Z.: RFES: a real-time fire evacuation system for mobile Web3D. Front. Inf. Technol. Electron. Eng. **20**(8), 1061–1074 (2019). https://doi.org/10.1631/FITEE.1700548

Automatic Fault Diagnosis of Smart Water Meter Based on BP Neural Network

Jing Lin[1,2,3(✉)] and Chunqiao Mi[1,2,3]

[1] School of Computer Science and Engineering, Huaihua University, Huaihua 418000, Hunan, People's Republic of China
LinjingL99@126.com
[2] Key Laboratory of Wuling-Mountain Health Big Data Intelligent Processing and Application in Hunan Province Universities, Huaihua 418000, Hunan, People's Republic of China
[3] Key Laboratory of Intelligent Control Technology for Wuling-Mountain Ecological Agriculture in Hunan Province, Huaihua 418000, Hunan, People's Republic of China

Abstract. The smart water meter in water supply network can directly affect water production and usage when faults occur. The traditional method of fault detection is inefficient with time lagging, which is not helpful for modernization of water supply system. The capability of automatic fault diagnosis of smart water meter is an important means to improve the service quality of water supply. In this paper, an automatic fault diagnosis method for the smart device is proposed based on BP neural network. And it was applied on Google Tensorflow platform. Fault symptom vectors were constructed using water meter status data and were used to train the neural network model. In order to improve the learning convergence speed and fault classification effect of the network, a method of weighted symptom was also employed. Experimental results show that it has good performance with a general fault diagnosis accuracy of 98.82%.

Keywords: Automatic fault diagnosis · Smart water meter · BP neural network · Tensorflow

1 Introduction

Smart water meters are located at the end of water supply network, and a self-organizing LAN is formed by those meters and data transmission units (DTU) through wired or wireless mode. They are important online flow monitoring equipment for intelligent water supply system at present and in the future. It can perceive the running status of the system in real time by integrating technologies including data acquisition, signal processing, data communication and flow sensing. In order to ensure the effective operation of the system and improve the service quality of the water company, it is a key issue that smart meter works well. So, the more water meters, the higher the fault probability becomes. It is helpful to improve the automation level and service quality if the system has the ability of automatic fault detection for smart meter. The causes of smart meter fault are complex, and different faults have different symptoms.

© Springer Nature Singapore Pte Ltd. 2020
P. Qin et al. (Eds.): ICPCSEE 2020, CCIS 1258, pp. 409–422, 2020.
https://doi.org/10.1007/978-981-15-7984-4_30

Therefore, it is a challenge task to clarify the relationship among the causes, symptoms and fault types, which is the basis of fault diagnosis and elimination.

In water meters' production process, the quality of the signal tube of base meter and the valve components of cold water meter [1, 2], and the correction method for measuring characteristics of smart water meter were studied to ensure that their performance met the requirement of application [3]. In some case, the data collected by the smart water meter can be used to detect the leakage of pipes in time according to the flow variation law of normal and abnormal water [4–6]. However, most of these studies were aimed at performance detection in the production of water meter, or the pipe leakage faults diagnosis by using them. There are few studies on automatic fault diagnosis for smart water meters in practice.

2 Related Work

Fault diagnosis technology on other smart devices has also attracted much attention. Sun et al. diagnosed the Roots flow meter fault of oil tanker in wireless sensor network according to the relationship between instantaneous flow and vibration frequency [7]. A neural network method for optimizing the selection of input voltage was proposed by Mahdavi et al. to diagnose analog and mixed signal circuit faults [8]. The status assessment method for fault location was proposed by Jamali et al. Through the measurement index of the smart meter in a distributed network and error data recognition technology of variable weight matrix, it could find the fault location, and had achieved good performance [9]. Guclu et al. proposed a detection method for fully distributed sensor fault based on the evaluation report of trusted aggregated data from adjacent nodes in intelligent space, which can detect stable faults and intermittent ones with high accuracy [10].

Artificial neural network is widely used in many complex tasks. It does not need to formulate working rules and pre-programming. It has the ability of self-organization, self-learning and self-reasoning by learning to comprehend the inherent laws of things. For example, in cyberbullying detection task, a model of character-level convolutional neural network with shortcuts was proposed to identify whether the text in social media contains cyberbullying [11]. In the field of mechanical manufacture, BP neural network is applied to predict the positioning accuracy error of feed servo system [12].

For the automatic diagnosis of multi-fault types for smart water meters, it is a non-linear classification problem in essentially. It is suitable to use neural network model to predict the fault types according to the symptoms. Therefore, in this paper, automatic diagnosis of multi-fault type for smart water meter based on BP neural network was studied. The status data of water meter is regarded as the fault symptoms. Then the symptom vector is used as the diagnosis sample for training BP neural network to find the non-linear mapping relationship between the fault symptoms and its types. At the same time, weighted symptoms are also used to improve the learning speed and classification effect of BP neural network.

3 Fault Diagnosis Model

3.1 Problem Description

(1) Fault types. Running status information of smart water meter is the original basis of fault diagnosis. In the self-organized LAN, the research on fault diagnosis is based on the following assumptions:

Hypothesis 1:The smart meter nodes can respond to the DTU instruction and return their running status information.

According to hypothesis 1, the communication unit and CPU of microcontroller of the smart nodes work well, and the faults are concentrated on the power supply unit, the valve-controlled unit, the metering unit and the data memory unit. The key units of the smart nodes are shown in Fig. 1.

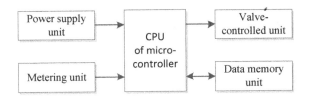

Fig. 1. The diagram of key unit structure of the smart node

The multi-unit concurrent failure is defined as the complex fault. Since the probability of complex fault in practice is small, in order to simplify the problems discussed, the faults are divided into four categories: the battery fault, the measure fault, the valve fault, and the memory fault. Thus, the fault space can be formalized into: $F = \{Y_j \in \mathbb{R}^m, j = 1, 2, \cdots, m\}$, where m indicates the dimension of the fault vector, which is the number of fault types.

(2) Fault symptoms. Symptoms are often their external manifestations when faults occur. A fault usually contains a set of symptoms, and a symptom may also belong to many faults. Several key symptoms are briefly described as below.

Bi-closure of Metering Signal Tubes: The metering signal tubes work in bi-stable mode, which can prevent metering errors caused by flutter and backwater. This ensures that the mechanical count is in conformance with the electronic count. Interfered by the outside magnetic-field, the signal tubes are closed simultaneously, which results in shutting the water valve. The flow sensor converts the mechanical measurement value to the digital value, and outputs it. This model can be described as Eq. (1):

$$f(t) = \alpha_1 \gamma(t) + \alpha_0 + \varepsilon(t) \tag{1}$$

Where α_1 is the magnification of sensors, α_0 is the sensor bias, $\gamma(t)$ indicates the mechanical measurement value of sensor, and $\varepsilon(t)$ is the measurement error of sensors, obeying Gauss distribution: $\varepsilon \sim N(0,1)$.

Battery Fault or Under-Voltage: The energy source of the smart water meter is from the battery, the power supply unit. The stable voltage output can ensure the water meter works well. Therefore, the supply voltage can be used as a symptom to indicate whether the battery is malfunctioning. The under-voltage function of the battery can be described as Eq. (2):

$$v_f(x) = \begin{cases} 1, V_n \leq x \leq M \\ 0, V_l \leq x < V_n \\ -1, 0 \leq x < V_l \end{cases} \tag{2}$$

Where M is the battery rated voltage, V_n, V_l are the lower limit and the upper limit of the battery under-voltage, respectively. When $v_f(.)$ is 1, the battery works normally. If $v_f(.)$ equals 0, it is in the low voltage transition status, and some measures should be taken to restore the voltage as soon as possible to avoid other faults caused by a long time of under-voltage. Otherwise, it is in the under-voltage status, meaning that the smart meter does not work normally and its valve is closed.

Meter's Arrearage: Detecting the insufficient remain water volume of the water meter, the system will automatically alarms and turns off the valve. The water meter arrearage can be formalized as Eq. (3):

$$g(t) = \alpha(t) - \beta(t) \tag{3}$$

Where $\alpha(t)$ and $\beta(t)$ are the cumulative purchase volume and cumulative consumption of water at time t; $g(t)$ is the available amount of water at the time t; $g(t) \leq 0$ represents the arrears of water meters.

Solenoid Valve Fault: Valve-controlled actuators cannot control the opening and closing of the water meter valves in a reasonable time. It may be that the closing valve or the opening valve cannot be in place.

Clock Error: The clock circuit error will lead to inaccurate timing, which affect the correct execution of water meter instructions and make the system data wrong.

Data Memory Error: All kinds of water meter information processed by CPU are stored in memories. Strong magnetic interference or clock error can also cause data loss or data errors.

   After the system status detection instructions are executed by the build-in software and hardware components of the meter, the above symptoms are returned in the form of the meter status information, written as $X = \{x_1, x_2 \cdots, x_N\}$, where N indicates the number of symptoms.

   (3) Problem formalization. Assuming that there are N training samples with fault symptoms, Let (X_i, Y_i) be sample set, $X_i \in \mathbb{R}^N$ is the input samples, $Y_i \in \mathbb{R}^m$ is the desired output of X_i, $i = (1, 2, \cdots, n)$.

   Let $S = \{X_i, i = 1, 2, \cdots, n\}$ be the symptom space, $X_i = \{x_{i1}, x_{i2} \cdots, x_{iN}\}$, $f(X_i) = Y_i$, $Y_i = \{y_{i1}, y_{i2} \cdots, y_{im}\}$, where f(.) is a vector map on S. For $\forall X_i \in S$, we calculate $f(X_i)$ and predict the possibility that it belongs to the label $t_L \in \{0, 1\}$ of the fault Y_i. The purpose of training model is to maximize the possibility $p(.)$ of all fault labels $T_L = (t_L^1, \ldots, t_L^m)$. Given input symptom S and model parameter θ, the problem can be formalized as Eq. (4):

$$\log p(T_L|S; \theta) = \sum_{i=1}^{m} \log p(t_L^i|S; \theta) \tag{4}$$

3.2 BP Neural Model for Fault Diagnosis

BP neural network (BPNN) is a multi-layer feed forward network trained by error back propagation algorithm. It has been widely used since it was proposed in 1986. Its learning process consists of two parts: the positive propagation of information and the backward propagation of errors. The BPNN model with single hidden layer is shown in Fig. 2, including the input layer, hidden layer and output layer. The neurons in input layer receive external fault symptoms as its input and transmit them to the neurons in hidden layer. The hidden layer, the internal processing layer, is responsible for information transformation. The output of the hidden layer neurons is delivered into the output layer neurons as its input. The output layer is in charge of post-processing information and report fault types, completing a learning process of forward propagation. When the actual output does not match the expected output, it begins the reverse propagation stage of error. The output layer error is propagated back to the hidden layer, then to the input layer one by one, according to error gradient descent function to update the value of each layer. The learning and training process of the BPNN continues to iterate until the output error of the model is reduced to an acceptable level, or the number of pre-learning times is reached. Finally, the trained model and the same parameters are used to predict the fault types of the symptom test samples, i.e. fault diagnosis.

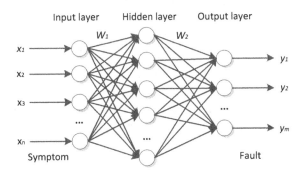

Fig. 2. The BPNN model for fault diagnosis

3.3 Input Layer

At first, the symptom $x = (x_1, \ldots, x_N)$ is embedded in the sample distribution space $D = (d_1, \ldots, d_N)$, where $d_j \in \mathbb{R}^f$ is a column of the Embedding matrix $M \in \mathbb{R}^{V \times f}$. By assigning the weight ($w = (w_1, \ldots, w_N)$, $w_j \in \mathbb{R}^f$) of the symptoms, the model is configured in a certain order. Then the two parts are combined to get the weighted Embedding $e = (x_1 + w_1, \ldots, x_N + w_N)$ of the symptom, as the input of the model. The weighted Embedding contains the importance of input or output of our model.

The purpose of initialization is to preserve the excitation variance in forward and backward propagation. Initializing W_i (i = 1,2) follows the uniform distribution of U (−0.5,0.5).

3.4 Hidden Layer

Determining the number of neurons in the hidden layer is a challenge task. It has an important impact on the generalization ability and training speed of the neural network to reasonably select the number of hidden neurons. For this reason, different solutions have been proposed, such as the estimation method based on the analysis of fuzzy equivalence relation [13], the pruning method using the generalized inverse matrix algorithm [14]. However, there is no particularly effective method to determine the number of hidden neurons in BPNN so far. Empirical Eq. (5) can be used to estimate the number of hidden neurons n_h in BPNN [15]:

$$n_h = \sqrt{n_i + n_o} + d \tag{5}$$

Where d is the constant between 1 and 10, n_i and n_o are the number of neurons in input/output layer, respectively.

Hidden neurons use RELU excitation function. When training, the neural network can automatically adjust the sparse ratio by the gradient descent method of error function to achieve faster training speed [16]. The incentive function, RELU, can be defined as following Eq. (6):

$$relu(z) = \max(0, z) \tag{6}$$

Given the input vector x, the hidden layer performs nonlinear combinatorial operations, which can be formalized as following Eq. (7):

$$h = relu(W_1 x + b_1) \tag{7}$$

Where $W_1 \in R^{H_1 \times N}$ is a parametric matrix, $b_1 \in R^{H_1}$ is a bias vector. H_1 is a hyper parameter, indicating the number of hidden nodes (n_h).

3.5 Output Layer

Similar to the hidden layer, the results of output neurons can be formalized as following Eq. (8):

$$y_o = softmax(W_2 h + b_2) \tag{8}$$

Where $W_2 \in R^{H_2 \times H_1}$ is a parametric matrix, $b_2 \in R^{H_2}$ is a bias vector. H_2 is a hyper parameter, indicating the number of output nodes.

Then the error propagates back from the output layer to adjacent layer in turn, and the weights of each layer are updated by gradient descent method, through Eq. (9) as following:

$$w = w - \eta \nabla E_p(w) \tag{9}$$

Where η (i.e. learning rate) is the step of each update, $\nabla E_p(w)$ is the partial derivative of the output error of the input sample p to the weight w. The error function E_p is defined as following Eq. (10):

$$E_p = \frac{1}{2} \sum_{j=1}^{m} \left(Y_{ij} - y_{oj}\right)^2 \tag{10}$$

And then $w_{ji}^{(l)}$ is considered as the connection weight between the neuron i in the layer $l - 1$ and the neuron j in the layer l. The recursive formula for error propagation from output layer to input layer is defined as following Eq. (11):

$$\delta_j^{(l)} = f'\left(z_j^{(l)}\right) \sum_{k=1}^{m_{l+1}} \delta_k^{(l+1)} w_{kj}^{(l+1)} \tag{11}$$

Where f(.) is incentive function, m_l is the number of neurons in the layer l, $z_j^{(l)}$ is the input of the neuron j in the layer l, $\delta_j^{(l)}$ indicates partial derivatives of error functions for the neuron j in layer l, namely, $\delta_j^{(l)} = \frac{\partial E_p}{\partial z_j^{(l)}}$. Then we can get the gradient $(\nabla E_p(w))$ of error for each layer.

Finally, the global error, E_T, is calculated by Eq. (12), where n is the total sample size. If the error reaches the preset accuracy or the times of learning is greater than the given maximum times, the learning process of the algorithm is completed; otherwise, the next sample is taken to continue to learn.

$$E_T = \sum_{p=1}^{n} E_p \tag{12}$$

4 Experiments

4.1 Experimental Environment

Settings for Experiment: Experiments run on a server, IBM xSeries 365, with Xeon MP 2.7G CPU, 8G DDR and SCSI 2 TB HD for developer. It also acts as the database server using MySQL server 5.6 DBMS, and as the web server on which Tomcat 6.5 is installed. Windows 7, JDK 1.7 and Python 3.5 are installed on the client with 2.0G CPU, 8G DDR and 600 GB HD for testing. Tensorflow 1.2.0+Keras 1.2.2, a flexible and cross-platform machine learning framework, is selected to implement

BPNN, and Matplotlib 2.0.0 is used to automatically draw the distribution map (i.e., Figure 4 and Fig. 6) of diagnostic or predictive score.

Data Set: Eight kinds of important status is selected as symptoms for the water meters fault diagnosis, and the status format of the water meter is defined as shown in Table 1, in which Y represents the fault type. In the traditional fault removal mode, the value of Y is filled in the system by repairman after troubleshooting. Therefore, Y, as the Gold value, is true and reliable. S_i (i = 1,2,...,8) is regarded as a symptom of the fault, its value is 0 or 1, 0 means normal, and 1 indicates a corresponding fault symptom.

Table 1. Information format of water meter running status.

Symptom	Description
S1	Opening of valve not in place
S2	Closure of valve not in place
S3	Meter's arrearage
S4	Bi-closure of metering signal tubes
S5	Clock error
S6	Battery under-voltage
S7	Cipher text error
S8	Forced valve-shut
Y	Actual fault, filled in after fault removal

The samples in experiment are from the status records of user's water meter in the water supply system. The 10-fold cross validation method is used to evaluate the model. 20% of the samples are used as the development and test set, and the remaining 80% are used as the training set. In order to improve the effectiveness of the model, all kinds of fault symptom samples in each dataset are made evenly distributed.

Evaluation Index: The accuracy (P) of fault diagnosis is used to evaluate our model in the experiment. P is calculated according to Eq. (13), where TP is the number of fault samples correctly classified, FP indicates the number of fault samples incorrectly classified.

$$P = TP/(TP+FP) \tag{13}$$

The comprehensive diagnosis accuracy (P_T) of the model takes the average of P for each type of fault. Generally, the larger P_T is, the better our method is.

Parameters: In view of the above-mentioned analysis of the fault symptom and fault type, our neural model has 8 neurons in the input layer (namely, $n_i = 8$), 4 in the output layer (namely, $n_o = 4$). The size of hidden neurons (n_h) is calculated according to formula (5). For example, let d = 7, then $n_h = 11$. Equation (14) is used to dynamically adjust the learning rate η to avoid slow convergence and local minimum in error propagation [17].

$$\eta(k+1) = \eta(k) - \tau \frac{\Delta E_T(k)}{E_T(k)}, 0 < \tau < 1 \tag{14}$$

Where $\eta(k+1)$ is the learning rate of output layer during the $k+1$ time of learning; τ is constant and takes the value of 0.5. The model parameters, W1 and W2, are randomly initialized according to the uniform distribution of $U(-0.5, 0.5)$. The expected accuracy of error is 0.0005, and the maximum number of iteration is 20000. For it is found that the learning process of our models ended after about 19000 iterations in our experiment at the latest.

Preprocessing: In the status samples of the water meter, all of the symptom values, Si (i = 1,2,···,8), are 0, which indicates that the water meter works well. When S3 and S8 are 1, and the others are 0, the water meter is in arrearage, not a fault. Therefore, the two kinds of "noise" data are removed from the samples. Finally, 400 symptom samples are selected. The sample distribution for the actual fault type is shown in Table 2.

Table 2. Distribution statistics of fault samples

Fault	Number of samples
Battery fault	140
Measure fault	120
Valve fault	80
Memory fault	60
Total samples	400

4.2 Analysis of Experimental Results

(1) Prediction Score. This score represents the probability of predicting the occurrence of a type of fault. After training our neural model on the training data set, 40 samples from each type of fault are selected as the test set. The effect of our model is measured by the score of predicting a certain kind of fault. The test results are shown in Fig. 3.

The battery fault, the valve fault, the measure fault and the memory fault are represented by red, green, blue and yellow lines, respectively. The highest scores predicted by our model have a range of [0.9953, 0.9999], which shows that our model has good prediction effect. However, because the fault types in each test set are the same, the predictive score cannot fully represent the accuracy of the prediction (i.e. fault diagnosis) of the model.

(2) Prediction accuracy. Next, the 10-fold cross validation method is used to predict the fault type with the trained model on the test set. The average value of five experiments is taken as the final prediction result, and the performance of our model is

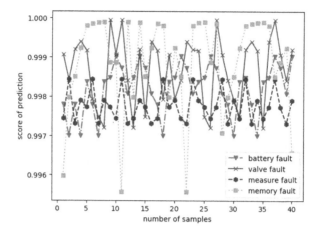

Fig. 3. The prediction scores of different test sets (Color figure online)

measured by the fusion matrix shown in Fig. 4. The numbers on the bottom X-axis and the left Y-axis of the fusion matrix represent the labels of the actual fault types and predictive ones. The numbers 1, 2, 3 and 4 are the labels of the battery faults, the valve faults, the measure faults and the memory faults, respectively. The right column and the bottom row (grey cell) of the matrix are the statistics of prediction accuracy and error rate for each type of fault. They are showed by upper blue number and lower red number respectively. The diagonal elements (green cells) of the upper left 4 × 4 matrix represent the number of samples predicted correctly and its percentage in all samples, while the other non-diagonal elements (pink cells) indicate the number and percentage of samples predicted incorrectly. The diagonal elements in the lowest right corner represent the overall accuracy of our model for fault diagnosis or prediction.

	1	2	3	4	
1	134 / 33.5%	6 / 1.5%	0 / 0%	0 / 0%	95.71% / 4.29%
2	3 / 0.75%	117 / 29.25%	0 / 0%	0 / 0%	97.5% / 2.5%
3	0 / 0%	0 / 0%	80 / 20%	0 / 0%	100% / 0%
4	0 / 0%	0 / 0%	0 / 0%	60 / 15%	100% / 0%
	97.81% / 2.19%	95.12% / 4.88%	100% / 0%	100% / 0%	98.23% / 1.77%

Prediction Type of the Model (left axis)

Standard Reference Type of the Model

Fig. 4. The confusion matrix of BPNN (Color figure online)

As seen in Fig. 4, the diagnostic accuracy of the measure fault and the memory fault is 100%. The reason is that the two kinds of fault symptoms are relatively independent and the symptom vectors are highly differentiated. The diagnostic accuracy of the battery fault and the valve fault is more than 95.12% and the error rate is less than 4.88%. The possible reason of the misdiagnosis is that the two kinds of faults have the same symptom (such as the forced valve-shut), which will lead to a smaller angle between the symptom vectors and a lower degree of discrimination, resulting in a diagnostic error. The overall accuracy of our model for fault diagnosis reaches 98.23%, which shows that the model predictive performance performs well in the experiment.

(3) Adaptive ability to complex fault diagnosis. In the actual application, the complex faults of water meters rarely occur, which means that the probability of simultaneous faults of multiple units in a single meter is very small. So far, in our system, there are no such cases in the symptoms and fault types of the smart meters. However, in order to test whether the neural model has adaptive ability to complex fault diagnosis, we combine different types of existing symptom samples into new inputs of the model to predict the faults, and observe the model's output, as shown in Table 3. Take generating the input, X, as an example, assuming that X1 and X2 are the samples for generating the input, then $X = X1 \wedge X2$, where $\wedge$ means bitwise AND operation.

Table 3. Examples of adaptive ability of the model to complex fault diagnosis

X (i.e., input)	Samples for generating X	Type of samples	Y (i.e., Output)
[0, 1, 1, 0, 1, 0, 1, 0]	[0, 1, 0, 0, 0, 0, 0, 0]	2	[0.0001, **0.5192**, 0, **0.9240**]
	[0, 0, 1, 0, 1, 0, 1, 0]	4	
[1, 0, 1, 1, 0, 1, 0, 1]	[1, 0, 1, 0, 0, 1, 0, 1]	1	[**0.9167**, 0.0006, **0.7916**, 0]
	[0, 0, 0, 1, 0, 0, 0, 0]	3	

The sample combination of fault type is described in Table 3, X is used as the input of our model to predict the fault, then the corresponding components of the output vector Y show the higher prediction scores (e.g., bold part). It shows that the model can predict the fault type of the combination, also has a certain adaptive ability to complex faults diagnosis. However, the prediction/diagnosis results of model are also random when the vectors outside the training sample set are used as input. In order to diagnose complex faults better, further researches are needed.

(4) Performance test of weighted BP neural network. In the above-mentioned BPNN, the corresponding fault symptoms of input vector elements are taken as 1 when symptoms appear, otherwise 0. In this case, we pay too much attention to the contribution of 1-value symptoms to fault diagnosis, and neglect the role of 0-value symptoms. In fact, the causes of the faults and symptoms are interlaced and complex. All kinds of symptoms can have an impact on fault diagnosis with different influence degrees.

Therefore, the influence degree is regarded as the weight, namely, the importance of symptoms. It is embedded into the input symptoms vector of BPNN to obtain the weighted symptoms Embedding, and the new neural model called wBPNN is trained by the weighted Embedding set corresponding to the original training set. The weight of symptoms, w_t, is calculated according to Eq. (15).

$$w_t(x) = \frac{(1 - \sigma) * I(x) - k}{k} + \frac{\sigma * (1 - I(x))}{N - k} \tag{15}$$

Where N is the number of elements in the symptom vector, and k is the number of 1-value elements. $I(.)$ is the indicator function. When the variable is 1, the function outputs 1, otherwise outputs 0. σ is an adjusting factor of the influence degree of 0-value elements, i.e., the overall contribution of 0-value elements to fault diagnosis. σ is often determined according to experience or experiment. The maintainer's suggestion was accepted in our experiment, and we let σ equal 0.2.

Fifty symptom samples were randomly selected, and then they were duplicated into two copies, one copy as A and the other weighted with Eq. (15) as B. Then trained BPNN and wBPNN were used to predict A and B respectively. The distribution of prediction scores was shown in Fig. 5.

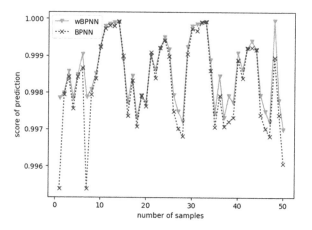

Fig. 5. Diagnostic score distribution map of wBPNN and BPNN

Experiment on prediction accuracy of wBPNN is similar to that of BPNN. It can be seen in the part (2) of this session. The fusion matrix of wBPNN is shown in Fig. 6.

As shown in Fig. 5, the highest score predicted by wBPNN model is distributed in [0.9969, 0.9999]. Its maximum lower limit score is 0.16% points higher than that of BPNN. Comparing Fig. 4 and Fig. 6, it can be seen that the number of misdiagnosis samples in wBPNN is 3 fewer than that in BPNN, and its overall accuracy of fault diagnosis is 0.59% points higher than that in BPNN, reaching 98.82%. It shows that the classification effect of hyper-plane in the wBPNN model is better.

In addition, during observing the error of the test set reaching to the expected error accuracy ($\varepsilon = 0.0005$), we found that the error dropped to 0.12%, at about 9000

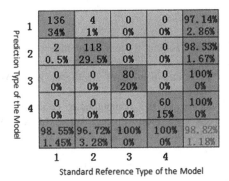

Fig. 6. The confusion matrix of wBPNN

iterations in BPNN, while only at 5000 iterations in wBPNN. It shows that the learning cycle of wBPNN is greatly shortened.

In conclusion, the weighted symptoms Embedding improves the learning convergence speed and fault diagnosis accuracy of wBPNN.

5   Conclusions

The ability of automatically diagnosing device faults is one of the important symbols of intelligence and automation level of water supply system. In this study, the status data of smart water meter in the system is first transformed into fault symptom vector. Then the vector is used to train BP neural model to realize the automatic diagnosis of various types of the faults of smart water meter. By adding the weight into fault symptoms and mapping them to the input vector space, the learning convergence speed and the fault diagnosis accuracy of wBPNN trained by weighted symptoms Embedding are improved. Its overall fault diagnosis accuracy is 0.59% points higher than that of BPNN, reaching 98.82%. The results show that the performance of the model is outstanding. This method can also be applied into automatic fault diagnosis of other intelligent meters, such as ammeters and gas meters.

Although the neural model has a certain adaptive ability to complex fault diagnosis on trial, the model prediction results become random when the input is outside the weighted symptom Embedding. Therefore, it is an important future work to do more effective diagnosis of complex faults. In addition, since the model parameters in the experiment are based on empiric, they also need a further experimental optimization.

Acknowledgements. This work was supported in part by the Huaihua University Double First-Class initiative Applied Characteristic Discipline of Control Science and Engineering, the Educational Cooperation Program of Ministry of Education of China (No. 201801006090), and the Hunan Provincial Natural Science Foundation of China (No.2017JJ3252).

References

1. Guobin, P.: Signal Tube Test Tooling System of Intelligent Water Meter. State Intellectual Property Office, PRC (2016)
2. Guobin, P.: Valve Test Tooling System of Intelligent Water Meter. State Intellectual Property Office, PRC (2016)
3. Ling, Y.: Risk and control of the computerized correction method for measuring characteristics of intelligent water meter. Process Autom. Instrum. **37**(2), 99–102 (2016)
4. Di, X.X., Guo, K., Tang, P., Tian, C., Wang, J., Xu, C., Yuan, Y.: Water leakage detection based on smart electricity meter. Patent Application Publication, United States (2017)
5. Hsia, S.C., Hsu, S.W., Chang, Y.J.: Remote monitoring and smart sensing for water meter system and leakage detection. IET Wirel. Sens. Syst. **2**(4), 402–408 (2012)
6. Xiyong, Z., Yue, P.: Research on the identification of real-time water leakage based on internet-based smart water meters and complex event processing. Group Technol. Prod. Modernization **32**(1), 23–28 (2015)
7. Sun, Y., Liu, X., Chen, X., Sun, Q., Zhao, J.: Research and application of a fault self-diagnosis method for roots flowmeter based on WSN node. Wirel. Pers. Commun. **95**(3), 2315–2330 (2017). https://doi.org/10.1007/s11277-017-4104-8
8. Seyyed Mahdavi, S.J., Mohammadi, K.: Evolutionary derivation of optimal test sets for neural network based analog and mixed signal circuits fault diagnosis approach. Microelectron. Reliab. **49**(2), 199–208 (2009)
9. Jamali, S., Bahmanyar, A., Bompard, E.: Fault location method for distribution networks using smart meters. Measurement **102**, 150–157 (2017)
10. Guclu, S.O., Ozcelebi, T., Lukkien, J.: Distributed fault detection in smart spaces based on trust management. Procedia Comput. Sci. **83**, 66–73 (2016)
11. Nijian, L., Guohua, W., Zhen, Z., Yitiao, Z., Kim-Kwang Raymond, C.: Cyberbullying detection in social media text based on character-level convolutional neural network with shortcuts. Concurrency Comput. Pract. Exp. (2020). https://doi.org/10.1002/cpe.5627
12. Chao, D., Yousheng, Q., Ju, W., Yao, X., Chaoqun, D.: A forecasting method of positioning accuracy for CNC machine tools feed system based on bp neural network. J. Vibr. Meas. Diagn. **37**(3), 449–455 (2017)
13. Yunjian, L.: A method to directly estimate the number of the hidden neurons in the feedforward neural networks. Chin. J. Comput. **22**(11), 1204–1208 (1999)
14. Fangju, A., Xiaofang, L.: A new feedforward neural network pruning algorithm. J. Sichuan Univ. (Nat. Sci. Edn.) **45**(6), 1352–1356 (2008)
15. Jiani, F., Zhenlei, W., Feng, Q.: Research progress structural design of hidden layer in BP artificial neural network. Control Eng. China **12**, 109–113 (2005)
16. Krizhevsky, A., Sutskever, I., Hinton, G.: ImageNet classification with deep convolutional neural networks. In: International Conference on Neural Information Processing Systems, pp. 1097–1105. Curran Associates Inc. (2012)
17. Guoyi, Z., Zheng, H.: Improved BP neural network model and its stability analysis. J. Central South Univ. (Sci. Technol.) **42**(1), 115–124 (2011)

Prediction of Primary Frequency Regulation Capability of Power System Based on Deep Belief Network

Wei Cui, Wujing Li[✉], Cong Wang, Nan Yang, Yunlong Zhu,
Xin Bai, and Chen Xue

Northwest Control Center of State Grid Corporation, Xi'an 710048, China
gyyx.bh@foxmail.com

Abstract. The primary frequency response ability plays a crucial role in the rapid recovery and stability of the power grid when the grid is disturbed to generate a power imbalance. In order to predict the primary frequency control ability of power system, a new model is proposed based on deep belief networks. The key feature of the proposed model lies in the fact that it considers three key factors, i.e., disturbance information, system state feature, and unit operation mode. Through this way, it predicts the primary frequency control ability of the power system accurately. The simulation results on real power system data verify the feasibility and accuracy of the proposed model.

Keywords: Primary frequency control · Power system · Deep belief neural network · Prediction

1 Introduction

In recent years, with the construction of ultra-high voltage transmission system, penetration of distributed power sources, and the commissioning of various electrical equipment, the complexity of the power grid has continued to rise. As a result, the safe and stable operation of power systems has been subject to various disturbances. When the power system is disturbed, there is an imbalance between the generation and load demand. Theoretically, primary frequency control scheme will quickly respond to restore the system frequency, thus keeping its safe and stable operation. However, the power system's primary frequency adjustment capacity is, in fact, limited. In this sense, if the primary frequency control scheme cannot quickly restore the power balance, then the system operator cannot make a correct prediction of the system disturbance and intervene in a timely manner. This may cause unit shutdown, load disconnection, and other undesirable consequences, which will develop into a chain failure, causing a large scale power outage. The primary frequency response capability of the power system mainly depends on the operating status and parameter settings of the units in the system, which cannot be directly measured. After the disturbance, the accident develops rapidly, and it is difficult for dispatchers to make correct judgments and interfere in a short time.

© Springer Nature Singapore Pte Ltd. 2020
P. Qin et al. (Eds.): ICPCSEE 2020, CCIS 1258, pp. 423–435, 2020.
https://doi.org/10.1007/978-981-15-7984-4_31

At present, research on primary frequency modulation mainly focuses on performance evaluation [1, 2], characteristic analysis [3–5], control strategy [6, 7], and participation of renewable energies [8–10]. The research on primary frequency regulation capability is less concerned. The research objects of the primary frequency response capability are mainly divided into two types: the primary frequency modulation capability of the unit and the primary frequency modulation capability of the power system. For the study of the unit's primary frequency response capability, reference [11] based on the information of a phasor measurement unit (PMU), measured the generation and frequency changes of the unit in real time and synchronously, calculated the parameters of the frequency modulation online, and evaluated the primary frequency modulation of the generator. performance. However, there are many units in the power system, and each unit's frequency regulation capability is different. The advantages and disadvantages of a unit's frequency regulation capability cannot represent the frequency regulation capability of the whole power grid. To address this, many scholars have conducted extensive research on the primary frequency regulation capability of the power system. Reference [12] defined the primary frequency control ability (PFCA) of the power grid, and proposed a mathematical expression of the primary frequency control ability of the power grid in combination with the variance. However, this method can only obtain the historical frequency response ability of the power grid, but cannot accurately reflect the power grid's real-time primary frequency regulation capability. Reference [13] was based on the historical data of the unit in the wide area measurement system (WAMS), combined with the time dimension and the frequency difference dimension, thus accurately predicted the primary frequency response capability of the single unit in the power grid. The predicted values of all the single machines are used to compute the prediction of the primary frequency response capability of the whole power grid. However, this method performs prediction according to a fixed frequency difference, and the prediction accuracy is limited. Furthermore, this method suffered from high computation complexities.

With the development of artificial intelligence, various algorithms are used in power systems [14]. Compared with the traditional neural network algorithm, the deep learning algorithm has more powerful feature extraction ability and can dig the deep connection between the data, which is especially suitable for the processing of massive data in the power system. Reference [15] applied deep belief networks to predict the frequency curve after power system disturbance, and the results show that it has more advantages than other shallow networks. Reference [16] applied deep belief neural network to predict and predict the transient stability of the power system. The results show that it has higher accuracy than conventional networks. The deep belief network has a flexible structure, is easy to expand, and has higher prediction accuracy for time series data [17].

Based on the above considerations, this paper proposes a new deep learning based model, which predicts the frequency response capability of the system quickly and accurately. To do so, we explore the application of deep neural network to the frequency response capability of the power system, and developed a prediction model based on deep belief network (DBN). Finally, the real data from actual power system is used to verify the effectiveness of the proposed model.

2 Deep Belief Network

In literature [18], a deep belief network was first proposed, which is a stack of multi-layer restricted Boltzmann machines (RBMs). It is often used for feature extraction and data classification. The standard RBM consists of a visible layer (input layer) and a hidden layer (output layer). There is no connection between nodes in the same layer, and nodes in different layers are connected to each other. The structure of a standard RBM is shown in Fig. 1.

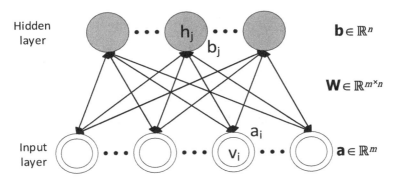

Fig. 1. Structure of a standard restricted Boltzmann machine.

Suppose there are m neurons in the visible layer of RBM, where v_i represents the i-th neuron, the hidden layer has n neurons, and h_j represents the j-th neuron. Its energy function is:

$$E(v, h|\theta) = -\sum_{i=1}^{m} a_i v_i - \sum_{j=1}^{n} b_j h_j - \sum_{i=1}^{m} \sum_{j=1}^{n} h_j W_{ij} v_i$$
$$= -\mathbf{a}^{\mathrm{T}} \mathbf{v} - \mathbf{b}^{\mathrm{T}} \mathbf{h} - \mathbf{h}^{\mathrm{T}} \mathbf{W} \mathbf{v} \tag{1}$$

In (1), $\theta = \{W_{ij}, a_i, b_j : 1 \leq i \leq m, 1 \leq j \leq n\}$, W_{ij} is the weight between the visible neuron i and the hidden neuron j, a_i is the bias of the visible neuron i, and b_j is the bias of the hidden neuron j, respectively. According to the energy function, the joint probability distribution of RBM can be obtained, as follows

$$P(v, h|\theta) = \frac{e^{-E(v,h|\theta)}}{\sum_{v,h} e^{-E(v,h|\theta)}} \tag{2}$$

The restricted Boltzmann machine obtains the maximum likelihood probability of the model through the given visible layer data, this way, the parameters of the RBM model are obtained. The specific formula is given by

$$\max P(v|h, \theta) = \sum_h e^{-E(v,h|\theta)} \tag{3}$$

The training procedure of deep belief network consists of two phases, i.e., unsupervised pre-training and supervised parameter fine-tuning. The whole training process of the DBN model is shown in Fig. 2 [19].

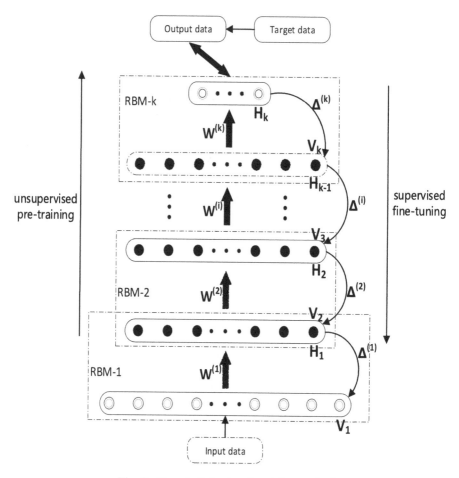

Fig. 2. Deep belief network training method.

Unsupervised pre-training initializes the network weight parameters layer by layer from low to high, this way, the network can get better initial value and avoid local optimization. First input the input data into RBM-1 for pre-training. Then use the activation probability of the RBM-1 hidden layer as the input for RBM-2, and so on to pre-train each RBM again, and finally get the neuron's bias matrix. Finally, the bias matrix $\mathbf{b}^{(k)}$ of the neuron and the weight matrix $\mathbf{W}^{(k)}$ between the neurons in each layer are obtained.

The objective of supervised parameter is to fine-tune the network weight parameters layer by layer from high to low. First compare the output data of the last layer of RBM in the unsupervised pre-training with the target data in the data set. Then use the back-propagation algorithm to fine-tune the network weight parameters from high to low. Finally, the DBN bias matrix $\tilde{\mathbf{b}}^{(k)} = \mathbf{b}^{(k)} + \Delta^{(k)}$ and Weight matrix $\tilde{\mathbf{W}}^{(k)} = \mathbf{W}^{(k)} + \Delta^{(k)}$, where $\Delta^{(k)}$ is the adjustment amount according to the k-th layer weight parameter after supervised learning.

3 DBN-Based Prediction Model

3.1 Network Model Design

The prediction of the primary frequency response capability of the power system is essentially a regression problem. Therefore, this paper extends the deep belief network by adding a single-layer BP neural network. The output layer uses a linear transfer function to output the primary frequency modulation capability of the power system for regression prediction. Its structure is shown in Fig. 3.

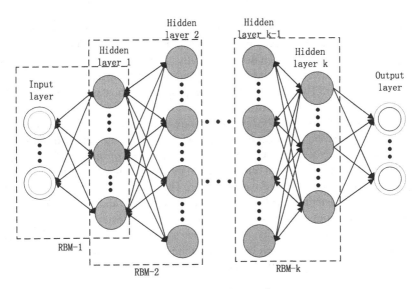

Fig. 3. Deep belief neural network structure.

After the power system is disturbed, there exists a power imbalance between generation and load demand. The grid will respond immediately after a frequency modulation. Then, under a frequency response action, the active power of the system will change with time, that is, the active power will change monotonically in a short time to restore the system frequency. Relevant characteristics such as the amplitude and speed of the active power change curve can reflect perturbation information, system characteristics, unit performance, and other related information, which can reflect the primary frequency regulation capability of the power system. Therefore, in this paper, the system 60 s active power variation curve, the steady state value of the active power, the maximum active power increase during the frequency modulation operation, and the time when the maximum active power increase occurs are used as output vectors to characterize the primary frequency modulation capability of the power system.

There are many factors that affect the primary frequency modulation ability of the system, such as disturbance degree, system state characteristics, unit operating state and parameter setting, etc. To ensure the accuracy of the system's frequency modulation capability prediction, this paper selects a variety of main variables that affect the system's primary frequency modulation.

The rotor speed of the generator is related to the balance of the unit's input and output active power, which is related to the unit's active power curve. The generator rotor motion equation is:

$$T_j \dot{\omega} = P_m - P_e - D\omega \tag{4}$$

In (4), T_j is the generator inertia time constant; P_m is the mechanical power; P_e is the electromagnetic power; D is the damping coefficient of the generator; ω is the angular frequency.

When a power disturbance occurs in the system, the electromagnetic power of each generator changes and the mechanical power remains basically unchanged, so the active power of the system responds dynamically. For different disturbance amounts, the electromagnetic power of each unit varies, and the resulting active power variation curve is also different. Therefore, the power gap representation perturbation information is selected as the input feature vector.

When a disturbance occurs, the operating mode of the power system also affects the active power response curve. It can be seen from (4) that both the generator inertia time constant and frequency will affect the power response curve of the unit. For the entire grid, the active power curve is not only related to the operation mode of the power grid, but also to the operation mode of the generator. Therefore, the grid frequency, active load, inertia time constant of each generator, standby power, and total system standby power of the system immediately after the disturbance are selected as the input vectors. Among them, the reserve capacity is equivalent to the maximum limit of the active power output of the system, which can increase the boundary conditions for the active power of the system. The specific formula is shown in formula (5):

$$\begin{cases} p_{r+}^i = p_{max}^i - p_0^i \\ p_{r-}^i = p_0^i - p_{min}^i \end{cases} \tag{5}$$

In (5), P_{r+}^i and P_{r-}^i are the reserve capacity of the system up and down respectively at the i-th event. P_{max}^i and P_{min}^i are the maximum and minimum active power that the system can quickly respond to at the ith event. P_0^i is the active power of the system at the i-th event.

3.2   Data Set Acquisition

The data in this study comes from an actual power grid. The amount of collected data is large. The original data has problems such as incorrect data and invalid data. Direct training using these data may yields large errors, slow training speed, and poor training results. Therefore, the data needs to be processed. To get a valid data set. Figure 4 shows the process of obtaining an effective data set.

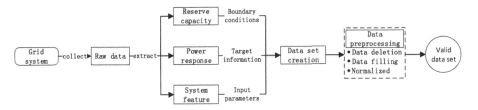

Fig. 4. Acquisition process of valid data set.

Perturbed event set acquisition

The historical data of the power grid system includes both the data at the time of disturbance and the data at the steady state. The historical data needs to be filtered to obtain the data of the unit and the entire network after the disturbance. Taking a large frequency event as a unit, the raw data of the power system and each generator set before and after the disturbance are collected, and then the characteristic parameters are extracted from the raw data to form a data set. However, historical data is susceptible to various influences, and there are many momentary disturbances, data bad points, high-frequency noise, etc. that the primary frequency modulation loop does not operate at one time. Therefore, it is determined that an effective disturbance is that the frequency offset exceeds the primary frequency modulation dead band and lasts for a period of time. That is, the selection rule of a valid disturbance is:

$$
\begin{cases}
\left| f_{i-1} - f_{ref} \right| \leq \Delta f \\
\left| f_i - f_{ref} \right| \geq \Delta f \\
\quad \vdots \\
\left| f_{i+N} - f_{ref} \right| \geq \Delta f
\end{cases}
\tag{6}
$$

Among them, f_i is the grid frequency at time i; f_{ref} is the standard frequency of the grid at 50 Hz; Δf is the effective disturbance frequency deviation, and the value is 0.07 Hz.

There are many factors that affect and characterize the frequency modulation capability of the power grid. All extraction into the data set will cause problems such as long model training time and over-fitting. Therefore, this article only extracts the main factors, that is, using system feature quantities as input parameters, power response as target information, and reserve capacity as boundary conditions. Among them, the system characteristic quantity includes the active load, inertia time constant, grid frequency and power shortage of the unit when the disturbance occurs, and reflects the static characteristics of the system and the unit when the disturbance occurs, and the degree of the deficit. The power response characteristics include three parameters: the average rate of active power change, the maximum active power increase, and the time when the maximum active power increase occurs, which reflects the response state of the system's active power after the disturbance occurs. The use of unit reserve capacity and total reserve capacity as boundary conditions reflects the limits of the frequency regulation capability of the power system in a short time.

Data processing

The collected data set has problems such as missing data, repeated data, and invalid data. Direct use will affect the prediction accuracy of the network model. Therefore, in order to ensure the accuracy and validity of the data, the data set needs to be pre-processed to obtain an effective data set. For the collected duplicate data and invalid data, these data can not help the training of the model, and even increase the error, so they are deleted directly. For the data missing problem, if the number of missing data in a disturbance event is too large, it is deleted directly; if the number of missing data is small, the data is filled to complete the data. This article uses the mean of the same indicator to fill in missing values, and fills in based on the mean of the series of data before and after the missing value.

The filled and deleted data sets have obvious differences in magnitude. Direct training will cause the network model to converge slowly and be prone to numerical problems. Therefore, the data set needs to be normalized to keep all the data in the same dimensions and intervals, remove the unit limitation of the data, convert it to a dimensionless pure value, and finally map it to the (0, 1) interval to improve the model. Accuracy and convergence speed.

4   Simulation Case Analysis

4.1   Simulation Setup

This paper uses the historical data from an actual power grid, and after processing to obtain a valid data set, it is divided into a training set, a validation set, and a test set according to the ratio of 6: 1: 1. The training set is used to train the network model and determine the parameters of the neural network. The validation set is used to judge the performance of the model after training, and to adjust the hyperparameters such as the number of hidden layers and the number of hidden layers in the network. The test set is used to evaluate the performance of the network model. After processing, 2400 valid data samples were obtained, of which 1800 were used as the training set, 300 as the validation set, and 300 as the test set.

For the evaluation index of model performance, this article uses mean relative error (MRE) and root mean square error (RMSE) to characterize the pros and cons of the model performance, and measures the effect of the model from unit metrics. The error calculation formula is shown in Eqs. (7) and (8):

$$MRE = \frac{1}{m} \sum_{i=1}^{m} \frac{|\hat{y}_i - y_i|}{|\hat{y}_i|} \tag{7}$$

$$RMSE = \sqrt{\frac{1}{m} \sum_{i=1}^{m} (\hat{y}_i - y_i)^2} \tag{8}$$

Here m is the amount of data involved in model training or verification in the data set, y_i is the model prediction value of sample i in the data set, and $\hat{y}_i$ is the actual target value of sample i in the data set.

4.2 Model Parameter Analysis

There is only one input layer and one output layer in our DBN, and the number of hidden layers needs to be set according to specific problems. For the number of neurons in the input layer and the output layer, set them to be equal to the dimensions of the input vector and the output vector, respectively. This paper selects 56 direct-tuning units that often participate in one frequency adjustment, so the input eigenvector is the reserve capacity and inertia time constant of the 56 units. There are 116 frequencies, active loads, power deficits, and total reserve capacity in the entire grid. The output characteristic vector is the active power change curve of the system for 60 s, the steady state value of the active power, the maximum active power increase during the frequency modulation operation, and the time corresponding to the maximum active power increase. Therefore, the number of neurons in the input layer and the output layer is set to 116 and 63 respectively. The contrast divergence algorithm is used in the unsupervised learning stage. The learning rate is 0.06, the momentum is 0.5, and the number of iterations is 300. The supervised learning stage uses back propagation. Algorithm, the activation function uses the sigmoid function, the learning rate is 0.03, and the number of iterations is 400. Among them, the selection of parameters such as the number of iterations is mainly through repeated simulation, and the value with the minimum error is selected. For the setting of the number of hidden layers, in this paper, under the condition that the number of hidden layers is not more than 5, each neuron is fixed at 50 neurons, and the network parameter structure containing different hidden layers is constructed in turn. Comparison is conducted to obtain a most effective network architecture. The comparison results are shown in Table 1.

Table 1. Maximum active power incremental error results for different hidden layer layers.

Number of hidden layers	Number of each hidden layer unit	RMSE/MW	MRE/%
1	50	64.7	4.79
2	50, 50	50.2	3.51
3	50, 50, 50	41.3	3.05
4	50, 50, 50, 50	32.6	2.31
5	50, 50, 50, 50, 50	25.1	1.76

It can be seen from Table 1 that when the hidden layer is 5 layers, the RMSE and MRE are lower than 1 to 4 layers, and the prediction effect is the best.This shows that when the number of hidden layers does not exceed 5, the five-layer structure DBN network model has the highest degree of fit to the data, and the model's prediction effect is relatively best. Therefore, this paper uses a network model with a hidden layer structure of 5 layers.

4.3   Analysis of Prediction Results

This paper compares the performance of the DBN model with the standard deep neural network model and the conventional shallow network model, and analyzes the performance of the prediction model. First select and construct three network models of DBN, DNN and BPNN. According to the determined network structure parameters, the number of DBN model layers is set to 6 and the number of neurons in each layer is 116, 100, 50, 50, 50, 20, 63. The structure of the DNN network model is the same as that of the DBN, but without unsupervised pre-training, only the back-propagation algorithm is used for supervised learning, the activation function uses the sigmoid function, the loss function uses the mean square error function, and the number of iterations is 300. The BPNN model has a hidden layer, the number of hidden neurons is 100, and the activation function, loss function, and iteration times are consistent with the DNN network. Through the learning and testing of the training and test sets, the specific prediction results of the three models are shown in Tables 2, 3 and 4 and Figs. 5 and 6.

Table 2. Incremental error results of maximum active power.

Predictive model	RMSE/MW	MRE/%
BPNN	58.1	4.36
DNN	26.5	1.83
DBN	18.3	1.18

Table 3. Time error results of maximum active power increment.

Predictive model	RMSE/s	MRE/%
BPNN	2.3	8.93
DNN	1.5	5.82
DBN	1.2	4.62

Table 4. Error results of active power steady state values.

Predictive model	RMSE/MW	MRE/%
BPNN	28.3	1.92
DNN	15.4	1.03
DBN	9.5	6.34×10^{-3}

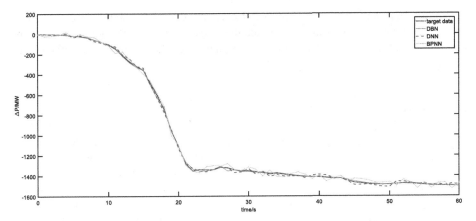

Fig. 5. Comparison of active power incremental curves of three models of a sample.

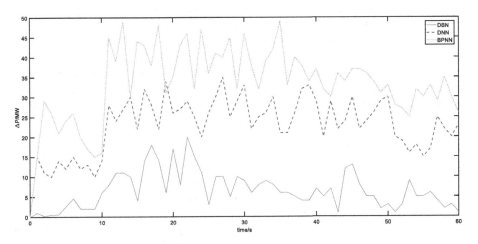

Fig. 6. Root mean square error of a sample over time.

According to Tables 2, 3 and 4, the errors of the DBN model on the steady state value of the active power, the maximum active power increment, and the time of occurrence are all smaller than the other two prediction models. Figures 5 and 6 are the 60 s active power change curve and the root mean square error change curve of the dynamic curve within 60 s for a sample. It can be seen that the DBN prediction curve has the smallest error and the BPNN prediction curve has the largest error. The historical data of the power system is stored in seconds, and the minimum unit of the predicted time is 0.1 s. At the same time, the deviation of the maximum active power increase prediction causes the three models to have large errors in the prediction of the maximum active power increase time. In summary, compared with the DNN and

BPNN models, the DBN model can better excavate the relationship between features under the same training sample, predict the active power curve within 60 s more accurately, and better characterize Primary frequency regulation capability of the power system.

5 Conclusions

In this paper, deep learning is applied to the prediction of primary frequency response capability of power system, and a prediction model of primary frequency response capability of power system based on DBN is established. Compared with other prediction models such as shallow neural networks, DBN is better at extracting features and mining hidden connections between data. In this paper, the disturbance information, state characteristics of the system and the unit during the disturbance are extracted as model input feature vectors to predict the dynamic changes and characteristics of the active power of the system after the disturbance. The simulation experiment was performed using historical disturbance data of a certain power system. By comparing the performance of the BPNN and DNN models, the prediction curve of the DBN model is closest to the actual curve with the smallest error. This model predicts the power grid's self-regulation ability in the event of a power system failure. It can estimate the impact of the faults to a certain extent, and plays a decision-making role for the dispatch control center. However, more factors need to be considered in practical application to improve the accuracy of model prediction, which will be taken into account in future studies.

References

1. Gao, L., Dai, Y.P., Wang, J.F., Zhao, P., Zhao, P.: An oline estimation method of primary frequency regulation parameters of generation units. Proc. CSEE **32**(16), 62–69 (2012)
2. Zhang, Y.J., Gao, K., Qu, Z.Y.: An evaluation method of primary frequency modulation performance based on characteristics of unit output power curves. Autom. Electr. Power Syst. **36**(7), 99–103 (2012)
3. Zeng, Q.D.: Dynamic characteristics of power system frequency and primary frequency regulation control. Guangdong Univ. Technol., 31–37 (2014)
4. Wang, X.: The analysis of primary frequency and study on simulation for thermal generator unit. Beijing Jiaotong Univ., 44–47 (2009)
5. Du, L., Liu, J.Y., Lie, X.: The primary frequency regulation dynamic model based on power network. In: 2006 International Conference on Power System Technology. IEEE (2006)
6. Tao, Q., He, Y., Pan, Y., Sun, J.J.: Characteristics of power system frequency abnormal distribution and improved primary frequency modulation control strategy. Power Syst. Prot. Control **44**(17), 133–138 (2016)
7. Zhao, C.H., Topcu, U., Li, N., Steven, L.: Design and stability of load-side primary frequency control in power systems. IEEE Trans. Autom. Control **59**(5), 1177–1189 (2014)
8. Liu, J.Z., Yao, Q., Liu, Y., Hu, Y.: Wind farm primary frequency control strategy based on wind & thermal power joint control. Proc. CSEE **37**(12), 3462–3469 (2017)

9. Li, X.R., Huang, J.Y., Chen, Y.Y., Liu, W.J.: Review on large-scale involvement of energy storage in power grid fast frequency regulation. Power Syst. Prot. Control **44**(7), 145–153 (2016)

10. Wang, Q., Guo, Y.F., Wan, J., Yu, D.R., Yu, J.L.: Primary frequency regulation strategy of thermal units for a power system with high penetration wind power. Proc. CSEE **38**(4), 974–984 (2018)

11. Zheng, T., Gao, F.Y.: On-line monitoring and computing of unit PFR characteristic parameter based on PMU. Autom. Electr. Power Syst. **33**(11), 57–61, 71 (2009)

12. Yu, D.R., Guo, Y.F.: The online estimate of primary frequency control ability in electric power system. Proc. CSEE **24**(3), 72–76 (2004)

13. Zhang, Q.B., Xu, C.L., Liu, D., Wang, B., Shan, X.: Ability of primary frequency regulation estimate based on wide area measurement system. Electr. Power Eng. Technol. **38**(2), 64–68 (2019)

14. Abdel-Nasser, M., Mahmoud, K.: Accurate photovoltaic power forecasting models using deep LSTM-RNN. Neural Comput. Appl. **31**(7), 2727–2740 (2019)

15. Zhang, Y.C., Wen, D., Wang, X.R., De Lin, J.: A method of frequency curve prediction based on deep belief network of post-disturbance power system. Proc. CSEE **39**(17), 5095–5104 (2019)

16. Zhu, Q.M., Dang, J., Chen, J.F., Xu, Y.P., Li, Y.H., Duan, X.Z.: A method for power system transient stability assessment based on deep belief networks. Proc. CSEE **38**(3), 735–743 (2018)

17. Zhou, N.C., Liao, J.Q., Wang, G.Q., Li, C.Y., Li, J.: Analysis and prospect of deep learning application in smart grid. Autom. Electr. Power Syst. **43**(4), 180–191 (2019)

18. Hinton, G.E., Salakhutdinov, R.R.: Reducing the dimensionality of data with neural networks. Science **313**, 505–507 (2006)

19. Zhu, C.Z., Yin, J.P., Li, Q.: A stock decision support system based on DBNs. J. Comput. Inf. Syst. **10**(2), 883–893 (2014)

Simplified Programming Design Based on Automatic Test System of Aeroengine

Shuang Xia[(⊠)]

SANYA Aviation and Tourism College, Snaya 572000, Hainan, China
6408718@qq.com

Abstract. Aeroengine is a highly complex and repairable multi-component system, and operates over a long time under the harsh conditions of high temperature, high pressure, high speed and high load, and any faults threatening the safety of the aircraft. Based on this, through the in-depth analysis of several common automatic fault detection methods used in aeroengines, an automatic test system based on association rules mining technology is proposed to realize automatic test of aeroengine fault. The system used association rule mining algorithm to deal with the database with a large amount of data. By improving the algorithm, the algorithm can reduce the size of the database and the number of programming. The test results show that the hardware design of the automatic test system is reasonable, the signal acquisition is accurate and the error can meet the requirements; the design of the fault detection process is applicable, the search algorithm is fast and accurate, the speed of detection is about twice as high as possible, and the service life of the engine is saved.

Keywords: Aeroengine · Automatic test system · Association rules · Apriori algorithm

1 Introduction

Aeroengine is the core of the whole aircraft, and its performance has a direct impact on the performance of the aircraft. Aeroengine is the heart of an aircraft. As the control center of the engine, the importance of the electronic regulator is self-evident. At present, the development of engine controller and test system is developing in the direction of digitalization and automation [1]. Automatic test system is an instrument which can make the measured device automatically measure, diagnose, process, store, transmit, and display or output the test results in an appropriate way [2]. At present, the manual inspection table is used to detect the engine by manual method. The detection operation of this method is complex, the fault location is difficult, and the requirement for the operator is high. At the same time, because of the complex environment of the controller detection, there are many other high-power communication, radar, and guidance equipment in the vicinity. The existing equipment can't meet the complex and bad electromagnetic environment [3].

With the use of modern microelectronics and computer technology, the complexity of aeroengine controller is increasing. The traditional manual way of testing has been unable to meet the requirements. Combined with bus technology, virtual instrument

© Springer Nature Singapore Pte Ltd. 2020
P. Qin et al. (Eds.): ICPCSEE 2020, CCIS 1258, pp. 436–444, 2020.
https://doi.org/10.1007/978-981-15-7984-4_32

technology, AI technology and information technology, modern automatic test equipment has become an effective guarantee for completing the test task of controller [4]. Therefore, it is very important to build an automatic detection system with data analysis and fault location function to liberate manpower, improve work efficiency and increase the stability and reliability of detection work [5].

2 State of the Art

Abroad has been the leader in the field of automatic testing. Since the middle of the 80 s, the US military has begun to develop a general automated test system consisting of reusable public testing resources for a variety of weapon platforms and systems [6]. At present, the US Army has formed a series of universal automatic test system for the internal military. The navy integrated automatic support system and the third echelon test system for the Marine Corps have been widely used. The integrated automatic support system is the largest automatic test system in the world at present, which is mainly used for 2 and 3 level testing [7]. The third echelon test system developed by MANTEC Company is a portable and universal automatic test system for the maintenance of field weapons system by the US Marine Corps, and it has good maneuverability and is able to diagnose for analog, digital and RF circuits [8].

The research and development of automatic test system in China has been following the trend of foreign countries. Many domestic airlines have developed an automatic test platform with high level of information and high test efficiency [9]. At present, the development technology of automatic test system in China is relatively late, and there is a big gap, which is mainly manifested in: firstly, the standard in the field of automatic testing is designated by foreign countries, and the development of ATS can only follow the foreign standards, and the development cost is high; secondly, the development of automatic test system is lack of unified planning, the product standards of various domestic companies and units are different, and the generality is poor [10].

3 Automatic Test System Modeling of Aircraft Engine

3.1 The Significance of Automatic Test System for Aeroengine

Aeroengine is usually in the harsh and complex working environment of high temperature and high pressure, and the fault occurs frequently, which makes up a large ratio of the whole flight system fault. Therefore, it is of great significance to study the automatic detection technology of aeroengine and improve the level of fault detection. The fault diagnosis and isolation of the air component of the aeroengine plays an important role in the safe flight of the aircraft. The accurate and timely detection and isolation of faults can avoid some significant losses. As shown in Fig. 1, the staff are manually testing the aeroengine. In recent years, great progress has been made in the research of aeroengine fault diagnosis, and many fault diagnosis algorithms with different structures and different platforms have emerged. At the same time, there are also some problems, such as the lack of original fault data in the stage of algorithm development, and it is difficult to make a direct comparison of the performance of different algorithms. The emergence of these problems has hindered the study of new fault diagnosis algorithms and the communication of existing fault diagnosis research results. The traditional fault diagnosis software and hardware simulation platform can only play the role of verification algorithm, but can't really solve the problem that the fault data is lack and the algorithm of diagnosis can't be compared horizontally.

Fig. 1. Aircraft engine fault detection

In traditional fault detection, the test data is usually compared with the data stored in the database, and then the fault is detected according to the result of comparison. There are unavoidable problems in traditional fault detection methods, such as too long training time and too many parameters. These problems will greatly reduce the practicality and effectiveness of the detection system, so that the detection effect is discounted.

3.2   Aeroengine Fault Detection Process Based on Association Rules

On the basis of fully studying association rule mining technology, automatic control system, aeroengine basic principle, aerospace flight system, fault detection technology and so on, and combined with the related characteristics of aeroengine, the basic flow of aeroengine fault detection is obtained by taking association rules as the core. The specific flow chart can be referred to Fig. 2.

The data used in this study is derived from the parameters of the engine. The parameters of the engine are related to the basic characteristics of the engine. When the engine is in a state of normal work, the data will show a certain relevance. By making full use of these associations, the underlying principles of engine work can be excavated through parameter records. The correlation between these parameters will fail when the engine fails. On this basis, the engine can be detected in the fault.

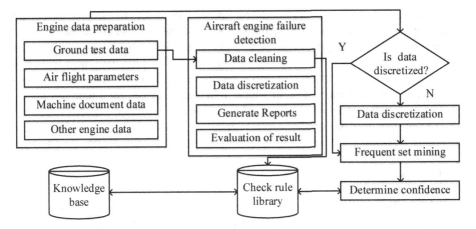

Fig. 2. Aircraft engine fault detection process based on association rules

3.3 Theory Definition of Association Rules Mining Algorithm

Item and item set: the inseparable minimum unit information in a database is called a item, which is represented by a symbol i. The set of items is called an item set. Setting the set $I = \{i_1, i_2, \cdots, i_k\}$ is an item set, the number of items in the I is k, and the set I is called the k item set.

Affairs: setting $I = \{i_1, i_2, \cdots, i_k\}$ is a collection of all items in the database, a set of items contained in a single process is represented by T, $T = \{t_1, t_2, \cdots, t_n\}$. The item set contained in each t_i is a subset of I.

Association rules: association rules are the implication of shape like $X \Rightarrow Y$. Where X, Y are the proper subset of the I, and $X \cap Y = \varphi$. X is called the premise of the rule, and Y is called the result of the rule. The association rules reflect that when a item in X occurs, the rules of the item in the Y are also followed.

Support of association rules: the support degree of association rules is the ratio of the transaction number of X and Y to the number of all transactions in the transaction concentration, and it is recorded as support $(X \Rightarrow Y)$, that is, support $(X \Rightarrow Y) =$ support $X \cup Y = P(XY)$.

Support reflects the frequency that the items contained in X and Y appear at the same time in the transaction concentration.

Confidence of association rules: the confidence degree of the association rule is the ratio of the transaction number of X and Y in the transaction concentration to the number of all transactions and the number of transactions containing X, and it is recorded as confidence $(X \Rightarrow Y)$, namely:

$$\text{Confidence } (X \Rightarrow Y) = \frac{\sup port(X \cup Y)}{\sup port(X)} = P(Y|X) \tag{1}$$

Confidence reflects the conditional probability of the occurrence of Y in a transaction containing X.

Minimum support and minimum confidence: in order to achieve a certain requirement, a user needs to specify the threshold of support and confidence that the rules must satisfy. When support $(X \Rightarrow Y)$ and confidence $(X \Rightarrow Y)$ are more than and equal to their threshold values, it is believed that $X \Rightarrow Y$ is interesting, then these two values are called the minimum support degree threshold (min_sup) and the minimum confidence threshold (min_conf). In which, min_sup describes the minimum importance of association rules, and min_conf specifies the minimum reliability that the association rules must satisfy.

Strong association rules: support $(X \Rightarrow Y) \geq$ min_ sup and confidence $(X \Rightarrow Y)$ $\geq$ min_ conf, the association rule $X \Rightarrow Y$ is called a strong association rule or $X \Rightarrow Y$ is called a weak association rule.

Setting X and Y are the item set in the data set D:

If $X \subseteq Y$, then support $(X) \geq$ support (Y); when $X \subseteq Y$, if X is a non frequent item set, then Y is also a non frequent item set, that is, the super set of any weak item set is a weak item set; when $X \subseteq Y$, if Y is an infrequent set of items, then X is also an infrequent set of items, that is, the subset of any large set of items is a large item set.

3.4   Improvement of Frequent Set Mining Algorithm Apriori

The basic principle of the Apriori algorithm is the calculation of the support degree of the item set. In the process of computing, the database needs to be processed, usually the database is scanned many times. In the process of scanning, if frequent item sets are found, which need to be recorded and the association rules are produced on the basis of the recorded data. In the first scanning, the set L_1 is obtained. When a data set is scanned in k $(k > 1)$ times, the results of the last scan are used, and the results of the previous scan are combined. The set C_k of the k $(k > 1)$ scan is obtained, and the support degree of the set C_k is determined. At the end of the scan, the set L_k of the frequent k set is calculated, and when the C_k is empty, the calculation process ends. The above discussion shows that if the improvement of the algorithm can be carried out from two angles, one is to reduce the size of the database, the other is to reduce the number of candidate set items.

The first step: in order to find L_k, a collection C_k of candidate k item set is generated by L_k-1 connecting with itself. Setting l_1 and l_2 are the set of items in L_k-1. $l_i[j]$ represents the j item of l_i. Apriori assumes that items in a affair or item set are sorted in a dictionary order. For $(k-1)$ of item set l_i, it means sorting items, which makes $l_i[1] < l_i[2] < \ldots < l_i[k-1]$. If the previous $(k-2)$ corresponding items of the elements l_1 and l_2 in L_k-1 are equal, then l_1 and l_2 can be connected, that is, if $(l_1[1] = l_2[1]) \cap (l_1[2] = l_2[2]) \cap \ldots \cap (l_1[k-2] = l_2[k-2]) \cap (l_1[k-1] < l_2[k-1])$, l_1 and l_2 can be connected. The condition $l_1[k-1] < l_2[k-1]$ can only guarantee no repetition. The result set generated by connecting l_1 and l_2 is $(l_1[1], l_1[2], \ldots, l_1[k-1], l_2[k-1])$.

The second step: the nature of the Apriori algorithm shows that any subset of frequent k item sets must be frequent item sets. The set C_k generated by the connection needs to be validated to remove the infrequent k item sets that do not meet the support degree.

3.5   Generation and Test Process of Association Rules for Engine Fault Detection

Association rules are related to two factors, one is confidence and the other is minimum support. If these two factors are different, the association rules will also be different. The main role of confidence is to measure the credibility of the association rules, which is very important to the fault detection. If there is a mistake in the detection process, it will cause a lot of loss. Therefore, in the process of detection, the minimum confidence is usually chosen to be 1. According to certain rules, whether there is a failure can be judged.

The above rules are converted to: if the rotating speed of the high pressure rotor increases, the residual fuel instantaneous value is reduced; if the rotating speed of the low pressure rotor increases, the oil pressure increases; if the rotor speed of the low pressure rotor increases, the vibration value of the casing increases; if the gas temperature in the rear of the turbine is constant, the rotor speed of the low pressure rotor increases and the remaining instantaneous fuel oil is reduced; if the gas temperature behind the turbine is constant, the remaining instantaneous fuel oil value decreases and the oil pressure increases; if the gas temperature behind the turbine is constant, the remaining instantaneous fuel oil value is reduced and the oil pressure increases and the rotor speed of the low pressure rotor increases; if the gas temperature in the rear of the turbine is constant and the rotor speed of the low pressure rotor increases, the remaining instantaneous fuel value is reduced and the oil pressure increases; if the gas temperature of the turbine is constant and the oil pressure increases, the rotor speed of the low pressure rotor increases and the remaining instantaneous fuel oil is reduced.

In the process of testing, in addition to the connection of the relay and the determination of the discrete quantity, the simulation manual adjustment is the key point. In static and dynamic testing, the signal is automatically introduced to adjust the engine or controller to the location of the detection point: for example, T_1 and other signals need to be input in the static testing; T_2 is introduced in the dynamic detection, and the output of the temperature controller is adjusted to reduce the speed to the corresponding detection point. In view of this problem, a sequential incremental search algorithm is designed with 0.618 methods of single factor test, and the artificial adjustment process is simulated. The basic principle is to take the symmetry point of the first test point and the second test point within the scope of the test, that is, 0.382; experiments can be carried out in the above 2 points and the test results can be compared, the above mentioned methods are repeated within the retaining scope until the best value is found.

4   Analysis of Examples

4.1   Experimental Test Results

The data used in this article is derived from the parameters of the engine. The parameters of these engines are related to the basic characteristics of the engine. When the engine is in a state of normal work, the data will show a certain relevance. By taking full advantage of these associations, the correlation between these parameters will fail when the engine fails. According to this, the fault detection of the engine can be realized. The test of the system was carried out in 2 steps on the test rig of an engine factory: the first was to test the accuracy of signal acquisition. Limited to the length, only the test results of T_1 are given, as shown in Table 1. The error of T_1 signal acquisition and measurement is within +0.5 C, which can meet the requirements of engineering application.

Table 1. Test result of T_1

Given temperature (°C)	−100	−77	−50	−20	0	20	65	100
Collect temperature of T_1 (°C)	−96.7	−76.5	−50.6	−20.8	−0.3	19.5	65.4	101.2

The second step was to verify the correctness of the software process and algorithm. This step connected the static and dynamic test for the engine. In order to verify whether the function and performance of the automatic tester meet the requirements of the design, the test was carried out. The test of automatic tester mainly included the hardware board level test, the hardware whole machine test and the software code test. In order to verify the reliability of the test function, a series of tests were carried out to verify the automatic test instrument.

4.2 Static Test Experiment for Engine

In the static state, T_6 temperature was introduced to adjust the response of a controller so that the tachometer voltage and the excitation voltage were zero, at this time, the temperature of the collection tile should be within the range of 3–5 °C of the maximum T_6 limit value, otherwise the temperature controller was not normal; in the static state, T_6 temperature was introduced to adjust the response of a controller so that the tachometer voltage and the excitation voltage were zero, the system automatically sent dead signals to the engine, and at the same time, it collected tile feedback signal and controller's nonlinear feedback response - excitation voltage and tachometer voltage signal, and the sampling interval was 0.1 s. The adjustment process of this section was extracted from the test results, as shown in Fig. 3. The adjustment process was increased from 3 s to 778 s, and the signal was gradually increased, and the precise adjustment process was from during 78 s to 100 s. After adjustment, the excitation

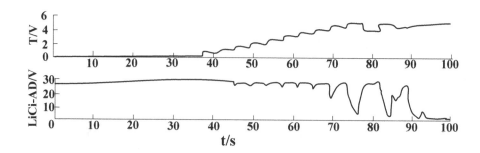

Fig. 3. Change curve of the self-searching process

voltage and tachometer voltage were 0.34 V and 0.018 V respectively, which satisfied the zero adjustment standard, and the search of the detection points was accurate, so that the artificial search process was simulated correctly.

4.3 Dynamic Test Experiment for Engine

In dynamic X_2 examination, T_6 temperature was introduced to adjust the response of a controller so that the tachometer voltage and the excitation voltage were zero, at this time, the location of the acquisition should be within the range of 1% of the design value, otherwise the engine was not normal. In the dynamic case, the test situation is shown in Fig. 4. As can be seen from Fig. 4, the rotor speed of the engine accurately followed the T_6 signal, and could quickly be reduced to the specified speed, and the value of X_2 was detected. In addition, the test system took only more than 20 min for completing dynamic and static testing. At present, the artificial detection takes more than 50 min, this method significantly improves the detection speed, which means shortening the engine ground start time and improving the service life of the engine.

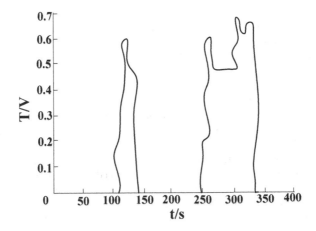

Fig. 4. Examine of slowdown process by importing T_6

5 Conclusions

Aeroengine is the center of an aircraft engine. In order to ensure the normal operation of the engine in use, there is an extremely strict standard for the automatic detection of the engine running state. Therefore, the performance and the speed of the test system directly affect the performance of the engine fault detection, and even affect the safety of the aircraft. In order to ensure that the aeroengine is always able to work normally, the engine is required to be automatically tested and repaired. Traditional manual testing can't meet the increasingly complex testing requirements in testing efficiency and measurement accuracy. Based on this, an automatic test system for aeroengine based on association rule algorithm was proposed. By improving the classical mining frequent set algorithm Apriori in two aspects of reducing the scale of the database and reducing the number of candidate item sets, the optimization of classical mining algorithm was realized. The example shows that the improved algorithm is more efficient and succinct than the original algorithm. The improved mining process was applied to the actual training data, which generated feasible and effective fault detection rule for aeroengine; and then the rules were used to excavate the actual test data, and the results were in accordance with the actual situation. Therefore, it is feasible and effective to apply the association rules to the fault detection of the aeroengine. Finally, the automatic test system was designed for an aeroengine, and the automatic test was realized; the design of the whole structure and the key signal was reasonable, and the measurement of the long distance small signal was realized; the test software process was correct, and the manual adjustment process was simulated successfully with the 0.618 method, and the rapid determination of the detection point was realized. The test results show that the designed test system meets the actual engineering requirements, and the detection speed is increased by about one time.

References

1. Li, Y.P., Li, W.: The design of the test control system of small piston aircraft engine. Adv. Mater. Res. **1061–1062**, 912–915 (2015)
2. Qiongwei, L., Li, A., Li, X., et al.: Research on built-in metrology system for aircraft integrated automatic test equipment. In: International Conference on Electronic Measurement & Instruments, pp. 340–343. IEEE (2011)
3. Liu, L., Wang, S., Liu, D., et al.: Entropy-based sensor selection for condition monitoring and prognostics of aircraft engine. Microelectron. Reliab. **55**(9–10), 2092–2096 (2015)
4. Anstead, J.W.: Automatic testing of electronic equipment for aircraft. Radio Electron. Eng. **34**(5), 269–275 (2010)
5. Korsun, O.N., Poplavsky, B.K., Prihodko, S.J.: Intelligent support for aircraft flight test data processing in problem of engine thrust estimation. Procedia Comput. Sci. **103**, 82–87 (2017)
6. Zang, X.J., Deng, J.N.: Aircraft pneumatic accessories automatic test system based on configuration software. Meas. Control Technol. **32**(7), 110–113 (2013)
7. Zheng, J., Xie, W.F., Viens, M., et al.: Design of an advanced automatic inspection system for aircraft parts based on fluorescent penetrant inspection analysis. Insight Non-Destr. Test. Condition Monit. **57**(1), 18–34 (2015)
8. Dimogianopoulos, D., Hios, J., Fassois, S.: Aircraft engine health management via stochastic modelling of flight data interrelations. Aeros. Sci. Technol. **16**(1), 70–81 (2012)
9. Côme, E., Cottrell, M., Verleysen, M., Lacaille, J.: Aircraft engine health monitoring using self-organizing maps. In: Perner, P. (ed.) ICDM 2010. LNCS (LNAI), vol. 6171, pp. 405–417. Springer, Heidelberg (2010). https://doi.org/10.1007/978-3-642-14400-4_31
10. Lv, Y., Gang, Z.: UG-based research and development of 3D pipe layout system of the aircraft engine. Procedia Eng. **17**, 660–667 (2011)

Intelligent Multi-objective Anomaly Detection Method Based on Robust Sparse Flow

Ke Wang[1], Weishan Huang[1], Yan Chen[1],
Ningjiang Chen[1,2(✉)], and Ziqi He[1]

[1] School of Computer and Electronic Information,
Guangxi University, Nanning 530004, China
chnj@gxu.edu.cn
[2] Guangxi Key Laboratory of Multimedia Communications
and Network Technology, Nanning 530004, China

Abstract. To meet the needs of transportation systems for smart scenic security services, real-time detection and identification of traffic anomalies with high accuracy is essential. Based on the multi-objective sparse optical flow estimation method based on KLT algorithm, an improved algorithm for robust sparse optical flow is designed. The Forward-Backward error calculation method was used to eliminate the error optical flow generated by the KLT algorithm and the robustness of optical flow was improved. The proposed algorithm was verified by the actual traffic scene monitoring example, and the anomaly detection accuracy is above 80%. Furthermore, it has good detection effect on the benchmark dataset.

Keywords: Traffic anomaly detection · KLT algorithm · Sparse optical flow · Forward-Backward error

1 Introduction

During the legal holidays in China, there are often some traffic anomalies in many popular scenic sites, such as the vehicle on the scenic road is retrograde, the vehicle is congested, and the state of the scenic visitors is suddenly changed. Traffic anomaly detection refers to the detection of pedestrians or vehicles on the road that do not comply with traffic rules to avoid traffic safety accidents, thus improving the security service system of the scenic sites. To detect an abnormal moving target, it is first necessary to acquire and accurately extract the positional relationship of successive video sequences to obtain the exact position of the target. Some unpredictable changes occur during the movement of the target, such as changes in the speed of the movement, changes in the brightness of the scene light, and occlusion of the background or target. Optical flow method [1] is very useful for analyzing images of continuous moving objects. Even if the object is occluded, the motion state of the target object can be judged and the target object can be detected and tracked. The optical flow has two types: a sparse optical flow and a dense optical flow. The dense optical flow calculates the offset of all points of the image, while the sparse optical flow is the image registration of the sparse points on the image. The relatively dense optical flow, the

© Springer Nature Singapore Pte Ltd. 2020
P. Qin et al. (Eds.): ICPCSEE 2020, CCIS 1258, pp. 445–457, 2020.
https://doi.org/10.1007/978-981-15-7984-4_33

calculation amount and the algorithm of sparse optical flow take less time, and the application of the algorithm in real-time optical flow computing can be implemented.

Lucas et al. [2] proposes an image registration technique that uses the spatial intensity gradient of the image and the Newton iterative method to search for the position that produces the best match. Tan et al. [3] proposes a fast anomaly detection algorithm for traffic monitoring video based on sparse optical flow. Adam et al. [4] proposes an anomaly detection based on multiple local monitors that collect low-level statistical information, which can improve the accuracy of detection in the crowded scene of the target object. Jodoina et al. [5] proposes an anomaly detection algorithm based on behavioral subtraction, which can reduce the computational amount and memory consumption. Lu et al. [6] proposes a sparse combinatorial learning framework that turns the original complicated problem to one in which only a few costless small-scale least square optimization steps are involved to improve real-time performance. Fan et al. [7] proposed an abnormal crowd behavior detection method based on improved statistical global optical flow entropy, which can improve the detection accuracy and robustness of crowd abnormal detection. Fan et al. [8] proposes a moving target detection algorithm combining single Gaussian and optical flow method, which can be used to solve the problems of high real-time requirement, moving background and easy change of ambient lighting in moving target detection in UAV scene. Nizar et al. [9] proposes a multi-object sparse optical flow estimation method based on KLT (Kanade-Lucas-Tomasi, Shi-Tomasi corner detection and Lucas-Kanade pyramid optical flow method) algorithm to achieve complete extraction of foreground mask and compute of the optical flow and optical flow field of the target, and it meets the real-time requirements.

However, these methods have a certain degree of error in the calculation of optical flow, such as the KLT algorithm proposed by Nizar et al. [9]. Kalal et al. [10] proposes a tracking fault detection method based on the Forward-Backward computational error method, which can reliably detect the tracking failure and select the reliable trajectory in the video sequence. Since the Forward-Backward computational error method is applicable to the KLT algorithm, this paper introduces the Forward-Backward computational error method after the KLT algorithm to realize the intelligent multi-objective anomaly detection based on robust sparse optical flow.

This paper is divided into five sections. The contents of each section are as follows: The first section is the introduction part, which mainly introduces the background and significance of the research topic, the research status, and the organizational structure of this paper. The second section is the working background and basic ideas. The third section introduces an improved method for multi-objective sparse optical flow estimation based on KLT algorithm. The fourth section carries out verification experiments on the algorithm. The fifth chapter summarizes the research results and the next work.

2 Work Background and Back Ideas

In [9], some phenomena have drifted in the process of solving the characteristic point optical flow by the KLT algorithm, so the optical flow calculated by the L-K (Lucas-Kanade) pyramid optical flow method is also the error optical flow. As shown in Fig. 1, the red frame is partially anomalous optical flow information, and it can be seen that the direction of the optical flow deviates from the actual direction.

Based on the KLT algorithm for the multi-objective sparse optical flow estimation method proposed in [9], this paper is inspired by the literature [10]. After the KLT algorithm, the Forward-Backward computational error method is introduced to correct the abnormal optical flow, and to realize traffic anomaly detection based on robust sparse optical flow.

Fig. 1. Abnormal optical flow obtained by L-K pyramid optical flow method

In this paper, the background difference method is used to extract the foreground mask. In order to adapt to changes such as illumination, the periodic update of the background model can be realized by calculating the Gaussian sliding average. If p(t) is the pixel value at time t, and u(t − 1) is the current average, then the new average would be:

$$u(t) = (1 - a)[u(t - 1)] + \alpha p(t) \tag{1}$$

Then, the motion foreground in the image sequence is extracted, and the extracted foreground is subjected to morphological filtering processing to remove noise.

The KLT algorithm is divided into two steps: using the Shi-Tomasi corner detection algorithm to find good feature points and using the L-K pyramid optical flow method to calculate the optical flow. The feature points in the image are called corner points. The Shi-Tomasi corner points detection algorithm performs the eigenvalue analysis on the autocorrelation matrix M, and the condition of the strong corner points is that the two eigenvalues λ_1 and λ_2 are larger than the minimum threshold. The scoring function of the Shi-Tomasi corner points detection algorithm is as follows:

$$R = \min(\lambda_1, \lambda_2) \tag{2}$$

The L-K pyramid optical flow method is based on local constraints. It is assumed that the optical flow maintains gray-scale conservation in a certain window range of the image. Different weights are assigned according to the pixel positions in the window range, and the following gray-scale conservation energy function can be obtained:

$$\sum_{(x,y)\in\Omega} W^2(x)\left(I_x u + I_y v + I_t\right)^2 \tag{3}$$

Ω is the neighborhood range centered at the (x, y) point, and $W(x)$ is the weight function of the window range, which determines the range in which the optical flow estimation in the neighborhood affects the overall result. Let $V = (u, v)^T, \nabla I(x) = (I_x, I_y)^T$, then get from Eq. (3):

$$A^T W^2 A V = A^T W^2 b \tag{4}$$

For the n points x_i in the neighborhood Ω, A, W and b are:

$$A = (\nabla I(x_1), \nabla I(x_2), \ldots, \nabla I(x_n))^T \tag{5}$$

$$W = \text{diag}(W(x_1), W(x_2), \ldots, W(x_n)) \tag{6}$$

$$b = -(I_t(x_1), I_t(x_2), \ldots, I_t(x_n))^T \tag{7}$$

The solution to the equation that can be solved is:

$$V = \left(A^T W^2 A\right)^{-1} A^T W^2 b \tag{8}$$

Where $A^T W^2 A$ is the following 2×2 matrix:

$$A^T W^2 A = \begin{pmatrix} \sum W^2(x) I_x^2(x) & \sum W^2(x) I_x(x) I_y(x) \\ \sum W^2(x) I_y(x) I_x(x) & \sum W^2(x) I_y^2(x) \end{pmatrix} \tag{9}$$

If the above matrix is not singular, the original formula has a solution.

Then, based on the feature extraction method of multi-scale optical flow histogram, the feature extraction and aggregation are carried out. Finally, the model is trained and detected based on the model theory of sparse reconstruction cost.

In order to eliminate the abnormal optical flow and improve the calculation accuracy of the optical flow field, this paper proposes an improved scheme of multi-objective sparse optical flow estimation method based on KLT algorithm: the intro-duction of the Forward-Backward computational error method adds a correction phase after the KLT algorithm. Thereby eliminating part of the abnormal optical flow to obtain a robust sparse optical flow.

3   An Improved Method for Multi-objective Sparse Optical Flow Estimation Based on KLT Algorithm

First, the premise of the consistency statement is given, that is, whether it is tracked forward in time or backward, the trajectory generated should be the same. Using the above premise, a flow for detecting Forward-Backward error is proposed as follows:

(1) When tracking by optical flow method, the feature points should be tracked forward in real time and a track should be generated;
(2) A verification track is initialized at the position of the feature point in the previous frame, which is the trajectory obtained from the back tracking of the previous frame to the first frame;
(3) Comparing the forward-tracking trajectory with the back-tracking trajectory, if there is a significant difference between them, the forward-tracking trajectory is considered to be erroneous.

When the background obscures part of the moving target, there will be inconsistencies between the forward trajectory and the backward trajectory. Figure 2 shows the two feature points marked: Point 1 is the head feature point of the tracking target, which is visible in both the front and back images, so the optical flow tracker can always achieve forward trajectory and the backward trajectory of the point. Point 2 is the feature point of the right foot of the moving target, and is blocked by the street sign in the right picture. A matching error occurs in the back tracking, so the tracking trajectory of the forward and backward directions is inconsistent.

Fig. 2.   Tracking process for a single feature point

Figure 3 shows the feature point trajectory for solving the Forward-Backward error. Let $S = (I_t, I_{t+1}, \ldots, I_{t+k})$ be a sequence of images, and X_t is the position of a feature point at time t. The k frames is tracked forward by the X_t point using the Lucas-Kanade optical flow tracker. The final trajectory is obtained as $T_f^k = (X_t, X_{t+1}, \ldots, X_{t+k})$, where f represents forward tracking and k represents the number of frames. In order to estimate the error reliability of a given image sequence S trajectory, a verification trajectory is first constructed. Point X_{t+k} is backtracked to the first frame and generates

$T_b^k = (\hat{X}_t, \hat{X}_{t+1}, \ldots, \hat{X}_{t+k})$, where $X_{t+k} = \hat{X}_{t+k}$. The Forward-Backward error is defined as the distance between the two tracks: $\mathrm{FB}\left(T_f^k|S\right) = distance\left(T_f^k, T_b^k\right)$. For simplicity, this paper calculates the distance using the Euclidean distance between the initial point and the last point in the verification trajectory: $distance\left(T_f^k, T_b^k\right) = \|X_t - \hat{X}_t\|$. Finally, the abnormal optical flow is eliminated by calculating the Forward-Backward error.

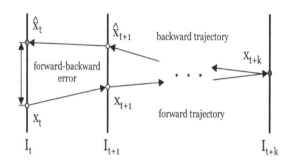

Fig. 3. Feature point trajectory diagram of Forward-Backward calculation error method

Based on the above principle, the integrated KLT algorithm and the Forward-Backward error detection optimization scheme are designed. By designing a new improved optical flow tracker to optimize the optical flow information extraction quality, and Fig. 4 shows the process of the proposed optical flow tracker.

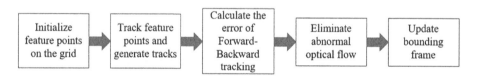

Fig. 4. The process of improvement of optical flow tracker

The tracker inputs two consecutive frames of images: the image $Image_t$, the image $Image_{t+1}$, and the bounding frame b_t of $Image_t$, and finally the bounding frame b_{t+1} of $Image_{t+1}$. A set of feature points is initialized on a rectangular grid within the bounding frame b_t. These feature points are tracked and extracted by the KLT algorithm, and these feature points generate sparse optical flow vectors in $Image_t$ and $Image_{t+1}$. Then, by calculating the Forward-Backward error, and assigning the Forward-Backward error values of these estimated optical flows, 50% of the worst error optical flow filtering is eliminated, and the remaining optical flow is used to detect the offset in the bounding frame. Finally, the bounding frame is updated by correcting the boundary, and the updated bounding frame is output for the next optical flow error detection.

Since the Forward-Backward error calculation method is applicable to the KLT algorithm, the flow of the intelligent multi-objective anomaly detection method based on the robust sparse optical flow as shown in Fig. 5 can be obtained.

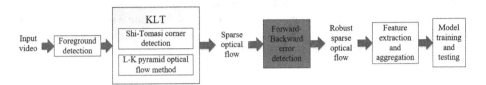

Fig. 5. The process of calculating robust sparse optical flow

By introducing the Forward-Backward error method after the KLT algorithm, the unreliable matching result and the error optical flow of the moving target are eliminated, and the robust sparse optical flow of the moving target is obtained, and the intelligent traffic anomaly detection based on the robust sparse optical flow is realized. This paper verifies the algorithm in the next section.

4   Experiments

The experimental hardware environment is: the processor is intel(R) Core(TM) i7-6700HQ CPU @2.60 GHz, the memory is 8 GB, the hard disk is 1 TB, the operating system is Windows10; the software development environment is: Microsoft Visual Studio 2010+OpenCV2.4.10. The dataset is provided by the literature [3] and the literature [11].

4.1   Experiment of Traffic Monitoring Video Detection

The dataset detected in this section is the real traffic monitoring video provided in [3]. First, the foreground detection is performed. The traffic scenario shown in Fig. 6 is the case where there is a traffic barrier in the middle of the road. The main traffic anomaly is the behavior of pedestrians crossing the road and crossing the traffic barrier. The background

(a) Original video image   (b) Background image of   (c) Extracted foreground mask
                                background modeling
                                construction

Fig. 6. Foreground for the extraction of traffic scenes

difference method is used to extract the motion foreground in the image sequence, and the extracted foreground is subjected to morphological filtering processing to remove the noise and obtain the foreground moving target.

The KLT algorithm is then used to extract the corner points and calculate the optical flow. Then the Forward-Backward error calculation method is introduced to improve the KLT algorithm. Figure 7 is a sparse optical flow and a sparse optical flow field respectively extracted, where in the direction and length of the arrow of the red arrow indicate the direction and velocity of the extracted optical flow.

(a) Extracted sparse optical flow   (b) Sparse optical flow field

Fig. 7. Experimental results of robust sparse optical flow calculation for traffic scenes

Table 1 shows the algorithm running time statistics and comparison results of the video sequence of the traffic scene (a total of 5495 frames). Among them, 10 frames of experimental data and 500 frames of incremental experimental data and the mean value of all frames are selected from 1000 frames to 5000 frames. It can be seen from the table that the time cost of calculating the robust sparse optical flow and optical flow field is relatively small, and the introduced Forwd-Backward error calculation phase takes about 8% of the total processing time of one frame, which ensures the real-time of overall algorithm.

Table 1. Optical flow operation time of traffic scene video sequence

Frames	Feature point extraction phase (s)	Optical flow solution stage (s)	Forward-Backward error calculation phase (s)	Total processing time of one frame (s)
1000	0.465	0.635	0.084	0.967
1500	0.472	0.655	0.089	0.978
2000	0.480	0.663	0.082	0.853
2500	0.451	0.630	0.076	0.898
3000	0.466	0.642	0.093	0.998
3500	0.468	0.645	0.091	0.974
4000	0.460	0.594	0.066	0.937
4500	0.457	0.589	0.063	0.943
5000	0.468	0.621	0.079	0.951
5500	0.479	0.654	0.086	0.980
Average	0.463	0.631	0.082	0.965

Then, based on the feature extraction method of multi-scale optical flow histogram, feature extraction and aggregation are carried out, and finally the model is trained and detected. Figure 8 is the experimental effect of traffic scene anomaly detection. Traffic anomalies are accurately detected and marked in real time with a red mask.

(a) Pedestrians crossing the road and crossing   (b) Retrograde non-motorized vehicles
the traffic barrier

Fig. 8. Experimental results of anomaly detection of traffic scenes

Figure 9 is a result of global traffic anomaly event detection of sample video from a traffic scene. There are two result bars under the picture, which are: the actual field of the abnormal situation on the ground and the result bar of the algorithm detection abnormality. Among them, the green part represents a normal frame, and the red part corresponds to an abnormal frame. Thus, it is possible to visually see the error between the actual abnormal situation and the detection of the abnormal situation. Due to the change of illumination or the swing of the camera, there is a certain degree of error in the detection of abnormal conditions, but the abnormal situation of the key frame is basically correctly detected, and the accuracy rate is over 80%.

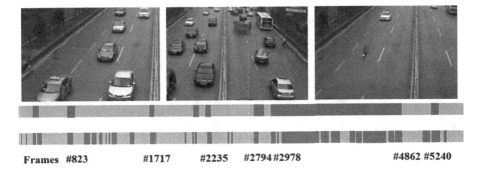

Frames #823   #1717   #2235   #2794 #2978   #4862 #5240

Fig. 9. Experimental results of global traffic anomaly detection in traffic scenarios

4.2   Experiment Based on Benchmark Dataset

This section examines the UCSD dataset provided by the literature [11], which mainly collects real campus surveillance video. In this paper, the Ped1 dataset and the Ped2 dataset are used. Figures 10 and 11 show the results of partial anomaly detection in

(a) Cycling in the crowd 1   (b) Skateboarding in the crowd 1

(c) Cycling in the crowd   (d) Skateboarding in the crowd 2

Fig. 10. Experimental results of anomaly detection in Ped1 datase

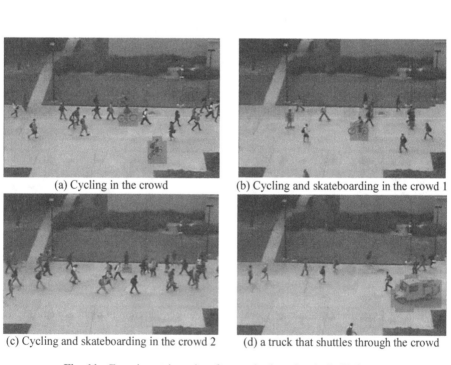

(a) Cycling in the crowd   (b) Cycling and skateboarding in the crowd 1

(c) Cycling and skateboarding in the crowd 2   (d) a truck that shuttles through the crowd

Fig. 11. Experimental results of anomaly detection in Ped2 dataset

which the cyclist and skateboarder are all detected by the improved method in this paper. It can be seen that the anomaly detection effect is better, and the abnormal motion targets can be accurately identified.

Table 2 shows the detection results of the improved algorithm running on the two sub-data sets of the UCSD dataset, and based on the number of real anomalies in the test video and the number of detected anomalies obtained by the algorithm, the accuracy of the algorithm is obtained.

Table 2. Accuracy of anomaly detection of the algorithm on the UCSD dataset

Dataset	Number of real anomalies	Number of detected anomalies	Accuracy
Ped1 dataset	48	40	83.3%
Ped2 dataset	20	17	85%

At the same time, in order to evaluate the performance of the improved method, the methods in Literature [4, 5] and [6] are compared, as shown in the table. It can be seen that the algorithm runs in real time on the UCSD Ped1 dataset, which is less time-consuming than other anomaly detection algorithms (Table 3).

Table 3. Comparison of run time on UCSD Ped1 dataset

Method	Literature [4]	Literature [5]	Literature [6]	This paper
A frame of processing speed (s)	25	3.8	0.007	0.094
FPS	0.04	0.26	142.86	25.64
Language platform	–	–	Matlab	C++
CPU(GHZ)	3.0	2.6	3.4	3.4
Mem(GB)	2.0	2.0	8.0	8.0

The comparison effect of introducing Forward-Backward error detection was tested. As shown in Figs. 12 and 13, the left side is the optical flow calculation result without introducing the Forward-Backward error detection, and the right side is the optical flow calculation result which introduces the forward and backward error detection. It can be seen from the results of the comparison experiments that before the introduction of the Forward-Backward error, there are more abnormal optical flows in the vicinity of the multiple moving targets; and after the introduction of the Forward-Backward error, the error optical flow near the moving targets is basically eliminated, and the original optical flow information is corrected to make the motion information expressed by the optical flow more accurate and improve the robustness of the optical flow information.

Fig. 12. Comparison of the Forward-Backward error introduced in the Ped1 dataset

Fig. 13. Comparison of the Forward-Backward error introduced in the Ped1 dataset

This section evaluates the traffic monitoring dataset captured in the real scene and the benchmark datasets UCSD Ped1 and Ped2. It can be seen that the intelligent traffic anomaly detection algorithm based on robust sparse optical flow with the introduction of the Forward-Backward error method basically satisfies the function of traffic abnormal detection, and the accuracy and real-time performance of optical flow calculation is higher than some existing methods.

5   Conclusions

In this paper, the Forward-Backward error method is introduced after the KLT algorithm to calculate the robust sparse optical flow to realize traffic anomaly detection. The improved algorithm comprehensively considers the real-time and accuracy, and the sparse optical flow is further improved in terms of robustness and efficiency. The algorithm validates its effectiveness through actual traffic monitoring and benchmark datasets. The next work will optimize the algorithm when the vehicle is seriously congested and occluded. With the improvement of theoretical basis and algorithm and the development of GPU computing, some defects of optical flow method will be overcome continuously, and some calculation and accuracy related bottlenecks will be broken, so that optical flow method can be carried out in a wider range of fields and scenes.

Acknowledgments. The work is supported by Xaar Network Next Generation Internet Technology Innovation Project(No.NGII20180901), and the Major special project of science and technology of Guangxi(No.AA18118047-7).

References

1. Horn, B.K.P., Schunck, B.G.: Determining optical flow. Artif. Intell. **17**(1–3), 185–203 (1981)
2. Lucas, B.D., Kanade, T.: An iterative image registration technique with an application to stereo vision. In: International Joint Conference on Artificial Intelligence, pp. 674–679. Morgan Kaufmann Publishers Inc. (1981)
3. Tan, H., Zhai, Y., Liu, Y., et al.: Fast anomaly detection in traffic surveillance video based on robust sparse optical flow. In: IEEE International Conference on Acoustics, Speech and Signal Processing, pp. 1976–1980. IEEE (2016)
4. Adam, A., Rivlin, E., Shimshoni, I., et al.: Robust real-time unusual event detection using multiple fixed-location monitors. IEEE Trans. Pattern Anal. Mach. Intell. **30**(3), 555–560 (2008)
5. Jodoina, P., Saligramab, V., Konrad, J.: Behavior subtraction. IEEE Trans. Image Process. A Publ. IEEE Sig. Process. Soc. **21**(9), 4244–4255 (2012)
6. Lu, C., Shi, J., Jia, J.: Abnormal event detection at 150 fps in matlab. In: IEEE International Conference on Computer Vision, 2720–2727. IEEE (2014)
7. Fan, Z.Y., Li, W., He, Z.H., et al.: Abnormal crowd behavior detection based on the entropy of optical flow. J. Beijing Inst. Technol. **28**(04), 756–763 (2019)
8. Fan, C.J., Wen, L.Y., Mao, Q.Y., et al.: Detection of moving objects in UAV video based on single gaussian model and optical flow analysis. Comput. Syst. Appl. **28**(02), 184–189 (2019)
9. Nizar, T.N., Anbarsanti, N., Prihatmanto, A.S.: Multi-object tracking and detection system based on feature detection of the intelligent transportation system. In: IEEE International Conference on System Engineering and Technology, pp. 1–6. IEEE (2015)
10. Kalal, Z., Mikolajczyk, K., Matas, J.: Forward-backward error: automatic detection of tracking failures. In: IEEE International Conference on Pattern Recognition, 2756–2759. IEEE (2010)
11. Mahadevan, V., Li, W., Bhalodia, V., Vasconcelos, N.: Ucsd ped dataset[DB/OL]. http://www.svcl.ucsd.edu/projects/anomaly/dataset.htm. Accessed 10 May 2018

Empirical Analysis of Tourism Revenues in Sanya

Yuanhui Li$^{(\boxtimes)}$ and Haiyun Han

Sanya Aviation and Tourism College, Sanya 572000, Hainan, China
576735855@qq.com

Abstract. Tourism industry has become a pillar industry, which is used to measure the general economic development of Sanya. According to the latest data related with Sanya tourism (2008–2019), a linear regression model is established in this paper. With statistical method, it takes the annual tourism revenue of Sanya as the explanatory variable and the number of domestic tourists received by Sanya as the explanatory variable. Based on the model, the results of econometric analysis on the factors influencing Sanya tourism revenue show that the number of domestic tourists is the main influence factors of tourism revenue in Sanya.

Keywords: Tourism revenue · Number of domestic tourists · Linear regression model

1 Introduction

Tourism revenue, as a direct reflection of overall condition of Sanya's tourism economic operation, is an indispensable comprehensive index to appraise the efficiency of tourism economic activities and an important indicator to measure the city's economic development. In recent years, many domestic scholars have made extensive and in-depth exploration on tourism income and its influencing factors.

Wang Ning (2019) used Eviews software to conduct a quantitative test and analysis on the factors influencing tourism income in Anhui Province, and the results showed that the total number of domestic tourists and GDP per capita were the main factors influencing tourism income in Anhui Province [1]. Wang Zhong-ke et al. used Eviews software to conduct regression analysis on statistics related to tourism revenue in Guangxi Province, and found that the total tourism revenue in Guangxi was significantly affected by the number of domestic tourists and the disposable income of domestic residents [2]. Zhang Guang-hai et al., using 4 layer BP neural network model to recognize the key factors influencing Shandong tourism revenue, found that there are significant differences of influence factors among the different cities in the province. The main factors are "the number of domestic tourism", "town residents' consumption expenditure", "number of museum" [3]. Song Hui-juan used statistical software to analyze tourism income and GDP in Sichuan Province, and the results showed that there was a long-term co-integration relationship between total tourism revenue, domestic tourism revenue, tourism foreign exchange revenue and GDP [4]. Liu Hui-xiang et al. used the grey correlation analysis method to analyze the tourism industry in Henan

© Springer Nature Singapore Pte Ltd. 2020
P. Qin et al. (Eds.): ICPCSEE 2020, CCIS 1258, pp. 458–466, 2020.
https://doi.org/10.1007/978-981-15-7984-4_34

Province, and the results showed that the major factors influencing the tourism industry in Henan Province were the total retail sales of consumer goods, the average salary of in-service employees in Henan Province and GDP per capita [5]. Liu Zhi-qian et al. used software programs such as SPSS, MATLAB and EVIEWS to solve the problem, and obtained the regression equation of domestic tourism market income, indicating that China's domestic tourism market income is mainly affected by the number of domestic tourists, tourism expenditure of rural residents per capita, highway mileage and other factors [6]. Long Dan used Stata software to analyze many factors affecting tourism income in Hainan Province, and the results showed that the number of tourists had the greatest impact on increasing tourism income, and the number of expressway miles, number of star-rated hotels and per capita tourism expenditure of domestic urban residents also had a positive effect on tourism income [7].

Although many scholars have conducted researches on the main factors affecting the tourism economy in different regions in recent years, there is still a shortage of researches on the factors affecting the tourism market in Hainan Province, especially Sanya city. Through the qualitative and quantitative analysis of the influencing factors of tourism revenue in Sanya, we can make good use of the local tourism resources and improve the tourism revenue. Based on the research results of several scholars and the actual situation of tourism industry in Sanya, this paper makes an empirical analysis of several representative factors affecting tourism income.

2   Data and Methods

2.1   Data Description

In the tourism industry, tourism revenue is an important indicator to measure the development of tourism. This model is to take Sanya tourism revenues as explained variable Y. Since there are many influence factors of tourism revenue, the domestic tourism income (X_1), the number of domestic tourists (X_2), Sanya in the number of domestic tourists (X_3), number of inbound tourists, Sanya (X_4), airport passenger throughput in Sanya (X_5), Haikou airport passenger throughput (X_6), the GDP per capita (X_7) are chosen as the explained variable.

Relevant data of 2008–2019 were obtained by looking up China statistical yearbook, Sanya statistical yearbook, official website—Sanya tourism bureau and official website of Haikou Meilan International Airport, as shown in Table 1.

Table 1. Statistics of tourism revenue and influencing factors in Sanya city from 2008 to 2019

Year	Y	X_1	X_2	X_3	X_4	X_5	X_6	X_7
2008	91.05	8749	17.1	553.01	51.15	600.62	822.20	22640
2009	103.77	10184	19.0	637.27	31.78	794.14	839.12	25125
2010	139.64	12580	21.0	841.14	41.51	929.39	877.38	29678
2011	160.71	19306	26.4	968.18	52.81	1036.18	1016.75	34999

(continued)

Table 1. (*continued*)

Year	Y	X_1	X_2	X_3	X_4	X_5	X_6	X_7
2012	192.22	22706	29.6	1054.08	48.14	1134.32	1069.67	38354
2013	233.33	26276	32.6	1180.21	48.19	1286.68	1193.55	41805
2014	269.7	30312	36.1	1313.90	38.86	1494.23	1385.39	46531
2015	302.31	34195	40.0	1459.91	35.82	1619.2	1616.7	49351
2016	322.40	39390	44.0	1606.69	44.89	1736.11	1880.38	53980
2017	406.17	45661	50.0	1761.69	69.28	1938.99	2258.48	59660
2018	514.73	51278	55.4	2160.89	81.68	2003.9	2412.36	64644
2019	633.19	57251	60.1	2305.70	90.63	2016.37	2421.66	70892

X_1: Annual revenue of domestic touris/100 million yuan.

X_2: Number of domestic tourists/100 million.

X_3: Number of domestic tourists in Sanya/Ten thousands of people.

X_4: Number of inbound tourists/10,000 person-times in Sanya.

X_5: Sanya airport passenger throughput/ten thousand passengers.

X_6: Haikou airport passenger throughput/ten thousand passengers.

X_7: National GDP per capita/yuan.

2.2   Modeling Procedures

Selection Dependent Variable. The goal of this study is to predict the annual tourism revenue of Sanya, so the "annual tourism revenue of Sanya" is chosen as the dependent variable.

Determine the Explanatory Power of the Independent Variable to the Dependent Variable. Scatter figures of the annual tourism revenue of Sanya and the number of domestic tourists in sanya, the passenger throughput of Sanya airport and the per capita GDP of the country are shown in Fig. 1, Fig. 2 and Fig. 3 respectively. As can be seen from the figures, there is an obvious linear dependency between the annual tourism income of Sanya and the three variables, so it is not suitable to use the linear regression equation of one variable to fit. This paper tries to introduce the multiple linear regression equation.

The idea and hypothesis of multiple linear regression: multiple linear regression is a statistical method for studying the dependence of multiple variables. Among the multiple variables, the dependent variable is represented by $\hat{y}$, the independent variable $x_1, x_2, \cdots, x_n$, and the linear regression equation is as Eq. (1):

$$\hat{y} = \beta_0 + \beta_1 x_1 + \beta_2 x_2 + \cdots + \beta_n x_n + \varepsilon \qquad (1)$$

$\beta_0, \beta_1, \beta_2, \cdots, \beta_n$ are the regression coefficient value, and ε is the random error of the Eq. (1) [8].

Multivariate linear regression needs to satisfy the following five conditions, none of which is indispensable [9]:

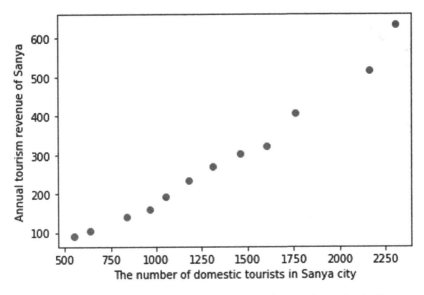

Fig. 1. Scatter chart of annual tourism revenue and domestic tourists in Sanya

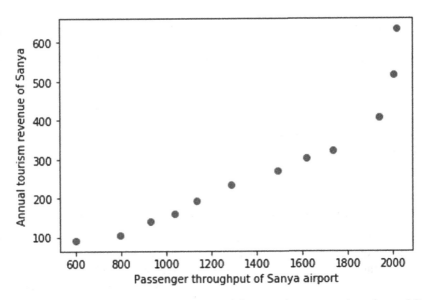

Fig. 2. Scatter chart of annual tourism revenue of Sanya and passenger throughput of Sanya airport

(1) Linearity is reasonable. There is only a linear relationship between dependent variables and independent variables, and there is no non-linear relationship.

(2) The mean of random errors is zero, independent of each other, and the variances are equal, which satisfies the following Eq. (2):

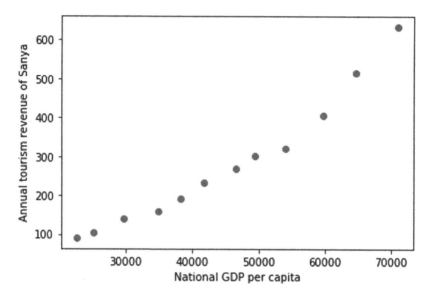

Fig. 3. Scatter chart of annual tourism income of Sanya and national per capita GDP

$$E(\varepsilon_i) = 0; \ \text{var}(\varepsilon_i) = \sigma^2; \ \text{cov}(\varepsilon_i, \varepsilon_j) = 0 \qquad i, j = 1, 2, \cdots, n, i \neq j \qquad (2)$$

(3) Random errors have the same distribution and are subject to a normal distribution, as Eq. (3):

$$\varepsilon_i \sim N(0, \ \sigma^2), \ i = 1, 2, \cdots, n \qquad (3)$$

(4) Random errors and independent variables are independent of each other in Eq. (4):

$$\text{cov}(\varepsilon_i, \varepsilon_j) = 0, \quad i, j = 1, 2, \cdots, n. \qquad (4)$$

(5) The independent variables are not correlated with each other [9].

SPSS software is used to substitute 7 independent variables into the multivariate linear regression model, such as annual tourism revenue of Sanya, annual domestic tourism revenue of China, number of domestic tourists of China, and number of domestic tourists of Sanya. The results are shown in Table 2, Table 3, and Table 4.

Table 2. Model summary

Model	R	R square	Adjusted R square	Std. error of the estimate
1	0.998	0.995	0.986	19.54696

Table 3. Anova

Model		Sum of squares	df	Mean square	F	Sig.
1	Regression	307199.952	7	43885.707	114.859	.000
	Residuals	1528.334	4	382.083		
	Total	308728.286	11			

Table 4. Coefficients^a

Model		Unstandardized coefficients		Standardized coefficients	t	Sig.
		B	Std. error	Beta		
1	(Constant)	−867.26	405.804		−2.137	0.099
	X_1	−0.062	0.032	−5.943	−1.947	0.123
	X_2	58.175	32.428	4.965	1.794	0.147
	X_3	−0.004	0.174	−0.012	−0.021	0.984
	X_4	−1.167	1.1	−0.128	−1.061	0.348
	X_5	−0.62	0.221	−1.815	−2.804	0.049
	X_6	0.158	0.116	0.584	1.365	0.244
	X_7	0.035	0.019	3.278	1.843	0.139

It can be seen from Table 3 that the significance test of the equation can be passed, and the regression coefficient is not all 0. In Table 4, if the significance level is assumed to be 0.05, only the regression coefficient of the variable "passenger throughput of Sanya airport" is significantly not 0, and the regression coefficient of the other variables is not significant. The possible reason for this is the existence of multicollinearity between the independent variables.

Eliminating Multiple Correlations of Independent Variables. Multicollinearity refers to the strong linear correlation between variables, which destroys the fifth condition of multiple linear regression: independent variables are irrelevant. The stepwise regression method is used to make SPSS automatically select the appropriate independent variables to establish the regression equation, and the variables that produce multicollinearity with the selected variables will be eliminated from the regression model, so as to obtain a linear fitting equation with less multicollinearity. Table 5 shows the two models obtained after the elimination of multicollinearity.

Table 5. Model summary

Model	R	R square	Adjusted R square	Std. error of the estimate	Durbin-Watson
1	0.986	0.972	0.969	29.45146	
2	0.993	0.985	0.982	22.54990	2.444

It can be seen from the goodness of fit shown in Table 5 that the second model is significantly improved over the first model, indicating that the second model is superior to the first model. Now let's look at the significance test of the equation.

Table 6. ANOVA[c]

Model		Sum of squares	df	Mean square	F	Sig.
1	Regression	300054.399	1	300054.399	345.928	0.000[a]
	Residuals	8673.888	10	867.389		
	Total	308728.286	11			
2	Regression	304151.806	2	152075.903	299.069	0.000[b]
	Residuals	4576.480	9	508.498		
	Total	308728.286	11			

It can be seen from Table 6 that both models have passed the equation significance test, indicating that the significance of the regression coefficients of both models is not 0.

Table 7. Coefficients[a]

Model		Unstandardized coefficients		Standardized coefficients	t	Sig.
		B	Std. error	Beta		
1	(Constant)	−107.270	22.529		−4.761	.001
	X_3	0.294	0.016	0.986	18.599	.000
2	(Constant)	−63.146	23.220		−2.719	.024
	X_3	0.437	0.052	1.464	8.448	.000
	X_5	−0.168	0.059	−0.492	−2.839	.019

Table 7 shows the coefficient significance test results. It can be seen from Table 7 that the coefficients of both models are significant. The second model increases the passenger throughput variable of Sanya airport compared with the first model, and the goodness of fit significantly increases. Therefore, this variable should be increased.

Fitting Linear Regression Equation. In summary, the multiple linear regression equation of annual tourism revenue of Sanya can be obtained, as shown in as Eq. (5):

$$\hat{y} = -63.146 + 0.437x_1 - 0.168x_2 \qquad (5)$$

Testing the Equations. The regression coefficient in the formula indicates that for every 10,000 more domestic tourists in Sanya, the annual tourism revenue of Sanya increases by 43.7 million yuan on average. However, for every 10,000 passengers increased in the throughput of Sanya airport, the annual tourism revenue of Sanya decreased by 16.8 million yuan on average.

Analysis of Residuals. Make a normal P-P diagram, as shown in Fig. 4.

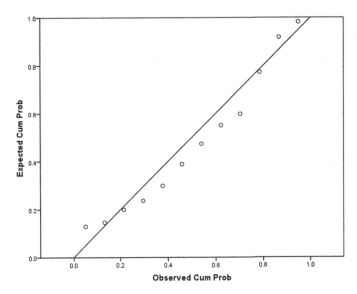

Fig. 4. Normal P-P plot of unstandardized residual

As can be seen from Fig. 4, the residual data is basically subject to a normal distribution. The autocorrelation analysis of the residuals is then performed by the Durbin-Watson statistic. The Durbin-Watson statistic is defined as Eq. (6):

$$D.W. = \frac{\sum_{i=2}^{n}(e_i - e_{i-1})^2}{\sum_{i=1}^{n} e_i^2} \tag{6}$$

The Durbin-Watson statistic is between $0 \sim 4$. When the residuals are positively correlated, that is, e_i, e_{i-1} are close, the Durbin-Watson statistic is close to 0. Similarly, when the residuals are negatively correlated, the Durbin-Watson statistic is close to 4. When the Durbin-Watson statistic value is close to 2, it indicates that there is no autocorrelation of the random error.

As can be seen from Table 5, the Durbin-Watson statistic has a value of 2.444, which is close to 2, indicating that there is no autocorrelation in the sequence.

Confirm the Model and Use it for Prediction. As can be seen from Table 5, the Durbin-Watson statistic value is 2.444, which is close to 2, indicating that the sequence has no autocorrelation. Therefore, the prediction model of annual tourism revenue of Sanya is now confirmed as Eq. (7):

$$\hat{y} = -63.146 + 0.437x_1 - 0.168x_2 \tag{7}$$

Now the data of 2019 is used as the test sample data to predict according to the multiple linear regression model. The error between the predicted value and the actual value of Sanya's tourism revenue is over 2.7 billion yuan, and the accuracy is over 95%, as shown in Table 8. It is shown that the model is valid.

· **Table 8.** Comparison of projected income with actual income

Year	Y	Predicted Y	Errors	Percentage error	X_3	X_5
2019	633.19	605.69	−27.50	−4.34%	2305.70	2016.37

3 Conclusions

(1) It can be concluded from the regression model that the number of domestic tourists has the most significant impact on tourism income and has a direct impact on tourism revenue in Sanya. Therefore, to attract more domestic and foreign tourists to sanya is the fundamental to improve the local tourism revenue.

(2) The impact of airport passenger flow on tourism income has reached an inflection point, and, therefore, it is important to increase the development of self-driving travel, vigorously develop rural tourism, and explore new tourism economic growth points.

Acknowledgments. This research was financially supported by Higher Education Scientific Research Program in Hainan Province of China (Hnky2019-100) and Hainan Provincial Natural Science Foundation of China (618QN258).

References

1. Wang, N.: An empirical study on the influencing factors of tourism income in Anhui Province. J. Chifeng Univ. (Philos. Soc. Sci. Chin. Edn.) **40**(03), 55–59 (2019)
2. Wang, Z.: Study on the influencing factors of tourism income in Guangxi. Bus. Econ. **512**(4), 40–43 (2019)
3. Zhang, G.: Identification of key factors of tourism income in Shandong Province based on bp neural network. J. Shandong Technol. Bus. Univ. **33**(02), 86–93 (2019)
4. Song, H.: Correlation between the tourist industry of Sichuan Province and its economic growth. J. Shenyang Agric. Univ. (Soc. Sci. Edn.) **21**(02), 129–135 (2019)
5. Liu, H.: Grey Relational analysis and application of uncertain problems in tourism of Henan Province. J. North China Univ. Sci. Technol. (Nat. Sci. Edn.) **41**(02), 112–117 (2019)
6. Liu, Z.: Weighted combination forecast of domestic tourism development based on multiple regression and GM (1, 1). J. Kashi Univ. **39**(03), 17–24 (2018)
7. Long, D.: An empirical analysis of the influencing factors of tourism income in Hainan Province. Mark. Mod. (25), 119–120 (2016)
8. Han, H.: Construction and empirical analysis of the prediction model of PRETCO-A scores. J. Shijiazhuang Univ. Appl. Technol. **31**(05), 76–80 (2019)
9. Xia, Y.: Essentials and Examples of SPSS Statistical Analysis, 1st edn. Electronic Industry Press, Beijing (2010)

Study of CNN-Based News-Driven Stock Price Movement Prediction in the A-Share Market

Yimeng Shang and Yue Wang[(✉)]

School of Information,
Central University of Finance and Economics, Beijing, China
yuelwang@163.com

Abstract. Stock market has always been an important research field of scholars in various industries, and short-term stock price forecasting is the focus of both finance and computer research. This paper applies the convolutional neural network (CNN) to short-term stock price movement forecasting by using daily stock news of a famous company called Guizhoumaotai in the Chinese wine industry. Two scenarios were taken into consideration: first, the news occurred in a day's transaction time was used to predict the day's stock price movement; second, the news occurred before a day's opening time and after the transaction time of the previous day was used to predict the day's stock price movement. In addition, the stock attentions of Baidu search index and/or media index were added into the model to explore whether they have significant improvement on prediction. The experimental results show that using the news data of the day can achieve better prediction performance. In addition, the introduction of Baidu Index improves the result of stock price prediction to some extent however, with a little effect.

Keywords: Stock price · Deep learning · CNN · Baidu index

1 Introduction

Stock market research has always been an important field of scholars' research. Stock market can represent a country's economic situation, and has received the attention of the government, managers and investors. Short-term stock price prediction is the focus of investor research. The stock market is affected by many factors, which has strong complexity and non-linearity, making the stock price have the characteristics of high noise and uncertainty, and the stock price prediction faces many difficulties. With the advent of the era of big data, we can use more and more online data to achieve better short-term prediction of stock movement. Can we have better prediction performance by combining news and attentions? This is the research motivation of this paper (Fig. 1).

Y. Shang and Y. Wang—contribute equally to this paper. This work is supported by National Defense Science and Technology Innovation Special Zone Project (No. 18-163-11-ZT-002-045-04).

P. Qin et al. (Eds.): ICPCSEE 2020, CCIS 1258, pp. 467–474, 2020.
https://doi.org/10.1007/978-981-15-7984-4_35

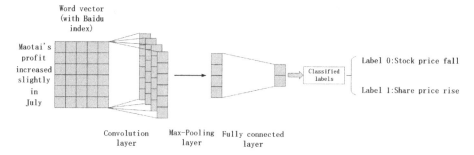

Fig. 1. CNN model of stock price forecast

In recent years, the rapid development of neural network promotes the development of in-depth learning theory. At present, deep learning convolutional neural network (CNN) is mainly used in speech recognition, image classification and other aspects, and has achieved unprecedented success. At the same time, as a common nonlinear dynamic system, neural network is also widely used in the prediction of economic and financial fields. Convolutional neural network is suitable for the classification and regression of complex data sets. Because of its internal structure and convolution and pooling operation, convolution neural network has the characteristics of local connection and weight sharing. By convolution operation, each neuron only corresponds to the inner region of convolution core, so the local connection mode ensures the most effective region to learn the input data. What the weight sharing brings is to reduce the scale of network model and the number of weights. Compared with the common neural network, the convolution neural network has lower network model complexity and higher calculation efficiency. In this paper, on the basis of in-depth study of the theory, we will explore the feasibility of convolutional neural network (CNN) to predict the stock price of China's stock market. In addition, the attention of stock is closely related to the volatility of stock price observed by the finance society. Then, on the basis of the model, combined with the search index and media index provided by Baidu, the corresponding model is improved and optimized, and the corresponding model is compared.

2 Related Work

Predicting the stock price is a common topic in the financial field and the computing field as well. With the rapid development of computer technology, machine learning and neural network is increasingly widely used in stock price prediction. Traditionally, people often use fundamental and technical parameters to predict stock prices. In recent decades, most of the stock market forecasts focus on the prediction of future stock prices by historical data. The researchers applied random forest [1], deep learning [2, 3], stepwise regression analysis [4] and other algorithms. Although these algorithms have achieved good results, due to the characteristics of the stock market, accident events can immediately affect the stock price. As a result, it is easy to over fit stock price forecasts using only historical data.

As told by Eugene Fama in his famous paper in 1965 [5], the financial market is informationally efficient, that is, stock prices you noted have reflected all known information, and the future price movement is in response to the future news or events. Recently, some works have begun to use events to predict stock price movements. Almost all of them use word embedding to represent text information [6]. X. Ding uses structured events extracted by Open IE and both linear and nonlinear models are employed to predict the S&P 500 index and three US companies [7]. X. Ding et al. then use a deep convolutional neural network to model the influences of events on stock price movements to get a further improvement on accuracy [8]. X. Ding leverage extra information from knowledge graph and their model obtains better event embedding and makes more accurate prediction on stock market volatilities [9]. Vargas et al. uses the RCNN classifier to predict intra-day stock price movement, and they conclude that the sentence embedding is better than the word embedding [10].

More recent works try to combine events and other information (e.g., traditional economical or technical parameters) to achieve better prediction accuracy. Peng in his master thesis combines financial news and historical data to predict stock price movement and achieve better results [11]. Xu et al. present a novel deep generative model jointly exploiting text and price signals [12]. Zhao et al. apply causality network to stock market movement prediction [13]. Zhang et al. extract the events from Web news and the users' sentiments from social media, and investigate their joint impacts on the stock price movements via a coupled matrix and tensor factorization framework [14]. Lei et al. bridge text-based deep learning models and end users through visually interpreting the key factors learned in the stock price prediction model [15]. Zhou et al. use multiple heterogeneous data sources, including historical trading data, technical indicators, stock posts, news and Baidu Index, to predict the direction of stock price trend [16].

Compared with the previous work, we choose the news headlines to predict the stock price movement. In addition, we introduce data such as Baidu Index and search index as the proxies of attentions to see if we can improve the accuracy of stock price prediction by adding these new features.

3 Methodology

3.1 Data

3.1.1 Stock Sample
We study the stock sample of GuizhouMaotai (600519) in the Chinese wine industry from January 2010 to March 2020. We get the historical stock price data from the CSMAR database.

3.1.2 Stock News
We use web crawlers to get historical news headlines and release dates from www. 58188.com, a comprehensive financial news website.

In the task of using news headline data to predict the trend of stock price, the first problem to be solved is how to make the computer understand text information in the headline. We use the method of word embedding to convert each word after word segmentation into a low-dimensional floating-point number vector. Compared with other methods (e.g., one-hot representation), this method requires less memory, and the obtained word vector can pack more information into a lower dimension.

In our experiments later, we use the Keras deep learning framework to build the word embedding layer neural network, and the word vector obtained from the word embedding layer is the input part of the subsequent part of neural network. There are two ways to get word embedding: 1) Learning word embedding while completing the task of predicting stock price of news headlines. In this case, the first step is to initialize random word vectors, and then these word vectors are learned in the same way as the weights of the neural network. 2) Using the word embedding matrix trained by previous researchers [17] and load it into the classification model. The word embedding matrix is set not to participate in the training of the neural network. We find the second way performs better by experiments.

To represent a news title, we first get the sequence number of each word in the news title from the dictionary, and then extract the word vector in the whole word embedding matrix according to the sequence number. The word vector matrix of a news headline can be obtained by putting the word vectors in the news title together. For example, a news headline is represented by a matrix of 37 rows and 300 columns where 37 is the predefined maximum number of words in a news title and 300 is the length of a word vector. If the length of a news title is less than 37, we pad zero word vectors into the matrix.

3.1.3 Search and Media Indices

We use the search index and media index provided by Baidu Index as the proxy variables of the public attention and media attention, respectively. The Baidu search index is the weighted sum of search frequencies of a keyword (i.e., the name of a stock here), and the Baidu media index reflects the number of keyword-related news that is included by Baidu News. Because the values of search index and media index are very large, and their ranges are very different, so they can not be used directly. The solution we take is to standardize them with the maximum value and the minimum value, so that they all change to the value of the fluctuation in the [0, 1] interval, as shown in Formula (1).

$$SI[i] = (SI[i] - \min(SI))/(\max(SI) - \min(SI))$$
$$MI[i] = (MI[i] - \min(MI))/(\max(MI) - \min(MI)),$$

$$(1)$$

where SI[i] and MI[i] represent the ith day's search index and media index, respectively. Min (SI), max (SI), min (MI) and max (MI) represent the minimum and maximum search index and media index in the time span, respectively. In this way, the ranges of the standardized search index and media index are both [0, 1].

3.1.4 Classification Tasks

In our classification tasks, we consider two types of time windows:

TW(a): Using the news occurred in today's transaction time (i.e., today's 9:30–15:00) to predict today's stock price movement (fall, if today's closing price < today's opening price; rise, otherwise). This time window is meant to explore how the current transaction time's information can predict the current stock price movement.

TW(b): Using the news occurred before today's opening time and after the transaction time of yesterday (i.e., yesterday's 15:30–today's 9:30) to predict today's stock price movement (fall, if today's closing price < yesterday's opening price; rise, otherwise). This time window is meant to explore how the information before the transaction time can predict the stock price movement.

There are two kinds of features: (i) a news headline of a date; (ii) a news headline with Baidu index of a date. We use the features in a time window to predict the rise and fall of stock prices. Note that in the implementation of (ii), the Baidu index (search index or/and media index) is repeated in each rows and is added to the columns of the word embedding matrix of a news title.

3.2 The Convolutional Neural Network Model

Convolutional neural networks (CNN) have made great progress in image classification, speech recognition and other fields. In addition, it has been proven to be equally effective in various fields of natural language processing (NLP). Its main method is to extract the features of the original target through convolution operation so as to reduce the data scale without affecting the accuracy. The convolution neural network is established by the Keras framework in this work.

We describe the process of extracting features from filters. The model uses multiple filters (with different window sizes) to obtain multiple features. These features form the penultimate layer and are transmitted to the full connection layer. The full connection layer has two kinds of output, which are divided into the corresponding stock price prediction of the rise and fall of the stock price. The softmax nonlinear activation function is used in the full connection layer. The function of this layer is to convert the result to the probability distribution on the label. In the training process, the predicted value is compared with the actual value to calculate the loss value, so as to train and optimize the model through back propagation.

4 Experimental Results

We use the Keras deep learning framework to implement the experiments. The programming language is Python 3.7. The experimental environment is Ubuntu 18.04 operating system, inter (R) core (TM) i5-7300hq CPU, NVIDIA 1050ti graphics card. Through the website described above, the total number of news headlines collected is about 6000. After that, the dataset is divided into three parts by using the Sklearn method in Python: 60% for training sets, 20% for validation sets and 20% for test sets.

4.1   Evaluation Metrics

The performance of the system is evaluated by calculating the confusion matrix. Table 1 describes the components of the confusion matrix, including TP, TN for the correct classification of corresponding classes, and FP and FN for the wrong classification of corresponding classes. After obtaining the values in the confusion matrix, the accuracy is used to evaluate the system performance, which refers to the percentage of samples correctly classified. In addition, in order to obtain more comprehensive results, we introduce two indicators, F1 and AUC. F1 takes into account both the precision and recall of the classification model, which is defined by the harmonic mean of precision and recall. Precision rate refers to the proportion of the number of correct positive samples in the number of positive samples determined by the classifier. Recall rate refers to the proportion of the number of positive samples with correct classification to the number of real positive samples. AUC (area under curve) is defined as the area under ROC curve, which can represent the ranking ability of a classifier (assuming a 'positive' instance ranks higher than a 'negative' instance) and can deal with the sample imbalance problem of two classes.

Table 1. Confusion matrix

		Predicted classes	
		Rise	Fall
Actual classes	Rise	True Positive (TP)	False Negative (FN)
	Fall	False Positive (FP)	True Negative (TN)

Note that we evaluate the rise and fall, respectively and some evaluation metrics are defined as follows.

$$
\begin{aligned}
\text{Precision}_{rise} &= \text{TP}/(\text{TP} + \text{FP}) \\
\text{Recall}_{rise} &= \text{TP}/(\text{TP} + \text{FN}) \\
\text{F1}_{rise} &= 2\text{Precision}_{rise}\text{Recall}_{rise}/(\text{Precision}_{rise} + \text{Recall}_{rise}) \\
\text{Precision}_{fall} &= \text{TN}/(\text{TN} + \text{FN}) \\
\text{Recall}_{fall} &= \text{TN}/(\text{TN} + \text{FP}) \\
\text{F1}_{fall} &= 2\text{Precision}_{fall}\text{Recall}_{fall}/(\text{Precision}_{fall} + \text{Recall}_{fall})
\end{aligned} \tag{2}
$$

And we define the average precision, recall and F1 for the rise and fall classes:

$$
\begin{aligned}
\text{Precision} &= \left(\text{Precision}_{rise} + \text{Precision}_{fall}\right)/2 \\
\text{Recall} &= \left(\text{Recall}_{rise} + \text{Recall}_{fall}\right)/2 \\
\text{F1} &= \left(\text{F1}_{rise} + \text{F1}_{fall}\right)/2
\end{aligned} \tag{3}
$$

4.2   Experiments

Experiment 1: Prediction based on News Headlines for Two Time Windows
We introduce two time windows in 3.1.4: TW(a) using the current information and TW (b) using the previous information.

Experiment 2: Prediction based on News Headlines + Baidu Index for Two Time Windows
Through experiment 1, we find that the news using TW(a) is better than the news using TW(b) in predicting the trend of stock price, as shown in Table 2, which shows that using the news data of the current day to predict the stock price can get more profits. In Table 3 of Experiment 2, Baidu index (media+search index) and search index are introduced to investigate whether the task of stock price prediction was improved by adding new features. Finally, it is found that the introduction of Baidu Index improves the result of stock price forecast, but the effect is not very significant. Note that [16] also gets similar result on adding Baidu index to news polarities. So we conclude that Baidu index has no significant improvement on *binary* stock movement.

Table 2. Prediction based on news headlines for two time windows

	AUC	Accuracy	Precision	Recall	F1
News, TW(a)	0.8994	0.5721	0.5724	0.5724	0.5724
News, TW(b)	0.7124	0.5660	0.5660	0.5660	0.5660

Table 3. Prediction based on news headlines + Baidu index for two time windows

	AUC	Accuracy	Precision	Recall	F1
News + media/search index, TW(a)	0.8498	0.6104	0.6119	0.6119	0.6119
News + media/search index, TW(b)	0.6976	0.5682	0.5671	0.5671	0.5671
News + search index, TW(a)	0.7892	0.5833	0.5748	0.5848	0.5848
News + search index, TW(b)	0.6675	0.5738	0.5742	0.5742	0.5742

5   Conclusion

Based on the convolution neural network model, this paper forecasts the stock price by the news headlines of the wine industry. In addition, through the use of two different kinds of data, we find that TW(a) has better performance in the stock price forecasting task. Finally, the news headline data and Baidu index data are combined. The experimental results show that the prediction accuracy is improved to some extent, but the effect is not very obvious.

References

1. Xin, F., Xu dong, L., Haiyan, C.: Stock trend prediction based on Improved Stochastic Forest algorithm. J. Hangzhou Univ. Electron. Sci. Technol. **039** (002), 22–27 (2019)
2. Chong, E., Han, C., Frank, C.P.: Deep learning networks for stock market analysis and prediction: methodology, data representations, and case studies. Expert Syst. **83**, 187–205 (2017)
3. Li, X., Huang, X., Deng, X., Zhu, S.: Enhancing quantitative intra-day stock return prediction by integrating both market news and stock prices information. Neurocomputing **142**, 228–238 (2014)
4. Jeon, S., Hong, B., Chang, V.: Pattern graph tracking-based stock price prediction using big data. Future Gener. Comput. Syst. **80**, 171–187 (2018)
5. Fama, E.: The behavior of stock-market prices. J. Bus. **38**(1), 34–105 (1965)
6. Mikolov, T., Sutskever, I., Chen, K., Corrado, G.S., Dean, J.: Distributed representations of words and phrases and their compositionality. In: NIPS'13, pp. 3111–3119 (2013)
7. Ding, X., Zhang, Y., Liu, T., Duan, J.: Using structured events to predict stock price movement: an empirical investigation. In: EMNLP'14, pp. 1415–1425 (2014)
8. Ding, X., Zhang, Y., Liu, T., Duan, J.: Deep learning for event-driven stock prediction. In: IJCAI'15, pp. 2327–2333 (2015)
9. Ding, X., Zhang, Y., Liu, T., Duan, J.: Knowledge-driven event embedding for stock prediction. In: COLING'16, pp. 2133–2142 (2016)
10. Vargas, M., Lima, B., Evsukoff, A.: Deep learning for stock market prediction from financial news articles. In: IJCNN'18, pp. 1–6 (2018)
11. Peng, Y.: Leverage financial news to predict stock price movements using word embeddings and deep neural networks. Master of Science Thesis, York University, Canada
12. Xu, Y., Cohen, S.: Stock movement prediction from tweets and historical prices. In: ACL'18, pp. 1970–1979 (2018)
13. Zhao, S., Wang, Q., Massung, S., Qin, B., Liu, T., Wang, B., Zhai, C.: Constructing and embedding abstract event causality networks from text snippets. In: WSDM'17, pp. 335–344 (2017)
14. Zhang, X., Zhang, Y., Wang, S., Yao, Y., Fang, B., Yu, P.: Improving stock market prediction via heterogeneous information fusion. Knowl.-Based Syst. **143**, 236–247 (2018)
15. Lei, S., Teng, Z, Wang, L., Zhang, Y., Binder, A.: DeepClue: visual interpretation of text-based deep stock prediction. IEEE Trans. Knowl. Data Eng. 1–14 (2018)
16. Zhou, Z., Gao, M., Liu, Q., Xiao, H.: Forecasting stock price movements with multiple data sources: evidence from stock market in China. Phys. A **542**, 1–15 (2020)
17. Mikolov, T.: Efficient estimation of word representations in vector space. arXiv: Computation and Language (2013)

Research on Equipment Fault Diagnosis Classification Model Based on Integrated Incremental Dynamic Weight Combination

Haipeng Ji[1,2(✉)], Xinduo Liu[3], Aoqi Tan[4], Zhijie Wang[3],
and Bing Yu[3]

[1] School of Mechanical Engineering, HeBei University of Technology,
Tianjin 300401, China
823801536@qq.com
[2] Tianjin JingNuo Data Technology Co., Ltd., Tianjin 300401, China
[3] School of Artificial Intelligence, HeBei University of Technology,
Tianjin 300401, China
[4] School of Economic Management, HeBei University of Technology,
Tianjin 300401, China

Abstract. This study proposes a classification model of equipment fault diagnosis based on integrated incremental learning mechanism on the basis of characteristics of industrial equipment status data. The model first proposes a dynamic weight combination classification model based on long short-term memory (LSTM) and support vector machine (SVM). It solved the problem of fault feature extraction and classification in high noise equipment state data. Then, in this model, integrated incremental learning mechanism and unbalanced data processing technology were introduced to solve problems of massive unbalanced new data feature extraction and classification and sample category imbalance under equipment status data. Finally, an equipment fault diagnosis classification model based on integrated incremental dynamic weight combination is formed. Experiments prove that the model can effectively overcome the problems of excessive data volume, unbalanced, high noise, and inability to correlate data samples in the process of equipment fault diagnosis.

Keywords: Neural network · Support Vector Machine · Integrated increment · Unbalanced data processing · Fault diagnosis

1 Introduction

In the actual production, it is very important for the mechanical equipment to operate safely and stably. In case of failure, it will affect the production and even endanger the life safety of relevant personnel. Therefore, it is necessary to carry out equipment state detection and fault diagnosis. The traditional methods of equipment fault diagnosis mostly rely on the accumulation of artificial experience. It is inefficient to collect and judge the fault information of equipment manually, which is not suitable for today's society [1]. This paper, based on a large number of domestic and foreign literatures, combined with the characteristics of mechanical equipment fault diagnosis data, by

© Springer Nature Singapore Pte Ltd. 2020
P. Qin et al. (Eds.): ICPCSEE 2020, CCIS 1258, pp. 475–489, 2020.
https://doi.org/10.1007/978-981-15-7984-4_36

studying the traditional Support Vector Machine (SVM), It is found that for SVM, its generalization ability is strong, only a small batch of sample data is needed for training to ensure good classification effect, but at the same time, support vector machine also has the problem of relying on parameter selection and weak fault tolerance; long-term and short-term memory Although the Long-Short Term Memory (LSTM) does not need to specify the prior probability in advance, it has good fault tolerance and non-linear processing ability. However, training the neural network requires a large amount of sample data, and it is easy to appear. Based on the above research, an Equipment Fault Diagnosis Classification Model Based on Integrated Incremental Dynamic Weight Combination (DWCMI model) is established, and experiments show that the model can not only effectively overcome The problem of large amount of fault data, unbalanced, high noise, and inability to correlate data samples during equipment fault diagnosis, and effectively saves time cost.

The rest of the paper is structured as follows: Sect. 2 summarizes the research methods of fault diagnosis methods, unbalanced data processing and incremental learning in the field of equipment fault diagnosis; the third section analyzes the equipment fault diagnosis classification model based on integrated incremental dynamic weight combination. The theoretical basis, method flow, and implementation steps of the DWCMI model are elaborated. In the fourth section, the DWCMI model is applied to the fault diagnosis process of rolling bearing equipment to achieve reliable feature extraction and fault classification. The validity of the DWCMI model is proved by experimental comparison. Finally, in the fifth section, the proposed method is briefly summarized, and the research direction that needs to be improved in the future is proposed.

2   Literature Review

The research status of equipment fault diagnosis, unbalanced data processing and incremental learning are summarized as follows.

2.1   Equipment Troubleshooting

In the current development of the industrial field, smart devices play an extremely important role. At the same time, the consequences caused by equipment failures become more and more serious. Therefore, how to carry out effective and accurate equipment fault diagnosis has been widely used by everyone. Pay attention to it. In the development of the fault diagnosis discipline, the most important and most critical issue is the extraction of fault feature information, which is directly related to the accuracy of fault diagnosis and the reliability of early fault prediction. In order to solve the problem of feature extraction of fault diagnosis, people mainly use signal processing, especially the theory and technical means of modern signal processing, to diagnose equipment by vibration signal, which is one of the most effective and commonly used methods in fault diagnosis [2]. In order to meet the needs of analyzing complex system faults, the use of deep learning to establish complex nonlinear fault diagnosis models based on the unique advantages of feature extraction and pattern recognition has begun to gradually show its characteristics [3].

2.2   Research Status of Unbalanced Data Processing in Equipment Fault Diagnosis

There are also many algorithms that have been improved for the lack of improved unbalanced data processing methods for classifiers. Qian et al. combine sampling methods with integrated learning theory, in which the sampling used is undersampling and SMOTE algorithm [4]. Han Hui et al. proposed to improve the AdaBoost algorithm and then combine it with the oversampling algorithm [5]. Chawla et al. combine Boosting and oversampling SMOTE algorithms. The basic idea is to incorporate the comprehensive sampling technique into each iteration process, making the following classifiers more concerned with a few classes [6].

2.3   Research Status of Incremental Learning in the Field of Equipment Fault Diagnosis

Incremental learning technology is a widely used intelligent data mining and knowledge discovery technology. The idea is that the learning accuracy should be improved when the samples are gradually accumulated. The incremental learning algorithm for pattern classification can be applied to applications with large data volumes and data streams. For incremental problems, Gauwenberghs G et al. proposed an exact solution for incremental training [7], adding one training sample or reducing the effect of a training sample on Lagrange coefficients and support vectors. In reference [8], an incremental solution to the exact solution of the global optimization problem is proposed.

3   Equipment Fault Diagnosis Classification Algorithm Based on Integrated Incremental Dynamic Weight Combination

This section introduces the DWCMI method in detail. This method firstly combines the long-short-term memory neural network and the support vector machine dynamically, and uses the support vector machine to dynamically adjust the weights of the respective classifiers in the combined model to improve the classification of rolling bearing fault diagnosis. Accuracy; secondly, the non-equilibrium processing model is used to solve the problem of fault category imbalance in massive equipment state data; finally, the fault learning feature extraction and classification problem of newly added equipment state data is processed in real time through the incremental learning mechanism.

3.1   Unbalanced Data Processing Model

Based on oversampling and undersampling, this paper proposes a new data sampling model, which is based on the oversampling and undersampling fusion data sampling model (New Kernel Synthetic Minority Over-Sampling Technique And Tomek links Based on K-Nearest Neighbors, Referred to as NKSMOTE-NKTomek model). Oversampling refers to the improved SMOTE algorithm, namely the NKSMOTE algorithm [9]. The algorithm mainly solves the effective. Reference [10] introduced the PSVM-Proximal Support Vector Machine, which is extremely fast.

In addition to the above methods for implementing incremental learning by improving the SVM algorithm, we can also implement incremental classification learning through multi-classifier integration. The multi-class integration method is divided into three categories [11].

- When a new set of data is encountered, a set of fixed classifiers updates the joint rule or voting weight.
- The existing classifier updates the parameters each time a new data set is encountered.
- Each time a new data set is encountered, a new classifier is trained with the new data set and added to the current integration.

The NKSMOTE-NKTomek algorithm steps are as follows:

1. New Kernel Synthetic Minority Over-Sampling Technique (NKSMOTE model): The sample set is divided into a few class samples and a majority class sample. For a few class samples x, look for x in the kernel space. k neighbor samples. Then, the sample types of x are divided according to the number of samples in the K samples and the number of samples in the majority.
   a. Safety samples: that is, the number of samples in a sample set is greater than or equal to the number of samples in most categories.
   b. Boundary samples: that is, the number of samples in the sample set is smaller than the number of samples in the majority.
   c. Noise samples: There are only a majority of samples in the sample set.

For the data of types a and b, 2 samples are randomly selected among the k samples, and n new samples are synthesized according to a certain rule among the 3 samples, wherein the n value is the upsampling magnification.

If the selected two samples y_1 and y_2 are majority samples, use the following formula to generate n samples:

a. Generate n temporary samples $t_j (j = 1, 2, \ldots, n)$ according to y_1 and y_2:

$$t_j = y_1 + rand(0, 0.5) \times (y_2 - y_1) \tag{1}$$

b. Generate a new minority class sample $X_j (j = 1, 2, \ldots, n)$ from t_j and x:

$$X_j = x + rand(0, 1) \times (t_j - x) \tag{2}$$

If there are a few samples in the selected two samples y_1 and y_2, use the following formula to generate n samples:

a. Generate N temporary samples $t_j (j = 1, 2, \ldots, n)$ according to y_1 and y_2:

$$t_j = y_1 + rand(0, 1) \times (y_2 - y_1) \tag{3}$$

b. Generate a new minority class sample $X_j (j = 1, 2, \ldots, n)$ from t_j and x:

$$X_j = x + rand(0, 1) \times (t_j - x) \tag{4}$$

For the c type of data, n is set to 1 in order to reduce the risk caused by the noise data. At the same time, a few classes y are randomly selected from a small number of samples, and a new sample is randomly generated using the following formula.

$$X = x + rand(0.5, 1) \times (y - x) \tag{5}$$

When obtaining the K-nearest neighbor, the nonlinear mapping function is first used to map the sample to the kernel space. The distance between the samples in the kernel space is called the kernel distance. The calculation formula is:

$$\begin{aligned} d(\varphi(x), \varphi(y)) &= \sqrt{\|\varphi(x) - \varphi(y)\|^2} \\ &= \sqrt{K(x,x) + K(y,y) - 2K(x,y)} \end{aligned} \tag{6}$$

Where $\varphi(x)$ is a nonlinear mapping function, $K(x,y)$ is a kernel function, and the kernel function used here is a Gaussian kernel function, and its calculation formula is:

$$K(x,z) = e^{-\frac{\|x-z\|^2}{2\sigma^2}} \tag{7}$$

2. Tomek links Based on K-Nearest Neighbors (NKTomeK model): For minority and majority samples, a small number of samples are divided into different categories of data based on k-nearest neighbors. Reduce the number of samples calculated by the Tomek links algorithm and increase the efficiency of undersampled samples.
   a. According to the minority class sample class in step 1, the sample set synthesized in step 1 is re-divided into a minority class sample and a majority class sample, and k neighbors are obtained for each sample in the minority class sample, according to k The sample categories of the majority and the minority samples in the nearest neighbor are the safety samples, the boundary samples and the noise samples. The category judgment criteria are as shown in step 1.
   b. Remove noise samples.
   c. Assume that the boundary sample set in the minority sample is D, and the majority sample set is U, where the number of U, is N. for $i = 1, 2, \ldots, N$.
   d. The distance between x_1 in the majority sample set U and x_2 in the boundary sample set D, is $d = d(x_1, x_2)$.
   e. The distance between x_1 in the majority sample set U and each sample in the synthesized sample set is the distance data set F. If $d < F$, the line where x_1 is located is returned.
   f. At the end of the loop, delete the sample set of the returned rows. Combine the majority class set U with a few class sets.

3.2   Integrated Incremental Learning Mechanism

The algorithm steps of the integrated incremental model based on learn++ are as follows:

1. Assume that the input is the training data set d^t to be processed at the t time, d^t is composed of the instance $x^t(i)$, and the examples $i = 1, 2, \ldots, m^t$, a total of m^t.
2. When $t = 1$, set instance weight w_i^1, penalty weight D_i^1 is equal weight:

$$D_i^1 = w_i^1 = \frac{1}{m^1}, \forall i \tag{8}$$

3. At the subsequent t time, w_i^1, and D_i^1 are determined according to the classification accuracy of the integrated classifier on the current data set. t time training on the new data set to generate a new classifier h_{i^t}. Then, all currently generated classifiers need to calculate the classification error rate on the new data set:

$$\varepsilon_k^t = \sum_{i=1}^{m^t} D_i^t [h_k^t(x^t(i) \neq y^t(i))], k = 1, 2, \ldots, t \tag{9}$$

Penalty weight is used to calculate the error rate. D_{i^t} weighting:

$$D_{i^t} = \frac{w_i^t}{\sum\limits_{i=1}^{m^t} w_i^t} \tag{10}$$

4. The base classifiers that produce different error rates in different periods are handled differently. For the currently generated base classifier h_k^t, if the error rate $\varepsilon_{k=t}^t > 0.5$, the classifier is invalid and needs to be re-learned to generate a new classifier. For the base classifier h_k^t generated at the previous k time, if the error rate $\varepsilon_{k=t}^t > 0.5$, the error rate $\varepsilon_{k<t}^t$ of the classifier is set to 0.5. h_k^t, indicates the use of the classifier generated at time k at the current t time. Finally, all base classifiers are weighted and integrated to form an integrated classifier:

$$H^t(x^t(i)) = \arg \max \sum\nolimits_k W_k^t [h_k^t(x^t(i) = c)] \tag{11}$$

The voting weight of each base classifier is determined by the weighted average error rate of the classifier:

$$W_k^t = -\lg \beta_k^t \tag{12}$$

3.3 LSTM-SVM Based Dynamic Weight Combination Classification Model

The LSTM-SVM dynamic weight model relies on SVM to dynamically adjust the weight between the LSTM and SVM classifiers. This section focuses on the construction process of the combined model.

1. The SVM model construction process is as follows
   a. Extract feature vectors from known training samples to build training sample sets $\{(x_i, y_i)|i = 1, 2, \ldots, n\}$;
   b. Select the corresponding kernel function and corresponding parameters;
   c. Under the condition that the condition $\sum_{i=1}^{n} y_i \alpha_{i=0}$ is satisfied, the optimal Lagrangian parameter α^* is found;
   d. The support vector is searched from the training sample set, and the weight coefficient w^* and the classification threshold b^* of the optimal classification hyperplane are solved to obtain the optimal classification hyperplane.
   e. End the training process and get the SVM classification model.
2. The LSTM model construction process is as follows
   a. Input layer

Let the training set $x \in R^{m \times n}$, m represents the number of samples, and n represents the data dimension. At the same time, the time dimension is added to the training set, and the training data is transformed into a three-dimensional matrix, that is $R^{m \times t \times i}$, where t represents the time dimension on the sample, that is, the sequence length, and i represents the input neuron at each moment. Dimensions. Map $x \in R^{m \times t \times i}$ to a linear input layer with weight $W^{(i)}$ and offset $b^{(i)}$, and change the data dimension i at each moment of the sample. Its formula is:

$$y^{(i)} = x \times W^{(i)} + b^{(i)} \tag{13}$$

among them: $W^{(i)} \in R^{i \times i_1}$; $b^{(i)} \in R^{i_1}$; $y^{(i)} \in R^{m \times t \times i_1}$.

   b. LSTM network layer

It can be seen from a that the input of the LSTM network layer is $y^{(i)} \in R^{m \times t \times i_1}$, and the number of neurons of the LSTM is d, assuming the hidden layer output of the last time of each sample as the LSTM. The output of the network $y^{(h)}$, then $y^{(h)} \in R^{m \times d}$.

   c. Output layer

Use the output of b as the input of the output layer, and use the softmax output layer to match the output dimension of the LSTM network layer with the last number of classifications.

$$y' = softmax(y^{(h)} \cdot W^{(o)}) \tag{14}$$

among them: $W^{(o)} \in R^{d \times q}$, q is the number of categories. y' is the output of the network architecture, $y' \in R^{m \times q}$.

d. Cost function

The output probability distribution of the training is compared with the real data distribution, and the cross-entropy cost function of the predicted output and the actual output is calculated.

$$H(y) = \sum_m y' \log_2 y \qquad (15)$$

Establish the basic structure of the LSTM as above, initialize the network parameters, and set the training times T. In each iteration, the cross-entropy cost function of the current iteration is obtained by forward propagation, and then the network parameters are updated by error backpropagation. Finally, iteratively T times, the cost function tends to converge gradually.

4 Experiment

This section mainly analyzes the application performance of the proposed DWCMI model in the fault diagnosis of rolling bearings, and realizes the feature extraction and fault mode classification of bearing equipment. Firstly, the experiment selects the appropriate model parameters and model structure through experiments, and then demonstrates the effectiveness of the DWCMI model by comparing the experiments with BP, LSTM, SVM and other algorithms without adding unbalanced data processing and without adding incremental learning.

4.1 Data Description

The experimental data comes from the bearing status data of the Electrical Engineering Laboratory of Case Western Reserve University (CWRU). The data acquisition device includes a 2 horsepower motor, a torque sensor, a power meter and electronic control equipment. Test bearing support motor shaft, bearing model is 6205-2RS JEM SKF deep groove ball bearing, using EDM technology to arrange a single point of failure on the bearing, the fault position is in the inner ring, outer ring and rolling element respectively, the fault diameter is 0.007, 0.014, 0.021 (1 in. = 2.54 cm). That is, the vibration signal in the four states of normal state (None), inner ring fault (IRF), outer ring fault (ORF) and rolling element fault (BF) is collected, and the sampling frequency is 12 kHz.

In order to verify the effectiveness of the DWCLS model and the DWCMI model in the field of bearing fault diagnosis, it is necessary to manually label the vibration signal, fault category classification 007 inner ring fault, 014 inner ring fault, 021 inner ring fault, 007 outer ring fault, 014 Circle fault, 021 outer ring fault, 007 rolling element fault, 014 rolling element fault, 021 rolling element fault, normal sample, for each type of fault data, randomly selected 40 samples as training samples, 20 samples as test samples Each sample contains 1024 sample points, that is, the training sample set data points are a total of 409,600, and the test sample set data points are 102,400. The specific bearing status data sample description is shown in Table 1.

Table 1. Bearing fault data description

DataSet	FaultType	FaultDiameter	TrainingExample	TestSample	SamplePoints
1	None	0	40	10	1024
2	IRF	007	40	10	1024
3	IRF	014	40	10	1024
4	IRF	021	40	10	1024
5	ORF	007	40	10	1024
6	ORF	014	40	10	1024
7	ORF	021	40	10	1024
8	BF	007	40	10	1024
9	BF	014	40	10	1024
10	BF	021	40	10	1024

To test the effectiveness of the DWCLS model in bearing fault diagnosis classification in the case of high noise data, Add random noise points based on the above samples. Add random noise points based on the above samples. The noise points account for 10%, 20%, 30%, 40% of the total number of training sample sets and the total number of test sample sets, the training sample set and the test sample set are then placed in the DWCLS model to record the classification results.

In order to test the effect of the NKSMOTE-NKTomek model on the classification of rolling bearing fault diagnosis in the case of unbalanced data sets, 80, 40, 20, 10 fault samples were randomly selected from the training samples and combined with 100 normal samples as 4 different. Training samples, 20 fault samples and 50 normal samples as test samples, each sample contains 1024 sample points, and four training samples are placed in the DWCLS model added to the NKSMOTE-NKTomek algorithm for training, get training results and test results. In order to verify the influence of the integrated incremental learning method on the classification accuracy of bearing fault diagnosis in massive data, each training set is divided into four groups. The first training set is used to train the DWCMI model, and the remaining three groups are used as new samples. Training the DWCMI model. The specific bearing status data sample description is shown in Table 2.

Table 2. Description of bearing status data

Group	Tilt rate	Status type	Number of samples	Sample points
DataSet1	1.25	Normal	100	1024
		Malfunction	80	1024
DataSet2	2.5	Normal	100	1024
		Malfunction	40	1024
DataSet3	5	Normal	100	1024
		Malfunction	20	1024
DataSet4	10	Normal	100	1024
		Malfunction	10	1024
TestSet	2.5	Normal	50	1024
		Malfunction	20	1024

4.2   Model Structure

1. The proportion of noise in the data set

The noise specific gravity added to the model is one of the important factors affecting the model's effect. Adding too much noise may cause the model to lose some useful information and increase the training time of the model. Therefore, we determine the non-equilibrium and wavelet transform of the model. After the structure, the LSTM model structure, and the SVM model structure, it is necessary to determine the magnitude of the noise specific gravity through experiments. Figure 1 shows the variation of the correctness rate with the number of training times when the noise specific gravity is 10%, 20%, 30%, 40%. It can be seen from the figure that when the noise specific gravity is 20%, the correct rate of the model varies with the number of trainings. The increase, the convergence speed is faster, and the learning efficiency is higher. Therefore, 20% noise specific gravity is added to the original sample to verify the effectiveness of wavelet denoising and reconstruction in the data set processing.

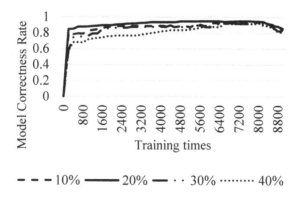

Fig. 1. Noise specific gravity line chart

2. LSTM model

When establishing a bearing fault diagnosis and recognition model based on long-term and short-term memory neural network, some parameters need to be set by themselves. After continuous iteration, the optimal parameters are selected to obtain appropriate data values. The parameters that have a large influence on the LSTM model include the training times T, the learning rate η, the sequence length step, and the number of neurons d in the hidden layer unit. Each parameter has a great influence on the training effect, training time, and computational complexity of the LSTM model.

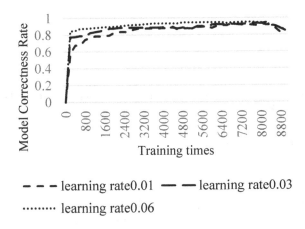

Fig. 2. Learning rate line chart

Figure 2 shows the relationship between the correct rate and the number of training times under different learning rates. It can be seen from the figure that when the learning rate is 0.001, 0.003, 0.006, the correct rate is roughly the same when the number of iterations is less than 5000, and the number of iterations is 7000. And when the learning rate is 0.006, the accuracy rate curve shows a significant drop. The learning rate generally determines the speed at which the parameter is iterated to the optimal value. In general, the larger the learning rate, the larger the step size of the gradient in training, and the easier it is to skip the optimal solution.

Figure 3 shows the correct rate curve for the sequence length s of 256 and 1024 (learning rate 0.006, training times 8000). In the LSTM network, the longer the sequence of the sample, the more the number of gradient iterations of the backward propagation of the error, and the larger the calculation amount, which affects the convergence speed and learning efficiency. As shown in Fig. 7, when the sequence length is 1024, the convergence speed is slow and the learning efficiency is low. When the sequence length is 256, the correct rate is significantly improved, and the convergence speed and learning efficiency can also be improved.

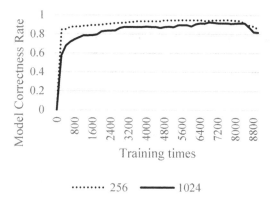

Fig. 3. Sequence length line graph

From the above experimental results, it can be concluded that the training frequency T of the LSTM model in the rolling bearing fault diagnosis experiment should be set to 8000, the learning rate η is set to 0.006, the sequence length s is set to 256, and the number of neurons is set to 200.

4.3   Experimental Results

After verifying the validity of the unbalanced data processing in the DWCMI model, a comparison experiment was performed on the performance of the integrated incremental learning algorithm in the DWCMI model, and four different training sets were divided into four groups, one of which was used as a training DWCMI model. The remaining three groups are used to add to the existing model for integrated incremental learning in three times. Use the DWCMI model proposed in this paper to compare the incremental learning with the DWCLS model, LSTM model, SVM model, and BP model, and use the test data set to test the model classification diagnosis effect, and record the accuracy of 20 experimental results for each set of incremental data. The G-mean value and the running time are averaged. The training average and test value comparison results of the four sets of incremental data are shown in Table 3.

It can be seen from Table 3 that the DWCMI model proposed in this paper is superior to the other four models in the training set of four different unbalanced tilt rates in terms of accuracy, G-mean and running time. From the final diagnosis of the model, the DWCMI model proposed in this paper maintains a high level of accuracy and G-mean values in different data sets of different tilt rates, which is significantly better than other BP models without incremental learning. Compared with the LSTM model, the DWCLS model and the SVM model have better training results, but the training time of the DWCLS model is significantly higher than that of other models. The SVM model has a large fluctuation in the test results of different tilt rate test sets. Therefore, the integrated incremental learning model in the DWCMI model can effectively incrementally train new data to obtain new models, and the DWCMI model can better adapt to changes in the environment through dynamic weights and penalty coefficients.

Table 3. Comparison of troubleshooting results

Train set	Method	Train precision (%)	Train G-mean	Test precision (%)	Test G-mean
1	DWCMI	91.38	0.9133	89.05	0.8774
	DWCLS	91.26	0.8991	88.23	0.8671
	LSTM	90.64	0.8923	87.05	0.8529
	SVM	89.83	0.8851	81.34	0.7745
	BP	90.25	0.8815	84.24	0.8306
2	DWCMI	91.03	0.8529	88.93	0.8441
	DWCLS	90.33	0.8485	87.36	0.7864
	LSTM	90.74	0.8513	86.85	0.7732
	SVM	87.24	0.7899	82.15	0.6892
	BP	89.71	0.8476	83.54	0.7014
3	DWCMI	91.56	0.8421	89.25	0.7436
	DWCLS	90.65	0.7585	87.93	0.6928
	LSTM	91.44	0.8402	87.87	0.6913
	SVM	89.19	0.7523	82.95	0.6708
	BP	90.63	0.7563	84.34	0.6864
4	DWCMI	91.91	0.7549	90.60	0.6892
	DWCLS	89.14	0.6819	88.24	0.6702
	LSTM	91.32	0.6981	88.05	0.6742
	SVM	92.24	0.7657	83.05	0.6257
	BP	91.34	0.6858	85.24	0.6454

5   Conclusion

In order to solve the problem of equipment fault diagnosis under the condition of massive unbalanced high-noise equipment status data, this paper proposes a classification model of equipment fault diagnosis based on integrated incremental dynamic weight combination. The model first uses the New Kernel Synthetic Minority Over-Sampling Technique And Tomek links Based on K-Nearest Neighbors (NKSMOTE-NKTomek model) to solve the category imbalance problem in the equipment operating state data. Secondly, the wavelet transform is used to remove the noise points in the vibration signal, and then the wavelet packet reconstruction is used to reconstruct the vibration signal after denoising into the original vibration signal; and the characteristic parameters of the vibration signal are realized by using the pole symmetric mode decomposition (ESMD). Extracting, normalizing feature parameters into feature vectors, using feature vectors to train long- and short-term memory neural networks and support vector machine models, and using support vector machines to dynamically adjust the weights of the respective classifiers in the combined model to achieve long-term and short-term The dynamic weight adjustment of the memory neural network and the support vector machine in the combined model. Finally, when there is new sample data, the feature vector of the newly added sample is put into the previously trained combination model as the test set. If the output result of the combined model meets the expected result, no processing is performed, if the output result does not meet the expected requirement. Integrate incremental learning for new sample feature vectors. After experimental verification, the DWCMI model makes the bearing fault diagnosis efficiency reach 89.45% on average, which is 3.93% higher than that of the non-incremental learning and unbalanced data processing process. Reliable classification of fault categories under massive unbalanced high-noise rolling bearing equipment status data.

Although the DWCMI model proposed in this paper has high classification performance for fault diagnosis of massive non-equilibrium and high-noise equipment operating state data, there are still some places where the model can be further improved. First, the model training time needs to be further shortened. In particular, the integrated incremental learning method can be further improved to obtain a faster integrated incremental learning model. At the same time, the unbalanced data processing in the DWCMI model is only for normal. Classes and fault classes are processed without further division of the different categories in the fault class. The next step is to refine the categories to obtain a more balanced data set and improve the classification of fault diagnosis.

Acknowledgment. This work is supported by Tianjin Science and Technology Project under Grant No. 18YFCZZC00060 and No. 18ZXZNGX00100, and Hebei Provincial Natural Science Foundation Project under Grant No. F2019202062.

References

1. Wang, T., Wang, T., Wang, P., Qiao, H., Xu, M.: An intelligent fault diagnosis method based on attention-based bidirectional LSTM network. J. Tianjin Univ. (Sci. Technol.) **53** (06), 601–608 (2020)
2. Min, Y.: Digital Image Processing, pp. 140–142. Mechanical Industry Press, Beijing (2006)
3. He, X.: Research on Theory and Method of Fault Intelligent Diagnosis Based on Support Vector Machine. Central South University (2004)
4. Batistag, E.A.P., Prati, R.C., Monard, M.C.: A study of the behavior of several methods for balancing machine learning training data. ACM SIGKDD Explor. Newslett. **6**(1), 20–29 (2004)
5. Qian, Y., Liang, Y., Li, M.: A resampling ensemble algorithm for classification of imbalance problems. Neurocomputing **143**(143), 57–67 (2014)
6. Wang, L., Chen, H.: Non-equilibrium dataset classification method based on NKSMOTE algorithm. Comput. Sci. **45**(9), 2–4 (2018)
7. Chawla, N.V., Lazarevic, A., Hall, L.O., Bowyer, K.W.: SMOTEBoost: improving prediction of the minority class in boosting. In: Lavrač, N., Gamberger, D., Todorovski, L., Blockeel, H. (eds.) PKDD 2003. LNCS (LNAI), vol. 2838, pp. 107–119. Springer, Heidelberg (2003). https://doi.org/10.1007/978-3-540-39804-2_12
8. Gauwenberghs, G., Poggio, T.: Incremental and decremental support vector machines. Mach. Learn. **44**(13), 409–415 (2001)

9. Sojodishijani, O., Ramli, A.R.: Just-in-time adaptive similarity component analysis in nonstationary environments. J. Intell. Fuzzy Syst. **26**(4), 1745–1758 (2014)
10. Gomes, H.M., Barddal, J.P., Enembreck, F.: Pairwise combination of classifiers for ensemble learning on data streams. In: Proceedings of the ACM Symposium on Applied Computing, New York, USA, pp. 941–946. ACM (2015)
11. Shao, H., Jiang, H., Zhang, X., et al.: Rolling bearing fault diagnosis using an optimization deep belief network. Meas. Sci. Technol. **26**(11), 115002 (2015)

Code Smell Detection Based on Multi-dimensional Software Data and Complex Networks

Heng Tong, Cheng Zhang$^{(\boxtimes)}$, and Futian Wang

Anhui Provincial Key Laboratory of Multimodal Cognitive Computation,
School of Computer Science and Technology, Anhui University, Hefei, China
cheng.zhang@ahu.edu.cn

Abstract. Code smell is the product of improper design and operation, which may be introduced in many situations. It will cause serious problems for further software development and maintenance. Currently, most code smell detection methods detect through a single type of software data. There are restrictions on detecting code smells with complex definitions and characteristics. In this paper, an approach of applying multi-dimensional software data is proposed. A complex network was built through structural data and historical version data, and code smell instances were determined by searching the network. We designed two smells detection strategies were designed and evaluated them in four open source projects. The results demonstrate that the proposed method has 23% and 15% higher F-measures on Shotgun Surgery and Parallel Inheritance Hierarchy than the existing mainstream detection ways. The code smell detection based on multi-dimensional software data and complex network is effective, and this method of processing multidimensional software data is also applicable for data-driven software research.

Keywords: Code smell · Detection technique · Software refactoring · Software maintenance

1 Introduction

High software demands make software systems have to provide more functions and fix more bugs [1]. However, it is difficult to meet all of them. In real development situations, developers may not be able to fully comply with development specifications [2]. And these unconscious negligence may lead to software aging and technical debt [3]. In this case, Fowler [4] proposed the concept of "code smell". It reveals a serious problem that some code does not directly affect the software system, but it can cause potential problems. Code smells are not an individual phenomenon, it is usually introduced due to the compromised handling in emergency situations [5] and leads to a series of serious problems. These issues include reducing the maintainability of the software [6], increasing the difficulty of understanding [7] and extending the development cycle [8]. Although

© Springer Nature Singapore Pte Ltd. 2020
P. Qin et al. (Eds.): ICPCSEE 2020, CCIS 1258, pp. 490–505, 2020.
https://doi.org/10.1007/978-981-15-7984-4_37

the impact of these problems is mostly indirect, they do reduce the quality of the software. Accurate detection of code smells and reasonable refactoring are necessary.

Researchers have always maintained high attention for the code smell detection. There are many methods to automatically detect code smells and alert developers to their risky development behavior [9–12]. Many detection tools are based on software metrics, which quantify some structural software data, and determine whether the object described by the metric information has the characteristics of the corresponding smells through specific rules. Machine learning is also widely used in metric-based code smell detection techniques [13]. In addition, historical version data for detection methods can complement metric-based detection techniques [14].

Existing studies show that there is a large research bias for code smell detection studies. Although existing methods show good performance, they are often limited to partial smells [13,15]. Previous methods are usually based on a single type of software data, but not all smells characteristics can be fully described by single types of data. Some code smells have both structural characteristics and historical characteristics. For example, if Shotgun Surgery is determined only from common changes in software historical version data, some methods that are not related to each other may be misjudged as Shotgun Surgery instances due to large-scale software refactoring.

Based on these considerations, we propose code smell detection method based on multi-dimensional software data and complex network. According to the method call and class inheritance relationship, we divide them into method level network and class level network. The purpose of this method is to detect the two smells defined by Fowler [4]. There are Shotgun Surgery and Parallel Inheritance Hierarchy. All two smells have obvious structural characteristics and historical characteristics.

Our study aims at addressing the below two research questions:

RQ1: How does our approach perform in code smell detection?

This research question is designed to determine the performance of our approach in detecting two code smells, namely Shotgun Surgery, Parallel Inheritance Hierarchy.

RQ2: How does our approach compare to the current code smell detection methods?

This research question is intended to compare our approach with other advanced code smell detection methods. The results of this study will demonstrate the effectiveness of complex network technology based on multi-dimensional software data in detecting code smells.

The organizational structure of this paper is as follows: In Sect. 2 we introduce the related works; In the Sect. 3, we introduce the specific method to detect code smells. In Sect. 4 we design the verification method. We report our results in Sect. 5. Section 6 summarizes our work and our future plans.

2 Related Work

Researchers take a lot of different approaches to extracting code smells from the software system [16]. Whether semi-automatic or automatic, they are primarily based on static code analysis or combining static analysis with dynamic analysis [15].

Manual-Based Code Smell Detection Method. This method is used early in the code smells study. Travassos et al. devised a "reading technique" that uses observation methods to find relevant information and identify shortcomings in software components by enumerating lists on paper [17]. The disadvantage of this type of approach is that it is inefficient for manual code smell detection for large systems.

Metric-Based Code Smell Detection. This method detects code smells by using metric such as the number of lines of code, the number of parameters and the number of methods. Traditional metric-based detection methods are generally based on specific detection rules. Moha et al. [9] proposed a metric-based detection method DECOR. This method designs a method for processing metrics and implements detection of three kinds of code smells. Rasool et al. [18] developed a rule-based lightweight detection method that proposes a solution that uses the same set of metrics to detect all of the original 22 code smells. Vidal et al. [10] proposed a semi-automatic detection method based on static code analysis. This method is currently limited by the detection of relatively simple code smells, and it is not possible to directly detect metrics for complex code smells.

Learning-Based Code Smell Detection Methods. They are largely based on metrics. Arcelli Fontana et al. [19] evaluated the use of machine learning techniques in the field of code smell detection and designed a set of methods for constructing code smells data sets. The performance of various machine learning algorithms is verified on 74 software systems included in Qualitas Corpus. In addition, Di Nucci et al. [20] further extended the research of Fontana. Liu et al. [21] also introduced deep learning and proposed an automatic generation method for training data based on neural network classifier markers. They believe that there may be some factors in the construction of code smells data sets, which may lead to the introduction of subjectivity. Palomba et al. [12] built a special bug prediction model for code smell based on previous research on bugs. This approach defines the severity of ten code smell instances based on a combination of three complementary metrics. For this method, there are challenges in getting a high-quality code smell data set of sufficient size and how to design a reasonable code smell data set.

The Historical Version Data-Based Code Smell Detection Methods. This method detects the code smells by analyzing the historical version data of the system. Fu et al. [22] analyzed the association rules mined in the history of software system changes and defined heuristic algorithms to complete the detection. Palomba et al. [14] uses changing historical version data to detect different code smells such as Divergent Change and evaluates its performance in 20 object-oriented open source projects. This method requires that the software system undergoes a period of evolution and places high requirements on the quality of the historical version data.

Cooperative-Based Code Smell Detection Methods. Boussaa et al. [23] propose to use a co-evolutionary search of competition to detect code smells, which allows for the generation of code smell instances to be combined with the generation of quality metric based detection rules. Kessentini et al. [24] performed evolutionary algorithms with different adapt abilities to detect code smells in a parallel collaborative manner. This method is still uncertain for detecting the generalization of genetic programming algorithms that detect code smells.

We conclude from the current studies that most code smell detection methods only use single type of data. It is difficult to simply extend the technology to provide smell instances that developers really have to notice. For some smells that not only have structural characteristics, it is difficult to find real instances with structural data detection methods. For methods that only use historical version data, the results of such methods are almost entirely determined by the quality of the software historical version data, which is affected by the subjectivity of the software developer. To our knowledge, there is currently no approach to design a combined detection technique based on the advantages of these two methods.

3 Method

3.1 Multi-dimensional Software Data

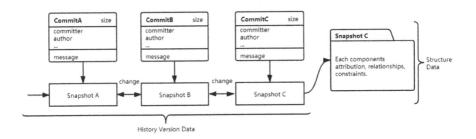

Fig. 1. Multi-dimensional software data

In the git repository, there are many commits that contain important data. This data is usually a separate change to a file (or set of files). The commit also saves revisions to the code and creates a unique ID. That is, the commit ID corresponds to an unique software snapshot. The set of commits at every two adjacent times in a period of time constitutes the data of the software system in the time dimension.

However, code smell detection requirements are usually raised in a specific software snapshot (usually the latest snapshot). The attributes of the internal classes and methods in this snapshot and their relationships are the usefully expression of the software structure. The software structured data represented by these attributes and relationships constitutes the data of the software system in the spatial dimension. The relationship between these two dimensions of software data is shown in Fig. 1. CommitA, CommitB and CommitC in the figure represent continuous commits of the software over a period of time. Each of these Commits corresponds to a software snapshot. The changes between every two continuous snapshots constitute the history version data. CommitC is the last commit in this period. The structure data is composed of software structure information such as attributes and relationships in the SnapshotC. These data constitute multi-dimensional software data.

3.2   Target Code Smell and Method Design

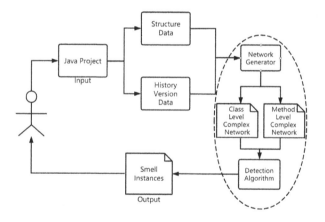

Fig. 2. The proposed code smell detection process our approach

Based on the data we applied and the definition of each code smell, we chose the following two code smells:

– **Shotgun Surgery**: When a software encounters some changes, developers have to make small changes in many of the classes in the software.
– **Parallel Inheritance Hierarchy**: Whenever developers create a subclass for a class, they will find themselves needing to create a subclass for another class.

The key idea of our approach is to identify code smell instances by analyzing complex networks generated from software historical version data and structured data. Figure 2 outlines the main steps of our approach. First, we extract data from the Java project. The Network Generator then processes the data and generates class-level and method-level networks. Finally, we set up the corresponding detection algorithm to search the network and generate smell instances. In next sections, we describe the steps of our approach to extract data and the rules for detecting code smells according to the network in detail.

3.3   Software Structured Data Extraction

We extracts software structured data for complex networks from software systems through a component called "network generator". The network generator first used java-callgraph[1] to extract the call relationship with software. Java-callgraph can read classes from Java, walks down the method bodies and prints a table of caller-caller relationships. After manually verifying the java-callgraph generated results, network generator builds method level complex network with method as nodes and method call relationships as direction edges. In a class level network, we use inheritance relationships to associate class nodes. This part of the inheritance relationship is extracted through CDA[2]. The purpose of this tool is to analyze Java class files in order to scan inheritance relationships between those classes.

3.4   Software History Data Extraction

For the selected code smells definition, analysis of historical version data is necessary. In this section, we designed a tool to extract relevant historical version data using GitPython[3]. GitPython is a Python library that interacts with Git libraries, including low level commands and high level commands. For consecutive snapshots in the selected time period, GitPython help us to get data between every two snapshots. The data extracted in this part is the records of all additions, deletions, and changes to the class/method in the evolution of historical versions. Then, we perform a text analysis on these records and get the relevant commit ID of the class/method as an attribute to the network node created in the previous step. At the end of this step, we have completed the complex network required for detection.

3.5   Complex Network Generator

Figure 3 is an example of a junit method level complex network and a node in the network. As can be seen from the figure, in addition to the relationship directed edge and node names that make up the network, each node stores the method name, formal parameter types and commit ID associated with it. This forms the search basis of our method. For class level networks, except that the relationship between nodes is a directed edge composed of inheritance. The parameter assigned to each node is the class name and the commit ID.

---

[1] https://github.com/gousiosg/java-callgraph.

[2] http://www.dependency-analyzer.org.

[3] https://gitpython.readthedocs.io/en/stable.

3.6   Code Smell Detector

We develop Code Smell Detector based on Networkx[4]. Networkx is a graph theory and complex network modeling tool developed in Python. The code smell detector takes the complex network as input. We have instantiated code smell detector and provide specific detection strategy for different code smells in the next. In the following we will detail the method we have designed to detect the different smells described.

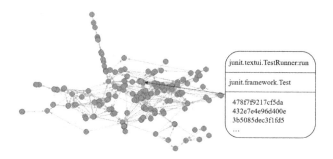

Fig. 3. The part of the jUnit method level network

Algorithm 1. Shotgun Surgery

Input: Result of $ComplexNetworkExtraction$
Output: Shotgun Surgery instances
1: $NodeRelatedClassSet \leftarrow \emptyset$
2: $CodeSmellInstances \leftarrow \emptyset$
3: **function** MAIN(MethodNetwork)
4:     **for** $Node \in MethodNetwork$ **do**
5:         $NodeRelatedClassSet = OutSameClass(Node.neighbor)$
6:         $NodeChangeIDList = Node.ChangeID$
7:         **if** $\sum NodeChangeIDList > \beta$ **then**
8:             **for** $NodeRelatedID \in combinations(NodeChangeIDList, \beta)$ **do**
9:                 **for** $neighborNode$ in $NodeRelatedClassSet$ **do**
10:                     **if** $NodeRelatedID \in neighborNode.ChangeID$ **then**
11:                         $CommonChangeNode \cup neighborNode$
12:                     **end if**
13:                     **if** $\sum ParentClassSum(CommonChangeNode) > \alpha$ **then**
14:                         $CodeSmellInstances \cup Node$
15:                         $Break$
16:                     **end if**
17:                 **end for**
18:             **end for**
19:         **end if**
20:     **end for**
21:     **return** $CodeSmellInstances$
22: **end function**

---
[4] http://networkx.github.io.

Shotgun Surgery. According to the definition of Shotgun Surgery, if an instance is affected by this smell, it will inevitably cause changes in some classes related to it. To detect this code smell, we searched the method level network. If a method is called by more than or equal to α methods from other different classes, and these methods have common changed more than or equal to β times in the historical version data, we define that the current instance is affected by Shotgun Surgery.

The detailed steps of our algorithm are shown in Algorithm 1. In the first and second lines, we created two lists that hold the results of the runs. The traversal of all nodes is started in the fourth line. In the fifth line, NodeRelatedClassSet collects other nodes that have edges with the target node. The OutSameClass function is to delete neighboring nodes that belong to the same class as the selected node. In the sixth line, NodeChangeIDList stores all the commitID that the target node has been changed. In the seventh to twelfth lines, we first select a combination of β IDs from NodeChangeIDList, and compare it with the change ID list of the neighboring node accordingly. In the thirteenth line, ParentClassSum is used to calculating how many different classes these nodes come from. When this last verification is true, we obtain a Shotgun Surgery instance.

Parallel Inheritance Hierarchy. According to the definition of Parallel Inheritance Hierarchy, if a group of instances is affected by this smell, they must have a parallel inheritance structure. To detect this smell, we traverse through the class level network and search for nodes (classes) generated in parallel in the network. If the number of nodes with parallel addition and same inheritance depth more than or equal to γ in the two inheritance substructure. We think that the current instance is affected by Parallel Inheritance Hierarchy.

The detailed steps of our algorithm are shown in Algorithm 2. After the first and second lines set up the list for storage, algorithm start traversing the network on the fourth line. For the selected two nodes, algorithm also traverse subclass trees of these nodes. In the sixth and seventh lines, algorithm match the addition ID and inheritance depth of the every two sub nodes from the subclass tree. The role of the Path method is to obtain the distance between two nodes. And record the matching successful combinations into the NodeMatchedClassSet. The twelfth line determines whether there are a sufficient number of matching node pairs (the root nodes of the two subtrees are also taken into account). That is, algorithm judge whether the number of these NodeMatchedClassSet reaches the limitation. When this last verification is true, we obtain a Parallel Inheritance Hierarchy instance.

Algorithm 2. Parallel Inheritance Hierarchy

Input: Result of $ComplexNetworkExtraction$
Output: Parallel Inheritance Hierarchy instances
1: $NodeMatchedClassSet \leftarrow \emptyset$
2: $CodeSmellInstances \leftarrow \emptyset$
3: **function** MAIN(MethodNetwork)
4:     **for** $Node1, Node2 \in MethodNetwork$ **do**
5:         **for** $SubNode1, SubNode2 \in Node1.Subclasstree, Node2.Subclasstree$ **do**
6:             **if** $Path(Node1, SubNode1) == Path(Node2, SubNode2)$ **then**
7:                 **if** $SubNode1.AddID == SubNode2.AddID$ **then**
8:                     $NodeMatchedClassSet \cup Node1 + Node2$
9:                 **end if**
10:             **end if**
11:         **end for**
12:         **if** $\sum NodeMatchedClassSet > \gamma + Node1.AddID == Node2.AddID$ **then**
13:             $CodeSmellInstances \cup (Node1, Node2)$
14:         **end if**
15:     **end for**
16:     **return** $CodeSmellInstances$
17: **end function**

3.7 Calibrating Strategy

Parameter settings have a significant impact on the performance of the search algorithm for a particular problem instance. Based on our limited parameter space, we use the brute force method to find the optimal setting of the parameters. We verified the results of using various parameter on Xerces[5]. Through this process, we determined the most appropriate value for each metric (the value used in the two detection rules). We show in Table 1 our selection range and the final determined values. SS and PIH are abbreviation for Shotgun Surgery and Parallel Inheritance Hierarchy, respectively.

Table 1. Selected value range

Metric	Experimental range	Best value
α(SS)	From 1 to 5 by steps of 1	3
β(SS)	From 1 to 5 by steps of 1	3
γ(PIH)	From 1 to 5 by steps of 1	2

4   Validation

To answer RQs, we use our approach on some software projects. In order to make the performance of our approach more informative. We not only consider the

---

[5] https://xerces.apache.org.

various sizes of software systems, but also select the system that is selected more times by the previous studies as possible [25]. We finally use four open source Java software systems to perform validation: Apache Tomcat[6], jEdit[7], jUnit[8], Android API (framework-opt-telephony)[9]. We list some information about these systems in Table 2. In particular, in order to simulate a real application environment, we selected a system snapshot from a previous time period, and applied our tools to the historical version data before this snapshot. For example, the history of Apache Tomcat from May 2006 to January 2020, we selected a 279d8563 snapshot from July 2013 and applied our method to all snapshots between July 2006 and July 2013. Subsequent historical version data can then be used to verify that we have found real instances of code smell.

Table 2. Selected projects

Project name	Number of classes	Git snapshot	Period
Apache Tomcat	1782	279d8563	2006.7–2020.1
jEdit	1226	692b051c	2002.1–2020.1
jUnit	764	8d26b450	2001.1–2020.1
Android API(framework-opt-telephony)	372	babd5ede	2011.1–2020.1

In order to report the detection performance of our method, we need to define a gold standard (GS) to use as a basis for comparison and evaluation. However, the definition of the gold standard in this area does not seem easy to implement. To our knowledge, no research has provided the number, size, and location of code smell instances in the software. Therefore, three software engineering master students from Anhui University were invited to manually determine the gold standard. They were asked to follow the code smell manual detection specification [17] to find smell instances in the four software systems. According to the definition of the code smell, they analyzed the corresponding software source code and version control repository. After obtaining the set of smell instances they provided, a software engineering doctor who also came from Anhui University verified these instances. At the beginning, the doctor believe that two of these instances were false positives. After discussion, they finally considered these two cases as true positives. And in order to avoid affecting the reliability of the experiment, the entire manual detecting process was carried out without knowing our detection strategy. The set of instances constitute our gold standard.

---

[6] http://tomcat.apache.org.

[7] http://www.jedit.org.

[8] https://junit.org.

[9] https://android.googlesource.com/platform/frameworks/opt/telephony.git.

Including the GS, the data used and defined in our research are publicly available[10]. We provide: (i) Inheritance (call) relationship between classes (methods); (ii) Commit ID for each method (class) when it is changed in (added to) the software system; (iii) Detect results of our method and competitive method; (iv) Each system's gold standard.

According to related studies, we have reproduced two detection methods, MLSD [18] and HIST [14]. Among them, HIST only extracts change history from the versioning system as input, but does not extract and analyze the structure of the software. It can be viewed as a detection tool that only uses historical version data. And MLSD uses software structured data to extract code smells in software source code. This is a typical code smell detection tool using only software structural features. We compared the results of our approach with the results of these two methods. After obtaining the oracle of code smells and the detection results of all the tools, we used the well-known recall and precision in the data retrieval as the performance evaluation indexes of our approach.

$$precision = \frac{TP}{TP + FP} \quad recall = \frac{TP}{TP + FN} \tag{1}$$

TP and FP refer to examples of true and false instances detected by the respective tools, respectively. To take into account the precision and recall rates, we also report F-measure, which is defined as the harmonic mean of the precision and recall rates:

$$F - measure = \frac{2 * precision * recall}{precision + recall} \tag{2}$$

5   Results

We report our findings in detail in this section. Tables 3 and 4 show the comparison results between our approach and two other recent detection methods. Specifically, we report (i) The number of smell instances (GS) for each system; (ii) TP and FP for each method; (iii) The corresponding precision, recall, and F-measure in the comparison results. We will then discuss the results of each code smell.

Table 3. Shotgun Surgery–comparison of results

Project	GS	MLSD					HIST					Our Approach				
		TP	FP	P	R	F	TP	FP	P	R	F	TP	FP	P	R	F
Apache Tomcat	2	0	3	0%	0%	0%	2	1	67%	100%	80%	2	0	100%	100%	100%
jEdit	1	1	5	17%	100%	29%	0	0	0%	0%	0%	0	0	0%	0%	0%
jUnit	0	0	6	0%	0%	0%	0	0	–	–	–	0	0	–	–	–
Android API (framework-opt-telephony)	0	0	1	0%	0%	0%	0	1	0%	0%	0%	0	0	–	–	–
Over all	3	1	15	6%	33%	10%	2	2	50%	67%	57%	2	0	100%	67%	80%

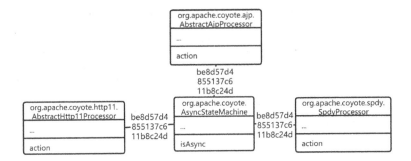

Fig. 4. A Shotgun Surgery instance

Shotgun Surgery (Table 3): We identified 3 Shotgun Surgery instances in 4 systems. Our approach detected 2 true positives of them (Precision: 100%) (Recall: 67%). MLSD did not produce any true positives on Apache Tomcat, and produced many false positives on these systems (Precision: 6%) (Recall: 33%). This result is expected. Structural data alone is not enough to match all the characteristics of this smell. HIST produced 2 false positives (Precision: 50%), and detected 2 instances (Recall: 67%). This method only uses the method of querying the common changes to complete the detection, and a typical false positive caused by this is that multiple refactor operations are recorded in the historical version, which makes the resulting common changes do not constitute Shotgun Characteristics of Surgery. A false positive by HIST in Android API (framework-opt-telephony) illustrates the problem. Although dispose method in GsmDataConnectionTracker get changed in several commit ID with some other methods, it has no relevance to these nodes that change together, which means that dispose is not an smells instance that caused these changes. In addition, this method is unavoidable because of the high threshold. Due to the structural association established by our method at the beginning of the search, not only the unrelated common changes are solved, but also the threshold setting is more relaxed.

An example from Shotgun Surgery in Apache Tomcat is shown in Fig. 4. The isAsync method in AsyncStateMachine is related to the existence of calls from three different classes of action methods, and common changes have occurred in all three commits.

Parallel Inheritance Hierarchy (Table 4): We identified 11 Parallel Inheritance Hierarchy instances in 4 systems. our approach's F-measure in this section is 87%, it detected 10 of them (Recall: 91%), and reported 2 false positives (Precision: 83%). The F-measure of HIST is 72% (Precision: 64%) (Recall: 82%), and the F-measure of MLSD is 6% (Precision: 5%) (Recall: 9%). HIST for detecting co-generated class pairs is an effective detection strategy in most cases, but this strategy will suffer losses when the inheritance subtree is complicated. This loss is mainly manifested in large-scale software refactoring and a different inheritance substructure. Although MIST has proved that it is difficult to determine

Table 4. Parallel Inheritance Hierarchy–comparison of results

Project	GS	MLSD					HIST					Our approach				
		TP	FP	P	R	F	TP	FP	P	R	F	TP	FP	P	R	F
Apache Tomcat	10	1	10	9%	10%	9%	8	4	67%	80%	73%	9	2	82%	90%	86%
jEdit	1	0	5	0%	0%	0%	1	0	100%	100%	100%	1	0	100%	100%	100%
jUnit	0	0	3	0%	0%	0%	0	1	0%	0%	0%	0	0	–	–	–
Android API (framework-opt-telephony)	0	0	3	0%	0%	0%	0	0	–	–	–	0	0	–	–	–
Over all	11	1	21	5%	9%	6%	9	5	64%	82%	72%	10	2	83%	91%	87%

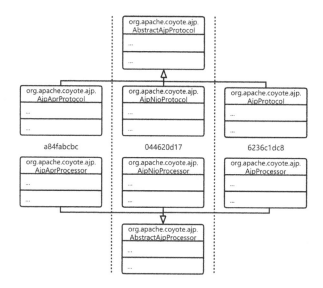

Fig. 5. A Parallel Inheritance Hierarchy instance

Parallel Inheritance Hierarchy only by structure, combining the structured data can still supplement the method of applying only historical version data.

An example of Parallel Inheritance Hierarchy from Apache Tomcat is shown in Fig. 5. The three subclasses of the AbstractAjpProtocol and AbstractAjpProcessor classes have been added to the software system in three different commit (Fig. 5 center number).

Time Complexity. Our algorithm are executed on machines with Intel I5-8400 processors and 16 GB RAM. Even for Apache Tomcat network, the most large complex network, our method only took 0.21 s to get the result. We believe that time consumed by the code smell detection algorithm itself is negligible. However, since our detection algorithm needs to use both source code and historical version changes as input, searching and extracting data may take some time. But in actual application scenarios, developers can easily obtain data from the software developed by themselves. Therefore, we believe that the time complexity of our code smell detection method is acceptable.

Summary for Results. In summary, for RQ1, our code smell detection method achieve high performance in detecting smells such as Shotgun Surgery and Parallel Inheritance Hierarchy. All two kinds of smells we detect have obvious structural and historical characteristics. The detection method based on multi-dimensional software data and complex network is an effective method to deal with this characteristics. Our experimental results also show that this method can specifically detect the code smell in software systems. For RQ2, we performed a detailed performance comparison of our method with other detection methods. The results show that our approach performs better than other detection methods. These competing methods usually only describe part of the characteristics of smells, especially the methods based on structural data are not applicable to the detection of these two types of smells. In addition, our method also provides a solution for combining multi-dimensional software data. It may support combining with other types of methods, such as machine learning.

6 Conclusion and Future Work

Most of the current code smell detection methods are based on a single software data for code smell detection, and mainly use software structured data. For more complicated code smells with multiple characteristics, their detection performance has limitations. Our approach is designed to address these issues. This paper proposes a new method of combining historical version data and structural data of software through complex network technology. This method focuses on two less popular code smells (Shotgun Surgery and Parallel Inheritance Hierarchy) that still pose a threat to software systems. Their characteristics are explained by both structural and historical version data. Based on these characteristics and definitions, we design two association rules.

Our method is evaluated in detail in four open source projects. According to the results compared with the other two mainstream code smell detection methods, our method improves the F-measure of Shotgun Surgery and Parallel Inheritance Hierarchy by 23% and 15%, respectively. This shows that our comprehensive multi-dimensional software data detection method is effective.

Our future plan is to further refine our approach and make them fully automated. We will also validate our method on more software systems. In addition, we will introduce a detection method based on machine learning or deep learning, thereby further improving the performance of code smell detection.

Acknowledgements. This work is supported by Anhui Provincial Natural Science Foundation (2008085MF189, 1908085MF206), National Natural Science Foundation of China (NO. 61402007), and the Scientific Research Foundation for the Returned Overseas Chinese Scholars, State Education Ministry.

References

1. Eick, S.G., Graves, T.L., Karr, A.F., Marron, J.S., Mockus, A.: Does code decay? Assessing the evidence from change management data. IEEE Trans. Software Eng. **27**(1), 1–12 (2001)
2. Li, J., Stalhane, T., Conradi, R., Kristiansen, J.M.W.: Enhancing defect tracking systems to facilitate software quality improvement. IEEE Softw. **29**(2), 59–66 (2019)
3. Kruchten, P., Nord, R.L., Ozkaya, I.: Technical debt: from metaphor to theory and practice. IEEE Softw. **29**(6), 18–21 (2012)
4. Fowler, M.: Refactoring: improving the design of existing code (1999)
5. Tufano, M., et al.: When and why your code starts to smell bad. In: 37th International Conference on Software Engineering, Firenze, Italy, pp. 403–414. IEEE (2015)
6. Yamashita, A., Moonen, L.: Exploring the impact of inter-smell relations on software maintainability: an empirical study. In: 2013 International Conference on Software Engineering, San Francisco, USA, pp. 682–691. IEEE (2013)
7. Abbes, M., Khomh, F., Gueheneuc, Y., Antoniol, G.: An empirical study of the impact of two antipatterns, Blob and Spaghetti Code, on program comprehension. In: 2011 15th European Conference on Software Maintenance and Reengineering, Oldenburg, Germany, pp. 181–190. IEEE (2011)
8. Khomh, F., Di Penta, M., Guéhéneucl, Y., Antoniol, G.: An exploratory study of the impact of antipatterns on class change- and fault-proneness. Empir. Softw. Eng. **17**(3), 243–275 (2012)
9. Moha, N., Gueheneuc, Y., Duchien, L., Le Meur, A.: DECOR: a method for the specification and detection of code and design smells. IEEE Trans. Software Eng. **36**(1), 20–36 (2010)
10. Vidal, S.A., Marcos, C., Díaz-Pace, J.A.: An approach to prioritize code smells for refactoring. Autom. Softw. Eng. **23**(3), 501–532 (2014). https://doi.org/10.1007/s10515-014-0175-x
11. Hall, T., Zhang, M., Bowes, D., Sun, Y.: Some code smells have a significant but small effect on faults. ACM Trans. Softw. Eng. Methodol. (TOSEM) **23**(4), 33:1–33:39 (2014)
12. Palomba, F., Zanoni, M., Fontana, F.A., De Lucia, A., Oliveto, R.: Toward a smell-aware bug prediction model. IEEE Trans. Software Eng. **45**(2), 194–218 (2019)
13. Azeem, M.I., Palomba, F., Shi, L., Wang, Q.: Machine learning techniques for code smell detection: a systematic literature review and meta-analysis. Inf. Softw. Technol. **108**, 115–138 (2019)
14. Palomba, F., Bavota, G., Di Penta, M., Oliveto, R., Poshyvanyk, D., De Lucia, A.: Mining version histories for detecting code smells. IEEE Trans. Software Eng. **45**(2), 194–218 (2015)
15. Fernandes, E., Oliveira, J., Vale, G., Paiva, T., Figueiredo, E.: A review-based comparative study of bad smell detection tools. In: Proceedings of the 20th International Conference on Evaluation and Assessment in Software Engineering, New York, USA, pp. 18:1–18:12. ACM (2016)
16. Rasool, G., Arshad, Z.: A review of code smell mining techniques. J. Softw. Evol. Process **27**(11), 867–895 (2015)
17. Travassos, G., Shull, F., Fredericks, M., Basili, V.: Detecting defects in object-oriented designs: using reading techniques to increase software quality. In: Conference on Object-Oriented Programming. Systems, Languages, and Applications (OOPSLA), Denver, USA, pp. 47–56. ACM (1999)

18. Rasool, G., Arshad, Z.: A lightweight approach for detection of code smells. Arab. J. Sci. Eng. **42**(2), 483–506 (2017)
19. Arcelli Fontana, F., Mäntylä, M.V., Zanoni, M., Marino, A.: Comparing and experimenting machine learning techniques for code smell detection. Empir. Softw. Eng. **21**(3), 1143–1191 (2015). https://doi.org/10.1007/s10664-015-9378-4
20. Di Nucci, D., Palomba, F., Tamburri, D.A., Serebrenik, A., De Lucia, A.: Detecting code smells using machine learning techniques: Are we there yet? In: 2018 IEEE 25th International Conference on Software Analysis. Evolution and Reengineering (SANER), Campobasso, Italy, pp. 612–621. IEEE (2018)
21. Liu, H., Jin, J., Xu, Z., Bu, Y., Zou, Y., Zhang: Deep learning based code smell detection. IEEE Trans. Softw. Eng. (2019, work in progress). https://doi.org/10.1109/TSE.2019.2936376
22. Fu, S., Shen, B.: Code bad smell detection through evolutionary data mining. In: 2015 ACM/IEEE International Symposium on Empirical Software Engineering and Measurement (ESEM), Beijing, China, pp. 1–9. IEEE (2015)
23. Boussaa, M., Kessentini, W., Kessentini, M., Bechikh, S., Ben Chikha, S.: Competitive coevolutionary code-smells detection. In: Ruhe, G., Zhang, Y. (eds.) SSBSE 2013. LNCS, vol. 8084, pp. 50–65. Springer, Heidelberg (2013). https://doi.org/10.1007/978-3-642-39742-4_6
24. Kessentini, W., Kessentini, M., Sahraoui, H., Bechikh, S., Ouni, A.: A cooperative parallel search-based software engineering approach for code-smells detection. IEEE Trans. Software Eng. **40**(9), 841–861 (2014)
25. Caram, F., Rodrigues, B., Campanelli, A., Parreiras, F.: Machine learning techniques for code smells detection: a systematic mapping study. Int. J. Software Eng. Knowl. Eng. **29**(2), 285–316 (2019)

Improvement of Association Rule Algorithm Based on Hadoop for Medical Data

Guangqian Kong, Huan Tian$^{(\boxtimes)}$, Yun Wu, and Qiang Wei

Guizhou University, Guiyang 550025, China
huan.tian22@qq.com

Abstract. Data mining technology and association rule mining can be important technologies to deal with a large amount of accumulated data in the medical field, and can reflect the value of large medical data. According to the characteristics of large medical data, aiming at the problem that the traditional Apriori algorithm scans the database too long and generates too many candidate itemsets, a method of digital mapping and sorting of itemsets is proposed. The method of the base model and generation model was used to generate superset, which can improve the efficiency of superset generation and pruning. By using open source framework Hadoop and transplanting the improved algorithm to the Hadoop platform combined with the MapReduce framework, the idea of parallel improvement was introduced based on database partition. Experimental results show that it solves the redundancy of large-scale data sets and makes Apriori algorithm have good parallel scalability. Finally, an example was given to demonstrate the possibility of improving the algorithm.

Keywords: Medical data · Hadoop · Apriori · MapReduce

1 Introduction

With the development of information technology, huge amounts of data have been accumulated in the medical field, and people pay more and more attention to medical data. Data mining can discover some potential rules from these data which are not easy to find by human beings and predict the behavior trend of the analytic objects, thus helping to make decisions or adjust strategies [1, 2]. Lin et al. [3] made a research on mining association rules of TCM symptoms-syndromes-drugs in chronic glomerulonephritis and Fang et al. [4] made a preliminary study on the utilization of cardiac architecture data in hypertension based on association rules and those studies show that association rules can be widely used in clinically choosing medicine, disease prediction, and analysis.

Data mining technology and association rule mining are important technologies that can reflect the value of large medical data. At present, data mining-based medical processing technology has been carried out for some time across the world, and a lot of research results have been achieved on the Apriori algorithm. In 2014, Zhang et al. [5] proposed an improved Apriori algorithm based on Map function to obtain candidate itemsets of successive transaction records. The candidate itemsets were encoded by a series of binary codes, and the final candidate itemsets were obtained by mathematical

© Springer Nature Singapore Pte Ltd. 2020
P. Qin et al. (Eds.): ICPCSEE 2020, CCIS 1258, pp. 506–520, 2020.
https://doi.org/10.1007/978-981-15-7984-4_38

logic operations. This method achieves some efficiency, but the process of manual coding in the process of re-coding is rather cumbersome. In 2015, Wang et al. [6] proposed an Apriori algorithm based on interest model, optimized the time complexity of the algorithm, and proposed a medical architecture. However, the selection of interest model does not effectively reduce the number of database scans and improve the efficiency of the algorithm. Wu et al. [7] proposed a new improved M-Apriori algorithm based on the original Apriori algorithm. In the process of searching frequent itemsets, this algorithm records transaction data sets containing K itemsets and verifies whether k + 1 itemsets are frequent itemsets. There is no need to scan the entire database, but only a part of the transaction. This improves the time efficiency of the algorithm, but it does not help the verification of frequent itemsets. And Chang et al. [8] proposed an effective Apriori Mend Algorithm, which generates item sets through hash functions. Compared with the traditional algorithm, the efficiency of the algorithm is improved, but the execution time of the algorithm is obviously increased, which results in the overall performance degradation of the algorithm.

In the past process of using Apriori algorithm shows obvious advantages, such as convenient application, simple structure, etc. but this algorithm still has limitations indeed [9]. Based on the feasibility of the application of the above-mentioned association algorithm in the medical industry, and the shortcomings of the existing association algorithm in medical data mining technology, such as too large scale of intermediate results and too long scanning time, this paper proposes a Hadoop-based algorithm for medical data association rules, which can improve the efficiency of medical data mining and provide support for doctors' decision-making. The improved method is applied to mining the correlation between different diseases, which is meant for active prevention of diseases and clinical auxiliary diagnosis [10].

This paper mainly includes the following parts. In the first part, the properties and flow of Apriori algorithm are introduced and the shortcomings of the algorithm are pointed out. In the second part, on the basis of detailed description of Apriori algorithm, aiming at the disadvantage of large scale of intermediate results and long scanning time, according to the idea of mapping and regulation, the method of digital mapping and sorting of Apriori algorithm itemsets is proposed, which facilitates data transmission and itemsets matching. Using the base model and generative model to generate supersets can improve the efficiency of the superset generation, and this method can also effectively improve the efficiency of pruning. In order to adapt to the high concurrent environment, the idea of parallel improvement based on database partition is proposed. The improved algorithm is transplanted to Hadoop platform and combined with MapReduce framework. The third part is the analysis of the improved algorithm and the design process. The feasibility and effectiveness of the improved algorithm are illustrated. And the experimental results show that the Apriori algorithm has good parallel expansion ability.

2 Apriori Algorithm

Apriori algorithm is proposed by Dr. Rakesh Agrawal and Dr. Ramakrishnan Srikan in 1994. For the first time, it proposes support-based join and pruning technology to control the growth of candidate itemsets. The purpose of association rules is to find out the relationship between items in data sets. Therefore, this algorithm is very suitable for us to analyze the relationship between diseases in medical data [11, 12]. However, the performance of the algorithm determines the performance of frequent itemsets mining [13].

2.1 The Nature and Process of Apriori Algorithm

Property 1: If I_i is a frequent itemsets, then any non-empty subset I_j of I_i must also be a frequent itemsets.

Property 2: If I_i is a non-frequent itemsets, then any superset I_j of Ii must also be a non-frequent itemsets.

Apriori algorithm is a hierarchical iteration-based method to generate test candidates. It scans data sets to calculate the support of each candidate itemsets, and determines whether each candidate set is frequent or not. Its basic idea is: Firstly, the transaction database is scanned to generate frequent itemsets L_1, then L_1 is used to generate L_2 through self-connection and pruning, and frequent itemsets L_2 is used to generate L_3. By analogy, iterate layer by layer until new frequent itemsets cannot be generated. Then, according to the set minimum confidence threshold, association rules are generated using the generated frequent itemsets. The flow chart of the algorithm is shown in Fig. 1.

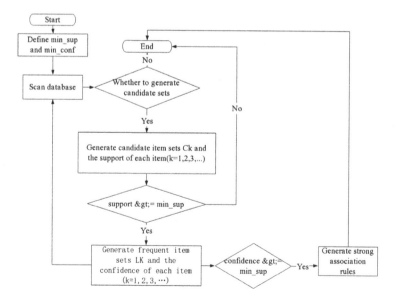

Fig. 1. Apriori algorithm flow.

The graph shows that the algorithm has two stages: The first stage is to generate frequent iterations through continuous iteration, which is obtained by counting the candidate itemsets and comparing them with the minimum support. The second stage is to generate strong association rules.

2.2   The Shortcomings of Apriori Algorithm

Apriori algorithm improves the efficiency of association rules to a great extent, through connection and pruning strategy. This can greatly reduce the size of candidate itemsets. However, when the data scale is large and there are many frequent items in the data or the support and confidence are small, the efficiency of the algorithm is obviously reduced, which seriously restricts the speed of the algorithm. The shortcomings are reflected in the following aspects:

1) A large number of supersets are generated. The exponential growth (especially C_2, $C_2 = C_{L1}^2$) occurs when the frequent itemset L_{k-1} is self-connected to generate the supersets $C_{k..}$.
2) When generating candidate $(k + 1)$-itemsets from $(k + 1)$-supersets, it takes a lot of time to match frequent k-itemsets many times.
3) Because Apriori algorithm is based on the idea of layer-by-layer search, every candidate k-itemsets must be scanned again to determine frequent itemsets. This increases the burden of I/O and is very time-consuming [14].

Because of the above shortcomings of Apriori algorithm, there are some problems in performance and scalability when using Apriori algorithm to process massive medical data. Therefore, this algorithm is further improved and transplanted to make it suitable for concurrent environment.

3   Improvement and Transplantation of Apriori Algorithm Based on Hadoop

Based on the introduction of Apriori algorithm mentioned above, this section will combine MapReduce with Apriori algorithm to improve the algorithm according to its shortcomings and the possibility of parallelization.

3.1   Improvement of Apriori Algorithm

Aiming at the three problems of Apriori algorithm, this paper improves the algorithm based on MapReduce.

For the first problem, the frequent itemset L_{k-1} is digitally mapped and sorted, because there are many items in medical data records and digital operations can reduce data transmission. When generating L_k supersets, only the last different item sets in the frequent itemsets need to be connected. Therefore, at this time, the same frequent L_{k-2} itemsets are prefixed, and the last item of each item in frequent L_{k-1} itemsets is combined with the prefix to generate the L_k supersets. We call the same frequent L_{k-2}

itemsets as the base model, and finally we call it the generation model which consists of different two terms.

For the second problem, it is necessary to verify whether the subset of each item in (k + 1)-supersets is in frequent k-itemsets during pruning. Since we use the method of base model + generation model to generate supersets, any item that does not end with two items in the generation model meets the requirements.

For the third problem, the database needs to be scanned many times. The essence of each scanning operation is to calculate the number of occurrences of each item in the candidate k-itemsets. When MapReduce is implemented, data sets are divided and distributed to multiple task nodes, which can reduce the number of scans.

Aiming at the improvement scheme of three problems of Apriori algorithm, the process of the algorithm is improved below.

Firstly, the data is initialized. The items of the frequency 1-itemsets are sorted according to word frequency, and the items are digitally mapped. And a sort function is added to the original Apriori algorithm. At the same time, when generating 2-supersets, the generative function is further improved. The superset is generated in the form of base model + generative model, and candidate 2-itemsets are generated by pruning operation.

The second step is to generate all frequent itemsets by iteration on the basis of initializing the generated frequent 1-itemsets and candidate 2-itemsets.

1) Generate frequent k-itemsets

Frequent k-itemsets are generated by loading candidate k-itemsets C_k and transaction databases. Each item in the data set is digitally mapped according to the mapping generated by data initialization, and then the k-item mapping is combined and sorted. By comparing with the candidate k-itemsets C_k, we can determine whether the mapping belongs to the C_k. The required k-item mapping is counted by word frequency, and then the frequent k-itemsets is obtained according to the set minimum support

2) Generate candidate (k + 1)-itemsets

The process of generating candidate (k + 1)-itemsets includes two steps: The k (k + 1)-supersets is generated, and then the candidate (k + 1)-itemsets is generated by pruning the (k + 1)-supersets. In the process of generating (k + 1)-supersets, the base model and generative model of frequent k-itemsets are first intercepted, and then the (k + i)-supersets is generated by arranging and combining the generated models. In the pruning process, because the superset is generated in the form of base model + generative model, it is only necessary to judge whether there is an item in the frequent k-level item set that ends with the generation mode. If there is an item, it belongs to the candidate itemsets, and if not, delete the itemsets.

3) Generate Association rules

For each frequent itemsets L, find out all its non-empty subset S. The confidence of each non-empty subset S is calculated and compared with the minimum confidence threshold set beforehand.

The basic process of the improved algorithm is shown in Fig. 2.

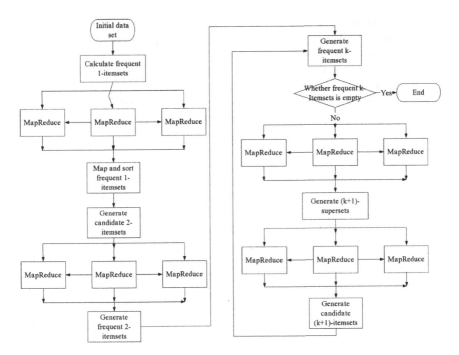

Fig. 2. The improved algorithm flow based on MapReduce.

3.2   Transplantation of Apriori Algorithm Based on MapReduce

In addition to the above methods to improve the Apriori algorithm, there are also improved methods based on hash table, Boolean matrix, array and so on [15]. However, the effect of improving the algorithm on a single computer is not very obvious, because of the computing ability and storage performance. With the advent of distributed computing, it is easy to think of transplanting the original algorithm to the distributed computing model. According to the improved idea of Apriori algorithm mentioned above, Apriori algorithm is transplanted into MapReduce framework of cloud computing to realize distributed parallel processing.

1) The first stage is data initialization.

   i. Data segmentation and distribution. The transaction database D is divided into N data blocks, and then input into the MapReduce framework to perform the Map task [16].

   ii. The Map task processes each record of the allocated transaction data block and outputs the result as <key, value> where key is each item in the record and value is 1.

   iii. Execute the Combiner function. The reason for introducing Combiner function is that: When the transaction database is large, the data records in each data block are similar, and the intermediate result key value produced by the Map task will repeat proportionally. If these intermediate results are sent directly to Reduce task, it will cause great pressure on network bandwidth.

iv. Map tasks generate candidate 2-itemsets in order of sorting, and output results like <key, value> in which key is the base model for generating the item and value is the generated item.

v. Reduce tasks aggregate Map tasks and aggregate Map tasks according to the base pattern.

2) In the second stage, frequent itemsets are generated iteratively.

    i. The data is loaded, and then the candidate k-itemsets data set generated in the previous stage is loaded. At the same time, the mapping table of frequent 1-itemsets is read..

    ii. The input data block of Map task is formatted and converted into <key, value> format, in which key refers to the data segment of each row in the loaded transaction database and value refers to the mapping table in frequent 1-itemsets.

    iii. Each record in the transaction database is digitally mapped and sorted according to the mapping table, and then a digitalized set is obtained.

    iv. Each item in the candidate k-itemsets is extracted to determine whether it belongs to the corresponding data set of each item in the transaction database. If it belongs to the data set, the output is in the form of <key, value> in which key is the item of the candidate k-itemsets and value is 1.

    v. Execute the Reduce task. The local frequent itemsets output from Map task are merged. Each item in the candidate k-itemsets is counted and compared with the minimum support threshold to generate frequent k-itemsets.

3) The third stage generates candidate $(k + 1)$-itemsets

    i. Generate $(k + 1)$-supersets and input frequent k-itemsets. The Map task inputs each row of data in the frequent k-itemsets.

    ii. The base model and generation model are intercepted from the frequent k-itemsets. Map task output is in the form of <key, value> in which key is the base model and value is the generation model.

    iii. Reduce tasks specify the output of Map tasks. The generated models are organized into a set of generated models. The key of Reduce output is selected by traversing the set and the value of Reduce output is the base model.

    iv. Pruning k-supersets. In the pruning process, according to the above optimization strategy, besides reading k-supersets records, it also needs to load the generating schema table of frequent k-itemsets.

    v. Map task reads the superset record ending in the generative mode, traverses all subsets of the superset item, determines whether it belongs to one item in the frequent itemsets, and deletes the superset record if any subset does not belong to one item in the frequent itemsets. The output is in the form of <key, value> where key is the base mode and value is the generative model.

    vi. Reduce task aggregates the results of Map task to get the final candidate $(k + 1)$-itemsets.

4) The fourth stage is the generation of association rules.

    i. Segmenting the generated frequent itemsets and distributing the items in the frequent itemsets to each task node to form a key-value pair <key1, value1>. Among them, key1 represents the offset of the row record, and value1 represents a row record, which is an item in the frequent itemsets.

ii. Map task calls rule generation function to generate all rules for each frequent itemsets, and outputs intermediate results in the form of <key2, value2>. Key2 represents an item in the frequent itemset, and Value2 represents the generated rule results.

iii. Reduce task merges <key2, value 2> and stores it in HDFS.

The flow of this phase is shown in Fig. 3.

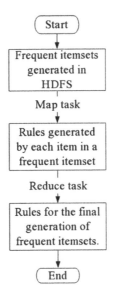

Fig. 3. Association rules generation process.

4 Examples of Improved Algorithm and Analysis of Experimental Results

In this section, an example of MapReduce-based algorithm is analyzed, and the scalability and computational efficiency of the improved algorithm are verified by experiments with different size data sets and number of nodes under different support levels.

4.1 Example Analysis of MapReduce-Based Algorithm

The improved algorithm still uses transaction database D in Sect. 3.2 to analyse the feasibility. Before the analysis, the minimum support threshold is determined to be 2, and min_sup is 0.2. The data initialization phase is shown in Table 1.

Table 1. Initialization data.

Data item	Sorting	Mapping	Generate the second level candidate set	
I1, I2, I5			1 12	(1, [12, 13, 14, 15])
I2, I4			1 13	
I1, I2, I3 I5	I2 7	1 I2 7	1 14	
I1, I3	I1 6	2 I1 6	1 15	
I2, I3	I3 6	3 I3 6	2 23	(2, [23, 24, 25])
I1, I2, I3	I5 5	4 I5 5	2 24	
I2, I5	I4 2	5 I4 2	2 25	
I1, I3			3 34	(3, [34, 35])
I3, I4, I5			3 35	
I1, I2, I5			4 45	(4, [45])

When generating the second layer frequent itemsets, the data processing process is shown in Table 2.

Table 2. Processing of generating second layer frequent itemsets data.

Item list	Map phase		Reduce phase	
	Mapping	Output	Input	Output
I1, I2, I5	1 2 4	([1, 2], 1) ([1, 4], 1) ([2, 4], 1)	[1, 2] 1 [1, 2], 1 [1, 2], 1 [1, 2],1	([1, 2], 4)
I2, I4	1 5	([1, 5], 1)	[1,3],1 [1,3],1 [1,3],1	([1, 4], 4)
I1, I2, I3, I5	1 2 3 4	([1, 2], 1) ([1, 3], 1) ([1, 4], 1) ([2, 3], 1) ([2, 4], 1) ([3, 4], 1)	[1,4],1 [1,4],1 [1,4],1 [1,4],1	([1, 3], 3)
I1, I3	2 3	([2, 3], 1)	[1, 5], 1	([2, 3], 4)
I2, I3	1 3	([1, 3], 1)	[2, 3], 1 [2, 3], 1 [2, 3], 1 [2, 3], 1	([2, 4], 3)
I1, I2, I3	1 2 3	([1, 2], 1) ([1, 3], 1) ([2, 3], 1)	[2, 4], 1 [2, 4],1 [2, 4], 1	
I2, I5	1 4	([1, 4], 1)	[3, 4], 1 [3, 4], 1	([3, 4], 2)
I1, I3	2 3	([2, 3], 1)	[3, 5], 1	
I3, I4, I5	3 4 5	([3, 4], 1) ([3, 5], 1) ([4, 5], 1)	[4, 5], 1	
I1, I2, I5	1 2 4	([1, 2], 1) ([1, 4], 1) ([2, 4], 1)		

When the third layer supersets are generated, the data are processed as shown in Table 3.

Table 3. Processing of generating third layer supersets data.

Initial input	Map phase		Reduce phase	
	Generative model	Output	Input	Output
([1,2], 4)	2	1, 2	1, 2	([2, 3], 1)
([1,4], 4)	4	1, 4	1, 4	([2, 4], 1)
([1,3], 3)	3	1, 3	1, 3	([3, 4], 1)
([2,3], 4)	3	2, 3	2, 3	([3, 4], 2)
([2,4], 3)	4	2, 4	2, 4	
([3,4], 2)	4	3, 4	3, 4	

The third layer supersets are pruned to generate three candidate itemsets, as shown in Table 4.

Table 4. Pruning generates candidate 3-itemsets

Initial input	Map phase		Reduce phase	
	Generative model	Output	Input	Output
[2, 3], 1	2 3	(1, [1, 2, 3])	(1, [1, 2, 3])	(1, [1, 2, 3])
[2, 4], 1	2 4	(1, [1, 2, 4])	(1, [1, 2, 4])	(1, [1, 2, 4])
[3, 4], 1	3 4	(1, [1, 3, 4])	(1, [1, 3, 4])	(1, [1, 3, 4])
[3, 4], 2	3 4	(2, [2, 3, 4])	(2, [2, 3, 4])	(1, [2, 3, 4])

Generating frequent 3-itemsets as shown in Table 5.

Table 5. Generating frequent 3-itemsets

Data item	Map phase		Reduce phase	
	Mapping	Output	Input	Output
I1, I2, I5	1 2 4	([1, 2, 4], 1)	([1, 2, 3], 1)	([1, 2, 3], 2)
			([1, 2, 3], 1)	([1, 2, 4], 3)
I1, I2, I3, I5	1 2 3 4	([1, 2, 3], 1)		
		([1, 2, 4], 1)		
		([1, 3, 4], 1)		
		([2, 3, 4], 1)		
I1, I2, I3	1 2 3	([1, 2, 3], 1)	([1, 2, 4], 1)	
I1, I2, I5	1 2 4	([1, 2, 4], 1)	([1, 2, 4], 1)	
			([1, 2, 4], 1)	

The 4-itemsets {1, 2, 3, 4} obtained by self-joining frequent 3-itemsets indicates the end of the algorithm if the relevant subset is not empty in frequent 3-itemsets. Finally, association rules are generated.

According to the examples in this section, we also set the minimum confidence min_sup = 55%, and send the records in frequent itemsets to map tasks in the form of <key, value>.

The Map task parses the input <T1, {1, 2, 3}>, <T2, {1, 2, 4}> and outputs intermediate results such as <frequent item sets, rules>. Among them:

Task Node 1: <T1, {1, 2, 3}> <{1, 2, 3}, {1, 3}, 3_2" Support 0.2, Confidence 67%>.

Task Node 2: <T2, {1, 2, 4}> <{1, 2, 4}, {1, 2}, {1, 4}, {1, 2, 4}, {2, 4}, {4, 0.3, reliability = 75%>; <{1,2,4}, {2,4}, {1, 4, support = 0.3, reliability = 100%>; <{1, 2, 4}, {4, 1, 2, 2, 4}, {1, 2, 2, 2, 2, 4}, support = 0.3, reliability = 60%>;

Reduce tasks specify the intermediate results generated by Map tasks:

<13 2, 67%>, <12 4, 75%>, <2 4 1, 100%>, <4 12, 60%>

Through the above example analysis, it shows that the Hadoop-based frequent itemsets mining and rule generation method improvement is effective.

4.2   Analysis of MapReduce-Based Algorithm

Through the above analysis of the example of Apriori improved algorithm and the design process, the following conclusions can be drawn:

1) The improved method shows that Apriori algorithm and MapReduce can be combined.
2) The improved Apriori algorithm relies on MapReduce framework to perform parallel operations efficiently and improve the efficiency of operation. Especially when the amount of data is large, the improved Apriori algorithm has more obvious advantages.
3) By digitizing mapping, sorting and deleting redundant items, the efficiency of Apriori algorithm is improved.

4.3   Experimental Results and Analysis

Comparison of Serial and Parallel Algorithm. In order to verify the performance of the improved algorithm, both the improved algorithm without MapReduce and the MapReduce algorithm are simulated on the same computer. Because of the confidentiality of medical data and electronic medical records, data mining training data *accidents.data* is used for simulation. The simulation hardware environment is: Intel (R) Core (TM) i5-4250 dual-core 1.3 Hz with 4 GB of memory. Software environment: operating system is 64-bit Win7; JDK version is 1.7.0_55; Hadoop version is 2.4.0; SSH client and server are used for communication between nodes; IDE platform is Eclipse.

Firstly, the serial algorithm is simulated. The improved Apriori algorithm runs under Win7 and records the specific performance of mining data *accidents.data*. Secondly, when virtual machines are enabled and configured to be pseudo-distributed, the node is

Name Node as well as Data Node. Finally, the improved MapReduce Apriori algorithm in Sect. 3.2 is run on Hadoop platform.

Serial and parallel algorithms mine the same transaction set, and the scale of transaction set is increasing. The two algorithms set the same threshold: min supp = 0.3, min conf = 0.7 and 6 experiments were conducted with different transaction set sizes. The experimental results are shown in Table 6.

Table 6. Performance Comparison of Serial and Parallel Algorithms.

Size of transaction set/KB	Number of association rules	Serial algorithm mining time/s	Parallel algorithm mining time/s
1260	14674	31	129
2410	15003	68	210
4590	15012	119	282
9310	15076	254	399
18300	14967	610	620
36500	14931	N/A	1148

According to the above experimental results, we can draw the broken line diagrams of the two algorithms, as shown in Fig. 4.

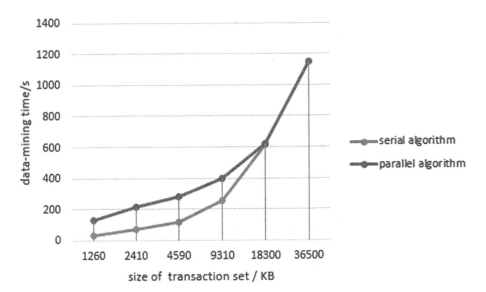

Fig. 4. Performance of serial algorithm and parallel algorithm

Comparisons of the Operation Efficiency of the Algorithm with Different Number of Node. In order to verify the scalability and computational efficiency of the improved transplanted algorithm, the experiment uses 5G, 10G and 15G data on 4 and 6 nodes with support of 10%, 15% and 20% respectively. Hardware environment: There are 7 PC machines with 7 nodes. One of them has 2 GB of memory as Name Node (master) and 200G of hard disk. The other six have 1G of memory as Data Node (node 1-node 6) and 100G of hard disk. Software environment: The operating system is 64-bit CentOS 6.5; kernel version is 2.6.32; JDK version is 1.7.0_55; Hadoop version is 2.4.0; SSH client and server are used for communication between nodes; IDE platform is Eclipse.

Comparing the efficiency of the algorithm with different number of nodes:

The first group of experiments is running on 4 or 6 nodes with 5G data, as shown in Fig. 5.

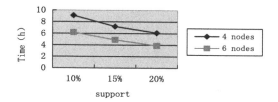

Fig. 5. A experimental results of 5G data.

The second group of experiments is running on 4 or 6 nodes with 10G data, as shown in Fig. 6.

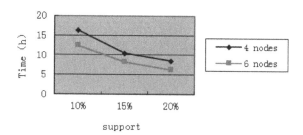

Fig. 6. Experimental results of 10G data.

The third group of experiments run on 4 or 6 nodes with 15G data, as shown in Fig. 7.

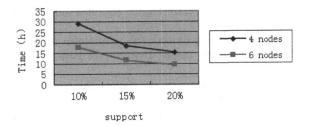

Fig. 7. Experimental results of 15G data.

According to the three groups of experimental results, it can be seen that the processing time of the improved algorithm increases linearly with the increase of the amount of data processed, which also shows the parallel processing characteristics of Hadoop framework.

5   Conclusions

In this paper, digital mapping and sorting of itemsets, base model and generation model are used to remedy the shortcomings of Apriori algorithm for medical data processing. It is adapted to MapReduce Parallel Framework by transplanting, and the improved method is proved to be effective and feasible by an example demonstration. Finally, through two comparative experiments, the results show that the algorithm has great advantages in processing large-scale data, and the improved transplantation has good expansibility. This algorithm saves a lot of time and memory space, improves the performance and scalability when the analysis of relationship between diseases is implemented, which gives great support to the research on the relationship between medical diseases.

Acknowledgments. This work was supported by the national natural science foundation of China ([2018]61741124) and the science planning project of Guizhou province (Guizhou science and technology cooperation platform talent [2018] no. 5781). What's more, we thank the anonymous reviewers sincerely for their significant and valuable feedback.

References

1. Zou, Y.: Analysis and prospect of data mining technology in hospital informatization construction. Digit. Technol. Appl. **36**(01), 233–234 (2018)
2. Wang, L.H.: Application of data mining in medical systems. Digit. Technol. Appl. (08), 96–98 (2017)
3. Lin, G.: Research on mining association rules of TCM symptoms-syndromes-drugs in chronic glomerulonephritis. University of Electronic Science and Technology (2016)
4. Fang, Y.Y., Zhu, X.M., Li, D.: Preliminary study on the utilization of cardiac configuration data of hypertension based on association rules. Sci. Technol. Innov. (06), 81–82 (2018)

5. Zang, W., Cao, B.X.: An improved parallel apriori algorithm with index structure. Electron. Technol. (2014)
6. Wang, D.M., Cui, X.Y.: Research on optimization of apriori algorithm based on cloud computing and big medical data. Beijing University of Posts and Telecommunications (2015)
7. Wu, X.Y., Mo, Z.: An optimization method for mining frequent itemsets based on APRI algorithm. Comput. Syst. Appl. **23**(06), 124–129 (2014)
8. Chang, R., Chen, Z.W.: A hybrid structural optimization algorithm. China Manuf. Inf. **40** (19), 49–53 (2011)
9. Yao, Q., Tian, Y., Li, P.: Design and development of a medical big data processing system based on Hadoop. J. Med. Syst. **39**, 23 (2015)
10. Zhou H.: Disease correlation analysis of big medical data. Electron. Technol. Softw. Eng. (18), 187–188 (2017)
11. Li, S.Q.: Analysis and research of association rule mining algorithms in data mining. Electron. Technol. Softw. Eng. (04), 200 (2015)
12. Savasere, A., Omiecinski, E., Navathe, S.: An efficient algorithm for mining association rules in large databases. In: Proceedings of 1995 VLDB International Conference, Zurich, Switzerland, pp. 432–433 (2015)
13. Liu, L.J.: Research and application of improved Apriori algorithms. Comput. Eng. Design (12), 3324–3328 (2017)
14. Goethals, B.: Survey on frequent pattern mining, HUT basic research unit, vol. 30, no. 10, pp. 1–43. Department of Computer Science, University of Helsinki (2013)
15. Yu, H.L., Wen, J., Wang, H.G.: An improved Apriori algorithm based on the boolean matrix and hadoop. Proc. Eng. **15**, 1827–1837 (2011)
16. Cui, L.: Application of MapReduce-based parallel association rule algorithms in social analysis. Hebei University of Engineering (2015)

An Interpretable Personal Credit Evaluation Model

Bing Chen, Xiuli Wang$^{(\boxtimes)}$, Yue Wang, and Weiyu Guo

School of Information, Central University of Finance and Economics,
Beijing, China
wangxiuli@cufe.edu.cn

Abstract. How to establish a personal credit evaluation model with both interpretability and high prediction accuracy is an essential task in the credit risk management of commercial banks. To realize interpretable personal credit evaluation with high accuracy, it proposes an interpretable personal credit evaluation model DTONN (i.e., Decision Tree extracted from Neural Network) that combines the interpretability of decision tree and the high prediction accuracy of neural network. Significant features were selected from raw features by a decision tree, and a four-layer neural network was constructed to predict personal credit by using the selected features. Therefore, the accurate credit evaluation was made through the neural network and associated decision process was intelligibly displayed in the form of a decision tree. In the experiments, DTONN was compared with four personal credit evaluation models: decision tree, neural network, support vector machine, and logistic regression, on give-me-some-credit credit dataset. The experimental results show that our proposed model is state-of-the-art both on the accuracy and interpretability.

Keywords: Decision tree · Neural network · Personal credit evaluation · Interpretability

1 Introduction

Due to the default of credit customers, the loans put by banks cannot be recovered on time, which is a significant focus for the loss of funds of commercial banks. Personal credit evaluation is a critical means of risk prevention and control for commercial banks. At present, commercial banks usually use logistic regression as a credit evaluation model. The advantages of logistic regression are interpretability and ease of maintenance. However, the prediction accuracy of such linear model is not high, especially when the model suffering from high-dimension inputs and noise, the performance will decline significantly.

This work is supported by National Defense Science and Technology Innovation Special Zone Project (No. 18-163-11-ZT-002-045-04).

P. Qin et al. (Eds.): ICPCSEE 2020, CCIS 1258, pp. 521–539, 2020.
https://doi.org/10.1007/978-981-15-7984-4_39

Compared with logistic regression, deep neural network can learn non-linear features more effectively, suppress noise, and generally have better generalization and prediction accuracy. However, due to the large number of parameters and the higher complexity of the model, the reasoning process is often not easy to understand. Such models are "black boxes" for ordinary users. In credit evaluation, the comprehensibility of the model is very important. Users need to understand the reason for the output of the model and the reasoning process in order to evaluate the reasonableness of its prediction and make the next action.

Those who work in the development and management of credit assessment models in banks believe that whether the credit score model is developed within the bank or developed jointly by the third party, it is always emphasized that the model that is finally launched can be fully explained, which needs to be understood from the perspective of bank supervision and internal management. From the perspective of bank supervision, the supervision bears the responsibility of supervising the bank to carry out credit services. It is necessary to check and confirm frequently whether the credit evaluation model is interpretable, so as to judge whether some unreasonable approval conditions are caused by social discrimination, vicious competition and other reasons. From the perspective of bank management, if the credit evaluation model is explicable, the relevant staff can quickly judge whether it is a process problem or other problem once the model results and business forecast are deviated in the process of credit business.

Comparing with deep neural networks, it is easy to explain how a decision tree makes any particular prediction because this depends on a relatively short sequence of decisions and each decision is based directly on the input data. However, the decision tree does not usually generalize as well as deep neural networks. Considering that the internal structure of neural networks is similar to that of decision trees, therefore, we can construct a personal credit evaluation model that has both the interpretability of decision tree and the high prediction accuracy of neural network for the healthy development of personal credit business. The overall idea of new model building in this paper is shown in Fig. 1.

Figure 1 mainly includes four parts. The first part is data preprocessing, including the processing of missing values, duplicate values, and discrete values. The second part is to build a neural network model, including using decision tree to select important variables and training an optimal decision tree according to important variables. The classification information obtained from the optimal decision tree is trained to build a neural network and use important variables to train the neural network. Then the important information is extracted from the trained neural network and displayed in the form of a decision tree as the new model for this article. Finally, using several model evaluation indicators evaluate the model.

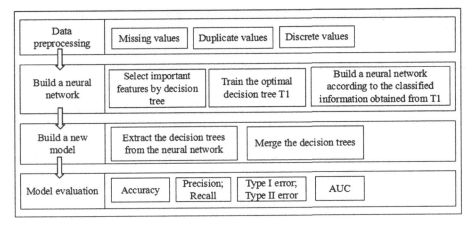

Fig. 1. The overall idea of new model building.

2   Related Work

2.1   Statistics in the Credit Score

[1] first used discriminant analysis in personal credit evaluation, classifying customers default categories according to the sample attribute values. [2] first used linear regression analysis in personal credit evaluation. Experiments have shown that the behavior characteristics of credit customers have a more significant impact than predicting credit customer dishonesty based on historical data. However, the linear regression method has significant problem in practical applications. Therefore, [3] compared logistic regression with linear regression, showing the advantages of less prior knowledge and high stability required for logistic regression. [4] first applied the decision tree to personal credit evaluation. Experiments have shown that the decision tree has better classification ability and higher prediction accuracy for credit evaluation data. [5] first applied the support vector machine in credit evaluation, and proved that the least squares support vector machine is superior to other models. [6] believed that support vector machines are more competitive in credit evaluation than traditional discriminant analysis and logistic regression.

Statistical methods such as logistic regression and decision trees are interpretable as well as the model is stable, but their prediction accuracy is generally not high. Although the prediction accuracy of support vector machines is relatively high, its disadvantages are unexplainable and subjective.

2.2   Neural Network in the Credit Score

[7–11] believed that the neural network can handle nonlinear problems presented between complex variables well, and has strong classification ability for credit data and high prediction accuracy. Neural network has been proved to be the model with the highest accuracy in data classification and prediction among the current single evaluation models. Although neural network has a good prospect in the personal credit score, it has been criticized by many scholars for its poor interpretability and poor model stability.

2.3   Combination of Decision Tree and Neural Network in the Credit Score

[12] proposed a combination model of personal credit evaluation based on decision tree and neural network. The model used decision tree to filter important indicators of credit data, then used the filtered important indicators as inputs to the BP neural network model. Experiments have shown that this method can reduce the interference of non-important indicators and enhance the stability of the model. [13] proposed an enterprise customer credit evaluation model based on decision tree and neural network. Using improved *C4.5* to train a high-precision decision tree, and then extracting rules from the decision tree for building a neural network. Experiments have shown that this method improves the prediction accuracy and stability of the model.

The predictive ability of the combined model is higher than that of the single statistical model, and its stability are better than that of the neural network. However, if the combined model is too complicated, it will reduce the understandability of the model and be difficult to maintain.

2.4   Related Research on the Interpretation of Neural Network

Aiming at the unexplainability of neural networks, scholars use the agent model method to explain neural networks. [14–16] proposed a quantitative method to explain each prediction made by pre-trained Convolutional Neural Networks (CNN), training a decision tree to clarify the specific reason for each prediction of CNN at the semantic level. [17] proposed a new algorithm DeepRED based on the CRED algorithm proposed in [18], which extended the three-layer BP neural network to the deep neural network and extracted the rules for the input layer, output layer and each hidden layer of the neural network, and finally merged all the rules into a decision tree containing only the inputs and outputs of the neural network. Use this decision tree to describe the behavior of the deep neural network.

When using the agent model method to explain the neural network, the predictive behavior made by the neural network can be understood directly through the interpretable model without understanding the complex relationships within it. However, the current agent model only provides an approximate explanation for the complex model, not the reconstruction of the internal knowledge of the complex model.

3   DTONN Modeling

The construction of the DTONN model contains two steps. The first step is to build a four-layer neural network. The second step is to extract rules from the trained neural network and display them in the form of a decision tree. The decision tree is named DTONN.

3.1   Neural Network Modeling

Although the neural network can produce high order nonlinear features and improve the prediction accuracy, training neural network with the important features selected by decision tree can effectively reduce the noise interference and the feature dimension, and improve the network training and reasoning speed.

To begin with, the decision tree is used to select important attributes from the preprocessed data to reduce the impact of non-core indicators on the personal credit assessment classification and prediction results, then the selected important attributes are used to train an optimal decision tree T1.

Because the decision tree and neural network are similar in structure and complementary in performance, building a neural network based on the classification information obtained by decision tree can get a more reasonable personal credit score result. Therefore, this paper creates a four-layer neural network based on the classification information obtained by T1.

The input layer corresponds to the input attributes of T1; the number of neurons in the first hidden layer corresponds to the number of distinct internal nodes of T1; and the number of neurons in the second hidden layer corresponds to the number of leaf nodes of T1 and the number of neurons in output layer corresponds to the categories in T1, where the connections between neurons are fully connected. Take Fig. 2 as an example of this four-layer neural network. After the neural network is built, it is trained using the selected features.

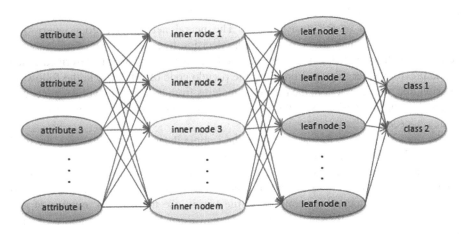

Fig. 2. An example of a neural network constructed from the classification information obtained by decision tree T1.

This four-layer fully-connected neural network uses sigmoid as the activation function $f(*)$. For convenience, taking a three-layer fully-connected neural network as an example, the calculation formula for the activation values of the hidden layer is shown in 1. Suppose there are L training examples, l neurons in the hidden layer and m neurons in the output layer.

$$H_j(x_p) = f\left(\sum_{i=1}^{n} w_{ij}x_{ip} - a_j\right) \tag{1}$$

Where x_p represents the p-th input data sample, $x_p = (x_{1p}, x_{2p}, \ldots, x_{np})^T, p = 1, 2, \ldots,$ L. w_{ij} is the connection weight between the i-th neuron in the input layer and the j-th neuron in the hidden layer, $i = 1, 2, \ldots n$. a_j is the bias of the j-th neuron, $j = 1, 2, \ldots, l$. The formula of the sigmoid function is shown in 2.

$$f(x) = \frac{1}{1+e^{-x}} \tag{2}$$

After calculating the activation value $H_j(x_p)$ of the hidden layer, the output value O of the output layer is calculated, and the calculation formula is shown in 3.

$$O_k(x_p) = f\left(\sum_{j=1}^{l} H_j(x_p)w_{jk} - b_k\right) \tag{3}$$

Where the connection weight between the j-th neuron in the hidden layer and the k-th neuron in the output layer is $w_{jk}, k = 1, 2, \ldots m$. b_k is the bias of the k-th neuron.

3.2   Extraction of Decision Tree Rules from Neural Network

Extract the decision tree rules from the trained neural network, i.e., the inputs, outputs and the activation values of the hidden layers in the trained neural network are presented in the form of the decision tree. The neural network has four layers. For convenience, this article starts to extract the decision tree rules in the neural network from back to front, which can ensure that the behavior of this layer is described based on the previous layer. Figure 3 provides the pseudo code of DTONN. In the following decision tree rules extraction, the extracted features from the hidden layer are generated by feeding the training data to the neural network and recording the outputs of the hidden neurons.

As shown in Fig. 3, for each class of the sample dataset, first initializing the total rule (line 2) from the last hidden layer to the output layer, and then traversing each layer of the network to initialize the total rules from the previous layer to the layer as an empty set. Based on the total rules extracted last time (line 5, including operations to remove duplicate rules), traversing each rule t in it, use the neurons' outputs of this layer as the inputs of the decision tree to be constructed (line 7), and apply t to the neurons' outputs of the next layer as the outputs of the decision tree (line 8), the C4.5 algorithm is used to construct the decision tree for the obtained inputs and outputs, and the newly obtained rules are merged with the existing rules (line 9). According to the above practices, traversing each layer of the network.

After getting the rules of each layer, these rules need to be merged (line 13, redundant and illogical rules also need to be removed (line 14)) layer-wise to get the tree rules that describe the neural network's outputs by its inputs. In this paper, first, substituting the terms in $R_{h_2 \to O}$ by the regarding rules in $R_{h_1 \to h_2}$ to get the rules $R_{h_1 \to O}$ and deleting the redundant and unsatisfiable rules. Then, replacing the terms in $R_{h_1 \to O}$

Algorithm (DTONN):

Input: training example$(x_1, x_2, \ldots, x_m)$, neural network (h_0, h_1, h_2, h_3)

1. **for all** class $\lambda_v \epsilon \lambda_1, \lambda_2$ do
2. $R_{k \to o}^v \Leftarrow IF\ h_{k+1,v} < 0\ THEN\ \hat{y} = \lambda_2$ //initial rule from kth layer to output layer
3.   **for all** hidden layer $j = k, k-1, \ldots, 0(k=2)$ do //go through every layer
4. $R_{h_{j-1} \to h_j}^v = \emptyset$ //initial rule from $(j-1)th$ layer to jth layer as none
5. $T \Leftarrow extractTermsFromRuleBodies(R_{h_j \to h_{j+1}}^v)$ //extract and remove rules
6.   **for all** $t \in T$ do
7. $x_1', x_2', \ldots, x_m' \Leftarrow h_j(x_1), h_j(x_2), \ldots, h_j(x_m)$ //obtain activation values for all neurons from each training example
8. $y_1', y_2', \ldots y_m' \Leftarrow t(h_{j+1}(x_1), h_{j+1}(x_2), \ldots, h_{j+1}(x_m))$ //obtain binary outcome in jth layer
9. $R_{h_{j-1} \to h_j}^v \Leftarrow R_{h_{j-1} \to h_j}^v \cup C4.5\ (x_1', y_1'), (x_2', y_2'), \ldots, (x_m', y_m'))$ //obtain the rule in jth layer
10.   **end for**
11.   **end for**
12.   **for all** hidden layer $j = k, k-1, \ldots, 0(k=2)$ do
13. $R_{h_{j-1} \to o}^v = mergeIntermediateTerms(R_{h_{j-1} \to h_j}^v, R_{h_j \to o}^v)$
14. $R_{h_{j-1} \to o}^v = deleteunsatisfiableTerms(R_{h_{j-1} \to o}^v)$
15.   **end for**
16. **end for**
17. **return** $R_{i \to o}^1, R_{i \to o}^2$ //get the total rules which describe output of NN by rules consisting of terms on input values

---

Fig. 3. Pseudo code of DTONN.

by the regarding rules in $R_{i \to h_i}$ to get the rule $R_{i \to O}$ and deleting the redundant and unsatisfiable rules. According to the above practices, until the rules contain only the attributes and categories of the neural network, i.e., the DTONN model with both interpretability and high prediction accuracy.

Figure 4 takes some rules extracted from the second hidden layer to the output layer as an example to show a small part of the intermediate rule extraction.

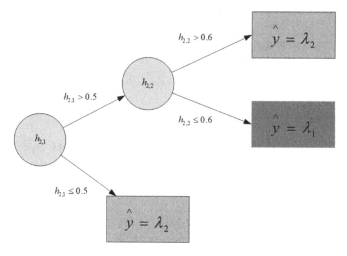

Fig. 4. A rule-term from the second hidden layer to the output layer, where h_{21} and h_{22} respectively represent the activation values of the two neurons in the second hidden layer of the neural network; $\hat{y}$ represents the output category of the input layer of the neural network, and $\hat{y} = 1$ for "good" customers, $\hat{y} = 0$ for "bad" customers. Therefore, if the classification result is λ_1, we can get a rule from the decision tree shown, that is, *IF $h_{2,2} \leq 0.6$ AND $h_{2,1} > 0.5$ THEN $\hat{y} = \lambda_1$.* If the classification result is λ_2, we can get a set of rules *IF $h_{2,2} > 0.6$ AND $h_{2,1} > 0.5$ THEN $\hat{y} = \lambda_2$ or IF $h_{2,1} \leq 0.5$ AND $h_{2,1} \leq 0.5$ THEN $\hat{y} = \lambda_2$.*

4   Experiments

4.1   Model Evaluation Indexes

Confusion Matrix. Set Y = 1 to indicate that a customer has better credit and belongs to the "good" customer, and Y = 0 means that a customer has poor credit and belongs to the "bad" customer. The confusion matrix is shown in Table 1.

Table 1. Confusion matrix, where TP (True Positive) indicates that a customer is actually a "good" customer, and the model predicts the customer as a "good" customer; FP (False Negative) indicates that a customer is actually a "bad" customer, and the model predicts the customer as a "good" customer; FN (False Positive) indicates that a customer is actually a "good" customer, and the model predicts the customer as a "bad" customer; TN (True Negative) indicates that a customer is actually a "bad" customer, and the model predicts the customer as a "bad" customers.

Confusion matrix		Actual results	
		Y = 1, "good" customer	Y = 0, "bad" customer
Predicted results	Y = 1, "good" customer	TP	FP
	Y = 0, "bad" customer	FN	TN

Type I Error and Type II Error. Type I error is the rate of misjudgment of "bad" customers classified as "good" customers, and Type II error is the rate of misjudgment of "good" customers classified as "bad" customers. The calculation formulas are shown in 4 and 5.

$$Type\ I\ error = \frac{FP}{TN + FP} \tag{4}$$

$$Type\ II\ error = \frac{FN}{FN + TP} \tag{5}$$

Accuracy. When evaluating the effect of the model, the higher the accuracy, the stronger the model's ability to accurately predict "good" and "bad" customers. The calculation formula is shown in 6.

$$Accuracy = \frac{TN + TP}{TN + FP + FN + TP} \tag{6}$$

Precision. When evaluating the effect of the model, the higher the precision, the more the number of actual "good" customers among the customers that the model predicts as "good". The calculation formula is shown in 7.

$$Precision = \frac{TP}{TP + FP} \tag{7}$$

Recall. When evaluating the effect of the model, the higher the recall, the more customers in the model are predicted to be "good" customers. The calculation formula is shown in 8.

$$Recall = \frac{TP}{TP + FN} \tag{8}$$

Receiver Operating Characteristic Curve (ROC curve). ROC is a comprehensive indicator reflecting continuous variables of True Positive Rate (TPR) and False Positive Rate (FPR) continuous variables. The calculation formulas of TPR and FPR are shown in 9 and 10.

$$TPR = \frac{TP}{TP + FN} \tag{9}$$

$$FPR = \frac{FP}{TP + TN} \tag{10}$$

AUC (Area Under Curve) represents the area under the ROC curve and is used to judge the quality of the model. The AUC value is usually between 0.5 and 1, and the relation between the AUC value and the model performance is shown in Table 2.

Table 2. The relation between AUC value and model performance.

AUC value	Model effect
0.95–1.00	Good
0.85–0.95	Better
0.70–0.85	General
0.50–0.70	Poor

4.2 Data Preprocessing and Analysis

Dataset. According to relevant laws in China, commercial banks keep customers' personal information confidential, so this article uses the classic credit dataset give-me-some-credit on *Kaggle*. Each sample in the dataset contains 11 attributes, the first 10 are related attributes describing the customer's credit status, and the 11th attribute refers to the credit evaluation of each customer tag after the dataset is comprehensively evaluated according to the first 10 attributes category. The names and types of indicators are shown in Table 3.

Table 3. Names and types of indicators.

Number	Attribute name	Attribute type
1	RevolvingUtilizationOfUnsecuredLines	Float
2	Age	Integer
3	NumberOfTime30-59DaysPastDueNotWorse	Integer
4	DebtRatio	Float
5	MonthlyIncome	Float
6	NumberOfOpenCreditLinesAndLoans	Integer
7	NumberOfTimes90DaysLate	Integer
8	NumberRealEstateLoansOrLines	Integer
9	NumberOfTime60-89DaysPastDueNotWorse	Integer
10	NumberOfDependents	Integer
11	SeriousDlqin2yrs	Integer

Data Preprocessing. *Missing Value Processing.* Table 4 shows that the *MonthlyIncome* in the raw data is seriously missing, and direct deletion may be affect the data distribution. This paper uses the random forest to fill in the missing values. In addition, the *NumberOfDependents* has slightly missing values, so deleted. Except for *MonthlyIncome* and *NumberOfDependents*, there are no missing values for other attributes.

Abnormal Value Processing. Figure 5 shows that a small number of customers' *age* are not in the normal range and so deleted directly. In addition, the loan overdue term is not in the range are considered as discrete values and deleted directly.

Data Conversion. Figure 6 shows that the marks of "good" customers are "0" and "bad" customers are "1" in the raw data. For convenience, the marks of "good" customers are converted to "1" and "bad" customers are converted to "0".

Table 4. Missing credit data.

Attribute name	Missing value amount
MonthlyIncome	29731
NumberOfDependents	3924

Data Analysis. Figure 7 shows that the customers' *age* is approximately normally distributed, which is in line with the assumption of statistical analysis. Figure 8 shows that the *MonthlyIncome* of customers is also generally distributed normally, which is in line with the assumptions of statistical analysis. Figure 9 shows that there is a very small correlation between any two variables.

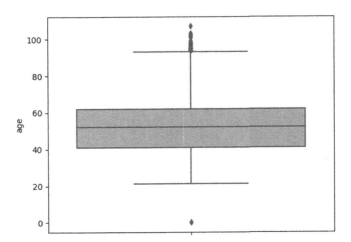

Fig. 5. Boxplot of customers' *age*.

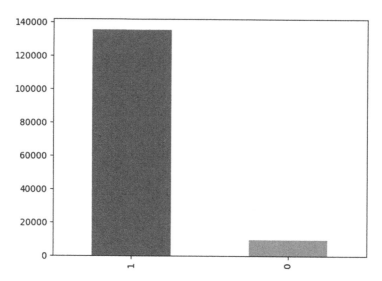

Fig. 6. Number of "good" and "bad" customers, where the blue bar represents the number of "good" customers and the yellow bar represents the number of "bad" customers. (Color figure online)

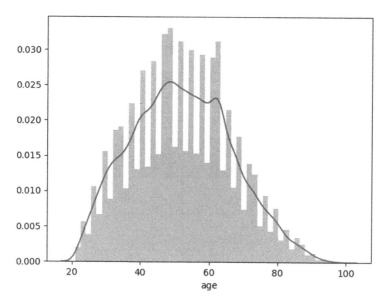

Fig. 7. Customers' *age* distribution.

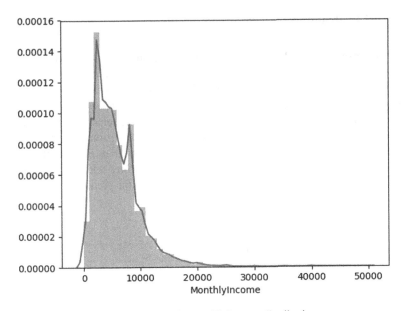

Fig. 8. Customers' *MonthlyIncome* distribution.

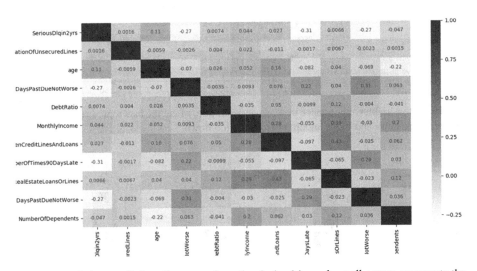

Fig. 9. Correlation coefficient diagram, where the decimal in each small square represents the correlation coefficient between the variables. The smaller the coefficient is, the weaker the correlation is. The different colors in the square represent the degree of correlation. Refer to the color band on the right of the figure, the closer the color is to blue, the higher the correlation is; the closer the color is to yellow, the smaller the correlation is. (Color figure online)

4.3   Model Evaluation Comparison

Selection of Important Indicators. As can be seen from Fig. 10, the importance of the attributes is ranked according to the internal mechanism of the decision tree. The importance of the six variables: *RevolvingUtilizationOfUnsecuredLines*, *DebtRatio*, *MonthlyIncome*, *NumberOfTimes90DaysLate*, *age*, and *NumberOfOpenCreditLines-AndLoans* are all greater than 0.05, so these six variables were selected to input the model. Figure 11 shows that the prediction accuracy of DTONN, decision tree (DT), and neural network (NN) that trained not using selected important variables is lower than the three models that using selected for important variables. Therefore, selecting important variables before the attributes is input into model can improve the prediction performance of the model.

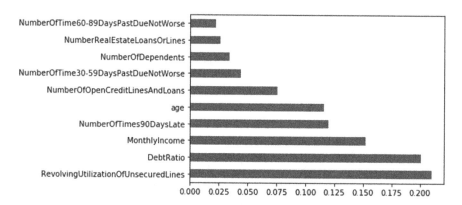

Fig. 10. The variables' importance rank via decision tree, where the bars represent that variables are becoming more and more important from the top down.

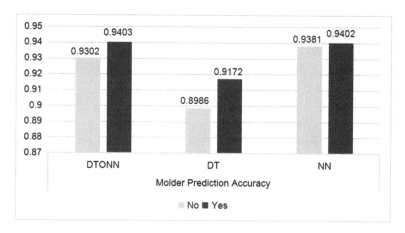

Fig. 11. Comparison of model prediction accuracy before and after selecting important variables using a decision tree, where the light blue bars represent the models' accuracy that trained with unselected variables, and dark blue bars represent the models' accuracy that trained with selected variables by decision tree. (Color figure online)

Comparative Analysis of Experimental Results of DTONN and Five Models. This section compares and analyzes the experimental results of the DTONN credit evaluation model with four other credit evaluation models: decision tree (DT), neural network (NN), support vector machine (SVM), and logistic regression (LR).

The prediction results of the five credit evaluation models on the test set: Accuracy, Precision, Recall, Type I error, Type II error, and AUC values are shown in Table 5. Based on these indicators, the performance of the model was analyzed through Fig. 12, Fig. 13, Fig. 14 and Fig. 15.

Table 5. The prediction results of the five models on the test set.

Model	Accuracy	Precision	Recall	Type error		AUC
				I	II	
DTONN	0.9403	0.9825	0.9552	0.4982	0.0447	0.9400
DT	0.9172	0.9384	0.9725	0.6239	0.0274	0.9100
NN	0.9402	0.9813	0.9561	0.5000	0.0438	0.9400
SVM	0.8417	0.9977	0.9426	0.4417	0.0573	0.8400
LR	0.8495	0.9807	0.9534	0.5620	0.0471	0.8500

Figure 12 shows that the classification accuracy of the DTONN model is higher than the other four models, indicating that the overall prediction performance of the DTONN model is better. At the same time, the precision of the DTONN model is second only to the SVM and higher than the other three models, indicating that the DTONN model has a better prediction ability for "good" customers. However, the recall of the DTONN model is lower than that of the DT model and NN model, and slightly higher than that of SVM and LR, indicating that the ability of DTONN model to correctly classify "good" customers as "good" customers is lower than that of DT and NN.

Figure 13 shows that the type I error of the DTONN is slightly higher than that of SVM and lower than that of the other three models. Figure 14 shows that the type II error is higher than DT and NN, and lower than the other two models. However, the risk of predicting "bad" customers as "good" customers is higher than the situation of predicting "good" customers as "bad" customers, so more attention needs to be paid to the type I error. Therefore, the DTONN model needs to be further optimized to reduce the type I error.

Figure 15 shows that the AUC value of DTONN is the highest among these five models, indicating that the DTONN model proposed in this paper is relatively reliable, however, the DTONN model needs to be further optimized to increase its AUC value.

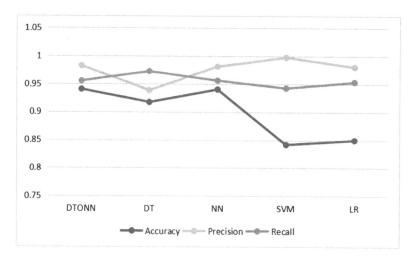

Fig. 12. Comparison of the accuracy, precision, and recall of the five models, where the blue line, orange line and gray line represent the accuracy, precision and recall of the five models respectively. (Color figure online)

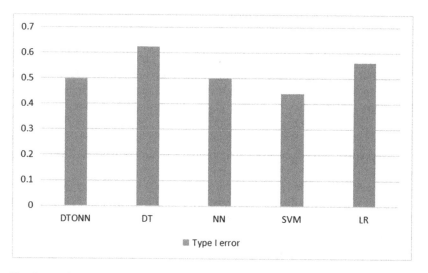

Fig. 13. Comparison of type I error of five models, where the yellow bars represent the type I error of the five models. (Color figure online)

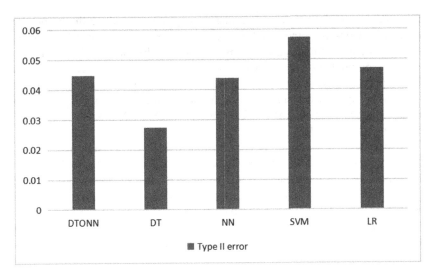

Fig. 14. Comparison of type II error of five models, where the blue bars represent the type II error of the five models. (Color figure online)

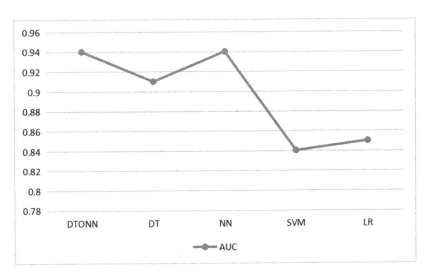

Fig. 15. Comparison of AUC values of five models, where every inflection point on the green line represents the AUC value of the five models. (Color figure online)

5   Conclusion

The personal credit evaluation model in commercial banks needs to be interpretable. The prediction accuracy of the neural network is high, and the interpretability is poor, while the decision tree is interpretable and the prediction accuracy is lower than the neural network. Considering that the decision tree and the neural network are complementary in performance and similar in structure, to reduce the probability of customer default risk in commercial banks, this paper proposes an interpretable personal credit evaluation model DTONN that combines the interpretability of decision tree and the high prediction accuracy of neural network. First, the decision tree is used to select important features for the preprocessed data. Then, the important features are used to train an optimal decision tree T1, and a neural network is built based on the classification information obtained by T1. Finally, the DTONN model is extracted from the trained neural network.

The experiment compares DTONN with four models of decision tree, neural network, support vector machine and logistic regression. Experimental results show that the DTONN model has the highest accuracy rate on the test set, the precision is second only to the support vector machine, the type I error is only slightly higher than the support vector machine, the recall and type II error value are all third, and the AUC value is the highest. Therefore, the DTONN model can be used for personal credit evaluation of commercial banks and reduce credit business risks.

References

1. Durand, D.: Risk Elements in Consumer Installment Financing, pp. 78–129. National Bureau of Economic Research, New York (1941)
2. Orgler, Y.E.: A credit scoring model for commercial loans. J. Money Credit Bank. **2**, 31–37 (1970)
3. Wiginton, J.C.: A note on the comparison of logit and discriminant models of consumer credit behavior. J. Financ. Quant. Anal. **15**(1), 757–770 (1980)
4. Makowski, P.: Credit scoring branches out. Credit World **75**(1), 30–50 (1985)
5. Baesens, B.: Bench marking state-of-the-art classification algorithms for credit scoring. J. Oper. Res. Soc. **54**(6), 627–635 (2003)
6. Bellotti, T., Crook, J.: Support vector machines for credit scoring and discovery of significant features. Expert Syst. Appl. **36**(2), 3302–3308 (2009)
7. Lucas, A.: Updating scorecards: removing the mystique. In: Credit Scoring and Credit Control, pp. 180–197. Oxford Clarendon Press, Oxford (1992)
8. Leonard, K.J.: Detecting credit card fraud using expert systems. Comput. Ind. Eng. **25**(1), 103–106 (1993)
9. Yobas, M.B., Ross, P.: Credit scoring using neural and evolutionary techniques. J. Math. Appl. Bus. Ind. **11**(1), 111–125 (2000)
10. West, D., Dellana, S.: Neural network ensemble strategies for financial decision applications. Comput. Oper. Res. **32**(10), 2543–2559 (2005)
11. Mariola, C., Dorota, W.: The individual borrowers recognition: single and ensemble trees. Expert Syst. Appl. **36**(3), 6409–6414 (2009)
12. Deng, S.F.: Personal Credit Evaluation Combined Model based on Decision Tree and Neural Network. Hunan University (2012)
13. Jiang, Y.H.: Construction of enterprise customer credit evaluation model based on decision tree neural network. J. Ningde Normal Univ. (Nat. Sci.) **30**(1), 8–14 (2018)
14. Zhang, Q.S., Wu, Y.N., et al.: Growing interpretable part graphs on ConvNets via multi-shot learning. In: Proceedings of AAAI Conference on Artificial Intelligence, pp. 2898–2906 (2017)

15. Zhang, Q.S., Shi, F., et al.: Interpreting CNN knowledge via an explanatory graph. In: Proceedings of AAAI Conference on Artificial Intelligence, pp. 4454–4463 (2018)
16. Zhang, Q.S., Yang, Y., et al.: Interpreting CNNs via decision trees. In: Proceedings of IEEE Conference on Computer Vision and Pattern Recognition, pp. 6261–6270 (2019)
17. Jan, R.Z., Frederik, J.: DeepRED-rule extraction from deep neural networks. In: Proceedings of International Conference on Discovery Science, pp. 457–473 (2016)
18. Sato, M., Tsukimoto, H.: Rule extraction from neural networks via decision tree induction. In: Proceedings of International Joint Conference on Neural Networks, pp. 1870–1875 (2001)

Education

Research on TCSOL Master Course Setting

Chunhong Qi[1(⊠)] and Xiao Zhang[2]

[1] Yunnan Chinese Language and Culture College, Yunnan Normal University, Kunming, China
3152@ynnu.edu.cn
[2] School of International Education, Yunnan University, Kunming, China

Abstract. Through questionnaires, discussions, interviews, classroom observations and other methods, this paper implements a study on the curriculum of Master of Teaching Chinese to Speakers of Other Languages (MTCSOL) at Yunnan Normal University. It was found that problems lied in many aspects, including the lack of teachers, teaching content being not cutting-edge and not meeting the teaching goals, and the shortage of teaching materials suitable for the training goals. Finally, relevant suggestions are proposed based on the analysis.

Keywords: Teaching Chinese to Speakers of Other Languages · Course · Training objectives · Teacher · Teaching material

1 Introduction

Since the setting of Teaching Chinese to Speakers of Other Languages in 2007, 151 colleges and universities in China (as of March 2020) have all complied with the "Guiding Training Program for Masters of Teaching Chinese to Speakers of Other Languages" to set up courses to train graduate students in the major. Various training colleges can fully use the program, and they can also set up special courses suitable for their school based on their own characteristics based on this program. Chen (2005), Li (2007), Jiang (2009), Li (2009), Ding (2009), Hu (2012), Wang (2012), Xu (2016), Liu (2018) et al. all have discussed about the curriculum setting of Teaching Chinese to Speakers of Other Languages at the macro level and pointed out that the curriculum content of the specialty has problems such as the theoretical curriculum and practical application courses, the proportion of language courses and cultural courses, the relevance of professional English and public courses. Yunnan Normal University is one of the first universities in China to offer a Professional Course in Teaching Chinese to Speakers of Other Languages. There are also many problems in the curriculum setting. In order to better promote the curriculum of Yunnan Normal University, and even provide some reference experience for the curriculum of this specialty in the country, we investigated and analyzed the rationality of overall curriculum of Teaching Chinese to Speakers of Other Languages of Yunnan Normal University.

© Springer Nature Singapore Pte Ltd. 2020
P. Qin et al. (Eds.): ICPCSEE 2020, CCIS 1258, pp. 543–560, 2020.
https://doi.org/10.1007/978-981-15-7984-4_40

2 Qualitative Analysis of Curriculum Setting of Teaching Chinese to Speakers of Other Languages of Yunnan Normal University

2.1 Overall Situation of Teaching Chinese to Speakers of Other Languages of Yunnan Normal University

In May 2009,[1] the "Teaching Chinese to Speakers of Other Languages Master's Degree Graduate Instructive Training Program" of the State Council Degree Office describes the curriculum system as: a core curriculum-oriented, modular expansion as supplementary, and practical training-focused curriculum system. The core curriculum aims to improve students' Chinese teaching ability, cultural communication skills, and cross-cultural communication skills. The core curriculum of the degree is set and credits: Teaching Chinese as a Second Language (4 credits), Second Language Acquisition (2 credits), Cases of Chinese Teaching in Foreign Countries (2 credits), Chinese Culture and Communication (2 credits) Intercultural Communication (2 credits).

The expanded course is divided into three modules: First, Chinese as a foreign language teaching (Chinese language element teaching, error analysis, contrast between Chinese and foreign languages, curriculum design, modern language teaching technology, Chinese teaching materials and teaching resources); the second is the spread of Chinese culture and cross-cultural communication (history of Chinese thought, national and regional cultures, special topics on cultural exchanges between China and foreign countries, etiquette and international relations); and the third is education and teaching management (foreign language education psychology, foreign primary and secondary education topics, instructional design and management, international Chinese promotion topics). The training courses include teaching investigation and analysis, classroom observation and practice, teaching testing and evaluation, and Chinese cultural talent display.

On the basis of the program of the State Council Degree Office "Guided Training Program for Postgraduates with a Master's Degree in Teaching Chinese to Speakers of Other Languages", Yunnan Normal University has formulated the first training program. And in September 2011, the second revised training program was implemented. In September 2015, the revised third training program was implemented. After the program is revised, the courses in this specialty are shown in Table 1. As can be seen from Table 1, the core curriculum of Yunnan Normal University has deleted "Teaching Chinese as a Second Language" in the 2009 Guidance Program of the Degree Office of the State Council and replaced it with "Chinese Language Analysis"; the "Chinese Culture and Communication" was classified as an extension course, and the 09 course teaching extension course "Course Design" was renamed as "Chinese Course Design" as the core curriculum. In addition, the "Chinese Classroom Teaching Method" not included in the plan is taken as a core course with two hours. Due to the limitation of the teaching staff, the teaching course of Teaching Chinese to Speakers of Other Languages at Yunnan Normal University did not provide statistics-related courses. An

---

[1] The Degree Office of the State Council: "Guidant Training Program for Full-time Chinese International Education Master's Degree Professionals", 2009.

18-h "Teaching Survey and Analysis" was opened. This course is also taught for teachers from the 2015–2019 school year, but the main research direction of the teachers is cultural, and there is not much research on statistics. In the course "Chinese Language Analysis", although the teacher introduced the ontology of Chinese language, the students' interest in learning was also very high. However, teachers did not combine Chinese ontological knowledge with Chinese teaching well, did not analyze language phenomena for teaching Chinese as a foreign language, and the knowledge spoken had little effect on teaching Chinese to non-native Chinese speakers. See Table 1 for the course setting of Teaching Chinese to Speakers of Other Languages at the Chinese Language Institute of Yunnan Normal University since 2015.

Table 1. Courses of teaching Chinese to Speakers of Other Languages by Yunnan Normal University from 2015 to present

Course type		Course code	Course name	Hours	Credits	Degree course	Term	Total credits
Core compulsory courses	Public core courses	9001	Research on Theory and Practice of Socialism with Chinese Characteristics	36	2	√	Term I	7
		9002	Introduction to Dialectics of Nature	18	1	√	Term II (1 optional)	
		9003	Marxism and Social Science Methodology					
		1068110020	Comprehensive English	36	2	√	Term I	
		150453001001	Thai	72	2	√	Term I (1 optional)	
		150453001002	Vietnamese	72	2	√		
	Degree core courses	150453002001	Introduction to Second Language Acquisition	36	2	√	Term I	12
		150453002002	Cases of Chinese Classrooms Abroad	36	2	√	Term II	
		150453002003	Cross-cultural Communication Case Analysis	36	2	√	Term I	
		150453002004	Analysis of Chinese Language	36	2	√	Term I	
		150453002005	Chinese Classroom Teaching Method	36	2	√	Term II	
		150453002006	Chinese Course Design	36	2	√	Term II	

(continued)

Table 1. (*continued*)

Course type		Course code	Course name	Hours	Credits	Degree course	Term	Total credits
Expanded selective courses	Teaching (at least 6 credits)	150453003001	Elements of Language and Teaching (1)	36	2		Term I	10
		150453003002	Elements of Language and Teaching (1I)	36	2		Term II	
		150453003003	Chinese Teaching Resources and Utilization	36	2		Term I	
		150453003004	Chinese Test and Evaluation	18	1		Term II (not set yet)	
	Culture (at least 2 credits)	150453003005	Chinese Culture and Communication	36	2		Term IV	
		150453003006	Comparison of Chinese and Foreign Cultures (Bilingual)	18	1		Term IV (not set yet)	
		150453003007	International Politics and Economics	18	1		Term IV (not set yet)	
	Education (at least 2 credits)	150453003008	Topics on Primary and Secondary Education Abroad	18	1		Term I	
		150453003009	Foreign Language Educational Psychology (Bilingual)	18	1		Term IV	
		150453003010	Introduction to Teaching Chinese to Speakers of Other Languages	18	1		Term I (not set yet)	
Training selective courses	At least 4 credits	150453004001	Teaching Survey and Analysis	18	1		Term II	4
		150453004002	Classroom Observation and Practice	18	1		Term II (not set yet)	
		150453004003	Teaching Chinese to Speakers of Other Languages Research	18	1		Term IV (not set yet)	

(*continued*)

Table 1. (*continued*)

Course type		Course code	Course name	Hours	Credits	Degree course	Term	Total credits
			Methods and Thesis Writing					
		150453004004	Chinese Cultural Talent Show and Display (I)	36	1		Term I (Calligraphy and Paper-cutting)	
		150453004005	Chinese Cultural Talent Show and Display (II)	36	1			
Internship		150453005001	Internship	240	6		Term III & IV	6
Paper		150453006001	Academic Degree Paper		2		Term V & VI	2
Make-up courses		150453007001	Foundation of Modern Chinese	36	N/A		Term I (not set yet)	No credit

3 Quantity Analysis of Questionnaires to the Overall Curriculum Setting of Yunnan Normal University

3.1 Questionnaire Design Theory Basis and Questionnaire Distribution

Questionnaire design theory basis In order to fully understand the teaching situation of the Yunnan Normal University Teaching Chinese to Speakers of Other Languages course, based on the existing research, this study designed the "Yunnan Normal University Teaching Chinese to Speakers of Other Languages Master Course Evaluation Questionnaire" according to the CIPP evaluation mode.

The CIPP evaluation mode is called the background, input, process, and result evaluation mode, and it was proposed by American scholar Stafford Beam in 1966. The evaluation mode is divided into four steps: background evaluation, input evaluation, process evaluation and result evaluation. Background assessment determines whether the goals of a plan are reasonable by evaluating the background of the plan's implementation. Input evaluation is to evaluate all aspects of the resources and conditions required to realize the plan so as to determine the advantages, disadvantages, feasibility and effectiveness of the plan. Process evaluation is the evaluation of the implementation process of the plan, the purpose is to monitor and feedback the implementation of the plan. Result evaluation is the evaluation of the results of the implementation of the scheme, and the interpretation and judgment of the results produced by the scheme.

The CIPP evaluation model includes both a diagnostic evaluation of course goals and teaching resources, and a formative evaluation of the teaching process and a result evaluation of the teaching results. This model overcomes the shortcomings of the traditional one-sided evaluation of teaching, and meets the multi-level requirements of diversity and comprehensiveness of university curriculum evaluation in the new era.

The author consults a large number of journals and literatures and concludes that the academic community's use of the CIPP education evaluation model is mainly concentrated in three aspects: evaluation of the curriculum, evaluation of teaching quality, and evaluation of talent training models. The research in various aspects can be roughly divided into the construction of evaluation system, evaluation mode, and evaluation index system.

Questionnaire Distribution. The questionnaire is divided into four parts. 1. Background evaluation: make an evaluation of the curriculum concept, curriculum goals, and curriculum structure; 2. Input evaluation: make an evaluation of the professional qualifications of the teacher, student conditions, teaching resources, and teaching management supervision; 3. Process evaluation: evaluate the content of the course, the teaching of the course, and the assessment of the course; three, evaluate the results: evaluate the effect achieved after the course. In the questionnaire, the Likert five-point scale method was used, and the options were "very inconsistent", "not consistent", "conforming", "relatively consistent", and "very consistent". The corresponding scores were 1, 2, 3, and 4 5 points. The score of "3" is the theoretical median value, so the closer the score is to "5", and the higher the evaluation, and the closer to "1", the lower the evaluation.

In June 2019, we issued a survey questionnaire for students majoring in Teaching Chinese to Speakers of Other Languages from Yunnan Normal University. A total of 102 questionnaires were distributed, of which 33 were from the first grade (2018), 33 were to the second grade (2017), 5 were to the third grade (2016), and 21 were to the 2015 grade. After 19 invalid questionnaires were excluded, there were 83 valid questionnaires. The courses studied by the respondents were all based on "Teaching Chinese to Speakers of Other Languages (Chinese Students) Training Program (Revised in 2015) (Professional Code: 045300)" by Yunnan Normal University. See the specific analysis of the curriculum of this major.

3.2   Reliability and Validity Analysis

Reliability Analysis. The reliability of the questionnaire in this study was tested using Cronbach's Alpha coefficient. Among them, "Yunnan Normal University Teaching Chinese to Speakers of Other Languages Master's Degree Program Evaluation Questionnaire", after SPSS reliability statistics, the Alpha coefficient is 0.971.

Validity Analysis. The validity of this research questionnaire was tested by SPSS for KMO and Bartlett's test. The results show that the KMO values are higher than 0.7, and the validity is high.

Questionnaire Data Analysis. This study has sorted the data of the indicators at all levels of the questionnaire, and now summarizes the scores of the indicators at all levels into Table 4 (Table 3).

Table 2. Alpha coefficient test of questionnaire about the curriculum setting of teaching Chinese to Speakers of Other Languages at Yunnan Normal University

Background evaluation reliability		Input evaluation reliability		Process evaluation reliability		Result evaluation reliability		Overall reliability of the questionnaire	
Cronbach's alpha	Items	Cronbach's alpha	Items	Cronbach's alpha	Items	Cronbach's alpha	Items	Cronbach's alpha	Items
.885	5	.929	11	.941	14	.789	4	.971	34

Table 3. KMO test of questionnaire about overall curriculum setting of teaching Chinese to Speakers of Other Languages at Yunnan Normal University

KMO & Bartlett test		Background evaluation	Input evaluation	Process evaluation	Result evaluation	Overall validity
KMO		.832	.900	.911	.764	.932
Bartlett's spherical test	About chi square	222.33	596.983	779.780	95.547	1857.053
	df	10	55	91	6	253
	Distinctiveness	0	.000	.000	.000	.000

Table 4. Evaluation data statistics table of the overall curriculum setting of teaching Chinese to Speakers of Other Languages at Yunnan Normal University

Level I	Level II	Level III	Min.	Max.	Average
Background evaluation (3.42)	Course concept (3.42)	Reasonable course positioning	2.00	5.00	3.4337
		Scientific and clear guidance	1.00	5.00	3.3976
	Course goal (3.42)	Curriculum goals in line with school running level	1.00	5.00	3.3735
		Clear course goals	1.00	5.00	3.4578
		Popularization of course goals	2.00	5.00	3.4217
Input evaluation (3.48)	Teachers (3.4)	Reasonable teacher-student ratio	1.00	5.00	3.1446
		Teacher level meets curriculum requirements	1.00	5.00	3.6627
	Students (3.57)	Reasonable enrollment	1.00	5.00	3.5783
		Students' professional background conforms to curriculum	1.00	5.00	3.5663
	Supporting resources (3.62)	Complete hardware facilities at school	1.00	5.00	3.7590
		Sufficient course funding	1.00	5.00	3.4819

(continued)

Table 4. (*continued*)

Level I	Level II	Level III	Min.	Max.	Average
	Management supervision (3.32)	Reasonable number of lessons	1.00	5.00	3.3373
		Improved school quality supervision system	1.00	5.00	3.3012
Process evaluation (3.51)	Course contents (3.48)	The course content is reasonable and practical	2.00	5.00	3.4819
		Moderate difficulty of contents	1.00	5.00	3.5301
		Course content reflects the characteristics of the school	1.00	5.00	3.5181
		Course content keeps pace with the times	1.00	5.00	3.3976
	Course teaching (3.54)	Correct teaching attitude of teachers	1.00	5.00	3.7108
		Flexible teaching methods	1.00	5.00	3.6867
		Reasonable teaching organization	2.00	5.00	3.4819
		Students have a good learning attitude	1.00	5.00	3.4819
		Students' learning methods and strategies are scientifically effective	1.00	5.00	3.3373
	Course evaluation (3.27)	Diverse and reasonable assessment methods	1.00	5.00	3.3012
		The assessment method is accurate and effective	2.00	5.00	3.2289
	Internship (3.64)	Internships can effectively train students	1.00	5.00	3.7831
		Instructor has effective guidance for student internship	1.00	5.00	3.6747
		Schools effectively monitor students' internships	1.00	5.00	3.4578
Result evaluation (3.35)	Goal achievement (3.35)	Improved second language teaching ability	2.00	5.00	3.4699
		Improved cross-cultural competence	1.00	5.00	3.7470
		Improved foreign language skills	1.00	5.00	2.8675
		Improved organizational capacity	1.00	5.00	3.3253

It can be seen from the scores of Table 4 that the overall score of the course setting is not high. The average of the four parts is less than 4 points. The scores of the four evaluation parts are process evaluation (3.51), input evaluation (3.48), Background evaluation (3.42) and result evaluation (3.35). We combined the data to summarize the problems in the curriculum setting of Teaching Chinese to Speakers of Other Languages at Yunnan Normal University as reflected in the questionnaire:

First, the background evaluation. In the curriculum setting indicators, the score of the "Curriculum Idea" is equal to the "Course Goal". The "Clear Curriculum Goal" in the "Curriculum Goal" has the highest score (3.46), and the "curriculum Goal in line with the school level" has the lowest score (3.37). According to the survey data, the participants believed that the orientation of Yunnan Normal University Teaching Chinese to Speakers of Other Languages was not clear, the course concept was not clear, the course goals were not clear, and they did not meet the school level.

Second, enter the evaluation aspect. The teaching resource index scores from high to low are "support resources" (3.62), "student source conditions" (3.57), "teacher conditions" (3.40), "management supervision" (3.32), and "teacher conditions" "teacher level and course requirements" meet the highest score (3.67), the lowest teacher-student ratio (3.14), and the lowest score in the teaching resources; in the "source conditions", the scores of "reasonable enrollment scale" and "student professional background and curriculum compliance" are both lower than 3.6; "supporting resources" score of "complete school hardware facilities" (3.76) is higher than "course funding" (3.48). "Management and supervision" is the second-level indicator with the lowest score in "Teaching Resources". It contains two items, "the number of courses is reasonable, and the time is arranged" (3.34). This shows that Yunnan Normal University has an unreasonable ratio of teachers, and the school has serious shortage of quality supervision and funding.

Third, process evaluation. The scores of the curriculum implementation indicators in order from high to low are "course practice" (3.64), "course teaching" (3.54), "course content" (3.48), and "course evaluation" (3.24). "Course assessment" is also the lowest score among all secondary indicators in the questionnaire. Among the "course contents", the lowest scoring item was "the course content is advancing with the times" (3.34), and the highest was "the course content is difficult to moderate" (3.53). "Course teaching attitude" in "Course Teaching" (3.71) is the item with the highest score in the course implementation; "Course Assessment" has the lowest score in the "accurate and effective assessment method" (3.23); "internship" in "Course Practice" "Position can effectively train students" (3.78) scored the highest, and also the highest score among all the items in the questionnaire. "Schools have effective supervision of students' internship" (3.46) scored the lowest. These reflect that the curriculum content of this specialty cannot keep pace with the times and is difficult to be moderate. The effectiveness of the curriculum assessment methods is not high.

Fourth, the results. In the course performance indicators, the scores in descending order are "improved cross-cultural competence" (3.75), "improved second language teaching ability" (3.47), "improved management organization ability" (3.33), and "improved foreign language level" (2.87). Among them, "improved foreign language proficiency" is also the lowest scoring item among all the items in the questionnaire. This shows that the participants are not very satisfied with the effect of the course, and the score of "improved foreign language level" is the lowest. This shows that the results of the two courses of Thai speaking and English speaking are not satisfactory.

4 Evaluation and Analysis of the 16 Professional Courses of Teaching Chinese to Speakers of Other Languages at Yunnan Normal University

This study refers to the previous research on the evaluation of curriculum settings, and produced the "Yunnan Normal University Teaching Chinese to Speakers of Other Languages Master's Course Evaluation Questionnaire" in order to evaluate the 16 courses offered by the Master of Teaching Chinese to Speakers of Other Languages of Yunnan University. The questionnaire is divided into four parts: course content, teacher lectures, course assessment and course results. The questionnaire uses the Likert five-point scale method. The options are "very inconsistent", "non-conforming", "conforming", "more consistent", and "very consistent". The corresponding scores are 1, 2, 3, 4, and 5, respectively. The questionnaire was issued together with a questionnaire survey of the overall situation of the curriculum.

According to SPSS reliability statistics, the Alpha coefficient is 0.943, and the reliability is high.

The validity of the questionnaire was tested by SPSS for KMO and Bartlett's test. The results show that the KMO values are all higher than 0.7, which has certain validity. We made the data of the questionnaire reliability test into Table 6 (Table 5).

Table 5. Alpha coefficient test of evaluation questionnaire to each professional course of teaching Chinese to Speakers of Other Languages at Yunnan Normal University

Course content		Teachers' teaching		Course assessment		Course result		Overall validity	
Cronbach's alpha	Items	Cronbach's alpha	Items	Cronbach's alpha	Items	Cronbach's alpha	Items	Cronbach's alpha	Items
.906	5	.875	11	.917	4	.895	3	.943	23

Table 6. KMO test of the evaluation questionnaire to each course of teaching Chinese to Speakers of Other Languages at Yunnan Normal University

		Course content	Teachers' teaching	Course assessment	Course results	Overall validity of the course
KMO		.821	.928	.844	.729	.932
Bartlett's spherical test	About chi square	262.477	902.671	230.775	146.033	1857.053
	df	10	55	6	3	253
	Distinctiveness	.000	.000	.000	.000	.000

We tested the homogeneity of variance for the mean of the 16 courses obtained from the questionnaire. The test result is p = 0.061 > 0.05, and the variance is homogeneous. The test result is shown in Table 7.

Table 8 shows that the variance is homogeneous, and one-way analysis of variance can be performed. The one-way ANOVA results are F(15,48) = 14.574, and p = 0.000 < 0.05. There is a significant difference in the evaluation value of each course. See Table 8 for the results of the one-way ANOVA.

Table 7. Test of variance homogeneity of the evaluation scores of 16 courses at Yunnan Normal University

	Levene statistics	df1	df2	Distinction
Evaluation value	1.812	15	48	.061

Table 8. Single-factor variance analysis of the average evaluation value of 16 professional courses of teaching Chinese to Speakers of Other Languages at Yunnan Normal University

ANOVA						
		Sum of square	df	Mean square	F	Distinction
Evaluation value	Between groups	5.730	15	.382	14.574	.000
	Within groups	1.258	48	.026		
	Total	6.988	63			

The average values of the 16 courses of the Yunnan Normal University Teaching Chinese to Speakers of Other Languages major are significantly different and the data are valid. We have made the average of the evaluation of 16 courses into Table 9, in order to more intuitively show the difference in the average of each course, we present the average of each course as Fig. 1.

Table 9. Statistics of the average value of 16 professional courses of teaching Chinese to Speakers of Other Languages at Yunnan Normal University

Course	Course content	Teachers' teaching	Course assessment	Course result	Average value
Intercultural Communication	4.13	4.21	3.75	4.11	4.05
Analysis of Chinese Language	4.03	4.21	3.84	3.81	3.97
Introduction to Second Language Acquisition	3.71	3.78	3.39	3.52	3.6
Oral Thai	3.09	3.3	3.26	3.1	3.19
Oral English	2.78	3.45	3.16	2.85	3.06
Topics on primary and secondary education abroad	4.01	4.2	4.1	4.05	4.09
Chinese Teaching Resources and the Application	3.56	3.74	3.53	3.6	3.61

(*continued*)

Table 9. (*continued*)

Course	Course content	Teachers' teaching	Course assessment	Course result	Average value
Case Analysis of Chinese Classrooms	3.95	4.22	3.64	3.99	3.95
Foreign Language Teaching Psychology	3.95	4.22	3.64	4	3.95
Language Elements and Teaching	4.06	4.05	3.73	4.06	3.98
Chinese Culture and Communication	3.81	3.88	3.76	3.96	3.85
Chinese Classroom Teaching Methods	3.72	3.71	3.69	3.64	3.69
Chinese Classroom Design	4	4.04	3.91	3.89	3.96
Classroom Observation and Practice	3.81	4.01	3.81	3.88	3.88
Teaching Investigation and Analysis	4	4.16	3.96	3.91	4.01
Chinese Cultural Talent Show and Display	3.93	4.19	4.17	4.12	4.1

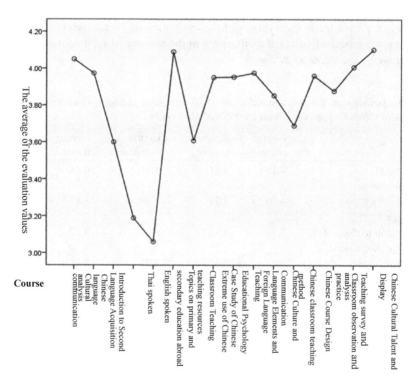

Fig. 1. Comparison of the average value in the assessment of 16 professional courses of teaching Chinese to Speakers of Other Languages at Yunnan Normal University

From Table 8 and Fig. 1, it can be seen that the courses with a mean curriculum evaluation higher than 4 include "Cross-Cultural Communication", "Topics on Primary and Secondary Education Abroad", "Teaching Survey and Analysis", and "Chinese Cultural Talent Show and Display". As the core courses, "Second Language Acquisition" and "Chinese Classroom Teaching Method" have lower scores, respectively 3.6 and 3.69. The score of "Chinese Teaching Resources and Application" as an extended course is also low; the average values of "Oral Thai" and "Oral English" in public core courses are very low, 3.19 and 3.06 respectively. From the evaluation results, the scores of the 16 courses of Yunnan Normal University Teaching Chinese to Speakers of Other Languages are not high. From the perspective of the judges, the only ones that can reach 80 points (out of 100 points) are "Cross-Cultural Communication", "Topics on Primary and Secondary Education Abroad", "Teaching Survey and Analysis", and "Chinese Cultural Talent Show and Display"; and the language courses closely related to Chinese teaching, such as "Chinese Language Analysis", "Language Elements and Teaching", and "Chinese Course Design", all of which scored close to 80 points. The core teaching theory course "Introduction to Second Language Acquisition" has just exceeded 70 points; the "Chinese Classroom Teaching Method" and "Chinese Teaching Resources and Use" related to teaching resources and teaching skills also just exceeded 70 points; and the teaching and working medium English is just passing. Therefore, the overall teaching situation is worrying.

5 Problems in the Curriculum Setting of Teaching Chinese to Speakers of Other Languages at Yunnan Normal University and Solutions

Based on interviews, discussions, and classroom observations, we analyze the curriculum and teaching implementation of Yunnan Normal University from the following aspects:

5.1 Problems in the Overall Curriculum Setting and Solutions

In order to understand the Teaching Chinese to Speakers of Other Languages professional curriculum and the specific issues of teaching each course at Yunnan Normal University, we held the graduate teaching and research meeting of the college in November 2019. There were 16 professional teachers at the conference, and most teachers thought that the core course of "Teaching Chinese as a Second Language" should be continued because Chinese knowledge is the foundation of teaching, although the course "Chinese Language Analysis" teaches ontological knowledge, it lacks pertinence. The Teaching Chinese to Speakers of Other Languages major should be oriented towards the most practical and practical teaching knowledge in teaching Chinese as a foreign language. The "Chinese Classroom Teaching Method" overlaps with "Chinese Course Design", "Chinese Language Elements Teaching", and "Classroom Observation and Practice", and appropriate deletion should be made. At the same time, we also held a student seminar to collect students' opinions. Students believe that the study background of students is more complicated, and many students have weak basic knowledge of Chinese. The college should provide supplementary courses such as Modern Chinese and Introduction to Linguistics. As some students' graduation thesis involves educational surveys, and statistical knowledge will be used in writing the graduation thesis, it is recommended that the college provide statistics-related courses to cultivate students' ability to scientifically process and analyze data.

5.2   Teacher-Student Ratio and Teacher Teaching

Yunnan Normal University Teaching Chinese to Speakers of Other Languages has 17 master's tutors, all with titles of associate senior or higher, including 16 with doctoral degrees. Looking at the enrollment situation of the Yunnan Normal University Teaching Chinese to Speakers of Other Languages in the past five years, the average number of Chinese students enrolled has remained at 51 and the number of international students enrolled has remained at about 28 in the past five years. Compared with the number of other masters, the number of students enrolled in Teaching Chinese to Speakers of Other Languages is relatively large, and there is a large class. This will lead to a series of problems. For example, a tutor not only brings Chinese students, but also foreign students, not only students of all grades of graduate students, but even undergraduates. An instructor teaches an average of more than 10 students, which not only puts a lot of pressure on the instructor, but the students cannot accept sufficient guidance, which affects the quality of the graduation thesis and leads to delays in graduation and other issues. This is also the reason why the average score of the "teacher-student ratio" in Table 2 is low (3.1446). It is recommended that schools introduce more professionals in Teaching Chinese to Speakers of Other Languages, so that the scale of the teaching staff can complete the training objectives of the specialty.

In the "course content", the lowest scoring items were "course content keeps pace with the times" (3.34), and the highest was "it is difficult to moderate course content" (3.53). Through our survey, we found that the teaching content of many subjects is not very relevant to the teaching Chinese to Speakers of Other Languages major. Students think that the curriculum content has no practical value. Moreover, the knowledge of the majors they have learned is quite outdated. Teachers fail to provide students with knowledge of cutting-edge theoretical subjects, which makes it very difficult for students to write graduation thesis.

Taking the "Chinese Teaching Resources and Its Utilization", which has a low score in the expanded course, as an example, the teaching content of this course is concentrated on the teaching materials. The teacher's teaching method is mainly to compare different teaching materials and summarize the advantages and disadvantages. The problem with teachers' teaching is that they don't pay much attention to the cross-integration with other disciplines, they just study the teaching materials, which is not conducive to expanding the students' knowledge. At the same time, the teacher's introduction of the cutting-edge dynamics of the subject is not enough. The study of the teaching materials studied in the classroom is still in 2000, which is also the deficiency of this course. This phenomenon also exists in courses such as "Chinese Culture and Communication", "Cross-Cultural Communication Case Studies", "Chinese Course Design", "Chinese Classroom Teaching Method", "Foreign Language Education

Psychology", "Language Elements and Teaching". Therefore, this issue should attract the attention of teachers. Some courses do not have teaching materials, such as the "Chinese Classroom Teaching Method". Teachers not only do not have teaching materials or courseware, and the lecture content is very random. In addition, in the course of the course, after the students gave a group report, they rarely receive teacher's comments and valuable feedback.

The teaching material of the course "Foreign Language Educational Psychology" is English, and students need translation to understand it. Moreover, this textbook itself is intended for working English teachers and researchers of English language teaching, especially for graduate students majoring in English education. It is not particularly relevant to the Teaching Chinese to Speakers of Other Languages major. Therefore, students think that the curriculum is not practical. Teaching Chinese to Speakers of Other Languages basically lacks teaching materials that can not only achieve the training goals, but also reflect the dynamics of cutting-edge research in the discipline.

5.3   The Setting of Second Language Classes

Currently in the Yunnan Normal University Teaching Chinese to Speakers of Other Languages course, in addition to public English, there is also a foreign language lesson, either Thai or Vietnamese. The first semester of the course is 36 semesters, two times a week. Most students choose Thai and only a few choose Vietnamese. For example, among students of Grade 2017, 38 people chose Thai, and only 4 people chose Vietnamese. Students from the beginning of Thai language think that in the Thai language class, the teacher just teaches pronunciation and word reading, and the students repeatedly follow it, but after one semester, they cannot say a complete sentence in Thai language. Teachers also have some problems in teaching. Due to the limited class hours, there are only two class hours per week. It is very difficult to complete a whole "Basic Thai" textbook. Teachers are too busy to catch up with the schedule and do not have enough time to practice. Therefore, it is recommended to increase class hours and give students sufficient practice, and teachers also have more time to practice sentence patterns to achieve better results.

In addition, students who have a basic knowledge of Thai language are recommended to pass the exam to exempt from Thai language. Many students hope that the College of Chinese Language and Literature will cooperate with the School of Foreign Languages to share resources so that students of Teaching Chinese to Speakers of Other Languages can take other foreign language courses, such as Korean, Japanese, and Russian.

5.4   Course Assessment

In the "Course Assessment", the item "Accurate and Effective Assessment Method" (3.23) has the lowest score, indicating that the students believe that there is a problem with the course assessment. At present, Yunnan Normal University Teaching Chinese to Speakers of Other Languages majors' course evaluation methods are "30% of normal grades + 70% of course thesis". Curriculum assessment methods are indeed not diverse enough, and it is not reasonable to take the final written curriculum thesis as an evaluation index, nor can it comprehensively test the students' true learning results, especially the teaching practice ability. Referring to the previous curriculum evaluation standards, there are drawbacks to the summary evaluation model. We should attach importance to procedural evaluation and change the assessment method based on course papers.

5.5   Establish a Sound Supervision Mechanism for Teaching and Practice

At present, the teachers of Yunnan Normal University Teaching Chinese to Speakers of Other Languages are more casual in teaching. Many teachers cannot teach around the training goals, some subjects have not even had textbooks for many years, or the textbooks are not suitable for the teaching of the specialty. After the student internship, the teacher just grades and writes reviews for the student's internship. In fact, the teacher and the college lack effective guidance for students' practical teaching practice. The college is equipped with instructors for some postgraduate internships. However, most of these teachers do not do research and have a master's or undergraduate degree. Most of their understanding of teaching comes from experience. They lack the support of theoretical knowledge and do not have the ability to guide graduate teaching practice.

5.6   Strengthen the Construction of Discipline

The college should train teachers regularly so that teachers can be charged in time and understand the latest developments in the discipline; provide appropriate financial support for graduate students and encourage them to participate in academic conferences related to the major.

6   Conclusion

This study explores the existing research on the evaluation of the Master's Program in Teaching Chinese to Speakers of Other Languages. For example, Wang and Li (2011) made a comparative study of the Teaching Chinese to Speakers of Other Languages master courses of Beijing Language and Culture University, Beijing Normal University, and East China Normal University. There are also problems of clarity, lack of teachers, and the deviation of teaching content from the research direction and content of teaching Chinese as a foreign language. After investigating 39 graduate students of Teaching Chinese to Speakers of Other Languages in 4 universities, Zhou (2014) found that: the proportion of theoretical courses is too high, the teaching methods of teachers are single, rigid, and the content of the courses is outdated. Zhang (2013) surveyed Sichuan University and Sichuan Normal University's Teaching Chinese to Speakers of Other Languages master's degree courses, and found that there are problems with the curriculum setting of unprofessional teachers, and the content of teachers' teaching does not match the actual needs of the major. Jiao (2017) surveyed the curriculum of the Master of Teaching Chinese to Speakers of Other Languages of Shaanxi University of Technology and found that there were problems with the curriculum setting, such as the imbalance between theoretical and practical classes, the single teaching method of teachers, and the disconnection between teaching content and reality. Wang (2018) surveyed the teaching curriculum of the four major universities in Tianjin, Teaching Chinese to Speakers of Other Languages, and found that the curriculum has problems such as inadequate combination of theory and practice, inconsistent practical teaching materials, and frequent changes in teaching materials. The author also investigated the curriculum of the Master of Teaching Chinese to Speakers of Other Languages of Yunnan University and found that there are problems such as: imbalance of teacher-student ratio, insufficient research on the frontiers and development dynamics of teachers.

Based on the above existing research, this study found that the imbalance in the ratio of teachers, the content of teaching and professional training goals are not consistent, the teaching is random, the teachers are insufficiently researched on the frontiers and development dynamics of the disciplines, and the lack of suitable teaching materials are general problems in Teaching Chinese to Speakers of Other Languages. It is recommended that the National Degree Office mobilize experts from across the country to write cutting-edge teaching materials for Teaching Chinese to Speakers of Other Languages, which are suitable for the training objectives of the major. And the teachers of this major should be regularly trained or designated to visit other universities or abroad to fundamentally solve the quality problem of personnel training in this major.

References

Chen, F.: On the knowledge structure of postgraduates in teaching Chinese as a foreign language. Lang. Appl. (S1), 124–128 (2005)

Li, X.: Construction of teachers in international promotion of Chinese language. J. Yunnan Norm. Univ. (5), 8–9 (2007)

Jiang, X.: Research on the Curriculum of the Master of International Education in Chinese Language. World Book Publishing House, Beijing (2009)

Ding, A.: Thoughts on the curriculum setting for the master of international education in Chinese language. Int. Chin. Educ. (2), 13–18 (2009)

Li, Q.: Discussion on the training objectives and teaching concepts of the master of Chinese international education. Lang. Appl. (03), 105–112 (2009)

Hu, B.: Study on the curriculum setting for chinese international education major. Master's Degree Thesis of Shenyang Normal University, Shenyang (2012)

Wang, S.: Research on the curriculum setting of Chinese international education. Shandong Normal University, Shandong (2012)

Xu, J.: Thoughts on the training program for the master degree of international education in Chinese language. Overseas Chin. Educ. (1), 27–35 (2016)

Liu, S.: Research on the curriculum of Chinese international education master (China). Liaoning Normal University, Liaoning (2018)

Wang, L., Li, X.: Research on the curriculum setting for Chinese graduate students in foreign languages. J. Chongqing Univ. Arts Sci. (Soc. Sci. Edn.) 30(04), 150–154 (2011)

Zhou, Y.: A review of researches on talent cultivation models for Chinese international education majors in colleges and universities. J. Liaoning Adm. Coll. 16(12), 96–97 (2014)

Zhang, L.: The inadequacy of master's program in Chinese International Education from the perspective of investigation. J. Chin. Stud. (05), 101–102 (2013)

Jiao, X.: A study on the curriculum of Chinese international education in local colleges and universities – taking Shaanxi University of science and technology as an example. Educ. Teach. Forum, (29), 80–81 (2017)

Wang, J.: Investigation and research on the curriculum of Chinese international education master programs in four universities in Tianjin. Tianjin Normal University, Tianjin (2018)

Analysis of Learner's Behavior Characteristics Based on Open University Online Teaching

Yang Zhao[✉]

Yunnan Open University, Kunming 650500, China
42755044@qq.com

Abstract. The analysis on the learning behavior characteristics based on big data is beneficial for improving the learning resource construction, teaching mode and interactive mode of online course platforms. Multiple aspects of analysis were conducted on nearly three million pieces of learning behavior data, which is from seven courses of 3,315 learners in the same major at a university. According to the quantity of course resources and policy of course scoring, four typical learning behaviors were selected, and the correlation between final exam results and learning behavior were analyzed. The analysis of behavior influences on the final exam results were also conducted. The analytical results give suggestions for online teaching and learning.

Keywords: Online course platform · Analysis of learning behavior · Learning characteristics · Online teaching of Open University

1 Introduction

The Open University of China has a running history of more than 40 years and a system based on online teaching. It has branches in every province in the country, each of them provides 'online teaching' to learners in the whole province by means of online course platform. So far, more than 20 million distance education college students graduated from the Open University. And the university achieved the national sharing of high-quality educational resources online, thus winning extensive recognition from people in China.

Learners are the main subjects of the online teaching of Open Universities. On the one hand, the Open University owns a large number of learners. In the same major, there are different student categories. Every course platform of branches has accumulated millions of learners and their learning behavior data. On the other hand, the traditional teaching and learning behavior evaluation method is not suitable for online learning. Compared with traditional education, most distance learners learn independently by course resources online provided on the platform, with their own time arrangement and favorite learning method. Therefore, traditional evaluation methods, such as 'sampling data evaluation method', are not suitable for the evaluation of online learning behavior.

Comprehensively learning behavior records of all learners were saved in the database of the online course platform, encompassing learner's information, types and times of learning behaviors, learning results, etc. So, based on the big data technology,

© Springer Nature Singapore Pte Ltd. 2020
P. Qin et al. (Eds.): ICPCSEE 2020, CCIS 1258, pp. 561–573, 2020.
https://doi.org/10.1007/978-981-15-7984-4_41

we can analyze the characteristics and trends of learning behaviors, so as to evaluate the learning effect and the value of course online.

The author selects the data of 7 compulsory courses learned by 3,315 learners who were enrolled in 2017 majoring in 'digital media technology'. Quantitative analysis and characteristics exploration were conducted on the data of online learning behavior. In this way, suggestions for the ODER (Open Distance Education Resources) construction and network teaching methods will be made.

2 Related Preparation

2.1 Researches on Online Learning Behavior Characteristics at Home and Abroad

Although the Open University of China enjoys an online teaching history of more than 20 years, it is not too early for online learning to attract public attention. In fact, the researches on behavior analysis, feedback and prediction of online learning attracted people's attention was that MOOC (Massive Open Online Course) rise in China. According to the queried literatures, the researches on online learning behavior characteristics at home and abroad are as follows:

It is Complex to Evaluate the Learning Behavior of the Online Course Platform. Since the rise of MOOC in 2013, more and more researchers have been engaged in analyzing the behavioral characteristics of online learning. In 2013, *Neutral Information Processing System* (NIPS) launched a seminar for data research in MOOC. Based on the analysis of courses and learner data on diverse online course platforms, including MOOC, Coursera, Edx and other platforms. Researchers found that the composition of learners, learning mode and evaluation of teaching quality are much more complicated than those in traditional teaching.

Statistics and Analysis By Single-Dimensional Data give Suggestions for Improvement. For example, the statistics and analysis of data of assignment submission in literature [6], of forum data in literature [7], of learning behavior in literature [8] and so on. However, there are few literatures on multidimensional data statistics and analysis.

Efforts are Made to Model Learners' Behaviors to Predict Learning Behaviors. For example, in literature [9–12], the distribution of learning records, the completion rate of courses, the data of watching videos and taking tests are studied.

There are Few Learner Behavior Analysis Results of Open University Based on Big Data Technology. Most of achievements of Open Education Research are in social science field such as macro policy research, teaching mode construction, teaching implementation methods, etc. There are few research achievements of learners' behavior characteristics analysis based on big data.

2.2   Learning Resources and Learning Requirements in Open University

Types of Learning Resources and Learning Behaviors. Open University provides learners with videos, texts, PPT, assignment, tests, etc. Each video has a button below for leaners to get in course forum, creating a communication between learners and teachers, learners and learners. Corresponding to these resources, learning behaviors include watching videos, browsing texts, browsing PPT, finishing and submitting assignment and tests, etc.

Minimum Course Requirements. In order to ensure the quality of teaching, in terms of the importance of course resource types, Open University, on the one hand, has set the minimum requirements for teachers to publish enough course resources, so as to ensure sufficient content for learners. On the other hand, there are minimum requirements for each learner in watching videos, submitting assignment and participating in tests, thus guaranteeing the quality of online learning (this stands out in the learner's learning behavior data, described in Sect. 4.4).

Video Requirements. Based on the concept of 'micro-course', the number of videos uploaded by teachers should not be less than 230% of the number of class hours, with each duration being mostly 5–15 min. The number of videos watched by the learners shall not be less than 60% of the total number of videos of the course, otherwise they will not be able to obtain online scores. When watching video, 'fast forward' is not available by default. Only by finishing watching the whole video can learners be marked as having watched one video.

Assignment Requirements. The number of assignments issued by teachers shall not be less than 1/6 of the number of hours. Learners must submit at least 60% of the total number of assignments, otherwise they will not be able to get online results. For the same task, students can submit multiple assignments. Among them, only the best result is adopted.

Test Requirements. The number of tests issued by teachers shall not be less than 1/10 of the number of hours. Learners must take at least 60% of the total number of tests, otherwise they will not be able to get online results. For the same test, students can submit multiple results. Among them, only the best result is adopted.

The Comprehensive Score. The comprehensive score consists of the final examination result and the online result. However, for the direct correlation of the data analysis, the final examination results are taken as the data of study results.

2.3   Several Aspects to Be Studied in This Paper

Based on the technical routes adopted by previous researchers in studying MOOC, Coursera, Edx and other platforms, in this paper, Yunnan Open University being an example, the author describe the learning behavior on the online course platform, and analyzes the four typical learning behaviors- watching videos, submitting assignment, participating in forums and participating in tests, which are associated with the final examination results, so as to judge the influence of learning behavior on the examination results, and to provide reference for advancing course resources and evaluation modes.

3 Analysis of Learner Data and Learning Behavior Data

3.1 Consideration About Select Data

In order to accurately judge the learning behavior characteristics of a large number of learners based on the same group and similar courses, these two aspects are considered in the selection of learning behavior data:

Different Types of Learners Having the Same Major and Enrolled at the Same Time. Different types of learners majoring in 'digital media' and enrolled at the same time are selected in this paper. They have similar knowledge background and respond to the same learning requirements. The data amount and the selected course samples are also the same, but they are different in age, learning time and online learning habits. There are 5 type of students: The Combine with *Secondary and Higher Vocational Education* (CSHVE), *Distance Open Education of Colleges* (DOEC), *Distance Open Education of Undergraduate* (DOEU), *Vocational Education in Campus* (VEC), *Education Promotion of Village Carders* (EPVC).

Seven Courses with the Same Number of Students, Number of Study Weeks and Study Requirements. The author counted and analyzed the seven selected required courses with 16 weeks of study. For these courses, the uploaded learning resources meet the requirements of the school and are of high quality, so their impact on learners is similar. The seven courses include: Open Education Learning Guide (O), Ideological and Moral Cultivation and Legal Basis (L), English (E), Advanced Mathematics (M), Video Production (V), Graphic Design (G), Animation Production (A).

3.2 Learner Types and Their Learning Motivation

Table 1 shows that there are as many as 12 kinds of motivation for learners, showing diversity, but in general, there are two points:

Most Learners Who Want to Get the Academic Certificate Have Strong Desire to Learn Online. Most learners aim to obtain a certificate or even a job. This means that in order to pass the course examination, they study according to the teaching requirements of the school. Some learners tend to improve their knowledge, skills and career. Although they don't just want to obtain certificates, they also have a strong desire to learn online.

There are Also Some Learners with Less Willingness to Learn Online. Such as for interest or fun, for experience of open courses, challenge myself, etc. They are likely to fall into a 'burnout period' after experiencing novelty. They would not keep on studying and even give up halfway.

Table 1. Survey about leaners' motivation

No.	Motivation	Ratio of agreement				
		CSHVE	DOEC	DOEU	VEC	EPVC
1	For diploma certificate	0.929	0.957	0.984	0.976	0.822
2	For promotion of career	0.323	0.902	0.916	0.521	0.755
3	Upgrade or expend related knowledge and skill	0.452	0.419	0.324	0.681	0.597
4	For getting a new job	0.876	0.533	0.498	0.925	0.121
5	For interest or fun	0.277	0.456	0.558	0.322	0.268
6	For experience of open courses	0.125	0.269	0.287	0.362	0.584
7	Challenge myself	0.103	0.247	0.355	0.122	0.664
8	Communicate with others with similar interest	0.098	0.225	0.279	0.064	0.325
9	Related to my major	0.003	0.315	0.334	0.048	0.128
10	Related to my job or business	0.002	0.198	0.214	0.002	0.412
11	For the instructor	0.158	0.225	0.321	0.113	0.523
12	Friends recommendation	0.036	0.147	0.122	0.056	0.009

3.3 Statistics and Analysis of Course Information, Learning Behaviors and Final Test Results

The Number of Each Type of Learning Resources is Positively Correlated with The Corresponding Number of Learning Behaviors. Table 2 shows the course information, the number of published resources and learning behavior data, including the teaching weeks, study hours, the number of student registration, the number of videos uploaded and videos viewed, the number of assignments published, assignments submitted and forums (active and passive participation), as well as the number of tests published and tests submitted. It can be seen that the learning behavior data corresponding to each type of learning resources is positively related to the number of resources released. Namely, the more videos, assignment and tests the teacher releases, the more the corresponding total number of learning behaviors, the more supervision on students' learning.

Table 2. Course resources and learning behavior records

Course	Week	Class hour	Registrants	Video (T|L)	Assignment (T|L)	Forum	Quiz (T|L)
O	16	36	3315	211547 | 96	25639 | 9	17212 | 5	10268 | 5
L	16	54	3315	256033 | 132	26114 | 12	22987 | 7	10023 | 7
E	16	72	3315	199784 | 96	59844 | 21	39885 | 12	15698 | 12
M	16	72	3315	378996 | 182	49556 | 16	27569 | 10	13225 | 10
V	16	72	3315	503372 | 211	91225 | 29	54269 | 16	11025 | 16
G	16	54	3315	316559 | 155	37884 | 15	24511 | 8	10935 | 8
A	16	72	3315	516988 | 228	102365 | 32	56122 | 16	12127 | 16

The Proportion of Different Score Grade That 0, Failed, Passed, and Excellent of Different Courses are Approximate. According to the final examination results (S), learners can be defined as: 0 (S = 0), Fail (0 < S < 60), Pass (60 $\leq$ S < 85), Excellent (S $\geq$ 85). It can be seen from Table 3, the score distribution of each course are approximately the same under same learning requirements, the proportion of 0 is 5%–7%, Fail is 18%–22%, Pass is 65%–73%, Excellent is 5%–8%, showing the great similarity with learning behavior.

Table 3. Courses requirements and score distribution

Course	Video	PPT	Text	Assignment	Forum	Quiz	Mid-term exam	Final exam	Ratio of			
									0	Fail	Pass	Excellent
O	Y	N	N	Y	N	Y	N	Y	0.06	0.18	0.73	0.8
L	Y	N	N	Y	N	N	N	Y	0.05	0.19	0.70	0.6
E	Y	N	N	Y	N	Y	Y	Y	0.07	0.22	0.69	0.6
M	Y	N	N	Y	N	Y	Y	Y	0.05	0.21	0.65	0.5
V	Y	N	N	Y	N	Y	Y	Y	0.07	0.19	0.71	0.7
G	Y	N	N	Y	N	Y	N	Y	0.06	0.18	0.70	0.8
A	Y	N	N	Y	N	Y	Y	Y	0.05	0.17	0.69	0.7

4   Data Analysis of Typical Learning Behavior

If learners' behavior was classified by learners' types and motivations, there would be too many categories and would make it difficult to build relevance. To simplify the issue is classify learners by the most direct empirical assumption, which means, more learning behaviors there are, stronger the learning motivation would be, and the performance (score) would be better correspondingly.

The number of videos watching, assignment submission, tests participation and forum participation, which was draw from the Open University learning platform data base, was analyzed by relating to the final exam results. Because these 4 types of behaviors affect the learning performance the most, and can speak for the learning initiative and the level of earnestness in the most efficient way.

4.1   The Relation Between Learning Behavior Frequency and Performance Distribution

Classifying Learning Performance to 4 Categories Based on the Data of Learning Behavior. Figure 1 shows the number of students with their frequency of 'watching videos + submitting assignment + participating in tests' in 7 subjects, classifying learners to 4 types according to the data of learning behaviors.

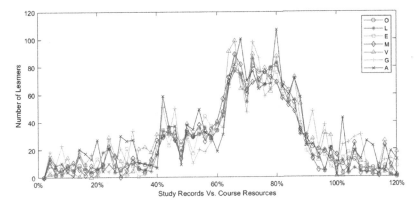

Fig. 1. Number of learners VS records of course.

Bystander: defining those whose learning records compose below 10% of the resources as Bystander, some of them even have a record of 0. Since most learners in Open University have strong intention of getting diploma, learners who have the record of 0 are few. Half of them are those who give up learning and testing after register. According to Fig. 1, there are around 5–7% bystanders in every subject.

Lazy: defining those whose learning records compose 10–59% of the resources as Lazy. Their characteristics are that their number of watching videos, submitting assignment and tests is lower than the requirements. There are 20% such kind of leaners.

Normal: defining those whose learning records compose around 50–90% of the resources as Normal, those learners fulfill the basic requirements of learning behaviors from the school and complete overall learning tasks basically, there are 65–70% such learners in every subject.

Hardworking: defining those whose learning records compose around 90–120% of the resources as Hardworking. They have watched all the course videos basically and submit almost all assignment and tests. Some of them even have studied the course resources multiple times. The motivation of those learners is mostly 'learning knowledge and skills' and they have stable learning records.

Final Tests Score Distribution. Figure 2 shows the final exam score distribution of 7 subjects. From which we can see that the final exam grade of every subject is in linear positive correlation with the learning behavior distribution of the subject. The proportion of the learning performances from 'Bystander, Lazy, Normal, Hardworking' are basically similar to the distribution of their final exam grades '0, Fail, Pass, Excellent'.

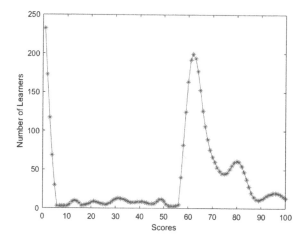

Fig. 2. Number of learners VS Course Score of course O *(Open Education Guidance).*

4.2   Analysis of Learning Behavior with Learning Time

All the courses expect learners to keep up with the courses closely and keep stable learning record from the beginning to the end. But based on Fig. 3, we can see that:

Week 1–2. The learning record surges from week 1 to week 2, and then decreases successively after week 2. This should be related to the 'freshness' to the courses from the learners in the first week, so most of the learners would look through some resources of the course;

Week 2–6. The record stays smooth from week 2 to week 6, it goes down and up but doesn't change much;

Week 6–8. Because of mid-term test, the record surges again from week 6 to week 8 (there are no mid-term tests for course O, L and G so there is no sharp increase of learning record of those courses), after then the record goes down sharply.

Week 9–13. The record keeps relatively gentle from week 9 to week 12, and begins to increase in week 13, indicating that some learners begin to review and prepare for the final exam.

Week 14–16. The record increases extraordinarily from week 14 to week 15 and even surpasses the total amount from the beginning, this should relate to the process of students reviewing, preparing for the final exam and filling learning gaps.

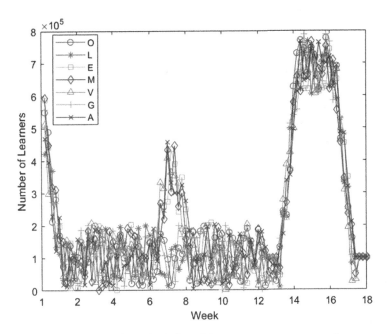

Fig. 3. Distribution of number of learners across course weeks.

4.3 The Records Distribution of Video Watching from Learners with Good Score and Poor Score

A Few Leaners Who Got 0 Score for Final Tests Have a Certain Amount of Learning Record. (Figure 4(a)), Which means that some learners are earnest and spend relatively much time to complete part of the learning tasks, they just don't meet the basic requirements; the other learners have good learning record on the internet but they don't participate in the tests. Taking the course O (Open Education Learning Direction) as an example, 97% of the learners watch the video at least once, 67% of the learners watch 70 videos, and the total number is 96. And there are similar distribution results in other courses too.

The Number of the Videos Watched by Learners Who Get High Scores in Final Test Generally Surpass the Total Amount of Videos Posted. (Figure 3(b)), which means that most of them maybe learn one video several times; some of them don't watch a lot of videos but they still pass the exam, we can say that they have good foundation or strong ability of learning. This also proves the importance of the course videos to an extent, and explains that comparing to traditional classroom, video course is more suitable for learners from different levels.

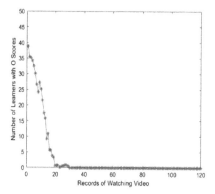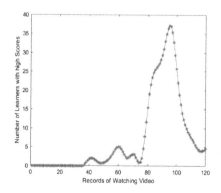

Fig. 4. Behavior of watching video of 0 score and high score learners in course O. *(Other courses have similar distribution)*

4.4   The Distribution of the Four Typical Learning Behaviors in Different Grade Scale

The 4 typical learning behavior are watching videos, participating in forums (viewing, posting, replying and making comments), submitting assignment and tests. The Fig. 5 shows the learning behavior distribution in different grade scale (0, Fail, Pass, Excellent). And the data reveals the differences between high-grade learners and low-grade learners in learning behavior. In total, most courses show a lineal relation that the higher grades the learners have, the higher level of participation in every learning behavior is.

3 Typical Behavior are Obviously Positive Correlation with Score. They are watching videos, submitting assignment and tests. Especially the behavior of watching video is the most significant.

Participation of Forum is Not so Positive Correlation with Score. Some high score learners don't have high forum records, including viewing, posting, replying and making comments. At the same time, learners who are above general participate in course forum the most. For example, the learners with lower grades participating in forum frequently means that the teaching videos fail to solve their problems, but instead, they should ask for solutions in the forum. Learners from every subject have the similar tendency of 'reading posts (passive participation)' and 'posting and replying posts (positive participation)', among the behaviors of positively participating in the forum, asking teachers questions and replying other learners (building learning cooperation) compose half each.

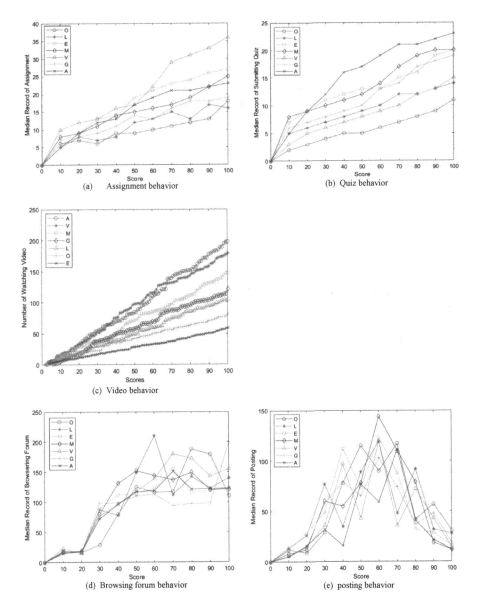

Fig. 5. Distribution of 4 typical learning behavior VS score of the 7 courses.

In words, the higher the grade is, the higher the participation level of every behavior is. Most courses have a lineal relation. Which means: taking efforts to participate in every learning behavior helps to obtain good grades. From the perspective of teacher and university, the videos and tests contribute the most to performance, more attention must payed to the video making and improving. The forum behavior isn't necessarily relevant to performance, and it really helpful for the learners of normal score, by learning collaboration with others.

5   Conclusion and Suggestion

In order to explore the regularity and relevance between the different learning behavior characteristics and score grades, multiple aspects analysis on nearly 3 million pieces of learning behavior data from 7 courses of 3,315 learners of the same major in Yunnan Open University are conducted. According to the learning behavior data and final exam score grade of the 7 courses, conclusion can be made that most type of learning behavior have linear correlation with final exam score. And according to the distribution of learning time, conclusion can be made that learning time is positive correlation to final exam score too.

General Characteristic. The learners with more learning behaviors study harder with better grade; learning behavior is largely related to the lowest learning requirement; video learning contributes most to grad, followed by test and assignment respectively. Besides, the performance in forum has non-obvious correlation with grade; the distribution of learning behavior in frequency and time is largely influenced by teaching management behavior.

Corresponding Suggestions for Online Course Teaching. Based on the above-mentioned analysis and description on data of video-watching, assignment and test, the teachers should increase video quantity, especially improve the quality of teaching videos to make learners obtain more high-quality teaching contents. And moreover, they should increase the quantity of tests properly. On the other hand, the forum behavior which is not included in scoring makes certain contribution to the intermediate-score learners, but still help many learner to solve their questions. Therefore, the teachers should log in the forum more to tutor learners, answer their questions and help them form the habit of learning collaboration.

Fund Project. (1) Humanities and Social Sciences Research and Planning Fund Project of Ministry of Education – 'On Training Mode of Academic Degree Linking Artificial Intelligence Applied Talents Based on '1+X' Certificate System', Project No. 20YJA880086; (2) Special Research Project of Open University of China: Research on the Training Mode of Modern Apprenticeship VR Technical Talents Based on Credit Bank; (3) Research and Cultivation Team of Yunnan Open University-'Research Team for Intelligent Programming, Production and Teaching Integration'.

References

1. Cao, M.: Exploring the learning behaviors of undergraduates with its impact on different achievement in blended learning. Mod. Dist. Educ. **2020**(1), 45–61 (2020)
2. Dorfsman, M.I.: The development of discourse in the online environment: between technology and multiculturalism. Int. J. Educ. Technol. High. Educ. **13**(7), 5458–5469 (2018)
3. Wang, P.: Research on online learning behavior analysis model in big data environment. Eur. J. Math. Sci. Technol. Educ. **13**(8), 5675–5684 (2017)
4. Bedenlier, S.M.: Culture and online learning: global perspectives and research. Int. Rev. Res. Open Dist. Learn. **17**(2), 358–371 (2016)

5. Li, X.: MOOC: showcase or shop? China Univ. Teach. **2016**(5), 73–79 (2016)
6. Evans, S., Myrick, J., et al.: How MOOC instructors view the pedagogy and purposes of massive open online courses. Dist. Educ. **2015**(3),77–89 (2015)
7. Wong, A.W., Wong, K., Hindle, A.: Tracing forum posts to MOOC content using topic analysis. Source code for research experiment **2019**(10), 89–97 (2019)
8. Ortega Arranz, A., Sanz Martínez, L., Álvarez Álvarez, S.: From low-scale to collaborative, gamified and massive-scale courses: redesigning a MOOC. Carlos Delgado Kloos **2017**(3), 77–87(2017)
9. Kaveri, A., Gunasekar, S., Gupta, D.: Decoding engagement in MOOCs: an indian learner perspective. In: IEEE Eighth International Conference on Technology for Education. IEEE (2017). vol. 7, pp. 536–544, Springer, Heidelberg
10. Isaac, C.: Four years of open online courses – fall 2012-summer 2016. SSRN Electron. J. **17**(2), 358–371 (2016)
11. Luaces, O., Diez, J., Alonso-Betanzos, A., et al.: Content-based methods in peer assessment of open-response questions to grade students as authors and as graders. Dist. Educ. **2017**(2), 77–89 (2017)
12. Coleman, A.: Identifying and characterizing subpopulations in massive open online courses. Cody **2015**(6), 39–52 (2015)
13. Talmon, R., Wu, H.T.: Latent common manifold learning with alternating diffusion: analysis and applications. Appl. Comput. Harmon. Anal. **2018**(1), 14–32 (2018)
14. Lwoga, E.T.: Critical success factors for adoption of web-based learning management systems in Tanzania. Int. J. Educ. Dev. Inf. Commun. Technol. **10**(1), 4–21 (2014)
15. Breslow, L., Pritchard, D.: Studying learning in the worldwide classroom research into edX's first MOOC. Res. Pract. Assess. **8**(1), 13–25 (2013)
16. Mcandrew, P., Scanlon, E.: Open learning at a distance: lessons for struggling MOOCs. Science **342**(6165), 1450–1451 (2013)

Empirical Study of College Teachers' Attitude Toward Introducing Mobile Learning to Classroom Teaching

Dan Ren[1(✉)], Liang Liao[2], Yuanhui Li[1], and Xia Liu[1]

[1] Sanya Aviation and Tourism College, Sanya, Hainan, China
340459345@qq.com
[2] Beijing Hollysys System Engineering Co. Ltd., Beijing, China

Abstract. With the rapid development of information technology and high broad reach of internet, mobile learning is becoming increasingly popular. In this paper, a mobile-learning method is proposed to explore college teachers' attitude toward introducing mobile learning to classroom teaching. An experiment was implemented to investigate teachers' attitude and classroom teaching activities. The results show that they hold an affirmative attitude toward mobile learning, and different mobile learning platforms were applied into almost every step of their classroom teaching. Compared to traditional teaching method, mobile learning has prominent advantages, such as instant and interactive way of teaching, fast updating knowledge, simple and convenient assessment function, various teaching approaches and improvements on students' learner autonomy. When introducing mobile learning to classroom teaching, teachers' main concern falls on that students use mobile devices to do other things in the name of study and can't balance their time spent on mobile learning. Finally, a suggestion is given that it is better for teachers to combine mobile learning with traditional teaching methods to achieve the best teaching effect.

Keywords: Mobile learning · College teachers · Attitude · Classroom teaching · Empirical study

1 Introduction

With the rapid development of modern information technology, more and more people have access to internet and learning resources online. The wide spread of Wi-fi and the popularization of mobile devices have turned mobile learning from theory into practice. College students, with peaked curiosity and sensitivity towards new things, use smart phones frequently in study. Mobile learning becomes not only an effective way for students to spend their spare time, but also becomes a choice for teachers' classroom teaching. It is confirmed that students are quite affirmative on mobile learning, but how is teachers' attitude toward introducing mobile learning to classroom teaching? Compared to traditional teaching method, is mobile learning superior? This paper aims to investigate college teachers' attitude toward introducing mobile learning to classroom teaching and try to find out how they actually apply mobile learning to classroom teaching, as well as the advantages and shortages of mobile learning compared to traditional teaching method.

© Springer Nature Singapore Pte Ltd. 2020
P. Qin et al. (Eds.): ICPCSEE 2020, CCIS 1258, pp. 574–587, 2020.
https://doi.org/10.1007/978-981-15-7984-4_42

2 Definition and Research Status of Mobile Learning

2.1 Definition of Mobile Learning

M-learning or mobile learning, is learning across multiple contexts through social and content interactions by using personal electronic devices [1]. According to Crescente, etc., m-learning is a form of distance education, and through it m-learners use mobile device to obtain educational technology at their time convenience [2]. In his article "Mobile Learning—A glance at the Future", Alexander Dye illustrated that mobile learning is a way of learning which can be conducted at any time and in any places with the help of mobile devices, and the mobile devices used by mobile learners must be capable of effectively presenting the learning content and providing interactive communication between teachers and students [3, 4]. In the 1970s, Alan Kay was the first to put forward the concept of mobile learning. He participated in a club to do research and development of portable personal computer, aiming to let kids have access to digital world. But they failed as a result of the lack of technological support at that time. In 1994, the first smart phone came into being, and it provided platform for mobile learning, and the afterward innovation and development of mobile devices pushed mobile learning to research hot spot for scholars and scientists.

The concept of mobile learning was first introduced to China in the year 2000 by a mobile learning laboratory of modern educational technology center in Peking University, and it has aroused great interest in the domestic academic world as well as research institutions. At present, mobile learning is authoritatively defined by domestic research institutions as the follows: mobile learning, is a learning platform based on modern information technologies, such as wi-fi, internet, multi-media, etc. and through this platform, teaching activities can be done with the assistance of mobile devices; meanwhile, both students and teachers can effectively communicate with each other [5, 6]. Although there are some differences among the definitions given by scholars from home and abroad, their illustrations on the approach, technological basis, instructional mode and final goal of mobile learning are almost the same. After reading their articles, it comes to me that mobile learning is a form of context-crossing learning, and it is quite significant to the idea of building a nationwide learning society and supporting informal learning.

2.2 The Research Status of Mobile Learning

The study of mobile learning in China can be divided into three periods. From 2000 to 2008, research on mobile learning was in an initial stage, with few scholars paying attention to it. From 2009 to 2016, as a result of rapid development of information technology and function improvement of smart phones, the study of mobile learning experienced fast growth. Since 2017, the study of mobile learning has stepped into a stage comparatively steady, because mobile devices, especially smart phone, have gradually become an irreplaceable part in people's life, and mobile learning also has become a significant way for study [7, 8]. Through these years, scholars and researchers mainly focus on studying from the following aspects, including: the feasibility analysis of mobile learning, the learners' attitude towards mobile learning, the factors affecting mobile-learners, mobile devices' influence on teaching, the model study of mobile learning, etc. According to the results of these studies, it is obvious that most students keep a positive attitude towards mobile learning and they have already introduced it to their study. However, there are few studies focusing on how the teachers use mobile devices in teaching, and almost no research concentrates on the teachers' attitude towards mobile learning [9, 10]. Thus, this research mainly aims to find out college teachers' attitude toward introducing mobile learning to classroom teaching, how they actually use mobile learning in their teaching, and try to figure out mobile learning's superiority to traditional teaching method from college teachers' perspective.

3   Research Methods and Process

3.1   Design Principles, Research Methods and Contents

Mainly based on attitude toward introducing mobile learning to classroom teaching, a sampling questionnaire survey is conducted on teachers in universities and colleges in Hainan province, China, and SPSS 19 is used to analyze the statistics. There are 26 items in this questionnaire, including twenty-five multiple-question items and one subjective item. The 25 multiple-question items are shown in Table 1. It can be divided into two parts. The first part include 21 items, from Q1 to Q21, mainly aiming to find out college teachers' attitude toward introducing mobile learning to classroom teaching as well as the advantages and shortages of introducing mobile learning to classroom teaching. This part adopts Likert Scale, and the format of each item is as follows: "strongly agree", "agree", "not sure", "disagree", "strongly disagree", with each option respectively scoring 5 points, 4 points, 3 points, 2 points, 1 point, and 3-point is the critical value. High scores represent high attention and consciousness of introducing mobile learning to classroom teaching. The second part includes four multiple choices, from Q22 to Q25, aiming to find out how college teachers actually use mobile learning in classroom teaching, and the mobile learning platform they prefer to use in every teaching step. At the end of the questionnaire, there is a subjective item "What is your main concern when introducing mobile learning to classroom teaching?", aiming to find out the biggest problem teachers are worried about when introducing mobile learning to classroom teaching.

Table 1. Multiple-question items of the questionnaire

NO.	Topic	Items
Q1	Attitude 1	You support the idea of introducing mobile learning to classroom teaching
Q2	Attitude 2	Colleges should provide necessary facilities to guarantee the introduction of mobile learning to classroom teaching
Q3	Advantage 1	Mobile learning can enrich teaching methods and improve students' participation

(continued)

Table 1. (*continued*)

NO.	Topic	Items
Q4	Advantage 2	The advantages of mobile learning outweigh its disadvantages on students' study
Q5	Advantage 3	Mobile learning can greatly improve students' learner autonomy
Q6	Advantage 4	Mobile learning can make up for the shortage of paper books in its slow updating of knowledge
Q7	Advantage 5	Students can choose resources according to their needs and interests on mobile-learning platform
Q8	Advantage 6	Mobile quiz is more convenient than traditional paper test
Q9	Shortage 1	Mobile quiz has its limits on question types
Q10	Shortage 2	The effectiveness of traditional paper test is better than that of mobile quiz
Q11	Shortage 3	Not all the resources on mobile-learning platform is fit for students They need to be distinguished and chosen
Q12	Shortage 4	Overuse of mobile-learning platform may degrade students' manual writing ability
Q13	Shortage 5	Overuse of mobile-learning platform may degrade students' face-to-face communication ability
Q14	Shortage 6	The frequent use of mobile-learning platform is not good for classroom discipline
Q15	Practice 1	You use mobile-learning platforms in every step of your teaching
Q16	Practice 2	You often prepare mobile-learning resources by yourself
Q17	Practice 3	You usually communicate with students on social media after class
Q18	Practice 4	You usually assign homework through mobile-learning platform
Q19	Practice 5	You usually give quiz to students through mobile-learning platform
Q20	Practice 6	You often recommend good resources to students according to their uniqueness
Q21	Practice 7	You will adjust your teaching according to the feedback on mobile-learning platform
Q22	Practice 8	What is the mobile-learning platform most frequently used in your preparation before class?
Q23	Practice 9	What is the mobile-learning platform most frequently used in your classroom teaching?
Q24	Practice 10	What is the mobile-learning platform most frequently used by you in review after class?
Q25	Practice 11	What is the mobile-learning platform most frequently used by you in classroom testing?

3.2   Research Participants

In the 100 participants, there are 72 females, accounting for 72%. The participants work on different subjects, and their age mainly range from 20 to 50, with 81% aging from 31 to 40. Through the frequency analysis of demographic variables, the result is shown in Table 2.

Table 2. Frequency statistic (N = 100)

Variables	Attribute	Frequency	Percentage (%)
Gender	Male	28	28
	Female	72	72
Age	21–30	10	10
	31–40	81	81
	41–50	7	7
	51–60	2	2
Post of duty (teachers)	Language	29	29
	Electromechanical engineering	10	10
	Hotel management	9	9
	Civil transportation	9	9
	General fundamental courses	33	33
	Flight attendant	3	3
	Economic management	7	7

3.3   Reliability and Validity Test

Reliability Test. Cronbach's α coefficient, commonly used by scholars and researchers, is used as an index to test the reliability. SPSS 19 is used to analyze the statistics of the survey, and Table 3 reports the reliability statistics.

As it is shown in Table 3, the Cronbach's α coefficient of the whole questionnaire is 0.852, and the Cronbach's α coefficient based on standardized items is 0.862, indicating that the reliability of the questionnaire is good.

Table 3. Reliability statistics

Cronbach's alpha	Cronbach's α based on standardized items	Case number	Number of items
.852	.862	100	27

Validity Test. Construct validity is used here as an evaluation index to test the validity of the questionnaire. As it is shown in Table 4, the KMO value is 0.797; the chi-square value of Bartlett's test is 1553.613; the degree of freedom is 378; and its significance level is about 0, suggesting that this sample is fit for factor analysis.

Table 4. KMO and Bartlett's test

Sampling sufficient KMO measures	.797	
Bartlett's test	Approximate chi-square	1553.613
	df	378
	Sig.	0.000

Table 5 shows rotated component matrix. As it is shown in Table 5, the variables of 13 items have respectively large loading in Factor 1, which means the 13 variables share common features. Five items contain variables with respectively large loading in Factor 2, and one item contains variables with respectively large loading in Factor 4. Meanwhile, the factor loading value of factor 3, 5, 6, 7 are all below 0.5, meaning there is a weak correlation between this variable and other variables, and these items need to be improved.

Table 5. Rotated component matrix

Item NO.	Factors						
	1	2	3	4	5	6	7
Q1	**.774**	−.115	.146	−.072	.300	.155	−.108
Q2	**.609**	−.143	.259	−.011	.306	.177	−.186
Q3	**.720**	−.269	.312	.017	.283	.157	−.048
Q4	**.846**	−.131	.201	.009	.025	.009	−.178
Q5	**.786**	−.162	.129	.236	−.180	−.075	.072
Q6	**.752**	−.105	.239	.031	.147	.150	.135
Q7	**.690**	−.078	.362	.203	.202	.056	.064
Q8	**.724**	−.112	.120	.048	−.036	.018	.066
Q9	.108	**.501**	.292	−.417	−.276	.230	−.142
Q10	.038	.489	.244	−.313	−.177	**.468**	.259
Q11	−.008	**.606**	.397	−.455	−.023	−.034	−.099
Q12	−.180	**.594**	.088	.298	.303	−.069	−.067
Q13	−.311	**.638**	.020	.463	.225	.087	.083
Q14	−.276	**.578**	−.086	**.501**	.073	.243	.103
Q15	**.650**	.157	−.175	.244	−.369	.042	−.137
Q16	**.584**	.054	−.362	.127	−.396	.078	.016
Q17	.318	.263	−.114	.005	**.314**	−.552	−.269
Q18	**.699**	.244	−.350	−.107	−.136	−.077	−.335
Q19	**.648**	.174	−.398	.092	−.377	.032	−.228
Q20	.476	.312	**.441**	−.064	−.042	−.085	.322
Q21	**.597**	.198	.055	.161	−.226	−.311	.260
Q22	.205	.246	−.488	−.311	.242	−.247	.339
Q23	.493	.055	−.546	−.202	.284	−.022	.175
Q24	.399	.218	−.578	−.236	.202	.185	.056
Q25	.448	.095	−.528	−.156	.217	.295	.166

3.4 The Present Implementation Mode of Introducing Mobile Learning to Classroom Teaching

According to college teachers' feedback, mobile learning platforms are used in almost each step of their teaching, which can be seen in Fig. 1. Before class, teachers can post message about study requirements and teaching goals to students through social media, such as WeChat, QQ, etc.; and they upload important materials through the new instant interactive teaching platform, such as mosoteach, for students to download and prepare for the class. In the class, teachers can use these new instant interactive teaching platforms to help fulfill each step of classroom teaching, from checking attendance to interacting with students. It breaks the normal procedure and enriches the classroom teaching forms. For example, in traditional classroom teaching, teachers usually call each student's name to check on work attendance, but now students can sign in on their own, and it is much easier than the traditional way for attendance checking and can also save time. After class, teachers can assign homework on the new instant interactive teaching platform, and students can leave message to their classmates and teachers at any time and any places. Through these platforms, they can have a discussion online and they can get teachers' help at any time and at any places. Mobile learning can provide instant access for every person to obtain knowledge, only if they have a wi-fi-covered smart phone or i-pad. Meanwhile, in addition to the traditional paper and pencil test, many new instant interactive teaching platforms, such as mosoteach, can also be used to do quiz and test. Teachers enter the test items and the correct answers to the platform, set the test time and time limits, and ask students to log in and finish the test according to the requirements. After the test, the system will automatically give students a score according to their performance, and it can automatically list the rank of the students as well as the percentage of pass rate, not only for the whole test, but also for each item. What's more, it provides different kinds of data maps, and it is quite convenient for teachers to do test analysis. Moreover, many mobile learning platforms can provide instant tests for students. These online test systems not only can offer an instant feedback for students to know if they have mastered the knowledge or not, but also greatly increase students' sense of competition and their autonomous learning ability. As it is shown in Fig. 1, teacher plays the role of guider, organizer and supervisor when introducing mobile learning to classroom teaching, while students become the center of the class, and this is good for them to fully play their subjective initiatives.

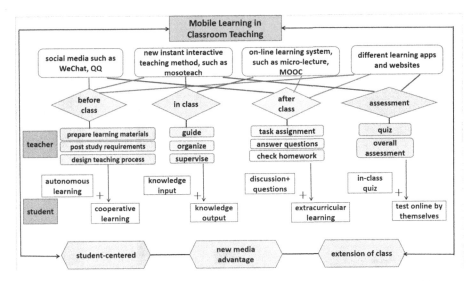

Fig. 1. The specific implementation mode of mobile learning in classroom teaching

4   Research Results

4.1   College Teachers' Attitude Toward Introducing Mobile Learning to Classroom Teaching

As it is shown in Table 6, the mean value of the item of supportive attitude is 4.37, while the mid-value is 3, it shows that teachers hold an obvious affirmative attitude toward introducing mobile learning to classroom teaching. According to Fig. 2, about 90% of the college teachers agree or strongly agree to the idea of introducing mobile learning to classroom teaching, and they also think that necessary conditions for mobile learning, such as basic equipment and network, etc. should be provided by schools to guarantee the practical use of mobile learning in classroom teaching.

Table 6. College teachers' attitude toward introducing mobile learning to classroom teaching

Attitude	M	SD
Introducing mobile learning to classroom teaching	4.37	0.825

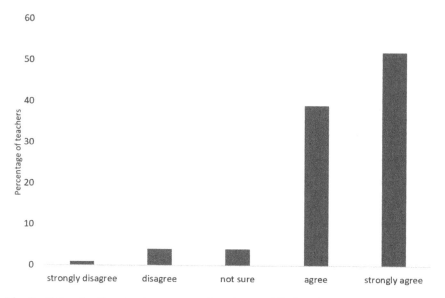

Fig. 2. Ratio of college teachers agree to introduce mobile learning to classroom teaching

4.2   College Teachers' Practical Use of Mobile Learning to Classroom Teaching

According to the survey result, it is obvious that most college teachers not only support the idea of introducing mobile learning to classroom teaching, they also actually use it in their classroom teaching.

As it is shown in Table 7, the mean value of the 7 items related to college teachers' actual practice of introducing mobile learning to classroom teaching are all above 3, which indicates that college teacher frequently use mobile learning in their classroom teaching. According to Fig. 3, more than half of the teachers use mobile learning platforms in almost every step of their teaching. About 80% of the teachers often recommend good learning materials and resources to students, and they often adjust their teaching program and teaching methods according to the feedback got from mobile-learning platforms. Moreover, about 70% of the teachers assign homework, tests and answer students' questions after class through mobile-learning platforms. However, according to the data of Practice 2 in Fig. 3, it can be seen that college teachers prefer to use ready-made learning resources online, and only about 45% of teachers choose to prepare learning materials by themselves.

Table 7. College teachers' practice of introducing mobile learning to classroom teaching

	M	SD
Practice 1	3.48	1.105
Practice 2	3.32	1.18
Practice 3	3.75	1.123
Practice 4	3.59	1.19
Practice 5	3.63	1.116
Practice 6	3.93	0.795
Practice 7	3.92	0.787

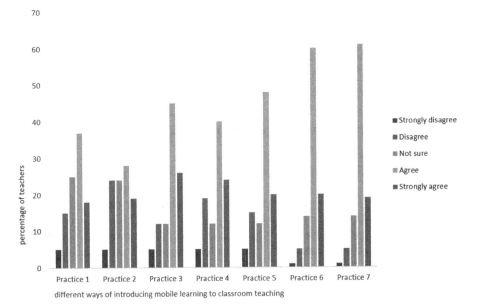

Fig. 3. College teachers' practical use of mobile learning in classroom teaching

A survey is also conducted to investigate the mobile-learning platforms preferred by college teachers in each step of their teaching, including: preparation, classroom teaching, after-class review, testing and evaluation. Seen from Fig. 4, the new instant interactive teaching method is the most popular mobile learning platform among college teachers, and it ranks the highest in almost every step of classroom teaching, followed by social media, online learning system, online learning apps, websites and others. This is due to the fact that the new instant interactive teaching method has powerful functions, and teachers can carry out each teaching step by using it, from uploading learning materials for preparation to check attendance and organize classroom activities, even to after-class review, as well as testing and evaluation. Besides, it is simple and useful. It keeps a record of students' study and their performance can be clearly traced according to the record.

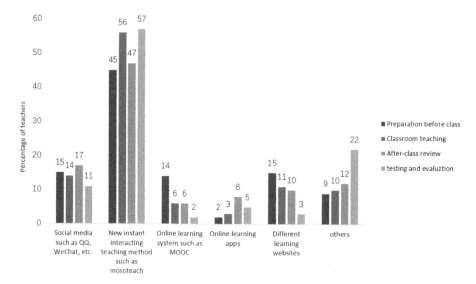

Fig. 4. Mobile learning platforms most frequently used in different stages of teaching

4.3   The Comparison Between Mobile Learning and Traditional Teaching Method

College teachers hold a supportive attitude towards introducing mobile learning to classroom teaching, as they find out that mobile learning is superior to traditional teaching method in some aspects. As it is shown in Table 8, the mean value of the 6 items related to mobile learning's advantage as well as the 6 items related to its shortages are all above 3. It means college teachers accept that compared to traditional teaching methods, mobile learning has its own advantages and shortages. More than 90% of the teachers think mobile learning can greatly enrich their teaching methods. More than 80% of the teachers think mobile learning can make up for the shortage of paper books in its low speed of updating knowledge, and test online is more convenient than traditional paper test. 77% of the teachers think that mobile learning platform provides a wide range of learning materials and resources, and students can choose these materials according to their own abilities and interests. Moreover, on the point whether mobile learning can improve students' learner autonomy, 60% of the teachers choose agree or strongly agree. It means that most of the teachers think mobile learning can help students improve their learner autonomy.

Although college teachers hold an affirmative attitude toward introducing mobile learning to classroom teaching, and they also accept that there are many advantages of mobile learning, it doesn't mean mobile learning is perfect. As it is shown in Fig. 5, 90% of the teachers think the wide range of learning resources online is a double-edged sword. Although it offers students a wide choice, it is also not easy for them to accurately choose materials suitable for themselves from the mass information online. About 75% of the teachers think mobile quiz has limits on question types and overuse of mobile-learning platforms may degrade students' manual writing ability. About 60%

of the teachers think that overuse of mobile-learning platforms may degrade students' face-to-face communication ability. On the point that traditional paper test is more effective than mobile quiz, only 49% of the teachers choose agree or strongly agree, while 36% of them choose not sure. It means that there is still an argument on whether traditional paper test has better effect than mobile quiz, and it needs to be further studied.

Table 8. The advantages and shortages of mobile learning

	M	SD
Advantage 1	4.41	0.805
Advantage 2	3.89	0.898
Advantage 3	3.67	0.9
Advantage 4	4.21	0.856
Advantage 5	4.05	0.88
Advantage 6	4.11	0.942
Shortage 1	4.02	0.91
Shortage 2	3.52	0.99
Shortage 3	4.21	0.808
Shortage 4	3.96	0.84
Shortage 5	3.58	1.027

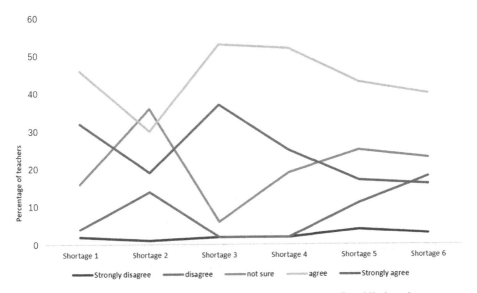

Fig. 5. Ratio of college teachers accepting the shortages of mobile learning

4.4   Teachers' Main Concern When Introducing Mobile Learning to Classroom Teaching

At the end of the questionnaire, a subjective item "What is your main concern when introducing mobile learning to classroom teaching?" is given to teachers, and the answers are gathered as it is shown in Fig. 6. Seen from Fig. 6, teachers' main concern focuses on students using mobile devices to do other things instead of study, or being unable to balance their time and overuse mobile devices, which takes up 59% in all. About 12% of the teachers worry that mobile learning may lead to degradation of students' manual writing ability and face-to-face communication ability. And 12% of the teachers are worried that mobile learning may become a mere form instead of an effective teaching method. 10% of the teachers think it is not good for classroom discipline. According to Fig. 6, it is clear that most teachers' main concern is largely connected with students' self-discipline and self-control. And the problem can be solved only if the students can reasonably use mobile devices and balance their time spent on these devices.

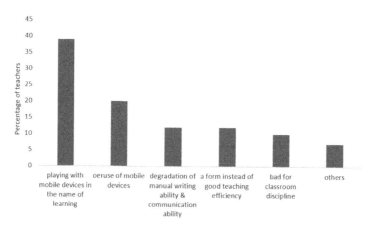

Fig. 6. Teachers' main concern when introducing mobile learning to classroom teaching

5   Conclusion

Through this empirical study, it can be concluded that most of the college teachers keep an affirmative attitude toward introducing mobile learning to classroom teaching, and more than half of the teachers adopt different mobile learning platforms in almost every step of their classroom teaching. The mobile platform most frequently used by them is the new instant interactive teaching method, such as mosoteach. Compared to traditional teaching method, mobile learning has apparent advantages in its fast updating of new knowledge, its simple and convenient quiz and assessment system, its various teaching approaches [11], its high requirement and tough training on students' learner autonomy, etc. However, although mobile learning is quite useful, it is not perfect.

When affirming the advantages of mobile learning, college teachers also think it has some shortages and may lead to some side effects, for example, degradation of students' manual writing ability and face-to-face communication ability, improper time distribution of using mobile devices, disorder of classroom discipline, etc. What's more, college teachers are also worried that students may use mobile devices to do something else in the name of mobile learning, and can't balance their time spent on mobile learning. This poses a great challenge for students' self-discipline and self-control.

In order to keep pace with the times and greatly improve classroom teaching effect, mobile learning is quite necessary and significant in the future classroom teaching. Since both mobile learning and traditional teaching methods have their own advantages, it is better if teachers can combine them effectively. But, how to effectively combine these two together is a new challenge for college teachers, and it needs to be further studied.

Acknowledgment. R. L. L. L. thanks the Education Department of Hainan Province for their support in the research project of education and teaching reform in colleges and universities of Hainan province "The Research of Applying Mobile Technology to Professional Basic Courses in Higher Vocational Education" (Hnjg2017-87).

References

1. Crompton, H.: A historical overview of mobile learning: toward learner-centered education. In: Berge, Z.L., Muilenburg, L.Y.: Handbook of Mobile Learning. Routledge, Florence (2013)
2. Crescente, M.L., Lee, D.: Critical issues of m-learning: design models, adoption processes, and future trends. J. Chin. Inst. Ind. Eng. **2**, 111–123 (2011)
3. Dye, A.: Mobile education—a glance at the future. http://www.dye.no/articles/a_glance_at_the_future/introduction.html. Accessed 15 Jan 2003
4. Liu, X., Jing, G.: An exploration on college english reform based on mobile learning. Chin. Vocat. Tech. Educ. **29**, 90–96 (2016)
5. Hu, J., Wu, Z.: The research of college english flipped classroom teaching mode based on MOOC. Technol. Enhanc. Foreign Lang. Educ. **6**, 40–45 (2014)
6. Fei, W.: Construction of mobile learning platform for college English. J. Shenyang Norm. Univ. (Nat. Sci. Edn.) **32**(4), 561–564 (2014)
7. Chen, Z., Jia, J.: The 20 years of mobile-assisted language learning research in China: review and outlook. Foreign Lang. World **1**, 88–95 (2020)
8. Tang, Y., Fu, X., Pu, C.: A review on the research status of mobile learning in recent 10 years. Dist. Educ. China **1**, 36–42 (2016)
9. Wang, D., Gao, D., Ning, Y.: The summary and review of mobile learning research abroad. Digit. Educ. **1**, 81–85 (2019)
10. Zhang, Q.: A study of application of smart phones to mobile learning of college English. J. Mudanjiang Univ. **26**(1), 154–156 (2017)
11. Chen, Y.: The practice and research in flipped classroom with the support of the mobile platform. Zhengjiang Normal University (2015)

Blended Teaching Mode Based on "Cloud Class + PAD Class" in the Course of Computer Graphics and Image Design

Yuhong Sun[1], Shengnan Pang[2(✉)], Baihui Yin[1],
and Xiaofang Zhong[1]

[1] School of Data and Computer Science,
Shandong Women's University, Jinan 250300, China
[2] Macau University of Science and Technology,
Avenida WaiLong, Taipa, Macau
103474533@qq.com

Abstract. As a teaching tool, Cloud class is specially developed for teaching and learning in mobile circumstance, and the PAD (Presentation Assimilation Discussion) class acts as a creative teaching mode. The combination of Cloud class and PAD class helps build online-offline blended teaching model. In this paper, a Cloud + PAD class mode is proposed. An experiment of teaching is implemented and the practical effectiveness is evaluated and verified. The experimental results show, compared with the conventional teaching method, the proposed teaching mode is capable of creating more high-level, innovative and complex teaching content, and thereby noticeably enhancing students' learner autonomy and motivating them to develop creativity.

Keywords: Cloud class · PAD class · Design of Computer Graphics Image · Blended teaching mode

1 Introduction

In October 2019, China released the implementation opinions of the Ministry of education on setting up first-class undergraduate courses, stipulating that we should vigorously boost the deep integration of modern information technology and teaching, proactively guide students to carry out exploratory and personalized learning, and enhance the high-level, innovative and challenging degree of courses. "Design of Computer Graphics Image" refers to a course of art and computer integration, while the conventional teaching mode cannot meet the needs of leading students for exploration and personalized learning. "Mobile network technology + PAD classroom" blended teaching mode, based on mobile network technology, starts with enhancing the high-level, innovative and challenging degree of the curriculums, while training talents exhibiting innovative and creative abilities.

Campus Technology conducted "Teaching with Technology" survey in 2016, and with seven in 10 survey respondents (71%) using a mix of online and face-to-face for their teaching environments. Among the remainder, more are using the traditional

© Springer Nature Singapore Pte Ltd. 2020
P. Qin et al. (Eds.): ICPCSEE 2020, CCIS 1258, pp. 588–600, 2020.
https://doi.org/10.1007/978-981-15-7984-4_43

course approach (19%) compared to the online-only mode (10 percent) [1]. According to NMC Horizon Report (2017 Higher Education Edition), effective collaborative activities can lead to higher-order thinking, better self-esteem, and increased leadership skills [2]. Mobile devices and collaborative learning environments are common tools in education but not all collaborative learning is structured the same [3]. Cloud class refers to a teaching tool specially developed for teaching and learning in mobile circumstance, and the PAD class acts as an innovative teaching mode based on collaborative activities. In this study, we build online-offline blended teaching model base on "cloud class + PAD class", while verifying the practical effectiveness and assessing the effects of them.

2 The Nature of the Course "Design of Computer Graphics Image" and the Problems Existing in Teaching

2.1 Nature of the Course

- The course "Design of Computer Graphics Image" refers to a fundamental course for digital media technology and digital media art majors; it also acts as a basic skill that should be nurtured by students [4].
- The course "Design of Computer Graphics Image" aims to nurture students' ability to use computer, while training their artistic design ability. This course is practical and creative, laying the basis for art design for students' follow-up courses (e.g., web page making, system development and data visualization design).

2.2 Problems in Teaching

The course "Design of Computer Graphics Image" is considered a novel, high-tech modern information technology; it has been extensively employed in life and work. This is a practical and creative course, not only requiring students to have software operation skills and solid knowledge base of design theory, but also requiring them to exhibit strong innovation and creation ability [5]. Through years of teaching this course, by analyzing students' learning effects and problems existing in the learning of this course, three problems are primarily identified:

The Lacks of Knowledge System and Knowledge Content. The Knowledge system structure of the course is unreasonable, and the teaching content is insufficiently high-level, innovative and challenging. The conventional curriculum knowledge system was organized based on the functions of computer image design software. When finishing the course, the students can master the scattered knowledge. They gain no insights into the application of knowledge in the workplace, as well as how to use knowledge to address sophisticated problems.

The Rigid Teaching Pattern. The teaching model is outdated, and students are less motivated to learn. On the one hand, learning initiative is impacted by learning motivation, and students have unclear learning goals since they do not understand the application value of knowledge. On the other hand, the conventional teaching mode is led by the teacher. Teacher will present the established content and guide students to think in accordance with his thinking. Students always cooperate with teachers in everything, so they are in a passive position, which gradually leads to the lack of initiative and enthusiasm of some students.

Lack of Creative and Innovative Awareness. Conventional classroom teaching is primarily to present knowledge and technology, whereas students less frequently apply knowledge and receive creative thinking training. They finished their homework easily. Over time, they will lose their creativity and innovative.

3   Cloud Class and PAD Class

3.1   Cloud Class

Recently, the usage of mobile technology has become crucial for higher educational institutions worldwide due to the wide spectrum of its benefits [6]. "Cloud class" refers to a teaching tool specially developed for teaching and learning in mobile circumstance. It supports various course resources (e.g., text, audio and video). After installing "cloud class" APP on the mobile phone, teachers are able to create virtual classes and publish courseware, pictures, videos, homework, documents, course subscription and other types of information. Students exploit mobile devices to log on to the virtual classroom and carry out independent learning. Meantime, they can get involved in interactive teaching activities (e.g., voting, thumb up, questionnaire survey, brainstorming, discussion, question-answering and photo display), while receiving immediate feedback and comments from teachers or other students. Teachers can exploit the cloud service function of cloud class platform to gain insights into the learning behavior and learning duration of all students, identify students' learning problems in time, and modify the teaching plan [7, 8].

3.2   PAD Class

PAD (Presentation Assimilation class) is a novel model of classroom reform proposed by professor Zhang Xuexin, PhD supervisor of Fudan University, in 2014. It has been employed in considerable courses in hundreds of universities; besides, it has been stressed as a teacher training project of the Ministry of Education of China. "PAD classes" split teaching into three interrelated processes, which are as follows:

- Teachers' classroom teaching: teachers elaborate the teaching content in class to help students gain insights into the framework of knowledge, as well as the vital points and difficulties in learning.
- Students' internalization and absorption after class: students internalize knowledge into personal experience with the use of reading textbooks or other materials after receiving classroom teaching, which refers to a process to combine learning and thinking.
- Students' classroom discussion: with the results of self-study and thinking after class, students can communicate with other students of the class, which can be split

into group and inter-group discussions. Students are allowed to achieve this via collaborative learning [9].

"PAD Class" requires teachers to give students enough space to fill content, construct their cognitive structure, encourage independent thinking, ask questions, show individual character when constructing the personal internalization knowledge, arrange the homework in the form of a "Sparkles Test Help", and nurture the students' ability of critical thinking and innovation to create, instead of ending everything. At the stage of class discussion, "PAD class" carries out discussions based on in-depth independent learning, as an attempt to nurture students' innovative creativity, language communication and teamwork ability [10].

4   Based on "Cloud Class + PAD Class" Blended Teaching Model Construction and Implementation

In this study, a novel model of student-centered curriculum teaching organization is explored, and the dominant position of teachers is clarified. Reference worldwide successfully combines online and offline blended teaching model. This study focuses on learning before learning, classroom teaching, after class and process assessment of the four major link, cloud class, classroom teaching and interactive learning environment. In accordance with the theory of constructivism learning, we construct the blended teaching mode based on "cloud class + PAD class", which is shown in Fig. 1.

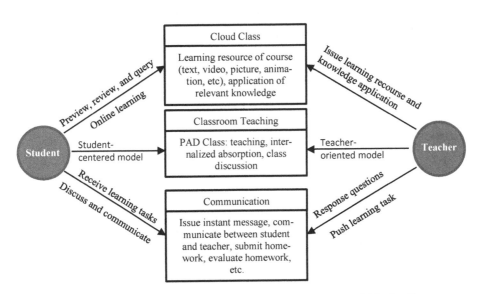

Fig. 1. A blended teaching model based on "cloud class + PAD class"

4.1   Elaborated Design of Cloud Class Network Learning Resources

Online learning resources underpin online and offline blended teaching mode. College students are yearning for acquiring the latest knowledge and technology in their study, and a single textbook can no longer satisfy their needs. Accordingly, when constructing network resources, new knowledge, novel technology and practical content are introduced, so the learning content and social needs are closely linked. In the production of resources, students' learning mentality, time and mobile devices are fully considered, and learning resources are set to comply with the "granitization, systematization and structure".

Reconstructing the Knowledge System and Teaching Content. On the one hand, the function of graphic design software is integrated with practical applications; on the other hand, the design principle involves the cultivation of innovation and creativity. To construct modular knowledge architecture, the relative integrity and continuity of knowledge points are required, so the teaching content is high-level, innovative and challenging.

The learning content is split into three modules, namely, design basis, software application and comprehensive practice. Before using the software, the basic design module introduces the preparatory knowledge. The software application module draws upon specific design cases to introduce the graphic design principles, while introducing the software functions. For instance, the sections of mask and channel are integrated into 5 simple cases (Table 1), so students can learn the software functions as well as the application of functions. The comprehensive practice module primarily consists of various posters, publicity pages, website advertisements, illustrations and drawings, etc. By comprehensive cases, multiple software functions and design principles are integrated to nurture students' ability to solve complex problems.

Table 1. Teaching content of mask and channel

Chapter	Case	Key knowledge points of this chapter
Mask and channel	Composite pictures	Principles of quick mask, relationship between selection and mask
	Enterprise publicity panels	Create a clipping mask
	Taobao website poster	Principles of layer mask
	Juicer promotion page	Alpha channels, filters in channel applications, computing channels
	Synthetic wedding photo	Color channel, selection, mask and channel relationship

Making Cloud Class Online Learning Resources. Learning resources fundamentally support students to conduct independent learning. Learning resources primarily consist of course resources, excellent cases, disciplinary frontiers and real-time push news.

As suggested from brain science research, the average person's effective attention span is about 10 min. To maintain learners' concentration time and ensure learners' keen interest in learning, the course content are divided into 3 modules, 10 sub-modules and 43 learning tasks based on course resources. This division is based on the "granulated resources, systematic design, structured curriculum" guidelines, and on the restructured teaching content. For each learning task, ac-cording to the fragmented learning needs, the knowledge points and skill points are segmented into small knowledge blocks suitable for learning in spare time, including knowledge points, implementation steps, design knowledge analysis, and independent practice. These small chunks of knowledge are presented in various forms: a short text, a short video, several pictures or PPT, etc., as shown in Table 2.

Table 2. Granular resources and presentation

Course content	Resource partitioning	Form of resources
Learning tasks	Learning objectives	Text, Picture
	Main points of knowledge	Text, Picture
	Implementation steps	Video, Text, PPT
	Knowledge analysis	PPT, Picture, Video
	Autonomous practice	Text, Picture

For excellent cases, disciplinary frontier and real-time push news, these three parts need to be well illustrated with appropriate pictures and accompanying text. It is not enough to focus on the content only, the students' reading interest (e.g., the selection of pictures, the size of the font, color collocation, the setting of the title, language, humor) should be considered as well. The mentioned factors will become the crucial factors affecting resources effectively. Suitable content and appropriate presentation mode are two critical factors of a high quality graphic information.

For the production of pictures and videos, we should not only refer to the requirements of China's ministry of education on MOOC teaching resources con-struction technology, but also consider mobile transmission speed and mobile screen size. In particular, the demo video should be taken into account. The soft-ware interface is relatively large, while the text of menu, label and panel is relatively small, so the presentation window needs to be adjusted to clearly and smoothly present the learning content on the phone.

4.2 "Cloud Class + PAD Class" Blended Teaching Process

Pre-class learning, primarily software operation knowledge. Teachers will push refer-ence materials and video files on the cloud class. Students learn online to master case operations. Cloud class provides learning in anytime anywhere set-tings. Such process can be conducive to nurturing their self-study ability and saving time in class.

Classroom teaching, which consists of two stages. The first stage is the discussion. Students first present their work in the group, appreciate and assess each other, as well as conduct discussion about the modification plan. Subsequently, the discussion between groups is performed in the class. At the discussion stage, students exchange information and knowledge to supplement personal information and knowledge. They progressively strengthen their ability of critical thinking, innovation and cooperative learning, as well as stimulate their internal learning motivation. The second stage refers to teaching. First, teachers summarize students' discussion results and problems existing during self-study. Subsequently, the new learning case is explained. Since students have taught themselves the operation steps of the new case after class, the teacher has more time to analyze the design principle of the new cases, and can introduce more novel technologies and ideas. Thus, the teaching content is high-level and innovative. Lastly, the teacher explains the critical operation steps of the next case, thereby helping students carry out self-study smoothly.

Internalized introspect, after classroom teaching, students should design works in accordance with the design principles and methods of classroom learning in a week. To complete the homework, students exploit the Internet to learn more de-sign methods and draw upon the cloud class to publish their design works and discuss the problems existing in the process. Students internalize knowledge into personal experience in the creation of works and gain more sense of learning effectiveness. This session can stimulate students' learning motivation and enhance their self-learning ability and innovative creativity.

Process Assessment. By using the function of check-in, assessment and learning time statistics of cloud class, students' attendance, homework quality and network resource use in the learning process are statistically tracked to ascertain the performance of process assessment. Process assessment can help teachers understand students' learning situation and identify existing problems, while motivating students to work harder to conduct later studies and boost their learning initiative.

5   Research Design

5.1   Research Questions

This study primarily discusses the issues below:

- Whether students are satisfied with the hybrid teaching mode of "Cloud class + PAD class";
- Whether the teaching model can facilitate students' learning motivation;
- Whether the teaching model can nurture students' autonomous learning ability;
- Whether the teaching mode can boost the development of students' innovative thinking ability;
- Whether learning motivation and self-learning ability positively impact innovation and creativity.

5.2 Research Objects

80 students from two classes of Digital media technology, school of data science and computer science of Shandong Women's University were recruited for the experimental teaching of Design of Computer Graphics Image.

5.3 Research Methods

In one class, the Blended Teaching Mode of "Cloud class + PAD class" was adopted, while the other class complied with the conventional teaching mode. After the course, the questionnaire approach and interview method were adopted. First, students of the two classes underwent the questionnaire survey respectively to analyze into the differences in the survey results of the two classes. Second, in-depth interviews were carried out with teachers teaching other courses.

5.4 Questionnaire Design

According to the methods of "students' satisfaction to blended teaching" [11], learning motivation questionnaire [12], independent learning model [13], and creative self-efficacy scale [14], design of course "Computer Graphics Image" blended teaching effect survey scale was developed, consisting of the satisfaction of blended teaching and students' learning effect. The scale covers 4 dimensions and 21 items, and the assessment and judgment criteria fall into 5 levels, namely, 1 strongly disagree, 2 disagree, 3 are not sure, 4 agree and 5 strongly agree.

The reliability of the questionnaire was analyzed. Questions 1 to 9 are split into the latitude of satisfaction, and questions 10 to 13 fall into the latitude of impact of learning motivation. Questions 14 to 17 are split into the latitude of the impact of self-learning ability, and questions 18 to 21 are split into the latitude of the impact of self-learning ability. It is concluded that the Cronbach α of all four latitudes is greater than 0.8 (Table 3). Besides, the questionnaire is stable and reliable.

Table 3. Reliability analysis of questionnaire

Numbers	Dimension names	Cronbach α	Question number
A	Satisfaction	0.901	9
B	Impact of learning motivation	0.912	4
C	Impact of self-study ability	0.887	4
D	Impact of innovation and creativity	0.941	4
	BCD three items	0.963	12

6   Analysis of Experimental Results

6.1   Analysis of Blended Teaching Satisfaction

A questionnaire survey was carried out for the experimental class. Table 4 presents that the mean value of each index of the three modules "online learning", "classroom teaching" and "blended teaching" reaches over 4 and becomes statistically significant ($p < 0.01$). Students rated the content as higher order, innovative, complex, slightly lower than other indicators, whereas it is still more acceptable ($M = 4.28$). On the whole, students are significantly satisfied with the hybrid teaching mode of "cloud class + PAD class".

Table 4. Students' satisfaction with blended teaching

Module	Indicators	M	SD
Online learning	The simplicity of cloud class operation	4.43	.583
	The function of cloud class to assist self-study	4.41	.652
	Abundant learning resources for cloud class	4.46	.585
Classroom learning	The content is higher order, innovation and complex	4.28	.621
	Classroom learning is interactive	4.50	.624
	Positive classroom atmosphere	4.61	.493
	The rationality of learning assessment	4.50	.506
Blended teaching	The integration of cloud class and classroom teaching	4.50	.587
	The effectiveness of blended teaching	4.61	.493

6.2   Analysis of Influence on Learning Motivation

Table 5. Comparison results of investigation on learning motivation

Title	Data					p
	Experimental group		The control group			
	M	SD	M	SD		
The current classroom teaching can stimulate my interest in learning	4.21	.695	3.41	.774		.000
The current classroom teaching inspires my confidence	3.87	.801	3.20	.872		.001
I can actively complete the homework and preview tasks	3.87	.695	3.12	.954		.000
I can actively participate in class discussions	4.15	.540	3.24	.860		.000

It is suggested that the average scores of the experimental class in learning interest, self-confidence, learning enthusiasm and other aspects are noticeably higher than those of the ordinary class, and the P values of all items reach below 0.01; the difference is noticeable from the statistical perspective. In an interview for other courses teachers, teachers in Z: "experimental classes of classroom atmosphere is more active, students grades higher than in control class". As teachers Y mentioned: "in the organization of course contests, experimental classes of students registration than the comparison classes", teachers' feedback in another angle suggests that the experimental class teaching to boost students' learning motivation exerts a positive effect (Table 5).

6.3   Analysis of the Influence on Self-learning Ability

Table 6. Comparison results of the impact of self-learning ability

Title	Data				
	Experimental group		The control group		p
	M	SD	M	SD	
I have the ability to collect and process information and acquire new knowledge	3.97	.668	3.17	.946	.000
I have a detailed study plan	3.64	.811	2.76	.916	.000
I can make good use of my time	3.59	.966	2.93	.985	.003
I am good at finding problems in learning and taking the initiative to solve them	3.92	.580	3.17	.863	.000

Table 6 presents that the average score of the experimental class is noticeably higher than that of the ordinary class in three aspects, namely, collecting and processing information, acquiring novel knowledge, formulating learning plans and exploiting their own time rationally. However, for "I can use my time reasonably", both classes achieved relatively low scores, revealing that teachers should still strengthen the training of students in this aspect.

6.4   Analysis of Impact on Innovation and Creativity

Table 7. Comparison of the survey results on the impact of self-learning ability

Title	Data				
	Experimental group		The control group		p
	M	SD	M	SD	
I have confidence in my ability to solve problems creatively	3.72	.686	3.00	1.025	.000
I think I'm good at coming up with new ideas	3.54	.682	2.93	.905	.001
I'm very good at developing my own ideas from other people's ideas	3.79	.656	3.02	.908	.003
I'm good at figuring out new ways to solve problems	3.72	.686	3.00	1.025	.000

Table 7 presents that students in the experimental class score noticeably higher than the control class in innovation and creativity. However, we should also see that the average score of the four options is relatively low, revealing that the nurturing of students' innovative and creative ability cannot rely on a single course alone, which is a long-term work, and teachers should more thoroughly deepen the training of students in the follow-up courses.

6.5   Analysis of the Impact of Learning Motivation and Self-learning Ability on Innovation and Creativity

Table 8. Regression analysis results of learning motivation and self-learning ability on innovation and creativity

	Learning motivation	Self-study ability
β	.882	0.822
SE	1.477	1.785
F	273.142	162.301
Adjusted R2	.775	0.671
p	.000	.000

Table 8 illustrates that learning motivation shows positive correlations with creativity ($\beta = .882$, $p < 0.001$), and self-study ability links to creativity positively ($\beta = .822$, $p < 0.001$). As revealed from the results, learning motivation and self-study ability positively impact innovation and creativity.

7 Conclusion

Design of Computer Graphics Image is a course of cross fusion of art and computer. Under the context of China's first-class undergraduate course construction, it is practically significant and critical to explore the "high-level, innovative and challenging" of the course. In the present study, a blended teaching mode based on Cloud class and PAD class is established, through which a relatively ideal teaching effect is achieved. Its innovation lies in that, first, cloud class has abundant teaching resources, contributing to students' self-study before class and internalization of knowledge after class. Second, it stimulates the classroom activity and students' participation in the PAD class teaching mode. Third, the nurturing of students' learning motivation, leaner autonomy and creativity is high-lighted. The combination of cloud class and PAD class can remedy the defects of the current teaching mode of computer design related courses and provide ideas for the future computer design course teaching.

Acknowledgements. This research was partially supported by the Undergraduate Education Reform Project of Shandong Provincial Department of Education under grant number M2018X238, and the Shandong Women's University Teaching Reform Project under grant number 2017jglx25.

References

1. Campustechnology. https://campustechnology.com/articles/2016/10/12/55-percent-of-faculty-are-flipping-the-classroom.aspx. Accessed 10 Feb 2020
2. Becker, S.A., Cummins, M., Davis, A., Freeman, A., Hall, C.G., Ananthanarayanan, V.: NMC horizon report: 2017 higher education edition. J. Beijing Radio TV Univ. **2**, 1–20 (2017)
3. Heflin, H., Shewmaker, J., Nguyen J.: Impact of mobile technology on student attitudes, engagement, and learning. Comput. Educ. **107**, 91–99 (2017)
4. Teaching Steering Committee of the Ministry of Education: Ministry of education national standard for teaching quality of undergraduate specialty in general colleges, pp. 952–963. Higher Education Press (2018)
5. Ding, Y.S., Li, C.H.: Design of computer graphics image and visual communication. China Comput. Commun. **10**, 126–127 (2018)
6. Mostafa, A.E., Hatem, M.E., Khaled, S.: Investigating attitudes towards the use of mobile learning in higher education. Comput. Hum. Behav. **96**, 93–102 (2016)
7. Wang, M.: Tflipped classroom teaching practice of higher vocational public English based on Mosoink cloud class platform. J. Kunming Metall. Coll. **33**(02), 28–33+47 (2017)
8. Guo, X.M., Li, Q.L., Yin, X.Y., Zou, C.X.: Research on practical teaching mode of tourism management based on cloud class platform. Exp. Technol. Manage. **1**, 176–180 (2020)
9. Zhang, X.X.: PAD Class: A new attempt in university teaching reform. Fudan Educ. Forum **12**(05), 5–10 (2014)

10. Zhao, W.L., Zhang, X.X.: The bipartition of classroom: a new local teaching mode to promote in-depth learning. Theory Pract. Educ. **38**(20), 47–49 (2018)
11. Chen, C.G.: Empirical study of college English blended teaching based on mobile learning platform. J. Chongqing Univ. Educ. **33**(01), 112–116 (2020)
12. Qu, T.F.: An empirical study of the teaching mode with combination between rain class and PAD class in the teaching of college English writing. J. Neijiang Normal Univ. **35**(01), 89–94 (2020)
13. Zhang, Y., Pan, S.P.: Theory of zimmerman's self-regulated learning model and its enlightenment. High. Educ. Dev. Eval. **1**, 48–50 (2006)
14. Gong, Y., Huang, J.C., Farh, J.L.: Employee learning orientation, transformational leadership, and employee creativity: the mediating role of employee creative self-efficacy. Acad. Manage. J. **52**(4), 765–778 (2009)

Application of 5G+VR Technology in Landscape Design Teaching

Jun Liu and Tiejun Zhu[✉]

Anhui Polytechnic University, Wuhu, China
ztj@ahpu.edu.cn

Abstract. As the VR technology that has been applied in many fields in China, the arrival of 5G era provides a strong impetus for VR technology to take off again. And the in-depth application of 5G + VR technology will further promote higher education. In view of the problems and deficiencies of the current landscape design teaching methods in China's universities, the *Introduction to landscape design* course and teaching as an example, 5G + VR technology is introduced in this paper to carry out the innovative design and teaching application of the course system. The teaching effect feedback and evaluation were carried out. The research and application results show that the innovation of 5G + VR in the teaching content and form of landscape design makes the course teaching simple and vivid, has strong situational experience and learning atmosphere, and effectively stimulates students' interest in learning.

Keywords: 5G · Virtual reality · Landscape design · New learning model · Application · Empirical research

1 Introduction

With the progress of science and technology, the design education in universities is facing profound changes. It needs to use new technology and new teaching mode to redesign the curriculum. In China, with the arrival of 5G era, as a VR technology widely used in higher education, the "perfect virtual" presentation of 5G + VR has become an important path of current research and development, and the new learning mode and learning experience it brings will open up a new field of education and teaching. At present, due to the limitations of traditional teaching thinking and hardware facilities in China, teaching mode is still in the stage of textbook theory explanation and picture display in a large proportion. Few students go out to investigate and practice. Students do not experience and feel it personally. They only rely on the case given by teachers, and draw and evaluate the design scheme by means of manual drawing, manual model making and comment [1]. The loss of the sense of space cannot make students directly understand and feel the process of landscape design, thus cannot effectively stimulate the enthusiasm of students' active learning, and the teaching effect is greatly reduced. Therefore, the new technology of 5G+VR should be applied in landscape design teaching.

2 Background

2.1 Virtual Teaching Under the Background of 5G Technology

VR was first proposed by American Jaron Lanier in 1989 and has been used up to now. VR has been applied in a variety of industries after three upsurge of development, and actively explored and applied in aerospace, medical, education and other industries. VR system mainly relies on computer, application software system, input equipment and output equipment for combined operation, which has three characteristics, namely immersion, interaction and imagination. With the arrival of 5G era, the upgrade of computer hardware and the optimization and development of related software, the resolution and bit rate of VR panoramic video are directly improved, which brings the ultra-high definition standard picture quality while reducing the network delay. Because of the support of 5G high-speed rate, low delay and other key technologies, the VR related calculation and rendering are carried out in the cloud, including mobile phone, TV, computers will become display terminals, and in the cloud mode, users can operate VR activities at any time and place by using any device. 5G will promote the revival of VR industry [2].

Virtual teaching is a teaching approach based on the development of VR technology. It is a new teaching mode of education and learning in virtual space. It uses science and technology to build a virtual learning environment. Through the objective and real simulation and presentation of knowledge points, teachers and students can complete the whole process of teaching in the virtual environment. The development of virtual teaching course involves the reform and exploration of concept, relationship, organization and means in teaching. In particular, the teaching application of VR technology can not only enable students to interact with the teaching content in the virtual environment, solve the problems such as the lack of interaction in the traditional classroom, but also can be used to create learning situations and increase the image and interest of learning content [3] (see Fig. 1). 5G technology has a subversive impact on the application of VR in education and training, because 5G technology can optimize the experience effect of VR at this stage, improve the problems such as overweight of VR display terminal, dizziness experiences, fixed application site etc., making VR education and training from immersion experience to full sensory extension experience. In the future, 5G + VR teaching mode and education method will develop rapidly. It is an intelligent education mode with the characteristics of openness, cooperation and autonomy. It can realize the targets of autonomous learning, cross regional cooperative learning, outdoor learning, etc.

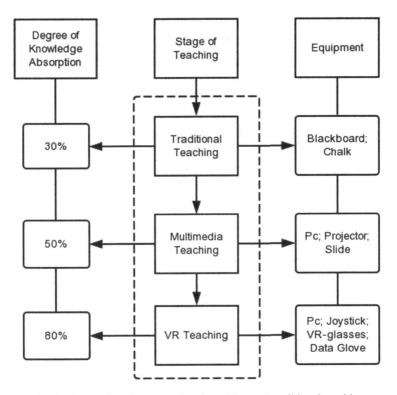

Fig. 1. Comparison between virtual teaching and traditional teaching

In the 2016 White Paper on Virtual Reality Industry Development, the Ministry of Industry and Information Technology of China clearly pointed out that the current VR technology needs to be explored in different industries in the consumption field, especially in various basic fields. In the same year, The Special Action for Innovation and Development of Intelligent Hardware Industry jointly published with the national development and Reform Commission proposed to build a virtual classroom with data resource base. On June 6, 2019, the Ministry of Industry and Information Technology of China officially issued 5G business license to China Telecom, China Mobile, China Unicom, and China Broadcast Network. China officially entered the first year of 5G business [4]. Taking this as a starting point, 5G + VR technology has opened a new chapter in China's education industry. Now many enterprises and institutions have invested in the research, development, promotion and application of the virtual teaching industry in the 5G era.

2.2   Current Situation of 5G + VR Teaching in China

Since the 1990s, there have been examples of VR application in classroom teaching activities in China. At that time, due to the lack of 5G technology, expensive equipment and software materials, the application in this area is still on the surface. Nowadays,

China's 5G + VR teaching activities are mainly focused on theoretical discussion and practical exploration, which are mainly reflected in the following three aspects.

For the Comparison of Teaching Advantages. Relevant scholars discuss the advantages of 5G + VR technology in teaching, such as summarizing the advantages of traditional classroom and 5G + VR classroom [5], or analyzing the current situation and typical cases of 5G + VR teaching research at home and abroad.

Put Forward the Principle and Requirements of Development. According to the concept and specific usage of 5G + VR system in the teaching subjects, with reference to the requirements of use, function and composition, get the practical significance of teaching activities [6] or formulate the relevant course system constitution [7] (see Fig. 2).

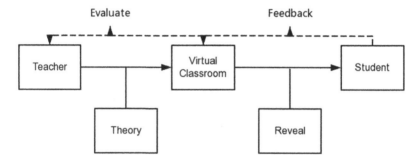

Fig. 2. Flow chart of virtual classroom teaching

The Specific Development of the Experimental System. The designer designed and developed a specific 5G + VR teaching experiment system, 5G + VR display and interaction system, and carried out relevant experiments.

According to the current research situation in China, there are few research cases of 5G + VR teaching practice, and the number of 5G + VR teaching software or system developed is relatively limited (see Table 1). Scholars mostly explore 5G + VR teaching ideas and theories, and the teaching practice is mainly focused on tour guides, geography and animation game, other disciplines are less involved. On this basis, it is necessary to explore the teaching mode of 5G + VR. In the key points of work in 2019, Ministry of Education of China put forward the action plan to launch and explore the implementation of educational informatization 2.0 and the pilot of new teaching mode based on information technology, research the development plan of China's intelligent education, and systematically promote the construction of the national virtual simulation experiment teaching project [8]. With the characteristics of 5G low latency and high network speed, the new connection mode promotes the improvement of "Wisdom + Intelligent" learning platform and learning environment, it is believed that VR technology will become more and more colorful in the future 5G era education stage [9].

Table 1. Overview of 5G+VR education projects in China (As of October 2019)

No.	Type	Project name	Main content of project
1	5G+VR teaching systems or products	Longhua Net	5G+VR 101 classroom, build 5G+VR editor
		Perfect illusion	Mobile 5G+VR live IP
		Allies cloud	Develop 5G+VR education game series products
		Sail	5G+VR education solution
		Black crystal technology	5G+VR super classroom
		Wisdom and sincerity	5G+VR education experience platform
2	5G+VR laboratories established in universities	5G+VR Laboratory of Tsinghua University	
		State Key Laboratory of 5G+VR new technology, Beijing University of Aeronautics and Astronautics	
		5G+VR and Visualization Technology Research Institute of Beijing Normal University	
		5G+VR and Multimedia Technology Laboratory of Southwest Jiaotong University	
		Human-machine interaction and 5G+VR research center of Shandong University	
		Visual computing and 5G+VR Sichuan Key Laboratory of Sichuan Normal University	
3	5G+VR early childhood education	Magic space	Application and development of 5G+VR early education
		Dragon Star	5G+VR education and entertainment for teenagers
		Visual +	5G+VR education for teenagers
4	5G+VR talent training	Dragon map Education	"Authorization + Certification + teaching support" integrated 5G+VR talent training system
		Turanka training	5G+VR industrial training + practical education curriculum system
		VR Star	5G+VR talent vertical training
		Crystal Education	Offering 5G+VR interactive display course
		The age of Mars	Offering 5G+VR game training course
5	5G+VR education website	5G+VR education network	5G+VR education research

2.3   Teaching Characteristics of Introduction to Landscape Design

Knowledge Points are Extensive. Landscape design, including natural landscape and human landscape, has different intersection with ecology, geography, religion, planning and other disciplines. The content of the study covers landscape plants, park design, architectural structure and design, public green space design and other aspects. The knowledge system has wide, multiple and miscellaneous characteristics. Such complex knowledge points need not only the teachers' timely answers, but also the students' subjective initiative to accept and digest, which makes the knowledge need to be effectively transmitted between them, or the feedback between them in time to produce interaction. At present, the interactive feedback mechanism of teaching is slow and biased due to the lack of situational experience.

Case Analysis. Case analysis is very necessary for the teaching of Introduction to landscape design. This method can let students directly understand the precautions and difficulties in the design process and leave a deep impression. Through excellent design cases, students can also feel the charm and value of landscape design. However, at this stage, in the course teaching of Introduction to landscape design in most universities in China, teachers mostly use PPT display, graphic and video as the way of case explanation.

Design Training. After case analysis, students are required to carry out design training. At this stage, the performance is mostly done by concept hand drawing, manual model building or computer modeling. Therefore, the quality of the work can only depend on the students' feeling on the two-dimensional plane, without the real feeling of being in the scene. Students at the beginning can only understand the structure and details of the landscape cases they have learned, and generally imitate them, lacking the direct experience of the landscape entities.

2.4   5G + VR Technology Makes up for the Shortcomings of Traditional Teaching

Today's Teaching Mode Structure Cannot Meet the Changing Landscape Design Knowledge Teaching. The traditional space-time constraints make it difficult for students to play the ability of spatial imagination, and the brain cannot quickly simulate the changes of various spatial elements in the curriculum. However, in the 5G + VR landscape design course teaching, students can experience the scene of landscape design with three-dimensional perspective and high-speed experience. The knowledge and perception transformed from the situational experience are easy to be understood and digested by students, and on this basis, they will also go to the subjective and dynamic practical application. As knowledge can be quickly fed back between the two, teachers can reasonably supplement knowledge points, so the quality of landscape design classroom teaching will be significantly improved.

As 5G + VR Classroom Can Simulate Any Landscape Scene, It Makes Case Teaching Easy in the Course. In the virtual landscape teaching, students step into the "real scene", and under the guidance of teachers, they can carefully understand and learn the design techniques, expression methods and landscape artistic conception. In the virtual reality environment, students can give full play to their imagination, put aside the constraints of reality and make bold innovation, and create their own works with new ideas and new design techniques. Moreover, 5G + VR teaching system can help teachers to observe students' learning activities instantaneously, so that students' feedback in teaching can be displayed to teachers with digital visualization for reference, and can be recorded in the cloud as one of the basis for course scoring, so that teachers can objectively understand students' class status and learning process.

The Intervention of 5G + VR Technology can Turn the Graphics and Texts in the Training Into a "Real Scene" that is Accessible. The computer simulation training environment allows students to carry out activities in the scene. The virtual nature of the training makes it unnecessary to worry about the risk of actual operation during the activities, so as to avoid students from being hurt. The VR system under the support of 5G technology is more stable and open, which can make team activities better and teachers can intervene to give guidance, and make the traditional training classroom become an organized, cooperative and sharing classroom.

By analyzing and examining the characteristics, current situation and development of landscape design teaching in Chinese universities, we can see that 5G + VR technology is imperative. Although the biggest pain point of the existing landscape design teaching VR technology is the delay of data transmission speed, the biggest difficulty is the high cost of large-scale application [10], but the arrival of the 5G era, it has completely changed (Table 2).

Table 2. Data propagating velocity comparison

Network type	Download speed rate	Actual download speed
2G	150 K/s	15 K/s–20 K/s
3G	1–6 M/s	120 K/s–600 K/s
4G	10–100 M/s	1.5 M/s–10 M/s
5G	20 G/s	2.5 G/s

3   5G + VR Teaching System Design of *Introduction to Landscape Design*

The course Introduction to landscape design is a basic course of landscape design major in Chinese universities. The teaching purpose of the course is to make students understand the development of landscape design, learn the basic theoretical knowledge of landscape design and master the basic principles and methods of landscape design [11]. The course is roughly divided into three parts: landscape basic knowledge, introduction to design analysis, and graphical plot. It mainly focuses on the cognition of landscape design performance and skill training, history of landscape design and introduction to design are auxiliary teaching contents. Therefore, in the design of teaching system, it also takes these three parts as the design subject, and tries to develop the existing VR landscape design modeling technology and 5G network combination.

3.1   Design Ideas

Because the course Introduction to landscape design is different from other landscape teaching courses, it has universality, generality and knowledge foundation, so it mainly focuses on the teaching of basic landscape theory, typical cases and drawing design operation. In the whole model teaching, it can be divided into several categories: introduction to landscape basic theory, introduction to regional landscape, introduction to plants, appreciation of landscape sketches and designer window. The contents of different models are placed in a large landscape environment, and displayed in the form of zones for teachers to teach and students to understand. The hardware equipment needed for class are mainly 5G network, computer host, display, VR glasses, joystick or personal mobile phone, etc. Students log in to the virtual landscape cloud platform through computer and enter the environment of teaching courseware, use VR glasses to watch the course content, use joystick and keyboard to give instructions to the project and roam and browse by themselves. Teachers can watch and guide through the screen or enter the virtual environment to give teaching.

3.2   Design Path

System Application. Desktop 5G + VR system: as the display viewport of the virtual environment, the display uses the mouse, keyboard, joystick, computer and force sensing equipment, tracks and calculates the data in real time, and carries out the activity interaction in the virtual environment. Compared with immersive VR system and network distributed VR system, this system is used for virtual environment demonstration, easy to use and popularize. As to hardware facilities, schools just need to be upgrade or purchase 5G network hardware, computer hardware and VR operation equipment, also the requirements for space size and use cost are low.

Landscape Display. At present, there are two common forms of virtual reality technology landscape display: one is to use panoramic camera to collect information from the actual environment to carry out the virtual scene; the other is to render the virtual scene through computer graphics software [12]. This system uses the first one to create a better experience effect.

Real Scene Scanning. After the satellite map is used to select the planar graph of the required design area, three-dimensional laser scanning is carried out to determine the main nodes of the plane, such as sign buildings, roads, flowers, trees, rivers, lakes and other corresponding landscape facilities. Then use the real photos of the area for PS mapping and lighting debugging. After the photos are processed, it can be imported into the virtual landscape simulation platform for system integration and material optimization. Finally, upload them to the cloud platform to log in the software, and then students can roam the landscape. The precision of the model made by this method is high, but the color display of the drawing is relatively poor, therefore, in the later revision of the picture, it is necessary to make the scene fresh and bright. Due to 5G's super fast network speed, the improvement of computer computing power and the innovation and improvement of algorithm, the cost of image modeling is short and the quality is high (see Fig. 3).

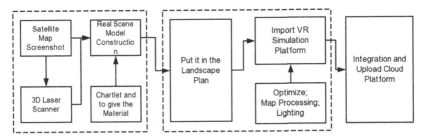

Fig. 3. Flow chart of live modeling

3D Modeling. Use 3DMAX, SU, RHINO and other modeling software to model the system, and build the scene of the selected area through the model. For example, to build a terrain model in SU, first import the CAD terrain map with contour set into the software, select the contour part in the original drawing, click "create according to contour" in the sandbox tool to build the first part of the model, then adjust the Y-axis scale according to the required height, then set up the models in the scene one by one and fill them with materials, finally, the plant blocks are placed on the terrain to complete the modeling. The completed model is imported into 3D development software, such as UNITY3D, and then redesigned to form a 3D landscape. This method performs well in the construction scene and can present an ideal virtual environment [13], but the final effect depends on the refinement of the model design (see Fig. 4).

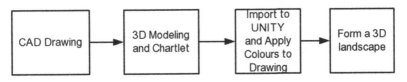

Fig. 4. 3D modeling flow chart

3.3   Virtual Landscape Construction

In the landscape design, the most commonly used design modeling software is 3DMAX, SU and other modeling software, which greatly enriches the designer's idea expression way, making the designer's idea expression fuller [14]. For example, the development of 3DMAX technology provides people with an opposite one-way receiving aesthetic view, changes people's cognition of the real world, expands the sense of visual experience and enables people to get virtual visual experience close to the real world. Therefore, in the design of this system, the scene modeling software such as 3DMAX is mainly used.

Physical Space Analysis. Virtual landscape design is based on space creation and experience. At the beginning of model design, it needs to analyze the physical space of the built model, which is mainly divided into two aspects: one is the analysis of base environment, which is the material basis of landscape design and has two connotations with the designed environment, that are the analysis of the relationship between designed landscape environment and surrounding areas environment, and topography and geomorphology characteristics of the landscape environment involved. Another one is the analysis of the environmental elements in the area, the location and quantity of the water body, mountains, roads, green plants, flowers and landscape sketches in the modeled landscape environment are determined, and the pedestrian flow dynamics and the main people stay area are investigated. In this way, the accuracy of data and drawing can be ensured, and the corresponding knowledge supplement can be provided for the later courseware design.

Model Construction and Output. Data acquisition is the premise and key of making model, and the accuracy, quantity and range of data directly affect the effect [15]. Therefore, before making the model, it is necessary to determine the plan area and altitude of the teaching contents, such as park landscape, city square, landscape sketch, etc., then the location of the geometric model and the behavior model of the virtual characters in the landscape environment are determined, and fill and mark the color in the imported CAD planar graph, so that students can quickly know and understand various projects and functions, and it is also convenient for them to feel different visual expressions of architecture, water body, flowers, trees, sketch roads, etc. in the landscape during post production.

The next operation is mainly divided into two stages: first, data pre-processing, which will delete the text and symbol in the CAD file to be imported, because these will produce position deviation and unknown dislocation after importing 3DMAX, and also need to delete the mark and filled pattern of some non main node elements, so as to keep the most concise but complete map foundation. The second is the modeling of the basic model. Generally, ARCGIS or SU is used to model the part of the surface and the area with uneven texture, and 3DMAX is used to model the other areas with clear shape. Import the finished CAD planar graph into 3DMAX, drag it into the top view work area, design the model in this work area, attach and arrange the similar model, assign the model material and set the large environment background, after debugging the light and camera, you can render with the help of 5G high-speed cloud (see Fig. 5).

Fig. 5. Making mountain landscape with 3DMAX

Landscape Generation and Editing. Save the 3D model landscape in FBX format, place it in the Assets folder, which belongs to UNITY3D project folder, tick the embedded media option when exporting the settings, and then you can see the model in UNITY3D working interface, then set the camera and light for the model, and add the third person perspective to browse the environment.

Perfect Details and Redesign. In theory, the more the modeling surface, the more delicate the modeling effect, the more real. Therefore, the model is checked in detail to ensure the integrity of the construction and the rationality of the collocation. The following is part of the scene experience display diagram of 5G + VR system in the course of Introduction to Landscape Design. The design of the interactive interface is mainly simple and easy to use. It is mainly divided into:

Homepage. Course teaching login interface of Introduction to Landscape Design (see Fig. 6).

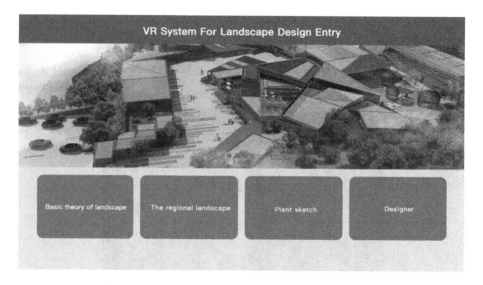

Fig. 6. Homepage

First level interface. It mainly includes four categories: landscape basic theory, regional landscape, plant sketch, and designer window (see Fig. 7 and 8).

Fig. 7. Designer interface

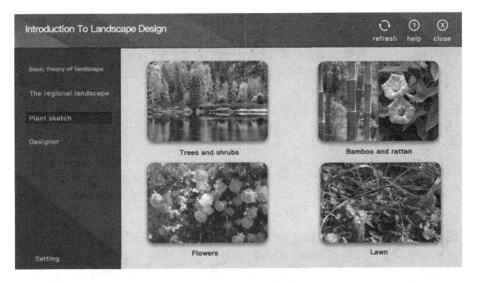

Fig. 8. Plantings Sketch interface

The second level interface. Taking regional landscape as an example, the second level interface is divided into four parts: park, square, waterfront and ancient garden (see Fig. 9).

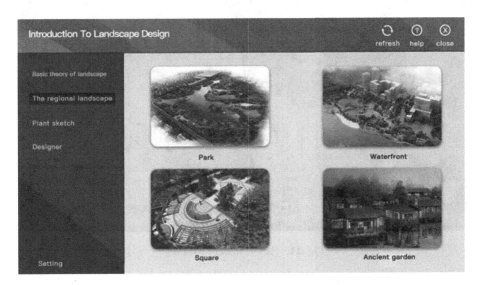

Fig. 9. Regional landscape interface

The third level interface. Taking the park landscape roaming as an example, students can roam in virtual scenes after wearing VR glasses (see Fig. 10 and 11).

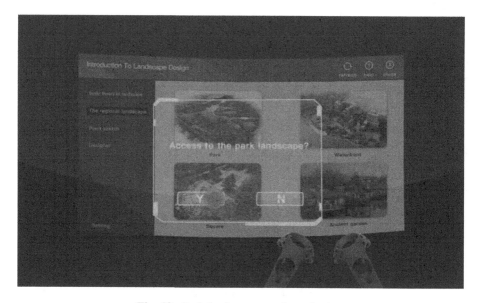

Fig. 10. Park landscape roaming selection

Fig. 11. Park landscape roaming

3.4 Scene Building and Management

After full situational experience and learning and cognition of landscape knowledge led by teachers, design training is another core content of the course Introduction to landscape design. The design thinking and design perception generated in the previous stage of learning need to be completed through design training. 5G + VR landscape design training session can provide an effective way for students' bold ideas and design practice. The system is developed and embedded in the "Practical Classroom" module to carry out scene building and management. Students can enter VR training class according to operating software. In virtual training, students can operate according to the software prompts, learn to design and build landscape model scientifically and programmatically, and complete the model building through computer virtualization, and carry out experimental design and operation in virtual environment. If there is any question during the design exercise, the system can record the text directly. The system is distributed to the client of the teacher through background processing to complete real-time feedback and tracking guidance. After the design training, teacher can upload to the Internet platform through this teaching system and carry out online scoring.

The design and application of the "Practical Classroom" module enable students' autonomy and imagination to be fully developed, strengthen the consolidation of knowledge points and deepen the image of landscape environment in their minds, and improve students' ability of landscape aesthetic cognition and understanding innovation. Due to the high transmission speed of 5G, the upload, download and analysis time of data is greatly reduced, which enables teachers to immediately carry out classroom teaching analysis, and at the same time plays an obvious burden reduction effect on teachers' classroom management and homework processing (Fig. 12).

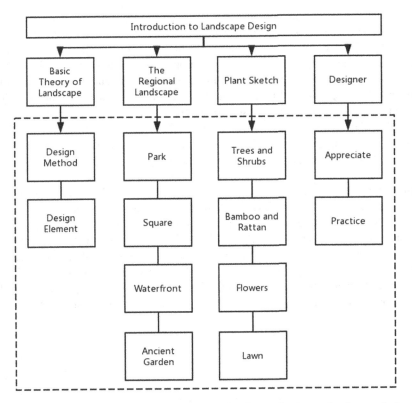

Fig. 12. Structure of 5G + VR teaching module of *Introduction to landscape design*

4   Effectiveness Analysis

4.1   Innovative Landscape Design Teaching Mode

5G + VR technology is involved in landscape design teaching activities, breaking the monotonous communication between teachers and students in the traditional landscape teaching mode, visualizing abstract concepts and deepening learners' understanding [16]. For example, the design of the park landscape environment roaming, students with helmet monitors, completely immersed in the virtual park, the model of the park can be Versailles Palace thousands of miles away or the New York Central Park.

In the virtual landscape environment, teachers and students are in the scene to conduct on-the-spot communication and case analysis. Such a classroom process is carried out in the way of games, forming an environment of independent learning in the classroom. Therefore, the traditional teaching method of teaching promoting learning is transformed into a mode of taking students as the main body and teachers as the assistant, and students carry out happy learning in the way of games in the situational experience.

As the part of teaching space is transferred to the network, the mobility, remoteness and interactivity make the landscape design classroom teaching flexible and free.

4.2   Significant Improvement of Practical Teaching Effect

In the teaching of Introduction to landscape design, students' design training and practice should occupy a large proportion, but usually many training projects are affected and limited by weather, geographical environment, practice site, safety, hardware and other factors. After the application of this system, the training projects that are difficult to be completed in reality can be carried out in VR environment, reducing the risk of students' practice and helping students understand and master the knowledge system of landscape design principle and aesthetic appreciation [17]. For example, in the teaching part of water plants, VR technology is used to simulate the water environment, so that students can deeply understand the growing habits and matching requirements of plants underwater, and avoid the unexpected risks in the real training.

In the virtual environment, the drawing of graphics and the building of models only need computer operation. The high-speed transmission of 5G broadband and the later cloud rendering shorten the design process time, greatly improved the quality and efficiency of students' practical training. Table 3 takes the practical training stage of park landscape roaming design as an example, and shows the comparison of learning efficiency in the two stages before and after the adoption of immersive teaching in virtual environment through the questionnaire on students' learning efficiency in four aspects: basic theory, case analysis, practical exercise and stage assignment.

Table 3. Stage teaching effect investigation: training stage of park landscape roaming design

Contents	Before use	After use
Basic theory	52%	77%
Case analysis	36%	85%
Practice exercise	53%	88%
Stage assignment	71%	92%

4.3 Rich and Diversified Teaching Achievements

5G + VR technology gives the new classroom a sense of immersion, presence and vividness, which makes the original boring landscape knowledge in a vivid and interesting form, and improves the enthusiasm of students in learning. As students get rid of the shackles of the traditional landscape classroom, making the knowledge transfer more effective, breaking the simple teaching routine, changing the way of controlled interaction between teachers and students [18] releasing the imagination space and creative thinking of students, and promoting students to be willing to learn and brave to create, many of their ideas can be presented in 5G + VR landscape and practiced in reality, therefore, the course teaching achievements have become rich and diverse (see Table 4).

Table 4. List of teaching achievements of 5G + VR *"Introduction to landscape design"*

Teaching achievements	Type	Number
Application for national patent	Invention patent	1
	Utility model patent	2
	Appearance patent	8
Competition Award (Person time)	National level	2
	Provincial level	5
	Municipal and school level	11
Model design and construction	Terrain model	13
	Main building model	27

4.4 Existing Problems

Causing Gamification Mood. Due to the real sense of situation and strong experience of the course teaching system of 5G + VR Introduction to landscape design, it is found that some students can't cooperate with the teaching teachers in the actual application process, and the main learning attention of some students is distracted. For example, at the beginning of the course teaching, students will not follow the teacher's guidance for virtual roaming, or students who have entered the next level still stay in the roaming interface of the previous stage, and they are excessively immersed in the mood of playing games.

Unit Teaching Time is Difficult to Control. Through the practical application of this system, students' high enthusiasm for learning and urgent experience psychology make the prescribed unit teaching time difficult to meet the students' experience requirements, thus making the teaching time control difficult to grasp, and teachers' time for knowledge supplement and further explanation is also insufficient.

Software and Hardware Need to be Further Developed and Upgraded. Currently in the initial stage of commercial 5G network, there are still some problems in the adaptation of 5G and VR system. In the later stage, 5G network needs further adaptation and hardware upgrading, while the development of teaching system resources is also insufficient, and the degree of interaction and integration needs to be further strengthened.

5   Conclusion

5G + VR new learning model can effectively make up for the shortcomings of the current traditional landscape design teaching, not only can make the course teaching become simple and vivid, the atmosphere of classroom learning is relaxed and pleasant, but also can make the classroom teaching get rid of the constraints of time and space, from two-dimensional conversion to three-dimensional space learning, high-definition and time free 5G virtual scene simulation on the one hand makes the traditional boring teaching cases and theoretical knowledge happy for students through situational experience, on the other hand, through the exercise and operation of the design training module, the design imagination of the students is fully stimulated, and the learning effect and quality are greatly improved.

Although the course teaching system of 5G + VR Introduction to landscape design has some deficiencies, it provides a reference for the reform of teaching mode in Chinese universities based on 5G + VR technology. We believe that with the comprehensive coverage of 5G network in the future, there will be more and more teaching design cases based on 5G + VR, and more universities will realize 5G + VR classroom, which will play an important role in interactive teaching, public welfare classroom, distance teaching, game and higher education industry integration and other fields, and make international teaching and experience exchange more frequent, and promote the international sharing of teaching resources. It will also play an active role in alleviating the imbalance of regional education resources in China.

Acknowledgement. This work is supported by the key online teaching reform research project of Anhui Polytechnic University "Research on online implementation and management of international students' intelligent classroom and practical teaching in case of outburst of large-scale epidemic "(2020xsjyxm08).

References

1. Wang, S., Wu, D.: The application of virtual reality technology in the teaching of "landscape architecture design" course. For. Educ. China **03**, 51–55 (2019)
2. Liu, Y.: 5G boosts virtual reality industry. Sci. Technol. J. **23**, 4 (2019)
3. Liang, L., Jiang, B., Jiang, W.: Research on the application of VR technology in practical teaching. Educ. Teach. Forum **16**, 50–52 (2019)
4. China has officially issued a 5G commercial license. http://www.sasac.gov.cn/n2588025/n2588139/c11432389/content.html. Accessed 6 June 2019
5. Jiang, F., Sun, Y., Li, R.: A brief analysis of the application of virtual reality in education and teaching. Course Educ. Res. **4**, 225–226 (2014)
6. Liu, L., Sun, H.: The application and design requirements of virtual reality technology in dance teaching. China Educ. Technol. **6**, 85–88 (2014)
7. Wu, M.: The application of virtual laboratory in middle school biology experiment teaching. Invention Creation **2**, 55–59 (2015)
8. The Ministry of Education. 2019 work guidelines. http://www.moe.gov.cn/jyb_xwfb/gzdt_gzdt/s5987/201902/t20190222_370722.html. Accessed 22 Feb 2019
9. Wang, S., Wang, Y.Y.: 5G+ education: connotation, key features and transmission model. Chongqing High. Educ. Res. **10**, 1–15 (2019)
10. Li, L.: Research on the integration of 5G+VR technology and ideological and political education - a case study of the first VR ideological and political education practice room in Jiangsu. Educ. Modern. **6**, 183–185 (2019)
11. Huang, Z., Zhang, S., Chen, Y.: Fundamentals of landscape design course teaching reform and practice. Light Text. Ind. Technol. **48**, 157–158, 169 (2019)
12. Wang, S., Wu, D.: The application of virtual reality technology in the teaching of "landscape architecture design" course. For. Educ. China **3**, 51–55 (2019)
13. Li, Y., Wang, H.: The design and realization of virtual tourism system based on VR technology. Autom. Instrum. **9**, 195–197, 201 (2019)
14. Sun, B., Zhang, J., Cai, Y., Guo, W.: Multiplayer online virtual experiment system based on kinect somatosensory interaction. Comput. Sci. **9**, 284–288 (2016)
15. Li, Q., Huo, S.: Modeling and virtual design of 3DMAX virtual landscape. Mod. Electron. Tech. **17**, 163–167 (2019)
16. Gao, H.: The summarize of VR educational application. China Comput. Commun. **2**, 231–232, 235 (2019)
17. Zhao, S.: Study on the theory and teaching practice of landscape planning and design specialty in universities - comment on landscape design teaching. Res. Educ. Dev. **21**, 85 (2018)
18. Liu, S.: Research and development of 3D virtual scene roaming and interaction system based on VR. China Comput. Commun. **15**, 71–73 (2019)

Feature Extraction of Chinese Characters Based on ASM Algorithm

Xingchen Wei[✉], Minhua Wu, and Liming Luo

Capital Normal University,
Information Engineering College, Beijing 100048, China
wxc_1218@foxmail.com

Abstract. In Chinese Calligraphy education, the computer-based evaluation on Chinese handwriting is one of the problems in the field of computer intelligent education. In this study, the method of feature comparison is first proposed in the process of computer-based evaluation on Chinese handwriting, focusing on automatically and accurately extracting the features of Chinese characters. Then, the key technologies applied in feature extraction of Chinese character were analyzed. It discussed the representation of features, aligns training samples, and reduces dimensions by principal component analysis, established local gray-scale model, and converged the gray-scale information of target feature points through statistical analysis. The experimental results show that the accuracy of the algorithm is 93.84%.

Keywords: Chinese character · Feature extraction · Active shape model

1 Introduction

When writing Chinese characters, the beauty and the neatness of writing is particularly important, which contributes to the art of calligraphy. Calligraphy is also the most unique and basic art category in Chinese art [1]. Moreover, calligraphy teaching is widely used in primary and secondary schools in China. However, there is a shortage of calligraphy teachers in these schools, which makes it necessary to study the computer evaluation of the aesthetic degree of Chinese characters writing [2].

There are a few pieces of research in the field of computer evaluation of Chinese character writing. At present, Zhuang Chongbiao is the leading researcher. By summing up the types of errors that often appear in the process of writing Chinese characters, Zhuang put forward an intelligent algorithm for judging the correctness and the neatness of Chinese characters written online [3]. The algorithm is based on online handwriting, so it can't judge Chinese character images. Moreover, there are only two levels of Chinese character evaluation in his research, correct or false. Therefore, the feature comparison method is helpful to evaluate the details of Chinese character writing. By extracting the feature information from the Chinese character images written by students and comparing it with the corresponding features of the standard template Chinese characters, the aesthetic degree of Chinese character writing can be quantified and evaluated.

© Springer Nature Singapore Pte Ltd. 2020
P. Qin et al. (Eds.): ICPCSEE 2020, CCIS 1258, pp. 620–637, 2020.
https://doi.org/10.1007/978-981-15-7984-4_45

In this process, how to extract the required features automatically and accurately is one of the key issues. However, there is a lack of research on Chinese character feature extraction. At present, Meng Yifei and others are leading in the related research. They study the feature extraction technology of characters of the kingdom of Xia and achieve a simple feature extraction of these characters [4]. However, they only focus on the simplest strokes of these characters, such as a dot, whether it can be applied to the feature extraction of complicated shapes in characters or even whole characters that remain to be explored.

It should be emphasized that the essence of this study is feature extraction, so Chinese characters do not have semantics but are regarded as an image with complex shape. Thus, based on the above background, the purpose of this study is to extract the features of Chinese characters, determine the representation of the features, and achieve the feature extraction of complicated shapes in characters or whole characters.

2   Research Ideas

In order to achieve the feature extraction of Chinese characters, this study analyzes and determines the representation of the features; then randomly selects some Chinese characters images as experimental samples; selects appropriate features of these characters images and carries out manual annotation; then uses the active shape model (ASM) algorithm to extract the features, and finally analyzes the experimental results. The experimental steps are shown in Fig. 1.

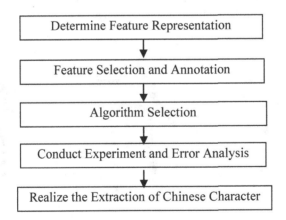

Fig. 1. Experimental steps

3   Feature Representation

In order to realize the feature extraction of the Chinese character image, it is necessary to determine the representation of the feature. Chinese character is a complex shape composed of a bunch of lines. The meaning of feature representation is to select a suitable representation method to record the obvious feature of lines in these images. There are many common methods for feature representation, such as region, line segment, and point set. As shown in Fig. 2, region, line segment and point set can be used to represent a part of the Chinese character "You" (the third tone of mandarin pronunciation) image, that is, one of the horizontal line shape.

The essence of using the region to represent feature is to record a rectangular region to represent one shape of the image contained in the region. Figure 2 (left) shows that there is one of the horizontal line shapes of "You" in the rectangular area. The advantage of this method is simple, but it can only record a shape that exists in a certain area, and cannot describe more information about this horizontal line.

The essence of using line segments to represent features is to record a line segment to roughly represent a shape in an image. This line segment can represent the slope and length of this shape. In Fig. 2 (middle), the line segment is used to represent a horizontal line shape of the Chinese character "You" image. It is more accurate than using the region to represent features, but it is difficult to describe more detailed information in this line segment, such as the small bends of this shape.

The essence of using point set to represent feature is to record the position of a series of feature points to represent all the obvious information of this shape. The advantage is that it describes more accurate and rich information, which can be expressed by recording the starting point, turning point, endpoint, and other points of all lines in the Chinese character image. Figure 2 (right) shows the horizontal line shape of the Chinese character "You" image represented by a set of 6 points. The number of point sets shows the richness of the described features.

Because of the advantages of the point set, this study chooses the point set to represent the features of the Chinese character image.

Fig. 2. Different feature representations of the horizontal line of character "you"

4   Feature Selection and Manual Annotation

4.1   Feature Selection

After choosing the point set as the feature representation of the Chinese character image, this study will explore which points are selected as feature points to represent the shape of the image, and these feature points should be able to fully reflect the shape features. Therefore, the selection of these feature points is mainly based on two purposes: 1. To represent more obvious features of the shape in the image of the Chinese character with as few points as possible. 2. The selected feature points are easy to find and annotate.

In order to achieve the above two purposes, we follow five principles when selecting feature points: 1. Select Clear turning point 2. Select T-shaped connection point 3. Select bisection point 4. Select high curvature point 5. Select the intersection point [5]. For the above reasons, taking the Chinese character "Ni" (the third tone of mandarin pronunciation) image as an example, we choose 32 feature points as shown in Fig. 3 to describe the features of the shape in this image. These 32 points can describe the endpoint position, turning point position, bisection position, etc. of the "Ni" character, basically depicting all the feature information of this shape. If needed, more feature points can be obtained by adding the midpoint of two adjacent points to cover all lines of the shape, so the 32 feature points selected meet the two purposes mentioned above.

Fig. 3. Feature point selection of character "ni"

In order to describe shape feature points, this study use parameter n to represent the number of feature points, and parameter x, y to describe the horizontal and vertical coordinates of each feature point. Thus, we can represent the vector (x_i, y_i) of each feature point, and the vector X with 2n elements, where $X = (x_1, ..., x_n, y_1, ..., y_n)^T$.

4.2 Manual Annotation of Feature Points

In this study, as mentioned later, the ASM algorithm is selected for research, which is divided into two stages: model building and shape matching. In the model building phase, some annotated images are selected as training sets to achieve the model building. So this paper introduces the manual annotation method of image feature points of training set needed in the model building stage in advance. In this study, the PTS file with the same format as Frank data set marked by Cootes is used, which has the advantages of simplicity, accuracy, and small size. The specific feature point set file representation is shown in Fig. 4. The file name is consistent with the name of the image to be annotated, but the file suffix is ".PTS". The first line in the file records the version of the annotation format, and the second line records the number of annotation point sets. Between the { }, from top to bottom, each line records the coordinate position of a feature point, the first number represents the abscissa value, and the second number represents the ordinate value. For example: in Fig. 4, the number of feature points is 3, and the coordinate positions of the three feature points are (20, 19), (20, 25), (17, 32) in turn. By using the program made by myself to annotate the feature points in all images, the corresponding PTS file is generated.

G4N005P01_H2_L1.pts

```
version: 1
n_points: 3
{
20 19
20 25
17 32
}
```

Fig. 4. PTS file

5   Algorithm Selection and Application

5.1   ASM Algorithm

The feature representation of this study uses the point set, because of the difference of each person's writing habits and the randomness of writing, the shape of the Chinese character image is deformed and uncertain. So we need an algorithm that can process the point set and adapt to the shape deformation to extract the feature points of the corresponding shape in the new image. Therefore, this study selects the ASM algorithm which can deal with certain deformation for feature extraction. ASM algorithm is a flexible model, which is mainly used to identify and locate the object image that may exist deformation [6]. This model can automatically eliminate the interference of a certain degree of deformation within a certain type of shape and is often used in image search to find the graphic structure represented by a batch of training samples [7]. The Chinese character image is the image with this characteristic because of the uncertainty of people's writing. Through modeling learning, the model represents the average position of points, and in the training process, the training results can be adjusted by modifying the relevant parameters of model training [8]. Moreover the feature representation required in this study is also the point set form. Considering that the ASM algorithm has some successful experience in the face recognition field, the ASM model is selected as the algorithm for research. The ASM algorithm used in this study mainly consists of two stages: model building and shape matching.

5.2   ASM Model Building

The process of building the ASM model is to use some annotated images as the training set for modeling, which includes three steps: alignment of training samples, principal component analysis (PCA), and building local gray-scale model.

1. Alignment of Training Samples

Through the manual annotation above, this study creates S (parameter, represents the number of samples) training samples, and generates S vectors x_i. Before the statistical analysis of these vectors, it is an important premise that the shapes expressed need to be in the same coordinate system. We want to eliminate the effects of displacement, scaling, and rotation as much as possible, which requires aligning all training samples.

There are many methods for aligning training samples, the most commonly used method is the Proclustes analysis [9]. By using this method, the Euclidean distance $(D = \Sigma |x_i - x|^2)$ between each shape and the average shape can be minimized [10]. In this process, the following iterative methods are adopted:

1. Displace, scale, and rotate each shape so that its center of gravity is at the origin.
2. Find a shape that is closest to the average shape.
3. Mark this shape as x_0 and use it as the standard shape.
4. Align all shapes with standard shapes.
5. Recalculation the average of all shapes after alignment.
6. Displace, scale and rotate the average value of all shapes after recalculation to align with x_0
7. Repeat step 4, and the average value gradually converges to x_0.
8. If the mean value does not change significantly and it below a certain threshold, the convergence iteration is finished.

The process of aligning the new shape with the standard shape is to make the Euclidean distance minimum by affine transformation through formula (1). Where t is the feature point matrix of the current shape to be transformed, and the goal is to make the feature point set of the current shape closest to the feature point set of the standard shape. x_t and y_t are the offset of the x-axis and y-axis respectively, r is the scaling factor, and θ is the rotation angle [11].

$$T\begin{pmatrix} x \\ y \end{pmatrix} = \begin{pmatrix} x_t \\ y_t \end{pmatrix} + \begin{pmatrix} r\cos\theta & r\sin\theta \\ -r\sin\theta & r\cos\theta \end{pmatrix}\begin{pmatrix} x \\ y \end{pmatrix} \tag{1}$$

2. Principal Component Analysis

More feature point information can provide more diverse information, which is conducive to convergence to more accurate shape, but it also increases the workload of operation. More importantly, there may be a correlation between many variables, which increases the complexity of problem analysis [12]. Moreover, it may cause overfitting when the shape converges, so it is necessary to find a reasonable method to reduce the parameters to be analyzed and the loss of information contained in the original parameters as much as possible, so as to achieve the purpose of accurate image convergence [13]. Therefore, this study adopts the PCA method.

By aligning the training samples, we can get n shapes, which are recorded as {F_1, F_2, ..., F_n}, their average shape $\bar{F}$ can be expressed as:

$$\bar{F} = \frac{1}{n}\sum_{i=1}^{n} F_i \tag{2}$$

According to the covariance formula:

$$Cov = \frac{1}{n}\sum_{i=1}^{n} (F_i - \bar{F})(F_i - \bar{F})^T \tag{3}$$

Next, the above covariance matrix is singular value decomposed to obtain the eigenvalue λ and eigenvector P. after arranging the eigenvalue λ from large to small, select n λ, where [λ_1, λ_2, λ_3,..., λ_n] meets the following requirements:

$$\frac{\sum_{i=1}^{n} \lambda_i}{\sum \lambda} \geq \eta \tag{4}$$

Among them, η is the weight threshold, usually above 94%.

The eigenvector corresponding to these n eigenvalues is P = [p_1, p_2, ..., p_n]. The statistical model of the training set is constructed by P: S = $\bar{S}$ + Pb.

In this case, a new Chinese character shape vector can be obtained whenever the value of the weighting coefficient vector b is changed. At the same time, in order to make the final shape vector change without excessive deformation and maintain the original basic shape, the ASM also has certain constraints on the value of the weighting coefficient. Through statistical analysis of the probability distribution of parameters, the value range of parameter b is obtained, $-3\sqrt{\lambda_i} \leq b \leq 3\sqrt{\lambda_i}$, where λ_i is the ith eigenvalue of Cov [4].

3. Build Local Gray-Scale Model for Each Feature Point

We determine the initial shape of all the feature points by principal component analysis. And we need to make some small adjustments. These adjustments can move each point to a better position, and establish the corresponding gray-scale model. Let the ith feature point of the jth picture be X_i, make a straight line L through the two adjacent points X_{i-1} and X_{i+1} of X_i, and move the feature point along the normal of L in a certain range, so as to adjust the feature point to the edge of the target image and better fit the image, as shown in Fig. 5. In the process of moving, the gray value contained in the vector is derived to get a local texture g_{ij}, and the changing trend of the value is shown in the figure. The maximum value of g_{ij} represents the position of the edge of the target image, as shown in Fig. 6 [14].

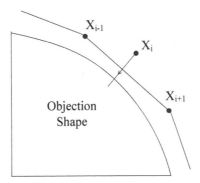

Fig. 5. Adjustment of feature point

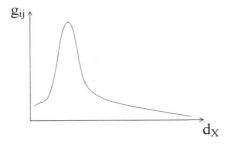

Fig. 6. Variation trend of gray value derivative

By the same operation for all training samples, the texture features of the ith point of n training samples can be obtained, and the average value can be calculated:

$$\bar{g}_i = \frac{1}{n} \sum_{j=1}^{n} g_{ij} \tag{5}$$

Variance:

$$S_i = \frac{1}{n} \sum_{j=1}^{n} (g_{ij} - g_i)(g_{ij} - g_i)^T \tag{6}$$

Finally, the gray model of the feature point is obtained.

5.3   Shape Matching of the ASM

Firstly, the trained shape of the ASM model is overlaid on the new image, and the shape of the new image is searched according to the established shape model of Chinese characters and the local gray-scale model of each feature point. The essence of the shape search is to find the best matching position of each feature point in the new image. The search method is shown in Fig. 7. Take the ith feature point as the center, take k points from left and right respectively, and form 2k + 1 candidate points. Each candidate point takes its left and right adjacent p points as the sub local feature of the candidate point, thus obtaining the sub local feature of 2(k−p) + 1 candidate point [15]. Suppose that the sub local feature of the jth feature point is expressed as g'_ξ, ξ = 1, 2, …, 2(k−p) + 1. Calculate the Mahalanobis distance D_ξ according to the following formula (7). After traversing all the above points, select the point with the smallest Mahalanobis distance as the best matching position of the feature point [16].

$$D_\xi = \left(g'_\xi - \bar{g}_j\right)^T C_j^{-1}\left(g'_\xi - \bar{g}_j\right) \tag{7}$$

Repeat the above search steps until the maximum number of iterations is reached.

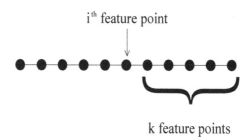

i^{th} feature point

k feature points

Fig. 7. Search method

5.4   Parameter Adjustment

6   Parameter Level

Before training with the training set image, this study hopes to find a suitable image size for training, which needs to adjust the resolution of the image. The method selected in this study is Gaussian pyramid. The principle is to convolute the original image with Gauss check, then delete one row or one column of pixel points every n rows or n columns, and finally form a new image. The image pyramid is shown in the Fig. 8.

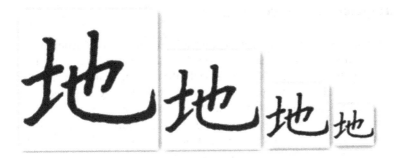

Fig. 8. Image pyramid

In the training stage, the higher the image resolution, the more image information is contained, but too many unnecessary details will affect the effect of model training. However, when the image resolution is too low, many important details will be lost. Therefore, for the training set image, through this subsampling method, the image resolution is gradually reduced to find the appropriate image resolution. In the experiment of the ASM algorithm in this study, the level of subsampling is 1–8. According to experience, for 70 * 70 resolution image, the training effect of 5-level subsampling is usually better. However, for different training sets of Chinese characters, when adjusting this parameter, we traverse 1–8 levels of subsampling to find the most appropriate subsampling level for the training set.

7   Parameter Width

In ASM shape matching stage, it is necessary to take the left and right adjacent P points of each candidate point as the sub local feature of the candidate point, calculate the Mahalanobis distance between all candidate points and the local gray-scale feature of the point, and select the candidate point with the minimum Mahalanobis distance as the final position. The selected P points are the calculated width of the local feature. If the width is too small to effectively represent the local feature of the feature point, the training effect is not good. If the width is too large, not only the calculation time is long, but also the pixels farther away from the feature point can not effectively represent the local features near the point. According to the experimental experience, even for different words and images with resolution of 70 * 70, the width setting of 12 generally has a better training effect, so in the early tuning, the initial width is set to 12. However, for different training sets of Chinese characters, when adjusting this parameter, the width is traversed from 1 to 30, so as to find the most suitable calculation width for the training set.

8 Parameter PCA Percentage

When the shape converges, in order to prevent the training model from overfitting and reduce the complexity of the problem, the high-dimensional information with high correlation is reduced to the low-dimensional information, so the principal component analysis is needed. In this process, the dimension reduction degree needs to be set, so the parameter PCA percentage is introduced. If the dimension reduction degree is too large, many useful dimension information will be lost, but if the dimension reduction degree is too low, useful principal components can not be selected to represent the shape information of the image. And for different training sets of Chinese characters, as well as different annotation feature points, sometimes we need to reduce the dimension to a greater extent, sometimes we need to reduce the dimension to a small extent or even not. Therefore, according to experience, first set the PCA percentage to 98.9%, then traverse the PCA percentage from 94% to 100%, each step is 0.1%, in order to find the most appropriate PCA dimension reduction degree for the training set.

9 Parameter n_iteration

This parameter represents the maximum number of iterations of the ASM algorithm in shape search on new images. It has little effect on experimental results. Generally, the number of iterations is only greater than 100 to get better search results. But for different training set samples, this parameter is traversed 100-3000 times, each step is increased to 100 times, in order to find the most stable image iteration times for the training set.

10 Application and Error Analysis of the ASM

10.1 Application of ASM

11 Experiment Process

To test the feature extraction effect of the ASM algorithm on Chinese character image, in this study, 960 standard Chinese characters were selected from 3500 commonly used Chinese characters. We made it into a book and printed by the publishing house. These books also distributed to primary and secondary school students in two schools in Beijing. The students wrote in imitation of the standard characters. The students' grades range from grade three to grade nine. The next step is to recycle the written books and use a scanner to convert the written books into electronic images for storage. As shown in Fig. 9 is one of the pages written by a student. In this study, 377 books were recovered and 1,085,760 Chinese character images were collected.

In this study, several Chinese characters were randomly selected for the experiment. They are "Guo" (the second tone of mandarin pronunciation), "Zhe" (the fourth tone of

Fig. 9. One page written by a student

mandarin pronunciation), "Ren" (the second tone of mandarin pronunciation), "De" (the fourth tone of mandarin pronunciation), "Shi" (the fourth tone of mandarin pronunciation), "Bai" (the second tone of mandarin pronunciation), "Zhong" (the first tone of mandarin pronunciation) and "Shang" (the fourth tone of mandarin pronunciation). And they have different shapes. ASM algorithm is used for feature extraction. In order to ensure the reliability of the results, 5 groups of cross-validation were carried out for each Chinese character shape extraction experiment. The specific experimental steps are as follows:

1. Randomly select 50 images of the Chinese character "Guo" with a resolution of 70 * 70 pixels from the data set mentioned above, and manually annotate these images. The location of the feature points follows the five principles described in Sect. 4.1, and finally, 21 feature points are determined. The annotation positions are shown in Fig. 10.
2. Denoise and binary preprocess these images to eliminate noise interference. All the images were randomly divided into 5 groups, 10 in each group.
3. 40 images of the other 4 groups are used as training sets and 10 images of this group are used as test sets. ASM algorithm is used to verify the results of feature point extraction. In order to ensure the reliability of the experimental results, 5 groups of cross-validation are performed successively. The experimental result is the average of 5 groups of feature extraction accuracy.

4. Replace the images with 50 randomly selected Chinese characters "Zhe" with a resolution of 70 * 70 pixels, and annotate these images manually. The location of the feature points follows the five principles described in Sect. 4.1, and finally, 42 feature points are determined. Repeat steps 2–3 for the experiment.

5. Replace the images with 50 randomly selected Chinese characters "Ren" with a resolution of 70 * 70 pixels, and annotate these images manually. The location of the feature points follows the five principles described in Sect. 4.1, and finally, 17 feature points are determined. Repeat steps 2–3 for the experiment.

6. Replace the images with 50 randomly selected Chinese characters "De" with a resolution of 70 * 70 pixels, and annotate these images manually. The location of the feature points follows the five principles described in Sect. 4.1, and finally, 23 feature points are determined. Repeat steps 2–3 for the experiment.

7. Replace the images with 50 randomly selected Chinese characters "Shi" with a resolution of 70 * 70 pixels, and annotate these images manually. The location of the feature points follows the five principles described in Sect. 4.1, and finally, 21 feature points are determined. Repeat steps 2–3 for the experiment.

8. Replace the images with 50 randomly selected Chinese characters "Bai" with a resolution of 70 * 70 pixels, and annotate these images manually. The location of the feature points follows the five principles described in Sect. 4.1, and finally, 12 feature points are determined. Repeat steps 2-3 for the experiment.

9. Replace the images with 50 randomly selected Chinese characters "Zhong" with a resolution of 70 * 70 pixels, and annotate these images manually. The location of the feature points follows the five principles described in Sect. 4.1, and finally, 16 feature points are determined. Repeat steps 2–3 for the experiment.

10. Replace the images with 50 randomly selected Chinese characters "Shang" with a resolution of 70 * 70 pixels, and annotate these images manually. The location of the feature points follows the five principles described in Sect. 4.1, and finally, 6 feature points are determined. Repeat steps 2–3 for the experiment.

Fig. 10. "Guo" annotation positions

12 Extraction Error Analysis

As mentioned above, in this study, a series of feature points are used to represent the features of the shape in the Chinese character image, so it is necessary to consider the allowable error range when extracting feature points. Because all feature points in the early stage are manually annotated, for the same feature point of the same image, every two different manual annotations cannot guarantee that the annotation position is the same pixel point. Especially the original image size is 1400 * 1400 pixels, it is almost impossible to annotate at the same pixel point twice. Therefore, we accept the annotation and extraction in a certain range of pixels. Any pixel point in this range can represent this feature point.

In determining the pixel range, 30 experiments were used in this study. One image of the Chinese character "De" is randomly selected each time, and the image resolution size is 1400 * 1400 pixels. In each experiment, three different people (P1, P2, P3) annotated the same image respectively until 30 Chinese images were annotated. Each image is annotated with 23 feature points, and the annotation location is shown in Fig. 11. 690 points were annotated by each person. The maximum and average distance of the same feature point in 30 groups of experiments were calculated as reference.

Fig. 11. "De" annotation positions

The experimental results are as follows (Table 1):

Table 1. Statistic of annotation errors

	Maximum distance			Average distance		
	P1, P2	P1, P3	P2, P3	P1, P2	P1, P3	P2, P3
Experiment 1	119.55	47.30	104.80	34.92	20.29	29.56
Experiment 2	81.27	55.47	94.26	37.62	26.40	28.42
Experiment 3	127.61	53.67	134.94	39.69	20.94	38.79
Experiment 4	253.21	57.43	242.20	69.32	21.25	59.71

(continued)

Table 1. (*continued*)

	Maximum distance			Average distance		
	P1, P2	P1, P3	P2, P3	P1, P2	P1, P3	P2, P3
Experiment 5	88.14	52.09	111.16	34.18	22.54	32.31
Experiment 6	101.83	78.34	102.00	34.59	25.98	30.86
Experiment 7	73.25	143.56	199.70	28.30	36.07	48.39
Experiment 8	89.02	56.08	83.22	33.74	22.32	28.89
Experiment 9	106.28	97.94	63.64	30.35	28.82	24.00
Experiment 10	99.30	54.08	81.39	36.08	27.59	25.63
Experiment 11	58.86	58.00	76.58	25.91	25.21	22.36
Experiment 12	98.49	108.67	95.63	35.55	32.19	30.24
Experiment 13	43.19	73.01	42.72	20.69	25.68	19.40
Experiment 14	138.68	52.39	133.60	34.92	20.40	33.58
Experiment 15	125.90	114.77	127.42	47.69	29.30	41.84
Experiment 16	70.21	55.90	47.80	23.77	23.20	18.35
Experiment 17	61.72	84.10	78.29	27.58	28.61	30.21
Experiment 18	40.02	83.63	86.61	19.86	30.94	27.69
Experiment 19	204.66	79.20	237.17	50.42	25.74	50.25
Experiment 20	55.47	82.29	103.02	23.93	30.72	30.51
Experiment 21	86.02	83.67	65.46	29.58	29.52	23.10
Experiment 22	91.83	70.41	77.01	29.69	26.41	26.00
Experiment 23	54.78	70.18	85.43	24.61	31.51	27.16
Experiment 24	77.03	84.88	54.78	29.67	31.83	19.37
Experiment 25	253.65	59.03	221.77	53.46	23.35	47.13
Experiment 26	72.20	56.22	65.80	25.86	21.89	21.91
Experiment 27	281.00	102.20	230.26	71.45	37.53	67.11
Experiment 28	155.79	126.06	180.85	40.79	50.79	55.98
Experiment 29	65.46	65.19	52.43	27.87	22.07	21.99
Experiment 30	65.00	46.69	80.06	27.61	19.67	26.35
Average	107.98	75.08	112.00	34.99	27.29	32.90

From the above table, for 1400 * 1400 pixel resolution images in 30 groups of experiments, the average value of the maximum distance annotated of the same feature point is 98.35 pixel points, and the average distance is 31.73 pixel points. Because this experiment finally uses 70 * 70 pixel resolution image, the error range correspondingly becomes the maximum distance is 4.92 pixels. In order to ensure the preciseness of the experimental results, when verifying the results of the algorithm, it must meet the requirement that the average distance of the results of feature points extraction for the same batch of images is also less than 1.59 pixels.

13 Experimental Results

According to the above steps, the accuracy of the experimental results is the number of feature points accurately extracted by the algorithm divided by the total number of feature points, the experimental results are as follows (Table 2):

Table 2. Different character feature extraction results

	Cross-validation 1	Cross-validation 2	Cross-validation 3	Cross-validation 4	Cross-validation 5	Average results
"Guo"	96.19%	96.67%	96.19%	93.81%	95.24%	95.62%
"Zhe"	92.86%	95.48%	94.05%	96.43%	90.48%	93.86%
"Ren"	92.35%	97.06%	92.35%	93.53%	92.94%	93.65%
"De"	93.83%	91.52%	92.83%	91.00%	86.57%	91.15%
"Shi"	93.81%	95.71%	94.29%	97.62%	97.14%	95.71%
"Bai"	97.50%	98.33%	95.83%	91.67%	95.83%	95.83%
"Zhong"	95.29%	92.35%	91.18%	92.94%	91.18%	92.59%
"Shang"	95.00%	98.33%	83.33%	93.33%	91.67%	92.33%

According to the experimental results of the above eight characters, the statistical chart of feature extraction accuracy is shown in Fig. 12:

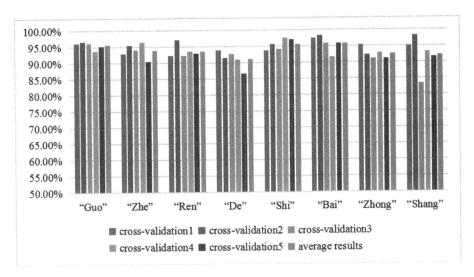

Fig. 12. Statistical chart of feature extraction accuracy

From the above results and statistical figures, it can be concluded that the accuracy of extracting feature points of different Chinese characters is basically the same, and all have good accuracy, the average accuracy of eight shapes is 93.84%. It should be emphasized that in the above experiments, the eight shapes of Chinese characters are selected randomly, and the experimental results of each shape are verified by five groups of cross-validation, which ensures the reliability of the experimental results. Therefore, the experimental results can be regarded as universal results.

13.1 Experiment Analysis

In the experiment, ASM has a good effect on feature point extraction of Chinese character shapes, but through the observation of the experimental results, it can be found that most of the feature points often converge to the appropriate position, but a few of the points can only change in a certain range due to the limitations of ASM model, resulting in these points cannot converge to the correct position. Moreover, because the shape of the Chinese character image is more changeable and arbitrary than face image, and ASM algorithm cannot adapt to the new image which is beyond the average deformation range of training set, the accuracy of feature point extraction still needs to be further improved.

14 Conclusion

Based on the ASM algorithm, through the selection of Chinese character shape feature points, eight Chinese character are taken as the experimental objects to annotate and extract the features. Finally, the experimental results show that the average accuracy is 93.998%. It has a certain contribution to solving the difficulty of feature extraction in the field of automatic evaluation on Chinese handwriting in primary and secondary schools.

However, some limits exist in this algorithm and experiment. First, the number of Chinese characters commonly used in primary and secondary schools is about 3500. Whether other Chinese characters can be well evaluated by this algorithm needs more experimental proof. Finally, future work will be implemented on the improvement of the algorithm.

References

1. Feng, T.: Modern value of calligraphy from the perspective of western "writing" theory. Chin. Calligr. **16**, 128–131 (2018)
2. Wu, L.: Strengthening the construction of calligraphy discipline in colleges and universities to change the backward situation of basic education calligraphy. China calligraphy daily, no. 007, 10 July 2018

3. Zhuang, C.: Intelligent evaluation algorithm for correct and neat writing of online Chinese characters. Signal processing branch of China Electronics Society and signal processing branch of China instrumentation society. In: Proceedings of the 12th National Annual Conference of Signal Processing (ccsp-2005). Signal Processing Branch of China Electronics Society and Signal Processing Branch of China Instrumentation Society: Signal Processing Branch of China Electronics Society, vol. 4 (2005)

4. Meng, Y., Zhang, X., Yang, X.: Application of feature extraction and matching based on ASM algorithm in character recognition. J. Guangxi Univ. **42**(06), 2183–2190 (2017). (Natural Science Edition)

5. Cootes, T.F., Taylor, C.J.: Statistical Models of Appearance for Computer Vision. The University of Manchester School of Medicine (2004)

6. He, J., Fang, L., Cai, J., He, Z.: Fatigue driving detection based on ASM and skin color model. Comput. Eng. Sci. **38**(07), 1447–1453 (2016)

7. Cootes, T.F., Taylor, C.J., Cooper, D.: Active shape models-their training and application. Comput. Vis. Image Underst. **61**(1), 38–59 (1995)

8. Chen, X., Li, W., Li, W., Zhang, W., Zhu, Y.: Multi feature fusion fatigue detection method based on improved ASM. Comput. Eng. Des. **40**(11), 3269–3275 (2019)

9. Goodall, C.: Procrustes methods in the statistical analysis of shape. J. Roy. Stat. Soc. B **53**(2), 285–339 (1991)

10. Gower, J.C.: Generalized procrustes analysis. Psychometrika **40**(1), 33–51 (1975)

11. Tang, F., Lu, X., Shen, L.: Extraction of facial proportional feature information based on ASM and ERT feature point location algorithm. J. Shenzhen Inst. Inf. Technol. **15**(03), 1–5 (2017)

12. Chen, Y., Chen, Y., Zheng, J.: An improved data preprocessing method based on PCA. Appl. Electron. Technol. **46**(01), 96–99 (2020)

13. Abdi, H., Williams, L.J.: Principal component analysis. Wiley Interdisc. Rev. Comput. Stat. **2**(4), 433–459 (2010)

14. Zeng, Z.: Analysis of personality traits of students based on facial features and deep learning. Jiangxi Normal University (2017)

15. Gong, J.: Research on facial expression recognition algorithm based on facial key points. Huazhong University of science and technology (2016)

16. Wang, Y., Li, J.: Improved face feature point location method based on ASM. J. Guilin Univ. Electron. Sci. Technol. **36**(06), 477–482 (2016)

Study of Priority Recommendation Method Based on Cognitive Diagnosis Model

Suojuan Zhang, Jiao Liu, Song Huang$^{(\boxtimes)}$, Jinyu Song, Xiaohan Yu, and Xianglin Liao

College of Command and Control Engineering, Army Engineering
University of PLA, Nanjing, China
kathyzhangs@aliyun.com

Abstract. In the context of personalized learning, the recommendation method aims to provide appropriate exercises for each student. And individualized knowledge status may give more effective recommendation. In this study, a priority recommendation method based on cognitive diagnosis model is proposed, and cosine similarity algorithm is applied to improve the accuracy and interpretability of recommendation. Then the performance of the methods was compared under cognitive diagnosis models. The experimental results show that the method proposed achieves more accurate results and better performance.

Keywords: Priority recommendation · Knowledge mastery · Cognitive diagnosis · Similarity

1 Introduction

With the gradual maturity of artificial intelligence, big data and other technologies, intelligent technology and education have been further integrated. More and more researchers have focused on how to use intelligent technology for individualized learning. However, it is a challenging research issue, how recommend questions text to meet the needs of the students from massive resource [1]. Cognitive diagnosis theory is one of the core theories in the field of educational measurement [2]. Through various models with the theory, the cognitive state of students can be identified. Therefore, what knowledge students have mastered, what they have not mastered should be the basis of the question recommendation.

Several researchers have been reported in the literature to recommend exercises for learning when using different methods. Current studies depend on either collaborative filtering methods or cognitive diagnosis models. Unfortunately, the traditional collaborative filtering recommendation method ignores the knowledge status of students, which only give recommendations according to students' preference. Moreover, cognitive diagnosis a fundamental task in education, which aims to discover the proficiency of knowledge for each student [3]. But the accuracy of recommendation based on cognitive models needs to be increased. Thus, the traditional recommendation method faces several challenges, such as: (1) the interpretability of recommended results; (2) the accuracy of personalized recommendation.

© Springer Nature Singapore Pte Ltd. 2020
P. Qin et al. (Eds.): ICPCSEE 2020, CCIS 1258, pp. 638–647, 2020.
https://doi.org/10.1007/978-981-15-7984-4_46

In this paper, we propose the PRM-CD method (Priority Recommendation Method based on Cognitive Diagnosis Model) and give priority to recommending more relevant questions. The DINA model (a kind of cognitive diagnosis model) is introduced into the PRM-CD method that illustrates the knowledge mastery (represents cognitive state of students) and uses a Bayesian method to form a diagnostic report. According to the report, we predict performance of the students; then combine with cognitive diagnosis and cosine similarity algorithm; finally give priority to the most suitable questions as candidates.

The rest of this article is organized as follows. In Sect. 2, we introduce the related work. Section 3 states basic concepts in cognitive diagnosis model-DINA. Section 4 details the whole implementation of PRM-CD method. Section 5 shows the experimental results. Finally, conclusions are given in Sect. 6.

2 Related Work

In this section, we introduce existing researches from two perspectives: Cognitive Diagnosis models and Recommended Method.

2.1 Cognitive Diagnosis Models

Cognitive psychology and psychometrics are the foundations of cognitive diagnosis. The cognitive diagnosis model can help describe the cognitive state of students [4]. Traditional test results only report a general score. Even if students who received the same score in an exam, they may have different cognitive state.

The cognitive diagnosis model can quantitatively reflect the cognitive level and structure of the students. In recent years, cognitive diagnostic models have emerged, greatly enriching the possibilities of modelling, analysis and interpretation. There are two main classifications of cognitive diagnosis models. They can be divided into discrete and continuous model [5]. Among the numerous cognitive diagnostic models, one of the most classic cognitive diagnosis methods are Item Response Theory (IRT) [6], which is one-dimensional continuous cognitive model and the DINA model (Deterministic Inputs, Noisy "And" gate model) of multidimensional discrete cognitive modelling.

To discover what makes a question easier or harder for students, predicting examinee performance has been considered as an important issue [7, 8] in the fields of cognitive diagnosis and education data mining. Cognitive diagnosis is applied to further specific educational scenarios, such as learning team formation and comprehension test validation, and difficulty estimation across multiple forms, student response behavior modelling.

2.2 Recommended Method

The recommendation system is one of the important applications in learning analysis. Individualized practice recommendation be roughly divided into collaborative filtering algorithm, matrix factorization method and the method based on cognitive diagnosis.

The recommendation system was first applied to mail filtering and news recommendation, and the personalized recommendation algorithm was also widely used in educational resource recommendation. Among them, personalized question recommendation has become a research focus to make up for knowledge deficiency and then to recommend questions for students. Collaborative filtering algorithms are widely used and are one of the most in-depth recommended algorithms in the current research [10]. However, the collaborative filtering algorithm is based on historical data for the recommendation. For novice, there is not enough historical records at the beginning, resulting in sparse data, poor quality of recommendation, and poor performance. At the same time, it ignores that each student as an individual does not have the same required knowledge points, so it is difficult to ensure that the recommended questions are really needed by the students. Matrix factorization method has better accuracy of recommendation but has certain limitations because of the weak interpretability [11].

In this article, we also address the following tasks, i.e., modelling the knowledge mastery for each student, cognitive diagnosis visualization and predicting students' performance, finally recommend effective questions text.

3   Cognitive Diagnosis Model

In this section, we describe the details of Cognitive Diagnosis model, such as DINA model, which is used widely as it is parsimonious, simple, compact and good intelligibility, and then present the basic principles and implementation of the model.

3.1   DINA Model

The DINA model aims to diagnose the potential knowledge mastery of the students through the history record of questions. It combines the R matrix (based on score) and the Q matrix (based on experts' knowledge) linked between questions an knowledge to set up the model for students. Meanwhile, we can assume a relationship between the students' knowledge mastery α and the skills required to solve a problem. Similarly, each item has a corresponding q-vector, which represents the skills that are required to solve that question. If the item j requires skill k, then $q_{ik} = 1$. Collecting all the q-vectors in a question results in a Q-Matrix [12] of dimension $J \times K$. In the case of the DINA model, it is assumed that students must possess all the skills required to effectively solve an item [13].

The interaction between the students' knowledge mastery and the question specification defines the latent response variable.

$$\eta_{ij} = \prod_{k=1}^{K} \alpha_{ik}^{q_{jk}} \tag{3-1}$$

Here, η_{ij} is the mastery of student i on question j computed by Eq. (3-1), which is also known as the ideal response [14]. Knowledge mastery vector of student i, denoted as $\alpha_i = \{\alpha_{i1}, \alpha_{i2} \cdots, \alpha_{ik}\}$. Each column corresponds to a knowledge point (or skill). When the value is 1, it means that student has mastered the knowledge. Otherwise, it equals 0, which presents the student has not mastered.

To account for the probabilistic nature of the observed response, slip and guessing parameters conditional on the ideal response [14, 15]. The slip and guessing parameters are denoted as:

$$s_j = P(r_{ij} = 0 \mid \eta_{ij} = 1) \tag{3-2}$$

$$g_j = P(r_{ij} = 1 \mid \eta_{ij} = 0) \tag{3-3}$$

respectively. Thus, the response function for an item (the correct response probability of student i on question j) is given by

$$P(r_{ij} = 1 \mid \alpha_i) = (1 - s_j)^{\eta_{ij}} g_j^{1-\eta_{ij}}, \tag{3-4}$$

where r_{ij} is defined as the answer to the question, with the correct answer being 1 and the wrong answer being 0.

3.2   Parameter Estimation of the DINA Model

In this section, the EM algorithm [16] has been introduced and programmed for the parameter estimation of the DINA model. After repeated iterations, the estimated value tends to be stable.

Then, based on the above parameter estimates and response, we estimate the mastery of knowledge for each student by Bayesian estimation [17] and maximize the posterior probability of students' response [18], which can be defined as:

$$\begin{aligned}
\hat{\alpha}_i &= \arg\ \max(\alpha \mid r_i) \\
&= \arg\ \max L(r_i \mid \alpha, \hat{s}_j, \hat{g}_j) P(\alpha) \\
&= \arg\ \max L(r_i \mid \alpha, \hat{s}_j, \hat{g}_j) \\
&= \arg\ \max \prod_{j=1}^{J} P(r_i \mid \alpha, \hat{s}_j, \hat{g}_j)
\end{aligned} \tag{3-5}$$

Thus, we can easily discover the knowledge mastery of students, which is assumed to be mastered (i.e. 1) or nonmetered (i.e. 0).

And the diagnosis results given in Fig. 1 and the visual diagnosis report is finally formed. Obviously, we can see the student have better proficiency of the knowledge 1, knowledge 2, knowledge 3 and knowledge 4, but was lack in knowledge 5.

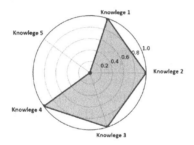

Fig. 1. A student cognitive diagnosis report

4   PRM-CD Method

Based on the DINA model in Sect. 3, we can further propose the PRM-CD method (Priority Recommendation Method based on Cognitive Diagnosis Model) using the Cosine Similarity method. Under the PRM-CD model, we prefer to recommend questions that is more suitable for students' cognitive state. Despite all wrong questions are need to be recommended, learners need more targeted test recommendations in a limited time.

4.1   Predicting Student Performance

It is not a trivial work that providing the appropriate exercises effectively during the massive learning resources. In vast test questions, students usually have only done a small number of questions. For those that students have not done before, that is, without any students' response to the questions, it is uncertain whether the commendation is suitable. If the recommendation are the questions that students have mastered, it is too easy for learners. Therefore, to improve the reliability and efficiency of the recommendation, it is necessary to predict the performance of undone and previously wrong questions based the cognitive diagnostic report.

First, the latent response about the question is obtained from Eq. 3-1, which represents whether the student can answer the question correctly. Then, the prediction result equals the probabilistic correct response, which can be obtained from the Eq. (3-4). Here, the result of the performance prediction is simplified. If the result is less than 0.5, it is regarded as 0, which means the answer is wrong, and vice versa. Finally, the questions with wrong answers, i.e., the predicted score of 0, are selected as the range of the recommended questions.

4.2   Calculation of Similarity Based on Cosine Similarity

With the cosine similarity method [19], we calculate the similarity between the knowledge mastery vector of students and q-vector, which represents the skills that are required to solve that question. For example, we selected a student randomly and calculate the knowledge mastery vector of student α_i and q vector as follows (Fig. 2). Thus, we get angle cosine between the vectors, which means similarity value.

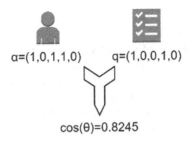

$a=(1,0,1,1,0)$ $q=(1,0,0,1,0)$

$\cos(\theta)=0.8245$

Fig. 2. Cosine similarity calculation

4.3 The Recommendation of Exercises

The cosine similarity reflects the matched degree between the recommendation and the students' knowledge status. The more similar they are, the more probability need to be recommended, and the more valuable they are to the students.

According to the similarity, the recommendations are sorted by the similarity degree, and finally a recommendation list is formed. The example given in Fig. 3 demonstrates a recommendation list for a student.

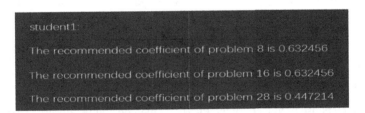

student1:

The recommended coefficient of problem 8 is 0.632456

The recommended coefficient of problem 16 is 0.632456

The recommended coefficient of problem 28 is 0.447214

Fig. 3. Recommendation list

5 Experiment

5.1 Data Set Introduction

The data source is the Online-Judge Platform of the programming club. The brief summary of these datasets is shown in Table 1. Among them, DataSet1 is used to validate the reliability of Cognitive Diagnostic; DataSet2 is used in chi-square experiment; DataSet3 and DataSet4 are used in comparative experiment.

Table 1. Datasets summary.

Data sets	Students	Items	Knowledge (skills)
DATESET1	226	50	5
DATESET2	40	80	5
DATESET3	15	20	5
DATESET4	20	30	5

5.2   Experimentation

Experiment 1: The Reliability of Cognitive Diagnostic

First, data set 1 are divided into training set and test set respectively. According to the training set, cognitive diagnosis was carried out; then the test set is predicted by the diagnostic results. The predicted results are compared with the observed response, then, the fitting degree is observed. Moreover, the ratio of training set to test set is adjusted and the proportion of training set and test set are gradually increased.

Table 2. Fit degree between predicted response and observed response

Data set	Fitting degree
A small proportion	87.5%
A large proportion	89%

From Table 2, it can be concluded that the fitting degree between predicted response and observed response is considerable, and with the increase of the proportion of the training set and the test set, the fitting degree is getting higher and higher. It can be seen that the prediction basis is more reliable, thus proving the effectiveness of the cognitive diagnosis results.

Experiment 2: Chi Square Test of Response Prediction

The effectiveness of performance prediction based the Priority Recommendation method PRM-CD method is proved from the view of statistics [20].
Hypothesis (Confidence $\alpha = 0.005$):

H0: The response predicted by the PRM-CD method and the observed response are independent. The prediction is unrelated to observed response.

H1: The response predicted by the PRM-CD method and the observed response are not independent. The prediction is related to observed response.

Table 3. Chi-square statistics

Groups	$p_{ij} = r_{ij}$	$p_{ij} \neq r_{ij}$	Total
$r_{ij} = 1$	497(420)	110(187)	607
$r_{ij} = 0$	56(133)	137(60)	193
Total	553	247	800

Note: The theoretical frequency in parentheses.

Here, the value of χ^2 can be calculated by Table 3:

$$\chi^2 = \frac{(497-420)^2}{420} + \frac{(110-187)^2}{187} + \frac{(56-133)^2}{133} + \frac{(137-60)^2}{60} = 189.2186.$$

Because $\chi^2_{pre} = 189.2186 > \chi^2_{0.005}(1)$, then $p_{pre} < 0.005$.

Therefore, we reject the original hypothesis H0 and accept the hypothesis H1, that is, through statistical chi-square test, it can be seen that the prediction is related to observed response. The effectiveness of RRM-CD method has better performance in response prediction.

Experiment 3: Contrast Experiment for Recommendation

To measure recommendation effectiveness, the traditional DINA method and PRM-CD method are both used to recommend questions. For the purpose of comparison, we record the performance of the traditional DINA method and PRM-CD method by tuning their parameters (Table 4) and Fig. 4 shows the results on two datasets.

Table 4. Comparison of the traditional DINA method and PRM-CD method

Evaluation index	Method	Data set3	Data set4
Precision	DINA	35%	25%
	PRM-CD	91%	69%
F1 value	DINA	38%	37%
	PRM-CD	53%	46%
Correct response rate	DINA	74%	76%
	PRM-CD	17%	31%

According to the experimental results (Fig. 4), the precision rate and the F1 score of the two methods are compared. It shows that the PRM-CD is more accurate. The correct answer rate reflects the difficulty of the recommended question, in other words, the knowledge is contained by the recommended question needs to be strengthened, which is in line with the original intention recommended by the questions. Under different test subjects, the PRM-CD method is better than the recommendation based on traditional DINA model, which proves that the former is more effective.

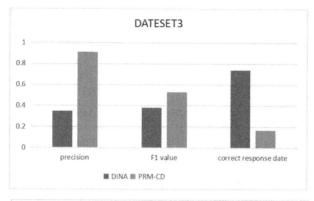

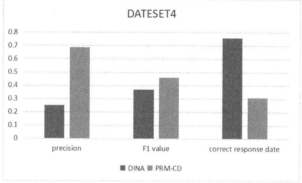

Fig. 4. Contrast experiment for recommendation

6 Conclusion

There has been increasing attention to recommendation in education. Personalized recommendation methods combining cognitive diagnosis and cosine similarity algorithm are proposed in this study.

The results of experiments on the real data set show that the prediction response by using the cognitive diagnosis result has little deviation from the observed response. Then, the evaluation index displays the accuracy of the recommendation results, and the difficulty of the recommended questions can be guaranteed. To sum up, the PRM-CD method proposed is effective in the recommendation and compared with the traditional DINA models, which considers the learning personality of students, thus enhances the guidance and interpretability of recommendation.

Acknowledgment. This research was supported by grants from the National Education Scientific Planning Projects (Research on learning paradigms in Adaptive Learning Space) (No. BCA190081).

References

1. Milicevic, A.K., et al.: E-Learning personalization based on hybrid recommendation strategy and learning style identification. Comput. Educ. **56**(3), 885–899 (2011)
2. Dandan, G.: Cognitive diagnosis theory and educational assessment. China Exam. **107**(3), 1008–1014 (2009)
3. Wang, F., et al.: Neural cognitive diagnosis for intelligent education systems. In: The Thirty-Fourth AAAI Conference on Artificial Intelligence (AAAI 2020), New York, USA (2020, accepted)
4. Leighton, J., Gierl, M.: Cognitive diagnostic assessment for education: theory and applications. J. Qingdao Tech. Coll. **45**(4), 407–411 (2007)
5. DiBello, L.V., Roussos, L.A., Stout, W.: 31a review of cognitively diagnostic assessment and a summary of psychometric models. In: Handbook of Statistics, vol. 26, pp. 979–1030 (2006)
6. Cheng, S., et al.: DIRT: deep learning enhanced item response theory for cognitive diagnosis. In: The 28th ACM International Conference on Information and Knowledge Management (CIKM 2019), Beijing, China, 3–7 November 2019, accepted
7. Hambleton, R.K., Swaminathan, H.: Item response theory: Principles and applications. Springer, New York (2013). https://doi.org/10.1007/978-1-4419-1698-3_100760
8. De La Torre, J.: The generalized DINA model framework. Psychometrika **76**(2), 179–199 (2011)
9. Toscher, A., Jahrer, M.: Collaborative filtering applied to educational data mining. In: KDD Cup (2010)
10. Liu, Q., et al.: Fuzzy cognitive diagnosis for modelling examinee performance. ACM Trans. Intell. Syst. Technol. (TIST) **9**(4), 1–26 (2018)
11. Vadivelou, G.: Collaborative filtering based web service recommender system using users' satisfaction on QoS attributes. In: 2016 International Conference on Inventive Computation Technologies (ICICT), vol. 3. IEEE (2016)
12. Koren, Y., Bell, R., Volinsky, C.: Matrix factorization techniques for recommender systems. Computer **42**(8), 30–37 (2009)
13. Tatsuoka, K.K.: Rule space: an approach for dealing with misconceptions based on item response theory. J. Educ. Meas. **20**(4), 345–354 (1983)
14. de la Torre, J., Minchen, N.: Cognitively diagnostic assessments and the cognitive diagnosis model framework. Educ. Psychol. **20**(2), 89–97 (2014)
15. Tatsuoka, K.K.: Architecture of knowledge structures and cognitive diagnosis: A statistical pattern recognition and classification approach. In: Cognitively Diagnostic Assessment, pp. 327–359 (1995)
16. De La Torre, J.: A cognitive diagnosis model for cognitively based multiple-choice options. Appl. Psychol. Meas. **33**(3), 163–183 (2009). Article in a Conference Proceeding
17. Culpepper, S.A.: Bayesian estimation of the DINA model with Gibbs sampling. J. Educ. Behav. Stat. **40**(5), 454–476 (2015)
18. De La Torre, J.: DINA model and parameter estimation: a didactic. J. Educ. Behav. Stat. **34**(1), 115–130 (2009)
19. Lahitani, A.R., Permanasari, A.E., Setiawan, N.A.: Cosine similarity to determine similarity measure: study case in online essay assessment. In: 2016 4th International Conference on Cyber and IT Service Management. IEEE (2016)
20. Zhu, T.Y., et al.: Cognitive diagnosis based personalized question recommendation. Chin. J. Comput. **40**(1), 176–191 (2017)

Correction to: A New Lightweight Database Encryption and Security Scheme for Internet-of-Things

Jishun Liu, Yan Zhang, Zhengda Zhou, and Huabin Tang

Correction to:
**Chapter "A New Lightweight Database Encryption
and Security Scheme for Internet-of-Things"**
in: P. Qin et al. (Eds.): *Data Science*, CCIS 1258,
https://doi.org/10.1007/978-981-15-7984-4_13

The originally published version of the paper starting on p. 168 contained a data error. The figures in Table 1. on p. 173 have been corrected.

The updated version of this chapter can be found at
https://doi.org/10.1007/978-981-15-7984-4_13

© Springer Nature Singapore Pte Ltd. 2020
P. Qin et al. (Eds.): ICPCSEE 2020, CCIS 1258, p. C1, 2020.
https://doi.org/10.1007/978-981-15-7984-4_47

Correction to: Effective Vietnamese Sentiment Analysis Model Using Sentiment Word Embedding and Transfer Learning

Yong Huang, Siwei Liu, Liangdong Qu, and Yongsheng Li

Correction to:
Chapter "Effective Vietnamese Sentiment Analysis Model Using Sentiment Word Embedding and Transfer Learning"
in: P. Qin et al. (Eds.): *Data Science*, CCIS 1258,
https://doi.org/10.1007/978-981-15-7984-4_3

In the originally published chapter 3 the reference to source [18] was erroneously omitted. The first sentence in section 4.2 has been revised by adding the reference [18], and the source has been added in the References section.

The updated version of this chapter can be found at
https://doi.org/10.1007/978-981-15-7984-4_3

Author Index

Printed in the United States
by Baker & Taylor Publisher Services